Applications of Big Data and Machine Learning in Galaxy Formation and Evolution

As investigations into our Universe become more complex, in-depth, and widespread, galaxy surveys are requiring state-of-the-art data scientific methods to analyze them. This book provides a practical introduction to big data in galaxy formation and evolution, introducing the astrophysical basics, before delving into the latest techniques being introduced to astronomy and astrophysics from data science. This book helps translate the cutting-edge methods into accessible guidance for those without a formal background in computer science. It is an ideal manual for astronomers and astrophysicists, in addition to graduate students and postgraduate students in science and engineering looking to learn how to apply data-science to their research.

Key Features:
- Introduces applications of data-science methods to the exciting subject of galaxy formation and evolution.
- Provides a practical guide to understanding cutting-edge data-scientific methods, as well as classical astrostatistical methods.
- Summarises a vast range of statistical and informatics methods in one volume, with concrete applications to astrophysics.

Tsutomu T. Takeuchi is Associate Professor, Division of Particle and Astrophysical Science, Nagoya University, Japan.

Series in Astronomy and Astrophysics

The *Series in Astronomy and Astrophysics* includes books on all aspects of theoretical and experimental astronomy and astrophysics. Books in the series range in level from textbooks and handbooks to more advanced expositions of current research.

Series Editors:
M Birkinshaw, University of Bristol, UK
J Silk, University of Oxford, UK
G Fuller, University of Manchester, UK

Recent books in the series

Dark Sky, Dark Matter
J M Overduin and P S Wesson

Dust in the Galactic Environment, 2nd Edition
D C B Whittet

The Physics of Interstellar Dust
E Krügel

Very High Energy Gamma-Ray Astronomy
T C Weekes

Numerical Methods in Astrophysics: An Introduction
P Bodenheimer, G P Laughlin, M Rózyczka, H W Yorke

An Introduction to the Physics of Interstellar Dust
Endrik Krugel

Astrobiology: An Introduction
Alan Longstaff

Fundamentals of Radio Astronomy: Observational Methods
Jonathan M Marr, Ronald L Snell, and Stanley E Kurtz

Stellar Explosions: Hydrodynamics and Nucleosynthesis
Jordi José

Cosmology for Physicists
David Lyth

Cosmology
Nicola Vittorio

Cosmology and the Early Universe
Pasquale Di Bari

Fundamentals of Radio Astronomy: Astrophysics
Ronald L. Snell, Stanley E. Kurtz, and Jonathan M. Marr

Introduction to Cosmic Inflation and Dark Energy
Konstantinos Dimopoulos

Physical Principles of Astronomical Instrumentation
Matthew Griffin, Peter A. R. Ade, Carole Tucker

A Guide to Close Binary Systems
Edwin Budding and Osman Demircan

An Overview of General Relativity and Space-Time
Nicola Vittorio

Applications of Big Data and Machine Learning in Galaxy Formation and Evolution
Tsutomu T. Takeuchi

Applications of Big Data and Machine Learning in Galaxy Formation and Evolution

Tsutomu T. Takeuchi

CRC Press
Taylor & Francis Group
Boca Raton London New York

CRC Press is an imprint of the
Taylor & Francis Group, an **informa** business

Designed cover image: Tsutomu T. Takeuchi

First edition published 2025
by CRC Press
2385 NW Executive Center Drive, Suite 320, Boca Raton FL 33431

and by CRC Press
4 Park Square, Milton Park, Abingdon, Oxon, OX14 4RN

CRC Press is an imprint of Taylor & Francis Group, LLC

ISBN: 978-0-367-61139-2 (hbk)
ISBN: 978-0-367-60823-1 (pbk)
ISBN: 978-1-003-10431-5 (ebk)

DOI: 10.1201/9781003104315

Typeset in Latin Modern font
by KnowledgeWorks Global Ltd.

Publisher's note: This book has been prepared from camera-ready copy provided by the authors.

Contents

Preface xv

SECTION I *Physics of Galaxy Structure, Formation, and Evolution*

Chapter 1   Introduction 3

    1.1   What is a Galaxy? 3
    1.2   Galaxy Morphology 4
        1.2.1   Galaxy Morphology: Tuning-Fork Diagram 5
        1.2.2   Peculiar and Other Special Galaxies 7
    1.3   Galaxy Size 7
    1.4   Surface Brightness 8
        1.4.1   Surface Brightness of Galaxies and its Structure 8
        1.4.2   Discovery of LSB Galaxies 9

Chapter 2   Properties of Galaxies 10

    2.1   Structure and Properties of Galaxies 10
        2.1.1   Structure of Spheroidal Galaxies 10
        2.1.2   Structure of Disk Galaxies 11
    2.2   Galaxy Morphology and Statistical Properties 12
    2.3   Galaxies and Environment 13
    2.4   Luminosity Function and Related Properties 15
        2.4.1   Schechter Function 16
        2.4.2   Integrated Quantities 17
        2.4.3   LFs at Other Wavelengths 18
    2.5   Scaling Laws 19
        2.5.1   Scaling Laws of Spheroidal Galaxies 19
        2.5.2   Scaling Laws of Disk Galaxies 22
        2.5.3   Photometric Scaling Relations for the Whole Population of Galaxies 23

Chapter 3   Interstellar Medium (ISM) 26

    3.1   Phases of the Interstellar Medium (ISM) 26
        3.1.1   Multiphase ISM Structure 26
    3.2   Neutral Hydrogen (HI) Medium 27
        3.2.1   21-cm Line Emission from HI 27

3.2.2   Collisions and Spin Flip 29
3.2.3   Measuring H I Mass 29
3.2.4   Radiative Transfer of the H I Line..............................29
3.2.5   Radiative Transfer Using Einstein Coefficients 30
3.2.6   Application to the Measurement of H I...................32
3.3   Molecular Clouds...33
3.3.1   Submm/mm Observations of Molecules 33
3.3.2   CO-to-H_2 Conversion Factor................................33
3.3.3   Mass and Dynamics of Molecular Clouds 34
3.4   Ionized (H II) Region..35
3.4.1   Physical Processes in H II Regions 35
3.4.2   Strömgren Radius...35
3.5   Photodissociation Region (PDR).......................................36
3.5.1   Self-Shielding...37
3.5.2   Heating and Cooling...38
3.6   Jeans Instability 39
3.6.1   What is Fluid Stability?..40
3.6.2   Formulation of the Jeans Instability 40
3.6.3   Jeans Criterion 42
3.7   Star Formation in the Context of Galaxies 42
3.7.1   Star Formation Criteria..43
3.7.2   Kennicutt–Schmidt Law 45
3.8   Dust..49
3.8.1   Basic Properties of Dust..49
3.8.2   Formation of Dust Grains..54
3.8.3   Attenuation in Galaxies...54
3.8.4   Dust Emission 55

Chapter 4   Chemical Evolution of Galaxies...59

4.1   Main Sequence Evolution of Stars 59
4.2   Post-Main Sequence Evolution...62
4.2.1   Very Low-Mass Stars ($\mathcal{M} < 0.45\,[\mathcal{M}_\odot]$)...............62
4.2.2   Low-Mass Stars ($0.45\,[\mathcal{M}_\odot] < \mathcal{M} < 2.0\,[\mathcal{M}_\odot]$)....63
4.2.3   Intermediate-Mass Stars ($2\,[\mathcal{M}_\odot] < \mathcal{M} < 8\,[\mathcal{M}_\odot]$)...65
4.2.4   Massive Stars ($\mathcal{M} > 8\,[\mathcal{M}_\odot]$) 66
4.3   Final Stage of Stellar Evolution...66
4.3.1   Thermal Pulse (Shell Flash) 66
4.3.2   Stars with $\mathcal{M} < 8\,[\mathcal{M}_\odot]$ 67
4.3.3   Stars with $8\,[\mathcal{M}_\odot] < \mathcal{M} < 10\,[\mathcal{M}_\odot]$ 67
4.3.4   Stars with $\mathcal{M} > 10\,[\mathcal{M}_\odot]$ 68
4.3.5   Supernova Explosion..69
4.4   Origin of Heavy Elements and Yield.................................71
4.4.1   Origin of Elements Up to Iron................................72

4.4.2 Origin of Elements Heavier Than Iron 72
4.4.3 Origin of Elements: Summary 73
4.5 Chemical Evolution of Galaxies 74
4.5.1 Initial Mass Function (IMF) 74
4.5.2 Chemical Evolution Model: Basic Equations 75
4.5.3 Instantaneous Recycling Approximation 77
4.5.4 Closed-Box Model 79
4.5.5 Numerical Example of a Closed-Box Model with Schmidt Law 80
4.5.6 The G-Dwarf Problem and Infall Model 80
4.6 Evolutionary Synthesis of Galaxy Spectra 83
4.6.1 Formulation of Spectral Synthesis 84

Chapter 5 Observational Star Formation Rate Indicator 89

5.1 Primary SFR Estimator 89
5.1.1 SFR and Observables 89
5.2 SFR Estimator Based on Dust Emission 91
5.2.1 SFR Estimator Based on Total IR Luminosity 91
5.3 Hybrid SFR Estimator 94
5.3.1 Physical Basis of the Hybrid SFR Estimator 94
5.3.2 UV/IR Ratio (IRX) as an Attenuation Estimator ...96
5.4 The IRX–β Relation 98
5.5 Secondary SFR Estimator 103
5.5.1 Optical Forbidden Lines 103
5.5.2 Metal Fine Structure Lines 104
5.5.3 X-ray Luminosity 106
5.5.4 Radio Continuum 106
5.6 Specific SFR and Related Quantities 108
5.6.1 Birthrate Parameter and Specific SFR 108
5.6.2 Star Formation Efficiency and Gas Consumption Timescale 109

Chapter 6 Clusters, Clustering of Galaxies, and the Large-Scale Structure 110

6.1 Clusters of Galaxies: Phenomenology 110
6.2 The Missing Mass Problem in Clusters 110
6.2.1 Mass From Galaxy Dynamics in Clusters 110
6.2.2 Mass Function of Clusters 111
6.3 Statistical Methods for Quantifying Galaxy Distribution ...112
6.3.1 The Large-Scale Structure in the Universe 112
6.3.2 Ensemble and Volume Average 113
6.3.3 2-Point Correlation Function 114
6.3.4 Power Spectrum Analysis 117
6.3.5 Higher-Order Moment Measures 119

Chapter 7   Structure and Galaxy Formation in the Universe 121

      7.1   Characterization of Fluctuation 121
      7.2   Initial Fluctuation: the Harrison–Zel'dovich Spectrum 122
      7.3   Deformation of the Harrison–Zel'dovich Spectrum 123
      7.4   Linear Theory of the Structure Formation 124
           7.4.1   From Physical Coordinates to Comoving Coordinates 125
           7.4.2   Jeans Instability in the Expanding Universe 128
           7.4.3   Peculiar Velocity Field 131
           7.4.4   Baryon Acoustic Oscillation (BAO) 134
           7.4.5   Press–Schechter Formalism 135
           7.4.6   Excursion Set Approach for the Mass Function... 138
           7.4.7   Merging in the EPS Formalism 141
           7.4.8   Bias 142
      7.5   Galaxy formation 144
           7.5.1   Cooling of Gas Clouds 145
           7.5.2   Stellar Feedback 147

SECTION II   Statistical Approach

Chapter 8   Basics of Statistics 153

      8.1   Probability and Statistics 153
           8.1.1   Mean, Dispersion, and Moments.......................... 153
           8.1.2   Limit Theorems 154
           8.1.3   Estimation and Hypothesis Testing 155
           8.1.4   Maximum Likelihood Estimation 156
      8.2   Information Theory and Statistics 157
           8.2.1   Information Entropy 157
           8.2.2   Maximum Likelihood Estimation in the Context of Information Theory 159
           8.2.3   Role of the Information Matrix 160
      8.3   Akaike's Information Criterion.. 162

Chapter 9   Expectation–Maximization (EM) Algorithm 164

      9.1   Introduction... 164
           9.1.1   Problem of Incomplete Data 164
           9.1.2   Slope of the Resolved Kennicutt–Schmidt Law... 165
      9.2   EM Algorithm with Gaussian Mixture Model.................... 167
           9.2.1   General Formulation... 167
           9.2.2   Derivation of the Likelihood Solutions in the EM Algorithm 169
      9.3   Data... 171
           9.3.1   CO Data.. 171

	9.3.2	Data for SFR		172
	9.4	Result and Discussion		174

Chapter 10 Copula and Luminosity and Mass Functions of Galaxies 178

10.1 Copula as a Tool to Construct a Multivariate Luminosity/Mass Function 178
  10.1.1 Copula 179
  10.1.2 Copulae and Dependence Measures Between Two Variables 181
  10.1.3 Estimators of the Nonparametric Dependence Measures ρ_S and τ 182
  10.1.4 Farlie–Gumbel–Morgenstern (FGM) Copula 183
  10.1.5 An Extension of the FGM System by Johnson & Kotz (1977) 185
  10.1.6 Gaussian Copula.................................... 187
  10.1.7 Application to Construct the Bivariate Luminosity Function (BLF) of Galaxies 188

10.2 Vine Copula for a Construction of Multivariate Distribution 195
  10.2.1 Vine Copula.. 195
  10.2.2 Likelihood for Vines................................ 202
  10.2.3 Model Selection with Vines 203
  10.2.4 Application to Construct a Multivariate Mass Function (MF) of Galaxies........................ 204
  10.2.5 Multivariate MF..................................... 204
  10.2.6 The Stellar–Atomic Gas–Molecular Gas Multivariate MF.................................... 205
  10.2.7 Data 206
  10.2.8 Result of the Vine Copula Analysis 207

Chapter 11 High-Dimensional Statistical Analysis....................................213

11.1 Introduction.. 213
11.2 Method: High-Dimensional Statistical Analysis 215
  11.2.1 High-Dimensional Statistical Analysis 215
  11.2.2 Geometric Representations of $\tilde{S}_D$ 216
  11.2.3 Geometric Representation of the Data 218
11.3 High-Dimensional Principal Component Analysis 220
  11.3.1 Limitation of the Traditional Principal Component Analysis (PCA)................... 222
  11.3.2 Noise-Reduction PCA (NRPCA): A New Method for HDLSS Data.................. 223
  11.3.3 Automatic Sparse PCA (A-SPCA)................ 226
11.4 Example of Data Analysis 227
  11.4.1 ALMA Map of NGC 253 227

11.4.2 Doppler Shifts Due to the Systemic Hubble Velocity and Rotation 227

11.4.3 Spectral Map as HDLSS Data 229

11.5 Preparatory Analysis 229

11.5.1 Eigenspectra 230

11.5.2 Distribution of the Contribution of PCs 230

11.5.3 Physical Meaning of First Two PCs 232

11.5.4 Spectral Features Corresponding to PCs: Original Data 233

11.6 Results and Discussion 235

11.6.1 Application of the High-Dimensional PCA to the Doppler Shift-Corrected Map of NGC 253 236

11.6.2 Map of PCs of the NGC 253 Spectra 240

11.6.3 Spectral Features of the Doppler-Corrected Data Corresponding to the PCs 240

11.7 Summary and Prospect 243

SECTION III Machine Learning Approach

Chapter 12   Basics of Machine Learning 247

12.1 Machine Learning 247

12.1.1 Unsupervised Machine Learning 248

12.1.2 Supervised Machine Learning 248

12.2 Neural Networks 249

12.2.1 The Concept of Neural Networks 249

12.2.2 Convolutional Neural Networks (CNN) 249

12.2.3 Learning Method 249

Chapter 13   Galaxy Face 252

13.1 Introduction 252

13.2 Data 253

13.2.1 Galaxy Catalogue 253

13.2.2 Galaxy Images 253

13.2.3 Image Preprocessing 254

13.3 Method 254

13.3.1 Principal Component Analysis 254

13.3.2 Autoencoders 255

13.3.3 Variational Autoencoders 258

13.4 Results and Diskussion 259

13.4.1 Principal Component Analysis 259

13.4.2 Autoencoders 262

13.4.3 Variational Autoencoders 265

Chapter 14   New Quantification of Galaxy Evolution by Manifold
Learning...277

14.1   Introduction...277
        14.1.1   Galaxy Evolution in the Era of Large-Scale
                    Galaxy Surveys..277
        14.1.2   Galaxy Manifold in the Multi-wavelength
                    Luminosity Space 278
14.2   Manifold Learning data 281
14.3   Method..282
        14.3.1   Galaxy Manifold in the Multi-wavelength
                    Luminosity Space 282
        14.3.2   Manifold Learning......................................283
        14.3.3   Isomap and UMAP Algorithms 284
14.4   Results and Discussion 286
        14.4.1   Results: Isomap and UMAP Galaxy Manifolds 286

Chapter 15   Topological Data Analysis of the Large-Scale Structure 292

15.1   Introduction...292
        15.1.1   Characterizing and Describing the Matter
                    Distribution in the Universe 292
        15.1.2   From Minkowski Functionals to Topological
                    Data Analysis 295
15.2   Topological Data Analysis (TDA) 296
        15.2.1   Persistent Homology Groups 297
        15.2.2   Inverse Analysis of Persistence Diagrams 307
15.3   Performance Evaluation of the Analysis Method Using
        Cosmological Simulations 308
        15.3.1   Data 308
        15.3.2   Analysis Results of the Simulation Data..............310
15.4   Analysis Using Sloan Digital Sky Survey Data
        Release 14...313
        15.4.1   SDSS Galaxy Survey Data.........................313
        15.4.2   SDSS Data Analysis Results 313
15.5   Summary and Discussion 315
        15.5.1   Difference of the Large-Scale Structure with
                    and Without Baryons.................................315
        15.5.2   Interpretation as the Baryonic Acoustic
                    Oscillation 315

Chapter 16   Radio Morphology of Galaxies with Machine Learning.............317

16.1   Introduction: Radio Galaxies and Their Classification.......317
        16.1.1   Radio Galaxies 317
        16.1.2   Fanaroff–Riley (FR) Classification 317
16.2   Morphology of Radio Galaxies with CNN 318

16.2.1   Data Acquisition...318
16.3 Data Processing 321
16.3.1   Sigma Clipping...321
16.3.2   Data Augmentation 322
16.4 Building the Classification Model 323
16.5 Results and Discussion 324
16.5.1   Class 1: Compact Galaxies...............................325
16.5.2   Class 2: FR I Galaxies 327
16.5.3   Class 3: FR II Galaxies.....................................327
16.5.4   Class 4: Bent-Tailed (BT) Galaxies....................328
16.5.5   Class 5: X-shaped Radio Galaxies (XRGs)..........328
16.5.6   Class 6: Ring-Like Radio Galaxies (RRGs).........329
16.5.7   Cross Validation 329
16.5.8   Comparison with Previous Studies....................333
16.6 Summary and Prospects..334

SECTION IV Appendix

Appendix A   Cosmological Basics 336

A.1 Homogeneous Universe..336
A.1.1   Friedmann-Lemaître-Robertson-Walker Metric...336
A.1.2   Friedmann Equations..337
A.1.3   Cosmological Parameters and Evolution of the
Universe..338
A.1.4   Dominant Epoch of Each Component..................339
A.2 Cosmological Redshift..341
A.2.1   Cosmological Quantities as a Function of
Redshift 343
A.2.2   Observable Properties.......................................344

Appendix B   Supplementary Information on Mathematics and Machine
Learning..349

B.1 Algebraic Preliminaries 349
B.2 Fundamentals of Manifolds 350
B.2.1   Topological Space 350
B.2.2   Riemannian Manifold..351
B.3 Methods for Dimensionality Reduction and Manifold
Learning...352
B.3.1   Multidimensional Scaling (MDS) 352
B.3.2   Isomap 353
B.3.3   Uniform Manifold Approximation and
Projection (UMAP) 355

Contents   **xiii**

Appendix C   Physical Constants and Units 360

        C.1   Physical Constants 360
        C.2   Units 361

Bibliography 363

Index 395

Preface

This book introduces the methods of using the rapidly developing field of data science to study the important topic of galaxy formation and evolution in astrophysics. The scope of the term "data science" is extremely broad and varies by interpretation. In this book, I aim at presenting new statistical methods, including computationally intensive techniques that have emerged alongside advances in computing power, as well as machine learning approaches for the era of big data science, using the research themes I have led as examples.

This book is structured into three parts: Part I, Part II, and Part III, with an additional Part IV to provide supplementary information. Part I begins with Chapter 1, introducing the concept of galaxies, followed by a discussion of their phenomenological and physical properties in Chapter 2, and an explanation of the physical processes within the interstellar medium in Chapter 3. Building on this foundation, Chapter 4 outlines the basics of chemical and spectral evolution, the most critical processes in the internal evolution of galaxies. The star formation rate in galaxies varies over time for many reasons and governs the subsequent stages of evolution through chemical evolution and feedback. Research connecting chemical evolution theory with observables fundamentally revolves around the star formation rate. Thus, in Chapter 5, I discuss the observational indicators of star formation rates in detail. Galaxies do not evolve in isolation but primarily interact through gravity and tend to cluster against cosmic expansion. Chapter 6 introduces the statistical properties of galaxy clusters and the methods for measuring them. A theoretical framework has been constructed to explore cosmic structure formation based on the statistical methods used to describe large-scale structures, the overarching patterns formed by galaxies. In Chapter 7, I discuss the theory of structure formation, which covers the process by which galaxies and large-scale structures emerge from uniform conditions in the early universe. This chapter also covers the cosmological aspects of galaxy formation.

With this preparation on galaxy formation and evolution, Part II introduces research topics on galaxies using newly developed statistical methods. Chapter 8 introduces the basic statistical concepts used in this book. In Chapter 9, I discuss the Expectation-Maximization (EM) algorithm, commonly used for statistical processing of incomplete data, starting from the general theory and demonstrating its application to analyzing the Kennicutt-Schmidt law, an important issue in galaxy physics. In recent galaxy astronomy, multi-wavelength observations have become the foundation of large-scale surveys, and studies that analyze many physical quantities simultaneously have become increasingly important. In Chapter 10, I explain copula theory, a systematic method for constructing multivariate distribution functions when several univariate distribution functions are given as marginal distributions. I also provide examples of applying copula theory to galaxy luminosity functions (or mass functions). Chapter 11 discusses high-dimensional statistical analysis, a field still largely unknown in astronomy. In the 21st century, new statistical systems for dealing with data where the number of samples n is much smaller than the dimension d of the data have developed significantly, prompted by genome analysis research. In this chapter, I demonstrate the effectiveness of applying high-dimensional principal component analysis (PCA), a representative method of high-dimensional statistical analysis, to galaxy radio spectroscopic maps.

Next, in Part III, I introduce more computer-intensive machine-learning techniques. Chapter 12 provides a concise review of the basic concepts of machine learning. Chapter 13 introduces PCA, autoencoders, and variational autoencoders. I present results from applying these methods to classify galaxy morphology and extract features from numerous galaxy images. In modern big data science, feature spaces handled by methods such as PCA often become high-dimensional. Furthermore, the physical quantities describing galaxy evolution are also highly multidimensional. If we try to handle all of these simultaneously, we encounter enormous nonlinear systems of hundreds of dimensions. In such cases, it becomes practically impossible to construct theories based on intuition, and new methods based on different concepts are required. Chapter 14 introduces manifold learning as an effective method in such cases. I present surprising results from applying manifold learning algorithms such as Isomap and UMAP to multi-wavelength survey data of galaxies, showing that galaxy features across multiple wavelengths can be described with only a small number of variables. Chapter 15 provides a detailed explanation of topological data analysis (TDA), a field of research that has gained much attention recently for its ability to extract features from point cloud data and irregular patterns. I introduce the fundamental concepts of persistent homology (PH), the main tool of TDA, and demonstrate the application of TDA to galaxy distributions obtained from simulations and observations, highlighting its effectiveness. In particular, I discuss the potential for physically characterizing differences in distributions caused by the presence or absence of baryons. Finally, in Chapter 16, I introduce the classification of radio galaxy morphology using convolutional neural networks (CNN). Radio galaxy morphology is more complex than optical morphology and is a challenging field for objective assessment. Using CNN, a machine learning technique well-suited to image analysis, I present results for six representative classes of radio galaxy morphology and discuss future prospects.

Part IV contains supplementary content covering knowledge that may be challenging to introduce within the main chapters. Appendix A covers the basics of cosmology, while Appendix B provides additional information on mathematics and machine learning algorithms. Appendix C lists physical constants and unit information, presented in CGS units.

The code and scripts used to process the data discussed in this book evolve so rapidly that including them in a written work would quickly render them outdated. Therefore, I have not introduced individual codes. They can be easily found through network searches. The chapters in Parts II and III are relatively independent of one another, allowing the reader to begin with any topic of interest. After reading Part I, I recommend selecting content based on individual interest. To ensure the book is as comprehensive as possible, I have explained the analysis methods and data handling in significant detail, even at the risk of being somewhat excessive. The topics covered in this book range from introductory to highly advanced and cutting-edge material still in development, making the content highly uneven in difficulty. I also note that the topics are rather inclined to unsupervised methods rather than supervised. The latter is introduced in various books, and in my personal opinion, not very easy to understand in the context of galaxy evolution. For the sake of reasonable interpretability, the current topics were chosen. I hope that readers will utilize this book in a way that suits their purposes.

Throughout the writing of this book, my colleagues in data science, Shiro Ikeda, Satoshi Kuriki, Kenji Fukumizu, and Shotaro Akaho, have provided invaluable contributions and assistance throughout our various collaborative research projects. I would like to express my sincere gratitude to them. My former graduate student Suchetha Cooray has made significant contributions to the research, analysis, and producing figures. Without him, this book would not have been completed. I offer my deepest gratitude to him. The content of Chapter 1 owes much to the significant and important contributions of Akihiko Tomita. I thank him. Ryosuke S. Asano and Aina May So provided invaluable assistance with figure creation and content review for Chapter 4. The data analysis in Chapter 9 is based on the COMING Legacy Project of Nobeyama Radio Observatory. I thank Kazuo Sorai, Moe Yoda, and all members of the COMING project for their support. Kai T. Kono contributed to the analysis and figure creation in Chaps. 10 and 15, which were part of his doctoral project. Yoh-Ichi Mototake has given a fundamental contribution to Chapter 15. We highly appreciate his profound knowledge of the subject. Makoto Aoshima, Kazuyoshi Yata, Kento Egashira, and Aki Ishii are actively involved in the ongoing joint research that underlies Chapter 11. Their contributions to this book are immeasurable. Chapter 14 is part of Suchetha Cooray's Ph.D. thesis project. His contributions to the ideas and analysis were invaluable. We thank Dillon Loh for his significant contribution to Chapter 13, based on his Bachelor Thesis at Nagoya University, Japan. We are grateful to Ayumi Wakida for her contribution to the explanation of neural networks and the main results in Chapter 16, which was the main theme of her Bachelor Thesis.

Writing this book has been a challenging process. Although the initial ideas were clear, several topics became outdated quickly, forcing me to replace them. The content of the book underwent several major revisions. Additionally, various responsibilities at my university, a considerable increase in administrative tasks due to the COVID fuss, and numerous large and small issues caused long interruptions in my work. Each time, it took me a while to regain the passion for writing. I sincerely apologize to my editors, Danny Kielty, Rebecca Hodges-Davies, and Kirsten Barr, for the huge inconvenience I caused. I extend my heartfelt gratitude for their endless patience, kind encouragement, and unwavering support over a long time of writing.

Last but not least, I offer my deepest gratitude to my family for their unwavering support and patience throughout this journey.

October 11, 2024
Tsutomu T. Takeuchi
from Nagoya where Autumn finally has come

Section I

Physics of Galaxy Structure, Formation, and Evolution

1 Introduction

1.1 WHAT IS A GALAXY?

A galaxy is a massive collection of stars, interstellar medium (ISM: a mixture of gas and dust), and dark matter, forming a complex system with intricate interactions between its components. The Milky Way, the giant galaxy where we reside, is one such astronomical object, composed of approximately 10^{11} stars and ISM. The stellar-to-ISM mass ratio is roughly 10 : 1. The diameter of the Milky Way is estimated to be about 30 kpc, with the Solar System located about 8.5 kpc from the center, although these values remain a topic of debate. The key components of the Milky Way Galaxy are:

- Bulge: A central spheroidal stellar system with dense ISM.
- Disk: A thin, rotationally supported system consisting of stars and ISM.
- Stellar halo: An extended spheroidal stellar system containing globular clusters and halo stars.
- Dark halo: A gravitationally bound system of dark matter responsible for the gravitational potential of a galaxy.

These components are illustrated in Fig. 1.1, where their scales and sizes are also indicated. The bulge is a spheroidal system at the center of the Milky Way, consisting of older stars (Population II) and gas. The disk is a flattened, rotating system of gas and stars, including the Solar System, where we reside. The Milky Way's disk features spiral arms. It is important to note that stars continue to form within the disk, which mainly consists of relatively young stars (Population I). Newly formed stars create ionized regions (HII regions) around them, due to the intense ionizing photons emitted by massive stars (Chapter 3). Over time, clusters of newly formed stars gradually diffuse and are observed as open clusters, sometimes accompanied by a reflection nebula. The stellar halo is an extended spheroidal system of stars, with globular clusters, dense groups of old stars, orbiting within it. Halo stars also belong to Population II. It is worth noting that Population II stars form a spheroid, while Population I stars form a disk. This may be due to the fact that when older stars were formed, the angular momentum of the system was not significant. Later, when the disk formed, the system likely had a much higher angular momentum. However, many questions about the formation of the Milky Way remain unresolved.

In addition to the stellar components, ISM is predominantly found in the Milky Way's disk. Atomic and molecular gas, mostly hydrogen, makes up the ISM, with hydrogen being the most abundant element in the Universe (see Chapter A). In lower-density regions, such as the outskirts of the disk, hydrogen exists as atomic gas (denoted as HI in older spectroscopic terminology). In contrast, in higher-density regions, typically the inner part of the Galactic disk, hydrogen is found in molecular form (H_2). As mentioned earlier, the region around massive stars becomes ionized, forming an HII region (HII is also derived from spectroscopic terminology). The ISM

DOI: 10.1201/9781003104315-1

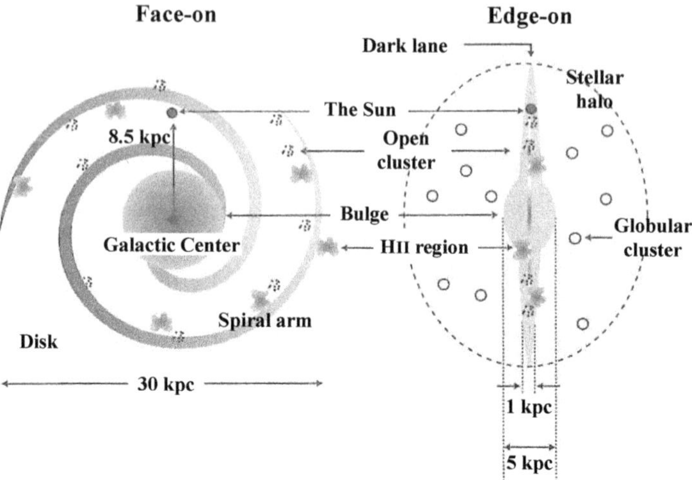

Figure 1.1 The structure and components of the Milky Way. Credit: Adopted from Tomita, A. 2010, Galaxies in Being Alive: Introduction to Galactic Astronomy, Fig. 2.4 with permission.

generally traces the structure of the Milky Way's disk, although a significant portion also exists in the regions between spiral arms. All these systems are enveloped in a much larger, extended dark matter halo (often called the dark halo), which governs the gravitational potential. Within this dark halo, a hot, diffuse ionized gas is present, sometimes referred to as the corona (analogous to the Sun's corona). The corona's gas distribution extends well beyond the disk, reflecting the larger structure of the dark halo. In general, we assume that the hot corona gas and the dark halo's gravitational potential are in dynamical equilibrium. Other galaxies share similar structural features. Before we dive into more detailed discussions, it's important to clarify the distinction between "galaxy" and "Galaxy" (capitalized) as used in the literature: the former refers to galaxies in general, while the latter specifically refers to the Milky Way. Throughout this monograph, we will consistently use "Milky Way" when referring to our own galaxy. Within the observable Universe, there are roughly 10^{11} galaxies. While their basic physical properties are somewhat similar, they exhibit a wide variety of appearances, with differences in size, mass, and luminosity spanning several orders of magnitude. Moreover, the diversity in their appearance is striking. Some galaxies have disks, some do not; some are flat, some are round, and others are irregular in shape. Given this diversity, it makes sense to classify galaxies according to their various properties.

1.2 GALAXY MORPHOLOGY

Astronomers have long recognized the diverse shapes of galaxies, even before they were confirmed as extragalactic objects. Large galaxies are typically classified into

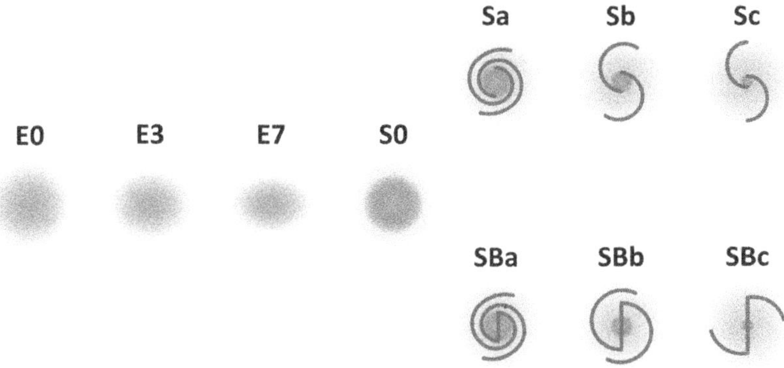

Figure 1.2 The classical tuning-fork diagram.

three main morphological types:

1. **Spirals**: Such as M31 and M81, these galaxies exhibit spiral arms.
2. **Ellipticals**: Like M87, they appear regular in shape, lacking significant substructure.
3. **Irregulars**: For example, the Magellanic Clouds, our closest neighbors, which show little organized structure.

Additionally, there are other exceptional cases.

1.2.1 GALAXY MORPHOLOGY: TUNING-FORK DIAGRAM

In natural sciences, classification is often the first step in studying a system. Early efforts by Jeans, Lundmark, and Reynolds in classifying galaxies led to Hubble's famous tuning-fork diagram (Fig. 1.2), which became the foundation of more intricate classification systems that followed. Spiral galaxies are divided into two subclasses: non-barred (S) and barred (SB) spirals, creating the tuning-fork shape.

The handle of the tuning fork represents elliptical galaxies. Although these galaxies appear featureless, detailed structures are often revealed through image enhancement or multiwavelength observations. The elliptical classification in Hubble's system is denoted by Ex, where

$$x = 10 \left(\frac{a - b}{a} \right) , \tag{1.1}$$

with a and b representing the major and minor axes of the ellipse, respectively. E0 galaxies are nearly spherical, while E7 galaxies are closer to lenticular (S0) galaxies, characterized by a central bulge and disk. These galaxies have absolute magnitudes ranging from $M_B \sim -24$ to $M_B \sim -8$, corresponding to luminosities of $L_B \sim 10^6$–10^{12} $[L_\odot]$.

Elliptical galaxies are further divided into subclasses, primarily based on their luminosity:

Normal ellipticals: Absolute magnitudes range from $M_B = -23$ to -15 mag.

- Giant ellipticals (gE)
- Ellipticals (E)
- Compact ellipticals (cE)
- S0 galaxies

cD galaxies: Absolute magnitude $M_B \simeq -25$ [mag]. These galaxies are typically located near the centers of galaxy clusters, possess extended diffuse envelopes, and exhibit high mass-to-light ratios ($\mathscr{M}/L$).

Dwarf ellipticals (dE): Absolute magnitudes range from $-15 \lesssim M_B \lesssim -10$ mag. These galaxies have lower surface brightness and are more metal-poor than compact ellipticals.

Dwarf spheroidals (dSph): Extremely low luminosity ($M_B \simeq -8$ mag) and surface brightness.

The prongs of the tuning fork represent spiral galaxies. After Hubble's classification, a subclass known as Sd galaxies, which lack bulges, was introduced. Sa galaxies have tightly wound spiral arms, while Sb, Sc, and Sd galaxies progressively exhibit more open and loosely wound structures. The size of the bulge also decreases from Sa to Sc galaxies, with Sc and Sd galaxies appearing almost chaotic. This gradual structural transition is depicted in Fig. 1.2.

The two prongs of the fork represent two branches of spiral galaxies. One prong includes galaxies where the spiral arms emanate from the central bulge (Sa, Sb, Sc), while the other consists of galaxies with spiral arms originating from the ends of a central bar (SBa, SBb, SBc). Over time, this distinction became less significant, and intermediate classes, such as SAa and SAb, were introduced by de Vaucouleurs.

S0 galaxies, which appear as lenticular structures with ellipticity $b/a < 0.3$, occupy the junction of the fork. These galaxies have smooth disks without internal spiral structures, likely resulting from environmental effects within galaxy clusters.

Irregular galaxies, placed at the end of the sequence, lack organized symmetry and central structure, often appearing chaotic. This class includes Magellanic Clouds (Irr I) and more disordered galaxies (Irr II), which are often rich in interstellar matter and young stars, sometimes experiencing bursts of star formation.

Irregular galaxies can be divided into:

Irr I: Similar to the Magellanic Clouds, these galaxies may display weak spiral structures.

Irr II (I0): These galaxies show no organized structure and are often associated with starburst activity.

Although not part of the tuning fork, blue compact dwarfs (BCDs) are a class of highly active, star-forming dwarf galaxies. BCDs are characterized by intense star-forming regions dominated by radiation from OB stars, and some are considered young galaxies.

1.2.2 PECULIAR AND OTHER SPECIAL GALAXIES

Many galaxies defy classification within the Hubble tuning fork. These galaxies, often disturbed by interactions with other galaxies, are labeled as peculiar galaxies (Pec). Features resulting from tidal interactions include:

- Strong deformations
- Double nuclei
- Tidal tails
- Polar rings
- Ripples and shells
- Ring-shaped disks

Other special types of galaxies include Seyfert galaxies, radio galaxies, quasars/QSOs, starburst galaxies, nucleated galaxies, and cD galaxies. These types often overlap and are collectively known as active galaxies.

Seyfert galaxies, first studied by Seyfert (1943), exhibit luminous nuclei powered by central massive black holes. Radio galaxies, powered by the same mechanism, emit strong radio waves, while quasars are the most powerful active galactic nuclei (AGN), with luminosities thousands of times that of the Milky Way. Despite their rarity, AGNs can have a significant impact on galaxy evolution due to their immense energy output.

Starburst galaxies have star formation rates exceeding $100 \, [\mathscr{M}_\odot \, \mathrm{yr}^{-1}]$, far higher than typical spiral galaxies. These galaxies often emit strongly in the blue and ultraviolet wavelengths but can also emit infrared radiation due to dust absorption. Luminous infrared galaxies (LIRGs) have infrared luminosities exceeding $10^{11} \, [L_\odot]$, while in extreme cases, ultraluminous and hyperluminous IR galaxies (ULIRGs and HyLIRGs), exceed 10^{12} and $10^{13} \, [L_\odot]$, respectively.

Nucleated galaxies, characterized by a dominant nucleus, belong to the Yerkes system of classification and may overlap with Seyfert and nuclear starburst galaxies. The cD galaxies, described by Matthews et al. (1964) and Morgan & Lesh (1965), are thought to result from sequential mergers within galaxy clusters.

1.3 GALAXY SIZE

Now, we turn to the measurable properties of galaxies and related considerations. Galaxies do not have a well-defined edge but gradually fade away at their outskirts. To describe the physical size of a galaxy, we often use its half-light or effective radius, r_e. When a galaxy is projected onto the sky, the light contained within a circle of radius r_e accounts for half of the total light of a galaxy. This radius provides a standardized measure of a size of a galaxy.

For instance, the Milky Way has an r_e of about 10 kpc, and large elliptical galaxies typically have a similar size. On the other hand, cD galaxies can reach r_e values as large as 100 kpc, while dwarf galaxies have much smaller r_e values, around 200 pc.

1.4 SURFACE BRIGHTNESS

1.4.1 SURFACE BRIGHTNESS OF GALAXIES AND ITS STRUCTURE

It is well established that luminous galaxies tend to have higher central surface brightness. Dwarf galaxies, on the other hand, generally exhibit faint surface brightness, which led to their being overlooked for a long time in galaxy studies. Physically, low surface brightness corresponds to a lower surface density of stars. Characterizing the overall surface brightness by the average within the effective radius, i.e., $L/2\pi r_{\rm e}^2$, both giant elliptical and spiral galaxies typically have $\mu_B \simeq 100\ [L_\odot\,{\rm pc}^{-2}]$. The brightness of the night sky is also comparable (e.g., Phillipps, 2006). Dwarf galaxies generally have surface brightness values about 10 times lower than that of giant galaxies. There is also a population of galaxies with the lowest surface brightness, known as low surface brightness galaxies (LSBs). In contrast, blue compact dwarfs (BCDs) and the recently discovered ultra-compact dwarfs (UCDs) exhibit much higher surface brightness than typical dwarfs.

We denote the surface brightness of the central region of spiral and lenticular galaxies as I_0. Freeman (1970) proposed that, despite significant differences in the brightness of galaxy disks, their central surface brightness remains remarkably constant ($I_0 = 21.67 \pm 0.3\ [{\rm mag\,arcsec}^{-2}]$ in the B-band). This finding had a substantial impact on studies of galaxy structure and evolution. However, at that time, it was assumed that galaxy luminosity and surface brightness were independent, and no account was made of selection effects—an assumption that later proved inaccurate. Figure 1.3 clearly demonstrates that the seemingly narrow distribution of surface brightness was a result of the luminosity–surface brightness selection effect.

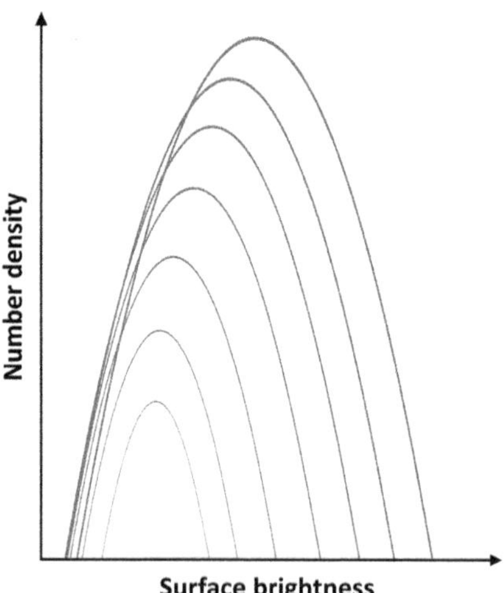

Figure 1.3 A schematic surface brightness distribution of galaxies at various luminosity intervals. From bottom to the top, the luminosity range increases. Generally, the distribution is narrow and constant for the brightest galaxies (Freeman's Law) and then broadens towards lower surface brightness for lower luminosity systems.

1.4.2 DISCOVERY OF LSB GALAXIES

Thanks to the great progress of recent observational techniques, a new population of galaxies not ever known was discovered, the above-mentioned LSB galaxies. The first discovered object of this type is Malin 1 (Bothun et al., 1987). Actually, it was the first and one of the most extreme examples. Then, a large number of LSB galaxies were found by subsequent observations. The LSBs include various morphological types, with a significant number of spirals. They have very low surface brightness, but otherwise very similar to that of high-surface brightness counterparts. A very important feature to be stressed is that they are not necessarily dwarfs, i.e., not small in mass nor faint in total magnitude at all. Also, they are as chemically evolved as high-surface brightness galaxies (Boissier et al., 2003).

2 Properties of Galaxies

2.1 STRUCTURE AND PROPERTIES OF GALAXIES

We begin by discussing the structure of elliptical galaxies, or more generally spheroidal systems, and disk galaxies. Notably, these galaxies exhibit fairly uniform surface brightness profiles on the celestial sphere.

2.1.1 STRUCTURE OF SPHEROIDAL GALAXIES

From a dynamical perspective, elliptical galaxies and the spheroidal components of galaxies can be treated similarly.

2.1.1.1 2D Structure de Vaucouleurs Profile

In general, the surface brightness profiles of spheroidal systems follow the de Vaucouleurs law

$$I(r) = I_e 10^{-3.3307\left[\left(\frac{r}{r_e}\right)^{\frac{1}{4}} - 1\right]}, \tag{2.1}$$

where I_e and r_e are the normalization and effective radius, respectively. This function declines more steeply with galactocentric radius r than an exponential function. The total luminosity of a spheroidal galaxy is obtained by integrating eq. (2.1) over $0 \leq r < \infty$

$$L_{\text{tot}} = 7.215 \pi I_e r_e^2 . \tag{2.2}$$

Several features specific to different subclasses of spheroids include

1. Normal ellipticals closely follow the de Vaucouleurs profile.
2. Higher- and lower-luminosity ellipticals deviate from this profile, declining more slowly or rapidly at large radii.
3. cD galaxies only follow this profile in their inner regions, with excess light from their outer regions.
4. Some studies suggest dwarf ellipticals are better described by an exponential profile, though this remains debated.

2.1.1.2 3D Structure

Since surface brightness profiles are projections of galaxies in the sky, we must consider their three-dimensional structure. Isophotes are generally elliptical. The possible shapes of the density distribution are

DOI: 10.1201/9781003104315-2

1. Oblate ($a = b > c$, a pancake-like shape)
2. Prolate ($a > b = c$, a lemon-like shape)
3. Triaxial ($a \neq b \neq c$, an ellipsoid resembling a box with smoothed edges).

Ellipticals are typically mildly triaxial, with axis ratios around $a : b : c \simeq 1 : 0.95 : 0.65$. Triaxiality is supported by observations of isophotal twists in some galaxies, which would not occur if the galaxies were purely oblate or prolate. This triaxial shape arises from anisotropic velocity dispersions that stretch the galaxy along its three principal axes.

Since ellipticity changes with radius, the major axes of the ellipses may appear rotated in projection, an effect known as isophote twist or axial twist.

2.1.1.3 Boxiness and Diskiness

Isophotes are not perfectly elliptical. Their deviations, termed "boxiness" or "diskiness", are quantified using the a_4 coefficient. First, the global ellipse $R_{\mathrm{e}}(\phi)$ is fitted to the isophote. Then, for each angle ϕ, we determine the distance

$$\delta(\phi) = R_{\mathrm{i}}(\phi) - R_{\mathrm{e}}(\phi) \tag{2.3}$$

between the radii of the corresponding points on the ellipse and isophote. The function $\delta(\phi)$ is then expanded in a Fourier series

$$\delta(\phi) = \bar{\delta} + \sum_{n=1}^{\infty} a_n \cos(n\phi) + \sum_{n=1}^{\infty} b_n \sin(n\phi) \,, \tag{2.4}$$

$$\begin{cases} a_4 < 0 & \text{boxy isophotes,} \\ a_4 > 0 & \text{disky isophotes.} \end{cases} \tag{2.5}$$

2.1.2 STRUCTURE OF DISK GALAXIES

We now turn to disk galaxies, including spirals and S0s. These galaxies have both disk and spheroidal (bulge) components, which are fit separately. The spheroidal component (bulge) is often described by the de Vaucouleurs $r^{\frac{1}{4}}$ law

$$I(r) = I_{\mathrm{e}} 10^{-3.3307 \left[\left(\frac{r}{r_{\mathrm{e}}} \right)^{1/4} - 1 \right]}, \tag{2.6}$$

while the disk is generally described by an exponential profile

$$I(r) = I_0 e^{-\left(\frac{r}{r_0} \right)}, \tag{2.7}$$

where r_0 is the disk scale length (e.g., $r_0 = 3$ kpc for the Milky Way), and I_0 is the central surface brightness. The total luminosity is given by integrating eq. (2.7)

$$L_{\mathrm{tot}} = 2\pi I_0 r_0^2 \,. \tag{2.8}$$

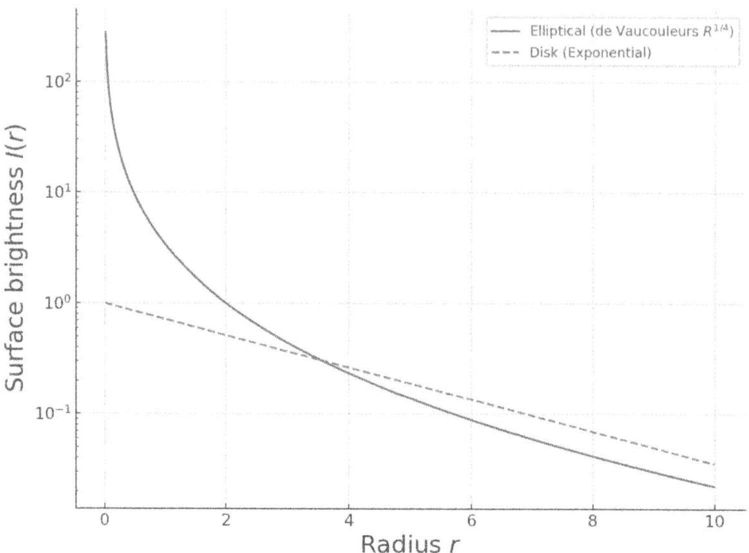

Figure 2.1 Surface brightness profiles of spheroids (elliptical galaxies and bulges of disk galaxies: de Vaucouleurs' law) and disks (exponential law).

The generalized Sérsic profile can describe both elliptical and spiral galaxies

$$I(r) = I_{\mathrm{e}} \exp\left\{ -b_n \left[\left(\frac{r}{r_{\mathrm{e}}} \right)^{1/n} - 1 \right] \right\}, \tag{2.9}$$

where n is the Sérsic index. We see that the de Vaucouleurs and exponential profiles are special cases of this profile

$$n = 4 \qquad \text{de Vaucouleurs profile}$$
$$n = 1 \qquad \text{exponential profile.}$$

Differences between surface brightness profiles of spheroids and disks can be seen in Fig. 2.1.

2.2 GALAXY MORPHOLOGY AND STATISTICAL PROPERTIES

Despite the wide variety of galaxy properties, they can be statistically characterized. One of the most basic statistical features of galaxies is their morphological dependence. The Hubble sequence is based on the optical appearance of galaxies, but morphology is closely tied to many physical properties.

We begin by examining the scale and mass of galaxies by morphology. As for giant galaxies, both luminosity and mass decrease for galaxies later than type Sc, with dynamical mass following a similar trend. The mass-to-light ratio of stars is roughly constant across different morphologies. The stellar mass $\mathscr{M}_*$ and luminosity L are related by

$$L \propto \mathscr{M}_*^a \, , \tag{2.10}$$

where $a = 3$–4. Averaging over stellar mass, the mean mass-to-light ratio is $\langle \mathscr{M}_*/L \rangle \simeq 1$–$2$ (see Chapter 4). However, the mass-to-light ratio is higher, reflecting the presence of dark matter. The latest-type galaxies show a variety of these properties, and it is due to their unique characteristics, like high gas fractions and shallow gravitational potentials.

Next, we summarize gas content in galaxies. Neutral hydrogen (HI) is easily observed in the local Universe, making it a representative measure of gas in galaxies (see Chapter 3). The gas mass peaks in Sb–Sc galaxies. Early-type galaxies have less gas, and while late-type galaxies are smaller, their total gas mass scales with size. When normalized by luminosity or dynamical mass, a trend emerges: later-type galaxies have higher gas fractions. For early types, gas mass fractions are $\sim$ a few percent, increasing to 10–20% for the latest types. This indicates that later-type galaxies are generally less evolved.

Galaxy colors also vary systematically with morphology. The $(B-V)$ color of local giant galaxies shows transitions from the reddest in ellipticals ($\langle (B-V) \rangle \simeq 0.9$) to the bluest in irregular galaxies ($\langle (B-V) \rangle \simeq 0.4$), with typical standard deviation of $\Delta(B-V) \simeq 0.1$. This reflects stellar populations and, to some extent, dust extinction. The colors primarily indicate the average age of stellar populations. Though not directly an evolutionary sequence, it reflects galaxy evolution. More detailed data and discussion can be found in Roberts & Haynes (1994).

When star formation occurs, short-lived massive stars ionize the surrounding medium, primarily hydrogen. The star formation rate (SFR) can be inferred from indicators related to massive stars, such as hydrogen recombination lines. Since galaxy color correlates strongly with morphology, there is a corresponding relationship between star formation and morphology. Figure 2.2 shows the distribution of the equivalent width of $H\alpha$ emission[1] by morphology (Kennicutt, 1998). The equivalent width measures star formation activity relative to galaxy luminosity, removing the scaling effect of total stellar mass.

The correlation in Fig. 2.2 is weaker than that with color. Since the timescale for SFR measured by $H\alpha$ is short, local variations can have a significant impact. Nevertheless, later-type galaxies exhibit higher $H\alpha$ equivalent widths, indicating more efficient star formation. The larger dispersion for irregular galaxies suggests they are more affected by internal and external disturbances.

2.3 GALAXIES AND ENVIRONMENT

A remarkable feature of galaxy morphology is its relationship not only with a galaxy's physical properties but also with its larger-scale structure and distribution. This is known as the morphology-density relation. Though this relation has been

---

[1] Strictly speaking, the equivalent width of $H\alpha$+[NII], as decomposing these lines is difficult with low-resolution spectra.

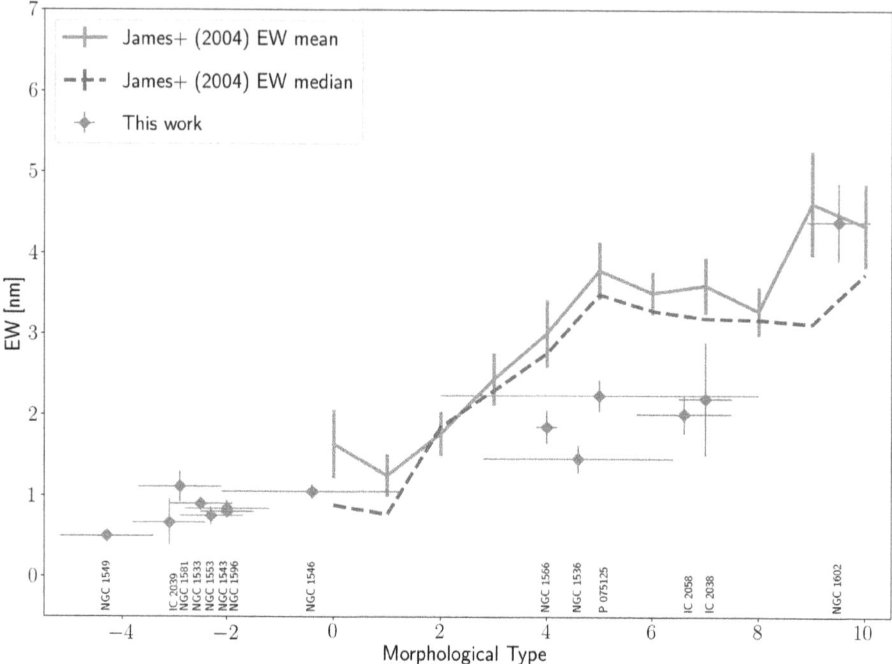

Figure 2.2 Morphological dependence of the Hα equivalent width of galaxies. A solid line with an error bar represents the mean of Hα equivalent widths of a galaxy sample of James et al. (2004), while a dashed line shows the median of the same galaxy sample. Solid symbols with error bars are data of Rampazzo et al. (2020). Galaxy morphology is represented by the index T. Credit: Rampazzo, R., et al. 2020, Astronomy and Astrophysics, 414, 23, Fig. 6.

known since the 1920s–1930s (see Biviano, 2000, for historical discoveries), A. Dressler (Dressler, 1980) was the first to quantitatively demonstrate it. He found that the fraction of elliptical galaxies increases in denser environments, like the centers of galaxy clusters, while the fraction of late-type galaxies (spirals and irregulars) decreases with increasing density. Conversely, late-types dominate in lower-density environments, and ellipticals and lenticulars become less frequent. This trend is clearly illustrated by Dressler et al. (1997) (Fig. 2.3). Houghton (2015) confirmed this trend by linking morphology with kinematic properties (fast and slow rotators). Sazonova et al. (2020) reported that the relation persists in clusters even at $1 < z < 2$.

Following Dressler (1980), Bhavsar (1981) and de Souza et al. (1982) extended the study to less dense environments, discovering that the morphology–density relation continues in small galaxy groups. This segregation is not confined to gravitationally bound structures like clusters or groups but also appears in large-scale structures (e.g., Giovanelli et al., 1986). However, two types of environmental effects must be considered:

1. The relation to local density during galaxy formation (nature).
2. Evolutionary effects that occur mainly in dense environments (nurture).

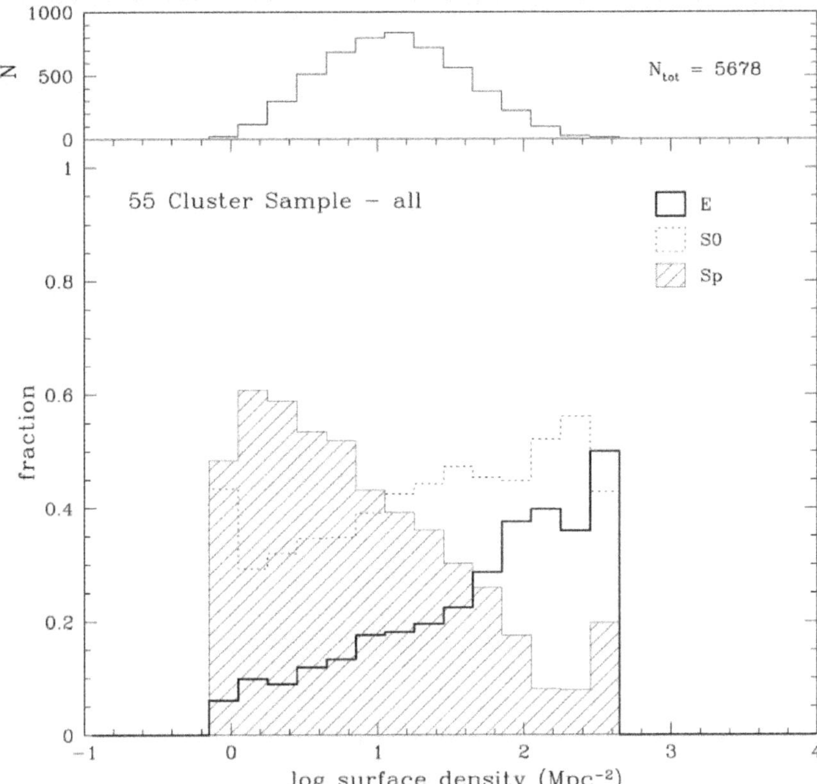

Figure 2.3 The morphology–density relation for the clusters. The sample galaxies are from Dressler (1980). The histogram at the top shows the number of galaxies in each bin of surface density, with the total number indicated in the upper right. Credit: Dressler, A., Oemler, A., Couch, W. J., et al. 1997, Astrophysical Journal, 490, 577, Fig. 1, reproduced by permission of AAS.

The distribution of galaxies likely results from both factors, making it difficult to disentangle them. Smaller-scale phenomena, such as those in clusters or groups, are primarily related to nurture, while larger-scale effects reflect conditions during galaxy formation.

2.4 LUMINOSITY FUNCTION AND RELATED PROPERTIES

The galaxy luminosity function (LF) is defined as the comoving number density of galaxies with luminosity $L \in [L, L + \mathrm{d}L]$, denoted as $\phi(L)$ in units of $[L_\odot^{-1}\,\mathrm{Mpc}^{-3}]$. Essentially, it represents the probability distribution function (PDF) of galaxy luminosity but also conveys the comoving number density of galaxies n_{gal}

$$\int \phi(L)\,\mathrm{d}L = n_{\text{gal}} \;. \tag{2.11}$$

The luminosity density ρ_L is given by

$$\int L\phi(L)\,\mathrm{d}L = \rho_{\text{L}} \;. \tag{2.12}$$

The LF, historically defined using physical density, is now typically expressed as a number density in a comoving volume in modern studies. Thus, the LF is crucial in extragalactic astronomy and observational cosmology. It serves as a fundamental descriptor of the galaxy population and is sometimes analyzed by color, morphology (e.g., Boselli et al., 2016; Marzke & da Costa, 1997; Marzke et al., 1998), spectral type (e.g., de Lapparent, 2003; de Lapparent et al., 2003), environment (e.g., Binggeli et al., 1988; Driver, 2004), and other factors.

The LF is essential for interpreting galaxy number counts (e.g., Koo & Kron, 1992) and analyzing galaxy clustering (Strauss & Willick, 1995, e.g., see Chapter 6). It also provides a key test for galaxy formation theories (e.g., Lacey et al., 2016). To address the 'faint blue galaxy' problem, the evolution of the LF became a focal point (e.g., Ellis, 1997) and has been linked to dwarf galaxy formation (e.g., Babul & Rees, 1992). Recently, the LF evolution has been studied in the context of cosmic star formation density (e.g., Madau & Dickinson, 2014).

2.4.1 SCHECHTER FUNCTION

The galaxy LF from UV to NIR is well described by a single family of functions. Schechter (1976a) proposed a fitting function for the optical galaxy LF

$$\phi(L)\mathrm{d}L = \phi_* \left(\frac{L}{L_*}\right)^\alpha \exp\left(-\frac{L}{L_*}\right) \mathrm{d}\left(\frac{L}{L_*}\right) \tag{2.13}$$

This function has three parameters: the normalization ϕ_*, characteristic luminosity L_*, and faint-end slope α. These parameters have clear physical meanings and are often used to evaluate galaxy populations. This form is inspired by the halo mass function of dark matter halos (Press & Schechter, 1974), discussed in Chapter 7.

The logarithmic definition of the LF is increasingly common, leading to the Schechter function in the form

$$\phi(L)\mathrm{d}\log L = (\ln 10)\phi_* \left(\frac{L}{L_*}\right)^{\alpha+1} \exp\left(-\frac{L}{L_*}\right) \mathrm{d}\log L \tag{2.14}$$

It is important to note that ϕ_* in this equation differs from the original by a factor of $\ln 10$, which often causes confusion in the literature.

In most optical and NIR studies, absolute magnitude M is used instead of luminosity L. In terms of M, the Schechter function becomes

$$\phi(M)\mathrm{d}M = 0.4(\ln 10)\phi_* 10^{-0.4(\alpha+1)(M-M_*)} \tag{2.15}$$

$$\times \exp\left[-10^{-0.4(M-M_*)}\right] \mathrm{d}M \tag{2.16}$$

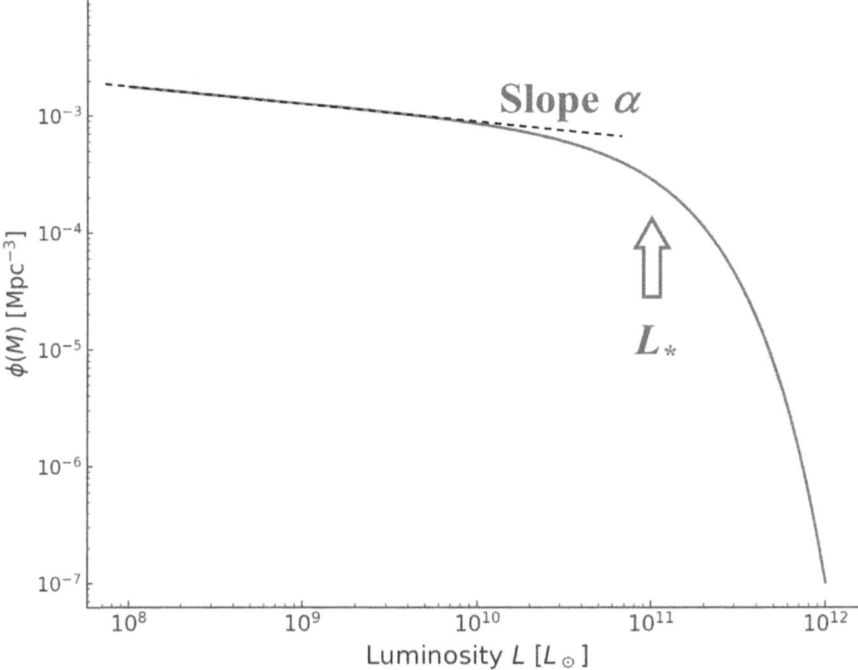

Figure 2.4   The Schechter luminosity function (LF) with its parameters.

Again, there is ambiguity in the normalization parameter in the literature. We follow the original definition of eq. (2.13).

The Schechter function fits various galaxy LFs, from UV (e.g., Wyder et al., 2005), NIR (e.g., Cole et al., 2001), CO (e.g., Kereš et al., 2003), stellar mass (e.g., Pérez-González et al., 2003), HI mass (e.g., Haynes et al., 2011), and baryonic mass (e.g., Eckert et al., 2016).

2.4.2   INTEGRATED QUANTITIES

The Schechter function has useful mathematical properties. Integrating eq. (2.13) over L gives the number density of galaxies with luminosity greater than L

$$n_{\mathrm{gal}}(>L) = \int_L^\infty \phi(L')\mathrm{d}L' = \phi_* \Gamma(\alpha+1, L/L_*) \qquad (2.17)$$

where $\Gamma(a,x)$ is the upper incomplete gamma function. The 1st-order moment with L yields the comoving luminosity density

$$\rho_L(>L) = \int_L^\infty L'\phi(L')\mathrm{d}L' = \phi_* L_* \Gamma(\alpha+2, L/L_*) \qquad (2.18)$$

Thus, $n_{\mathrm{gal}}(> L)$ and $\rho_L(> L)$ are primarily governed by ϕ_* and $\phi_* L_*$, respectively. These integrated quantities are essential for understanding galaxy populations and their evolution (e.g., Takeuchi et al., 2000).

2.4.3 LFS AT OTHER WAVELENGTHS

Equation (2.13) works well for many galaxy populations but not all. At certain wavelengths, LFs deviate significantly from the Schechter form, as seen in far-infrared (FIR), X-ray, and radio continuum LFs.

2.4.3.1 Far-Infrared (FIR)

Some galaxy populations have a heavy tail at the luminous end of the LF, deviating from the exponential decay. This is better described by a double-power law or the function proposed by Saunders et al. (1990b)

$$\phi(L) = \frac{\mathrm{d}n}{\mathrm{d}\log L} = \phi_* \left(\frac{L}{L_*} \right)^{\alpha+1} \exp\left[-\frac{1}{2\sigma^2} \log^2 \left(1 + \frac{L}{L_*} \right) \right] \tag{2.19}$$

Here, ϕ_*, L_*, and α are similar to those in eq. (2.13), with σ^2 characterizing the heavy tail. This form describes the LF of *IRAS* galaxies (Saunders et al., 1990b) and has been useful for IR galaxy LFs in general (e.g., Le Floc'h et al., 2005). As clearly seen in Fig. 2.5, the behavior of the Saunders function is very different at the high-luminosity end. This can be observed in Fig. 2.6, which compares the UV and FIR LFs and their evolution.

2.4.3.2 X-ray

Active galactic nuclei (AGN) emit energetic radiation efficiently detected in X-rays. Ranalli et al. (2005) showed that the X-ray LF fits the Saunders function [eq. (2.19)]. Local luminous IR galaxies (LIRGs) are often dominated by AGNs (Sanders & Mirabel, 1996), making the similarity between IR and X-ray LFs understandable. Miyaji et al. (2015) also found consistency with a double power-law function

$$\phi(L) = \frac{\phi_*^{\mathrm{DP}}}{\left(\frac{L}{L_*} \right)^{-\alpha} + \left(\frac{L}{L_*} \right)^{-\beta}} , \tag{2.20}$$

where ϕ_*^{DP} normalizes the double power-law (e.g., Dunlop & Peacock, 1990).

2.4.3.3 Radio Continuum

Radio continuum LFs, especially at 1.4 GHz, show a different shape. Machalski & Godlowski (2000) and Mauch & Sadler (2007) demonstrated that the radio LF is the sum of two populations: star-forming (SF) galaxies and AGNs. They used eq. (2.19) for SF galaxies and eq. (2.20) for AGNs. Thus, LFs at different wavelengths reveal a range of features that are crucial for understanding galaxy properties.

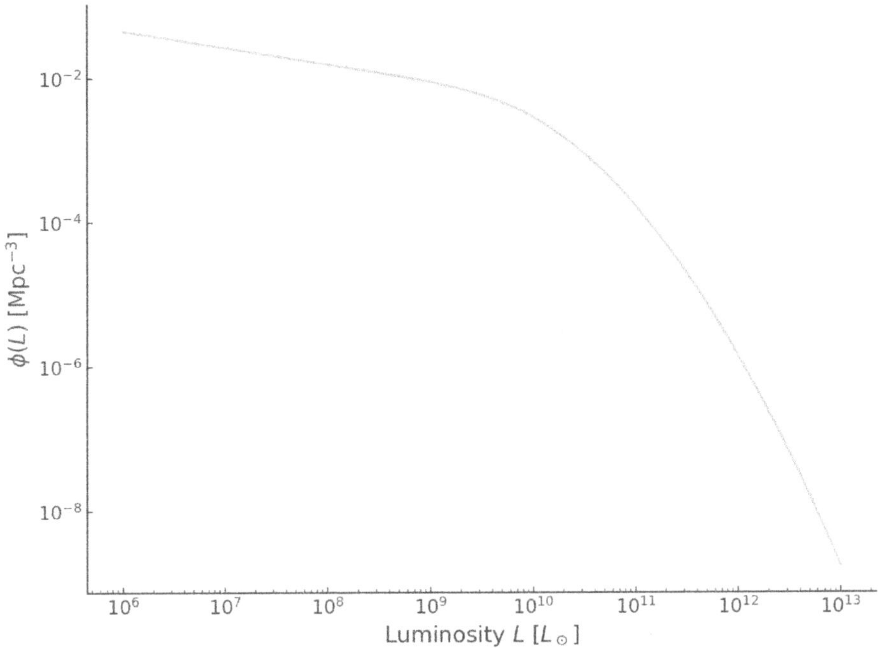

Figure 2.5   The infrared (IR) luminosity function. The parameters are taken from Takeuchi et al. (2003b) and Takeuchi et al. (2004).

2.5 SCALING LAWS

As discussed, galaxies cover a vast range in luminosity or mass (over 10^6 orders), from the largest cD galaxies to dwarf spheroidals (dSphs). Despite this wide range, galaxies follow well-defined scaling relations linking physical, structural, spectrophotometric, or kinematic properties such as size, luminosity, or mass. These relations often indicate a common origin for various galaxy types and help test models of galaxy formation and evolution.

2.5.1 SCALING LAWS OF SPHEROIDAL GALAXIES

2.5.1.1 Faber–Jackson Relation

Faber & Jackson (1976) discovered a relation between galaxy luminosity and central velocity dispersion, known as the Faber–Jackson (FJ) relation

$$L \propto \sigma^4 , \tag{2.21}$$

where σ^2 is the stellar velocity dispersion.

Elliptical galaxies are supported not by rotation but by velocity dispersion, acting as "pressure" to counteract gravity. More luminous galaxies, typically more massive,

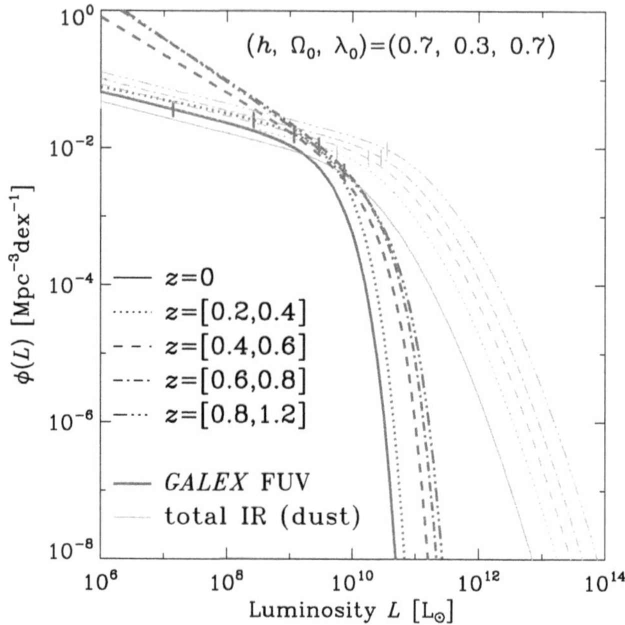

Figure 2.6 The ultraviolet (UV) and infrared (IR) luminosity functions (LFs) and their evolution. The former is described by the Schechter function, while the latter is better represented by the function proposed by Saunders (Saunders et al., 1990b). (Takeuchi et al., 2005a). The series of curves with exponential decline at the high-luminosity end describes the UV LFs, while that with a power-law-like decline shows the FIR LFs. See the ebook for the color version of this figure. Credit: Takeuchi, T. T., et al. 2005, Astronomy and Astrophysics, 440, L17, Fig. 1.

require higher velocity dispersion for dynamical stability. Assuming constant surface brightness μ, the virial theorem yields

$$2\frac{3}{2}\mathcal{M}\sigma^2 = \frac{G\mathcal{M}^2}{R} , \tag{2.22}$$

where $\mathcal{M}$ and R are the mass and size of the galaxy. Luminosity is related by

$$L = 4\pi R^2 \mu , \tag{2.23}$$

leading to the FJ relation after eliminating R. In practice, the power index in the FJ relation varies from 3 to 5, reflecting the complex structure of elliptical galaxies. The FJ relation is useful for estimating galaxy distances without requiring the Hubble constant, but it has a large scatter (~ 2 mag). The D_n–σ relation, where D_n is a parameter representing the radius where surface brightness reaches $20.75\ B[\mathrm{mag\,arcsec^{-2}}]$

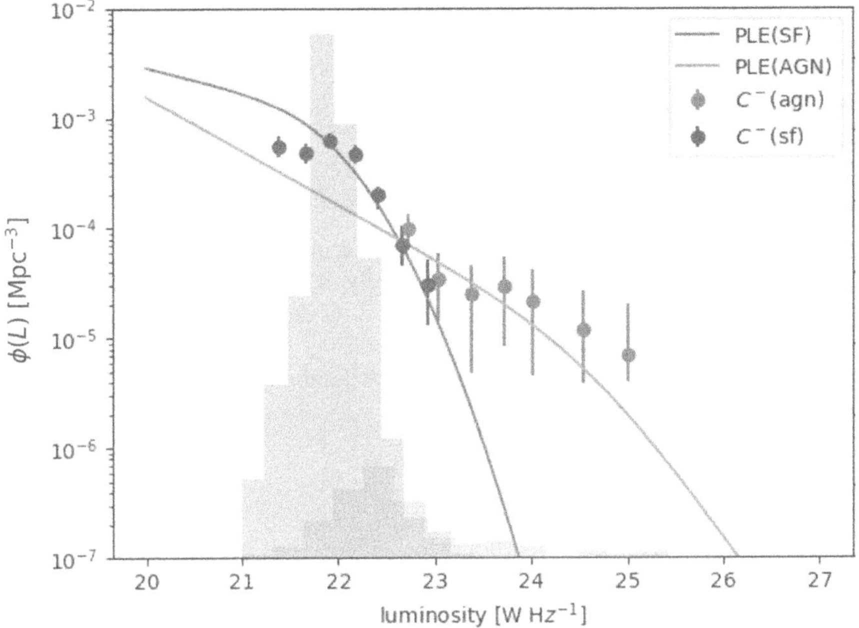

Figure 2.7 The local LF at 3 GHz estimated from VLA-COSMOS data at $0.1 < z < 0.4$. Symbols represent nonparametric estimate of the luminosity functions of star-forming galaxies (denoted as SF) and active galactic nuclei (AGN). The curves are parametric estimate of the luminosity functions of SF galaxies and AGNs. See ebook for color version of this figure. Courtesy: Kai T. Kono.

(Dressler et al., 1987), refines the FJ relation, reducing the scatter

$$D_n \propto \sigma^{1.33} \, , \tag{2.24}$$

2.5.1.2 The Fundamental Plane Relation

To reduce the scatter in the FJ relation, additional parameters were introduced (e.g., Djorgovski & Davis, 1987; Dressler et al., 1987). In 3D space, the relation becomes

$$L \propto \sigma^{3.45} I^{-0.86} \, . \tag{2.25}$$

Here, L is luminosity, I is surface brightness, and σ is velocity dispersion, all measured within the effective radius r_e. The FJ relation is a projection of this fundamental plane, and its scatter arises from projecting the 3D space onto 2D. The κ-parameter space, constructed from orthogonal combinations of L, I, and σ, provides better insight (Bender et al., 1992, 1993). Though the theoretical basis for the fundamental plane is not fully understood, it remains a useful tool for studying elliptical galaxies.

2.5.2 SCALING LAWS OF DISK GALAXIES

2.5.2.1 Tully–Fisher relation

For disk galaxies, the well-known Tully–Fisher (TF) relation links rotational velocity v_{max} to luminosity L_{band}

$$L_{\mathrm{band}} \propto v_{\mathrm{max}}^{\alpha} , \tag{2.26}$$

where v_{max} is the rotational velocity corrected for inclination, and L_{band} is the luminosity in a specific photometric band. Originally, $\alpha = 2.5$ from V-band data (Tully & Fisher, 1977), but later studies found $\alpha \simeq 3$–4, depending on wavelength, with H-band ($\lambda = 1.65\ \mu$m) giving $\alpha = 4.3$.

The TF relation is physically explained by assuming mass distribution follows the surface brightness profile

$$I(r) = I_0 \exp\left(-\frac{r}{r_0} \right) , \tag{2.27}$$

leading to a total disk mass proportional to

$$\mathcal{M} \propto 2\pi I_0 r_0^2 . \tag{2.28}$$

This results in the scaling relation

$$L \propto v_{\mathrm{max}}^4 . \tag{2.29}$$

2.5.2.2 Baryonic Tully–Fisher Relation

The classical TF relation depends on wavelength because it uses optical/NIR luminosity, which varies with star formation history and other factors. A deviation from a log-linear relation for lower v_{max} galaxies was noted by McGaugh et al. (2000). Lower-mass galaxies are less luminous than expected, but this issue is resolved by using the baryonic Tully–Fisher (BTF) relation, which replaces luminosity with total baryon mass. McGaugh et al. (2000) and McGaugh (2005) showed that the BTF relation restores log-linearity across a wide range of rotational velocities

$$\mathcal{M}_{\mathrm{B}} \simeq 50 v_{\mathrm{max}}^{\alpha_{\mathrm{BTF}}} , \tag{2.30}$$

where $\alpha_{\mathrm{BTF}} = 4.0$. Various representations of the Tully–Fisher (TF) relation are shown in Fig. 2.8. Classical TF relation as the relation between the B-band luminosity and the rotation velocity v_{max} (denoted as V_f in the plot) is shown in the top-left panel in Fig. 2.8. We see a slight downward deviation at the low rotational velocity end. In contrast, the relation between the gas mass $\mathcal{M}_{\mathrm{ISM}}$ ($\mathcal{M}_g$) and v_{max} shows a significant downward deviation at high v_{max} (top-right). The bottom left panel plots stellar mass $\mathcal{M}_*$ converted from luminosity, showing even more prominent nonlinear deviation at low v_{max}. Finally, if we see the baryonic TF relation, the relation between $\mathcal{M}_{\mathrm{B}} = \mathcal{M}_* + \mathcal{M}_{\mathrm{ISM}}$ ($\mathcal{M}_d$) and v_{max}, show a very tight single log-linear relation (McGaugh et al., 2000).

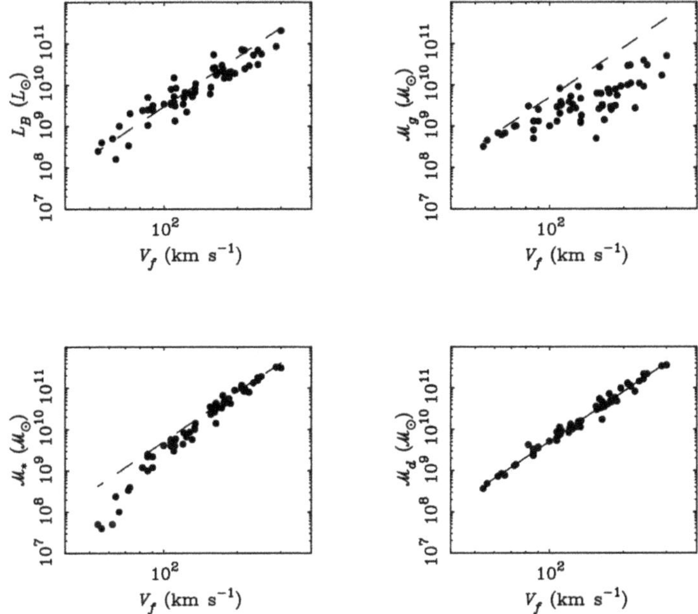

Figure 2.8 Four versions of the Tully–Fisher (TF) relation. The top left panel shows the B-band luminosity as a function of the rotation velocity $v_{\max}$ (denoted as V_f in the plot). The top right panel plots the gas mass $\mathcal{M}_{\mathrm{ISM}}$ ($\mathcal{M}_g$) instead of luminosity. The bottom left panel plots stellar mass $\mathcal{M}_*$ converted from luminosity. The bottom right panel shows the baryonic TF relation, with $\mathcal{M}_{\mathrm{B}} = \mathcal{M}_* + \mathcal{M}_{\mathrm{ISM}}$ ($\mathcal{M}_d$). The solid line in the bottom right panel is a fit for the data. Credit: McGaugh, S. S. 2005, The Astrophysical Journal, 632, 859, Fig. 1, reproduced by permission of AAS.

2.5.3   PHOTOMETRIC SCALING RELATIONS FOR THE WHOLE POPULATION OF GALAXIES

2.5.3.1   Color–Magnitude Relation

Since the 1970s, astronomers have studied the galaxy counterpart to the color–magnitude diagram used for stars. Visvanathan & Sandage (1977) identified a tight red sequence for early-type galaxies, while Visvanathan & Griersmith (1977) found a looser sequence for spirals, now referred to as the blue cloud. Both populations, shown in Fig. 2.9, are discussed within the unified color–magnitude diagram (Blanton, 2006). The less populated region between the red sequence and the blue cloud is sometimes called the green valley.

It has been demonstrated that the red sequence corresponds to ellipticals, while the blue cloud corresponds to spirals (e.g., Driver et al., 2006). The formation of these two distinct branches has been a subject of long-standing debate, but no definitive explanation has been found yet (e.g., Faber et al., 2007).

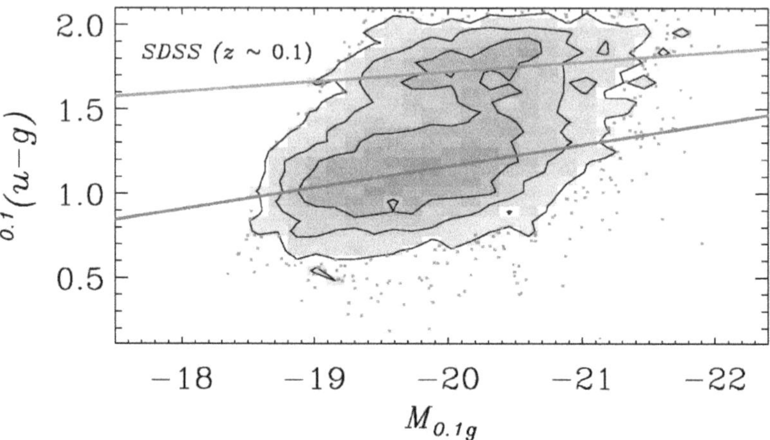

Figure 2.9　The Local ($z \simeq 0.1$) color–magnitude relation of the SDSS galaxies. The upper solid line indicates the red sequence, while the lower solid line indicates the blue cloud. Credit: Blanton, M. R. 2006, Astrophysical Journal, 648, 268, Fig. 3, reproduced by permission of AAS.

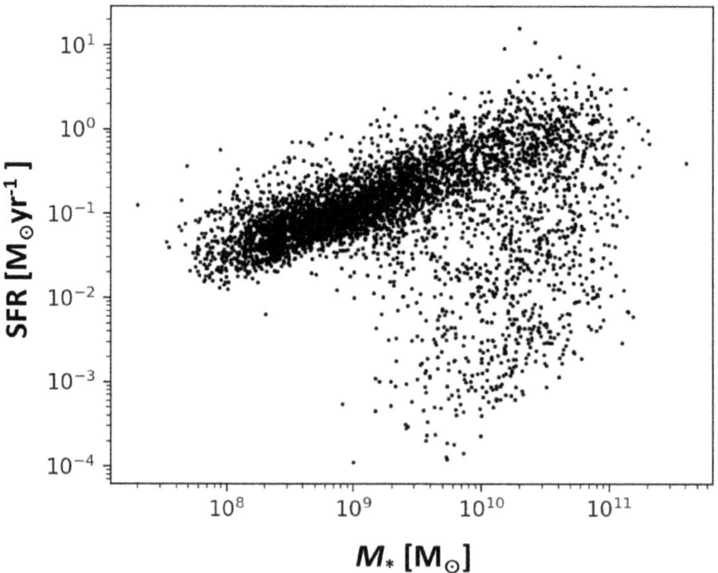

Figure 2.10　The correlation between stellar mass and SFR derived from the fitting of spectral energy distribution to RCSED sample galaxies (Chillingarian et al. 2017). Courtesy: Suchetha Cooray.

2.5.3.2 The Star-Forming Galaxy Main Sequence

Closely related to the color–magnitude relation is the stellar mass—SFR diagram. Red sequence galaxies are often referred to as "red and dead" because they have stopped forming stars, whereas blue cloud galaxies are generally still forming stars. Plotting galaxies on a stellar mass ($\mathscr{M}_*$)–SFR diagram shows that star-forming galaxies form a relatively tight sequence, often called the "main sequence of star-forming galaxies" (e.g., Elbaz et al., 2007; Noeske et al., 2007). This main sequence has been found to evolve with cosmic time and depends on various physical parameters, which continues to be a topic of extensive study (e.g., Magnelli et al., 2015; Pannella et al., 2015; Santini et al., 2014).

3 Interstellar Medium (ISM)

3.1 PHASES OF THE INTERSTELLAR MEDIUM (ISM)

The interstellar medium (ISM) is a mixture of gas and dust that fills the space between stars. Dust consists of solid particles made from heavy elements like carbon, silicon, oxygen, and iron. The gas-to-dust mass ratio in the Milky Way is around 100 but varies across galaxies, depending on their metallicity. The composition of gas in the ISM is approximately:

Hydrogen	92 % (by number)
Helium	8 %
Other elements (Oxygen, Carbon, etc.)	0.1 %

The ISM is predominantly composed of atomic hydrogen (HI), with typical conditions of $n = 1$ $[\mathrm{cm}^{-3}]$ and $T = 100$ [K]. The temperature is regulated by the balance between heating and cooling processes, while density is determined by forces like pressure and gravity.

3.1.1 MULTIPHASE ISM STRUCTURE

The ISM is made up of different phases (see Fig. 3.1 and Table 3.1), each characterized by its temperature and density:

- Molecular hydrogen (H_2): forms dense molecular clouds where stars are born.
- Atomic hydrogen (HI): the largest gas component in the Galaxy, serving as a reservoir for molecular clouds.
- Ionized hydrogen (HII or H^+): ionized by UV radiation from massive stars.
- Coronal (hot, ionized) gas: primarily ionized by supernova explosions, releasing energies around 10^{51} erg per explosion.

A common classification divides the ISM into three main phases: hot ($T > 10^6$ K), warm ($T \sim 10^4$ K), and cold ($T \sim 10^2$ K) (see McKee & Ostriker, 1977). These phases are generally in pressure equilibrium, with sharp boundaries between them.

The ISM reaches equilibrium through a balance of heating (from young stars and supernova shocks) and cooling (mainly through radiation). The cooling function Λ describes how efficiently the ISM radiates energy, depending on temperature and metallicity (see Fig. 3.2) (e.g., Gnat & Ferland, 2012; Revaz et al., 2009; Sutherland & Dopita, 1993). Higher metallicities significantly enhance cooling efficiency due to contributions from heavy elements like iron (for details, see, e.g., Gnat & Ferland, 2012).

DOI: 10.1201/9781003104315-3

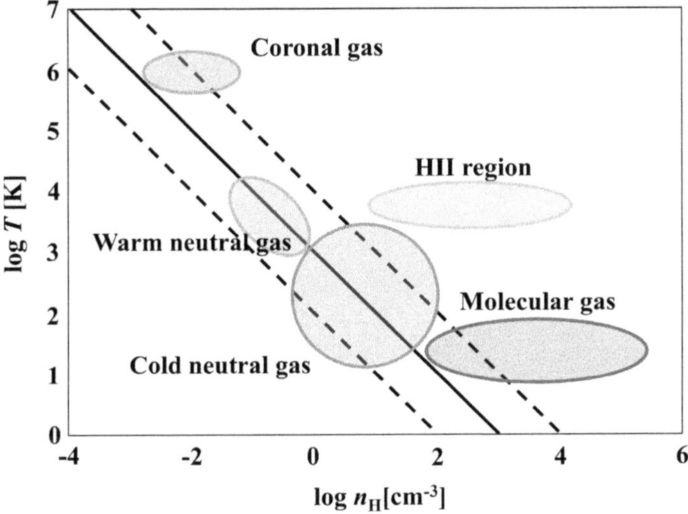

Figure 3.1 Phase diagram of the interstellar medium (ISM).

3.2 NEUTRAL HYDROGEN (HI) MEDIUM

The neutral phase of the ISM is dominated by atomic hydrogen (HI), which is studied primarily through long-wavelength radio observations, specifically the 21-cm line emission.

3.2.1 21-CM LINE EMISSION FROM HI

The 21-cm line arises from a quantum mechanical transition in hydrogen atoms, where the spin of the electron flips relative to the proton spin. The relative population of the upper and lower energy states follows the Boltzmann distribution

$$\frac{n_u}{n_l} = \frac{g_u}{g_l} \exp\left(-\frac{h_P \nu}{k_B T_{ex}}\right), \tag{3.1}$$

Table 3.1

Physical states of the multiphase ISM.

Component	Temperature [K]	Density [cm⁻³]	Volume Fraction
Diffuse ionized	10^6	0.01	50 %
HII regions	5000	0.5	< 1 %
Diffuse atomic	30–100	10–100	30 %
Diffuse molecular	30–100	10–500	30 %
Dense molecular	10–15	$> 10^4$	10 %

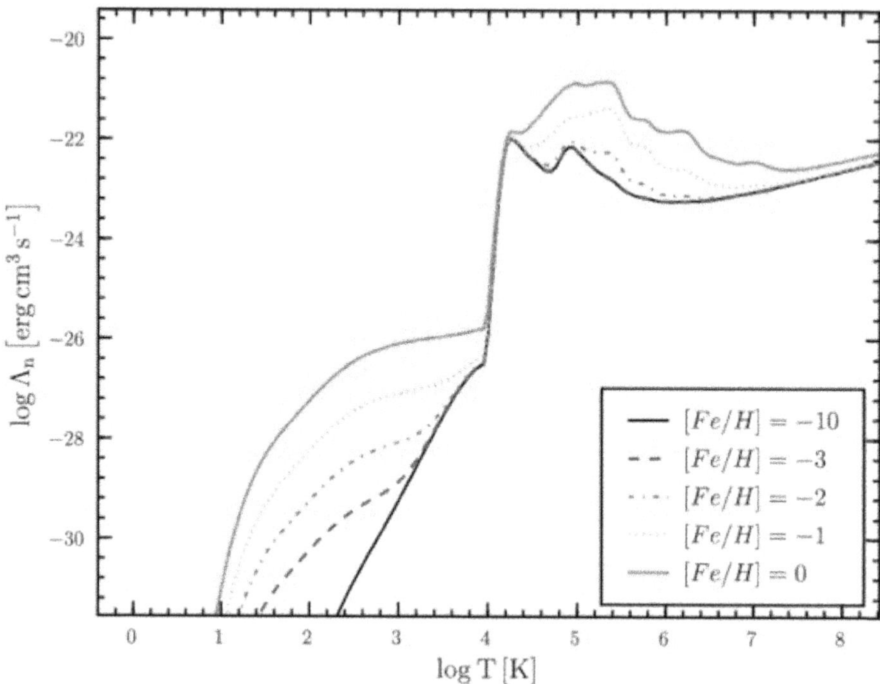

Figure 3.2 Cooling function of the ISM. Top: shows cooling rates for different metallicities ($\log Z$ from -10 to 0). Credit: Revaz, Y., Jablonka, P., Sawala, T., et al. 2009, Astronomy and Astrophysics, 501, 189, Fig. 1.

where g_u and g_l are the statistical weights of the upper and lower states, and T_ex is the excitation temperature.

The 21-cm line was discovered in 1951 (Ewen & Purcell, 1951). The energy levels in a hydrogen atom are

$$E = \frac{\mu_{ep}e^4}{2\hbar n^2} = 13.6 n^{-2} \ [\mathrm{eV}], \tag{3.2}$$

where n is the principal quantum number and μ_{ep} is the reduced mass of the electron-proton system

$$\mu_{ep} = \frac{m_e m_p}{m_e + m_p}. \tag{3.3}$$

The hyperfine splitting occurs due to the interaction between the spins of the proton and electron, creating a small energy difference between parallel and anti-parallel spin configurations.

The probability of spontaneous emission from the upper state is determined by the Einstein A coefficient, calculated from quantum mechanics

$$A_{21\,\mathrm{cm}} = \frac{64\pi v^3}{3h_\mathrm{P}c^3} |\mu_{21}|^2 = 2.85 \times 10^{-15} \ [\mathrm{s}^{-1}], \tag{3.4}$$

where $\mu_{21} = 8 \times 10^{-41}$ [erg^{2}G^{-2}] is the magnetic moment. This results in a long lifetime for the upper state ($\sim 10^7$ years), but the vast number of HI atoms in galaxies makes this radiation detectable.

3.2.2 COLLISIONS AND SPIN FLIP

In addition to spontaneous emission, collisions between HI atoms can also excite the transition. The rate of collisions, known as the C coefficient, is approximately

$$C \simeq n_{\rm HI}\sigma_{\rm HI}\langle v \rangle = 1.5 \times 10^{11} \left(\frac{n}{1\,[{\rm cm}^{-3}]} \right) \left(\frac{T_{\rm K}}{100\,[{\rm K}]} \right)^{\frac{1}{2}} [{\rm s}^{-1}], \qquad (3.5)$$

where $T_{\rm K}$ is the kinetic temperature, and the average thermal velocity of HI atoms is $\langle v \rangle = 1.1 \times 10^6\, T_{\rm K}^{1/2}\,[{\rm cm\,s}^{-1}]$.

The typical collision timescale is $C^{-1} \sim 2 \times 10^3$ [years], much shorter than the spontaneous emission timescale, making collisional spin-flip transitions common in the ISM.

3.2.3 MEASURING HI MASS

We now discuss how to measure physical properties, such as the mass of HI clouds, from observations. Generally, the optical depth of HI clouds is small, $\tau_{\rm HI} < 1$, which indicates that photons emitted from HI can escape without undergoing collisions. Therefore, the observed flux can be directly related to the column density (the density integrated along the line of sight). To express this quantitatively, we employ radiative transfer theory and the Einstein coefficients.

3.2.4 RADIATIVE TRANSFER OF THE HI LINE

The radiative transfer equation is defined as

$$dI_v = -\kappa_v I_v dx + j_v dx, \qquad (3.6)$$

where

$$
\begin{aligned}
\kappa_v \quad &: \text{absorption coefficient } [{\rm cm}^{-1}], \\
j_v \quad &: \text{emissivity } [{\rm erg\,s}^{-1}\,{\rm cm}^{-3}\,{\rm sr}^{-1}\,{\rm Hz}^{-1}], \\
dI_v \quad &: \text{energy per unit area, per unit time, per unit solid angle,} \\
&\quad\ \text{and per unit frequency } [{\rm erg\,s}^{-1}\,{\rm cm}^{-2}\,{\rm sr}^{-1}\,{\rm Hz}^{-1}], \\
dx \quad &: \text{path length } [{\rm cm}].
\end{aligned}
$$

We define the optical depth as

$$\tau_v(x) \equiv \int_0^x \kappa_v(x')dx', \qquad (3.7)$$

and the source function as

$$\mathscr{S}_v \equiv \frac{j_v}{\kappa_v} \,. \tag{3.8}$$

If $j_v(x)$ and $\kappa_v(x)$ are constant, the radiative transfer equation can be integrated as

$$I_v(x) = I_{v0}(x)e^{-\tau_v(x)} + \mathscr{S}_v \left[1 - e^{-\tau_v(x)}\right] \,. \tag{3.9}$$

The first term on the right-hand side represents the attenuation of the initial intensity through absorption, while the second term represents the net radiation from the cloud's emission and self-absorption.

When the intensity remains constant, i.e.,

$$\frac{\mathrm{d}I_v}{\mathrm{d}x} = 0 \,, \tag{3.10}$$

we find

$$I_v = \mathscr{S}_v \,. \tag{3.11}$$

Here, $\mathscr{S}_v$ is the intensity in equilibrium, where the rates of emission and absorption are balanced. For material in thermal equilibrium with radiation at temperature T for a given frequency v, i.e., when

$$I_v = B_v(T) \,, \tag{3.12}$$

the equation becomes

$$\mathrm{d}I_v = -\kappa_v B_v(T)\mathrm{d}x + j_v\mathrm{d}x \,. \tag{3.13}$$

Thus, the source function in thermal equilibrium becomes

$$\mathscr{S}_v = B_v(T) \,. \tag{3.14}$$

This implies that the source function behaves as a blackbody in local thermodynamic equilibrium (LTE). Under LTE conditions, a single temperature T sufficiently describes both the local gas and the radiation field, provided temperature variations are small. Additionally, the thermalization length must be shorter than the scale over which temperature changes.

3.2.5 RADIATIVE TRANSFER USING EINSTEIN COEFFICIENTS

The energy released by spontaneous emission from a volume element $\mathrm{d}V$ in time $\mathrm{d}t$, into a solid angle $\mathrm{d}\omega$, and within a frequency interval $\mathrm{d}v$, is given by the Einstein A coefficient as

$$\mathrm{d}E_{\mathrm{em}} = h_\mathrm{P}\, v n_\mathrm{u}\, \phi_{\mathrm{line}}(v)A_{\mathrm{ul}}\mathrm{d}v\,\mathrm{d}V\mathrm{d}t\,\frac{\mathrm{d}\Omega}{4\pi}\,[\mathrm{erg}] \,. \tag{3.15}$$

From this equation, we can derive the emissivity

$$j_\nu = h_P \nu n_u \phi_{\text{line}}(\nu) \frac{A_{ul}}{4\pi} \; [\text{erg}\,\text{s}^{-1}\,\text{cm}^{-3}\,\text{sr}^{-1}\,\text{Hz}^{-1}]. \tag{3.16}$$

Here, $\phi_{\text{line}}(\nu)$ is the line profile function, which characterizes the frequency distribution of the line emission from volume element dV, and is normalized as

$$\int_0^\infty \phi_{\text{line}}(\nu)\,d\nu = 1. \tag{3.17}$$

The broadening of spectral lines is caused by

- The natural spread in photon energies due to Heisenberg's uncertainty principle, $\Delta E \Delta t \geq \hbar$, where Δt is the lifetime of the excited state.
- Doppler broadening, which results from the thermal motion of atoms along the line of sight, causing shifts in the observed frequency.

In the actual interstellar medium (ISM), Doppler broadening typically dominates for the H$\textsc{i}$ line.

The energy absorbed by a hydrogen atom receiving isotropic radiation is described using the Einstein B coefficient

$$dE_{\text{abs}} = h_P \nu n_l \phi_{\text{line}}(\nu) B_{lu} I_\nu d\nu\, dV dt \frac{d\Omega}{4\pi} \; [\text{erg}]. \tag{3.18}$$

From this, we obtain the absorption coefficient

$$\kappa_\nu = h_P \nu n_l \phi_{\text{line}}(\nu) \frac{B_{lu}}{4\pi} \; [\text{cm}^{-1}]. \tag{3.19}$$

There is also another quantum mechanical effect known as stimulated emission, where a photon with energy equal to the energy gap between states can induce further emission. This stimulated emission sometimes referred to as "negative absorption," is described as

$$dE_{\text{se}} = h_P \nu n_u \phi_{\text{line}}(\nu) B_{ul} I_\nu d\nu\, dV dt \frac{d\Omega}{4\pi} \; [\text{erg}]. \tag{3.20}$$

When combining these processes, the radiative transfer equation can be written as

$$\begin{aligned}
\frac{dI_\nu(x)}{dx} &= -\kappa_\nu I_\nu + j_\nu \\
&= -\frac{h_P \nu}{4\pi}\left(n_l B_{lu} - n_u B_{ul}\right)\phi_{\text{line}}(\nu) I_\nu(x) + \frac{h_P \nu}{4\pi} n_u A_{ul} \phi_{\text{line}}(\nu).
\end{aligned} \tag{3.21}$$

The Einstein coefficients are not independent of each other. In thermodynamic equilibrium, the number of transitions from the upper state must equal the number from the lower state. This gives the condition

$$n_l B_{lu} I_\nu^{\text{TE}} = n_u A_{ul} + n_u B_{ul} I_\nu^{\text{TE}}. \tag{3.22}$$

Solving the intensity of I_v^{TE}, we get

$$I_v^{\text{TE}} = \frac{\dfrac{A_{\text{ul}}}{B_{\text{ul}}}}{\left(\dfrac{n_{\text{u}}}{n_{\text{l}}}\right)\left(\dfrac{B_{\text{lu}}}{B_{\text{ul}}}\right) - 1} \,. \tag{3.23}$$

In thermodynamic equilibrium, the ratio $n_{\text{u}}/n_{\text{l}}$ is governed by the Boltzmann law as shown in [eq. (3.1)]

$$\frac{n_{\text{u}}}{n_{\text{l}}} = \frac{g_{\text{u}}}{g_{\text{l}}} \exp\left(-\frac{h_{\text{P}} v}{k_{\text{B}} T_{\text{ex}}}\right) ,$$

leading to

$$I_v^{\text{TE}} = \frac{\dfrac{A_{\text{ul}}}{B_{\text{ul}}}}{\left(\dfrac{g_{\text{l}} B_{\text{lu}}}{g_{\text{u}} B_{\text{ul}}}\right) \exp\left(\dfrac{h_{\text{P}} v}{k_{\text{B}} T_{\text{ex}}}\right) - 1} \,. \tag{3.24}$$

In statistical equilibrium, the radiation intensity should follow that of a blackbody. Hence, we obtain the following relations

$$A_{\text{ul}} = \frac{2 h_{\text{P}} v^3}{c^2} B_{\text{ul}} \,, \tag{3.25}$$

$$g_{\text{l}} B_{\text{lu}} = g_{\text{u}} B_{\text{ul}} \,. \tag{3.26}$$

These equations link the quantum mechanical quantities A_{ul}, B_{lu}, and B_{ul}, and they hold regardless of whether the atoms are in thermodynamic equilibrium.

3.2.6 APPLICATION TO THE MEASUREMENT OF H$\textsc{i}$

For the hyperfine splitting of H$\textsc{i}$, where $g_{\text{u}} = 3$ and $g_{\text{l}} = 1$, the intensity can be expressed as

$$I_v^{\text{TE}} \simeq B_v(T)\tau_v = B_v(T) \int_0^{D_{\text{obj}}} \kappa_v \, dx \simeq \frac{2 v^2 k_{\text{B}} T}{c^2} \kappa_v \, dx \,. \tag{3.27}$$

We also have

$$\begin{aligned}
\kappa_v &= \frac{h_{\text{P}} v}{4\pi} \left(n_{\text{l}} B_{\text{lu}} - n_{\text{u}} B_{\text{ul}}\right) \phi_{\text{line}}(v) \\
&= \frac{h_{\text{P}} v}{4\pi} \frac{g_{\text{u}}}{g_{\text{l}}} \frac{c^2}{2 h_{\text{P}} v^3} A_{\text{ul}} \phi_{\text{line}}(v) n_{\text{l}} \frac{h_{\text{P}} v}{k_{\text{B}} T} \,.
\end{aligned} \tag{3.28}$$

Therefore

$$\int_{\text{line}} I_v \, dv = \frac{h_{\text{P}} v}{4\pi} \frac{g_{\text{u}}}{g_{\text{l}}} A_{\text{ul}} \int n_{\text{l}} \, dx \,. \tag{3.29}$$

In eq. (3.29), the integral of n_u gives the column density itself. This means that measurements of line intensity allow us to determine the column density of the HI gas. Furthermore, with a sky map of line intensity, we can calculate the total HI mass $\mathscr{M}_\mathrm{HI}$.

3.3 MOLECULAR CLOUDS

Next, we explore molecular clouds in the ISM. These dense clouds are also very cold, with temperatures around 20 K (see Table 3.1). Molecular clouds consist predominantly of molecules and are named accordingly. The formation of molecules requires high densities and sufficient dust to shield the gas from UV radiation. However, studying the physics of molecular clouds using optical or near-infrared (NIR) wavelengths is challenging due to heavy dust extinction. Instead, observations in the submillimeter (submm) and millimeter (mm) wavelength regimes are more informative.

3.3.1 SUBMM/MM OBSERVATIONS OF MOLECULES

Transitions between rotational states in the electric dipole of a molecule emit photons in the radio (submm–mm) frequency range. For a diatomic molecule with reduced mass μ, the energy level is given by

$$E_\mathrm{rot} = J(J+1)\frac{\hbar^2}{2\mu r^2} , \qquad (3.30)$$

where J is the rotational quantum number and r is the separation between the two atoms. Since radio-wavelength photons experience minimal absorption, molecular gas clouds can be detected even in cases where other gas and dust clouds are present along the same line of sight. This makes submm/mm observations highly effective for studying interstellar molecular gas.

These clouds are primarily composed of molecular hydrogen (H_2). However, since H_2 molecules do not have an electric dipole moment, they do not emit in the radio wavelength range. Instead, molecular clouds are typically observed through the mm-wave emission of CO molecules, specifically from the rotational transition from $J = 1$ to $J - 0$, which corresponds to $\lambda \simeq 2.6$ [mm]. Additionally, in very dense clouds, the isotope ^{13}CO—which is about 40 times less abundant than ^{12}CO—can still be observed in regions that are optically thick in ^{12}CO.

3.3.2 CO-TO-H_2 CONVERSION FACTOR

To determine the total molecular content of a galaxy, we primarily need to measure the mass of H_2. This requires converting the observed molecular line data into information about H_2. The strength of CO lines is known to correlate well with the amount of H_2, and the following relationship is often used

$$X(\text{CO}) = \frac{N(H_2)}{\displaystyle\int T_A \, dv}, \tag{3.31}$$

where T_A is the antenna temperature. For a homogeneous, self-gravitating cloud, eq. (3.31) simplifies to

$$X(\text{CO}) = \frac{k_B \lambda}{16\pi^2 \hbar c} \left(\frac{15 n_H}{2.8 G m_H}\right)^{\frac{1}{2}} \left[\exp\left(\frac{h_P \nu}{k_B T_{\text{ex}}}\right) - 1\right], \tag{3.32}$$

(Draine, 2011). For the CO $J = 1\text{–}0$ line ($\lambda = 0.26$ [cm]) with $n_H = 10^3$ [cm^{-3}] and $T_{\text{ex}} = 8$ [K], we obtain

$$X(\text{CO } J = 1\text{–}0) = 1.56 \times 10^{20} \ [\text{cm}^{-2}(\text{K km s}^{-1})^{-1}]. \tag{3.33}$$

However, it is important to note that there are significant uncertainties in this relationship. For instance, in dark molecular gas (Wolfire et al., 2010), hydrogen exists in molecular form, but CO does not. It is estimated that up to 30 % of a molecular cloud mass could reside in such gas.

3.3.3   MASS AND DYNAMICS OF MOLECULAR CLOUDS

Small dark clouds typically have masses ranging from 10^2 to 10^4 $\mathcal{M}_\odot$ and radii between 1 and 10 pc. In contrast, giant molecular clouds have sizes of 10 to 60 pc and masses between 10^5 and $10^{6.5}$ $\mathcal{M}_\odot$. The velocity dispersion estimated from CO emission line widths in molecular clouds is usually around 3 km s^{-1}, much larger than the thermal velocities expected at $T \simeq 30$ [K]. This suggests that the line width is primarily due to turbulent motion. The velocity dispersion σ_v^2 is found to correlate with the cloud size R_{cl} as

$$\sigma_v \ [\text{km s}^{-1}] \simeq 0.8 \left(\frac{R_{\text{cl}}}{1 \ [\text{pc}]}\right)^{0.5} \tag{3.34}$$

(Larson, 1981). Since the pressure from turbulence exceeds the thermal pressure, it must be counterbalanced by the gravitation of a cloud. Giant molecular clouds are likely in near-virial equilibrium. Assuming a cloud mass $\mathcal{M}_{\text{cl}}$, size R_{cl}, and density profile $\rho(r) \propto r^{-1}$, the virial theorem gives

$$3 \mathcal{M}_{\text{cl}} \sigma_v^2 = \frac{2}{3} \frac{G \mathcal{M}_{\text{cl}}^2}{R_{\text{cl}}} \tag{3.35}$$

From this, the cloud mass is

$$\mathcal{M}_{\text{cl}} = \frac{9}{2} \frac{R_{\text{cl}} \sigma_v^2}{G}. \tag{3.36}$$

Combining eqs. (3.34) and (3.36), we find

$$\mathscr{M}_{\mathrm{cl}} \simeq \frac{2.8 R_{\mathrm{cl}}^2}{G} , \tag{3.37}$$

and the column number density of a molecular cloud can be expressed as

$$\mathscr{N}_{\mathrm{H}} \simeq \frac{2.8}{\pi G \mu m_{\mathrm{H}}} = 1.5 \times 10^{22} \, [\mathrm{cm}^{-2}] , \tag{3.38}$$

which is nearly constant, independent of the cloud mass.

3.4   IONIZED (HII) REGION

In the region around a massive star, HI gas is ionized. The ionized hydrogen region is referred to as the HII region.

3.4.1   PHYSICAL PROCESSES IN HII REGIONS

In HII regions, photo-ionization and recombination take place, as

$$\mathrm{H} + h\nu \leftrightarrow \mathrm{H}^+ + e^- . \tag{3.39}$$

The right-hand side is the photoionization ,and the left-hand side is the recombination. In the recombination process and subsequent cascade of the energy levels, the recombination lines are emitted. Ionizing photons should have an energy of $h_{\mathrm{P}}\nu > 13.6$ [eV] ($\lambda < 912\,[\text{Å}]$). Massive stars have higher temperatures, emit UV photons that ionizes H (as well as He, Li etc.) in the uncollapsed parts of the molecular cloud. In HII regions, gas is in steady state, in the balance between photoionization by stellar photons and the inverse process of recombination that occurs when ions capture free electrons.

3.4.2   STRÖMGREN RADIUS

A spherical region of radius r forms around a hot star, within which all photons capable of ionizing hydrogen are absorbed by the surrounding gas. In steady-state conditions, the following balance must be maintained

[Total number of H recombinations]

= [Total number of photo-ionizations]   (per unit time)   (3.40)

Here, we define Q_* as the number of ionizing photons emitted by the star per unit time, and R_{rec} as the number of recombinations per unit of time and volume. This leads to the relation

$$Q_* = \frac{4\pi r^3}{3} R_{\mathrm{rec}} . \tag{3.41}$$

Recombination requires the interaction of two particles with relative velocity v and cross-section σ_{rec}. Assuming $n_e = n_p$ for pure hydrogen gas, the recombination rate is given by

$$R_{\mathrm{rec}} = n_p n_e \langle \sigma_{\mathrm{rec}} v \rangle \equiv n_p n_e \alpha(T) = n_e^2 \alpha(T) = x_e n^2 \alpha(T) \,, \qquad (3.42)$$

where $\alpha(T) = \langle \sigma_{\mathrm{rec}} v \rangle$ is the recombination coefficient, and x_e is the fraction of ionized gas with density n as

$$n_e = x_e n \qquad (3.43)$$

(as discussed in Appendix A). Assuming the gas is fully ionized ($x_e \simeq 1$), we obtain the Strömgren radius

$$r_{\mathrm{S}} = \left(\frac{3 Q_*}{4 \pi \alpha n^2} \right)^{\frac{1}{3}} . \qquad (3.44)$$

Now, we estimate the typical value of the Strömgren radius. A main-sequence O5 star emits ionizing photons at a rate of

$$Q_* = \int_{h_{\mathrm{P}} v = 13.6\,[\mathrm{eV}]}^{\infty} \frac{\mathscr{L}_v(\mathrm{O5})}{h_{\mathrm{P}} v} \, dv \simeq 3 \times 10^{49} \, [\mathrm{s}^{-1}] . \qquad (3.45)$$

Meanwhile, the recombination coefficient at $T = 10^4$ K is

$$\alpha(10^4\,\mathrm{K}) = 2.6 \times 10^{-13} \, [\mathrm{cm}^3 \mathrm{s}^{-1}] . \qquad (3.46)$$

Note that recombinations to the ground state ($n = 1$) result in photons with energies greater than 13.6 eV, which will quickly ionize nearby neutral hydrogen (HI) atoms. Therefore, these recombinations do not affect the overall ionization balance and can be neglected.

For gas with a number density of $n = 10^4$ $[\mathrm{cm}^{-3}]$, the Strömgren radius is calculated as

$$r_{\mathrm{S}} = \left[\frac{3 \times 3 \times 10^{49} \, [\mathrm{s}^{-1}]}{4\pi \times 2.6 \times 10^{-13} \, [\mathrm{cm}^3 \mathrm{s}^{-1}] \times (10^4 \, [\mathrm{cm}^{-3}])^2} \right]^{\frac{1}{3}}$$
$$\simeq 6.27 \times 10^{17} \, [\mathrm{cm}] \simeq 0.2 \, [\mathrm{pc}] . \qquad (3.47)$$

Although the static Strömgren radius model is somewhat simplistic, as HII regions typically expand over time, these formulae still provide a useful estimate of the ionization-recombination balance within HII regions. More related discussions will follow in Chapter 5.

3.5   PHOTODISSOCIATION REGION (PDR)

A photodissociation region (PDR) is the interface between an HII region and a molecular cloud. It is also referred to as a photon-dominated region, with the same

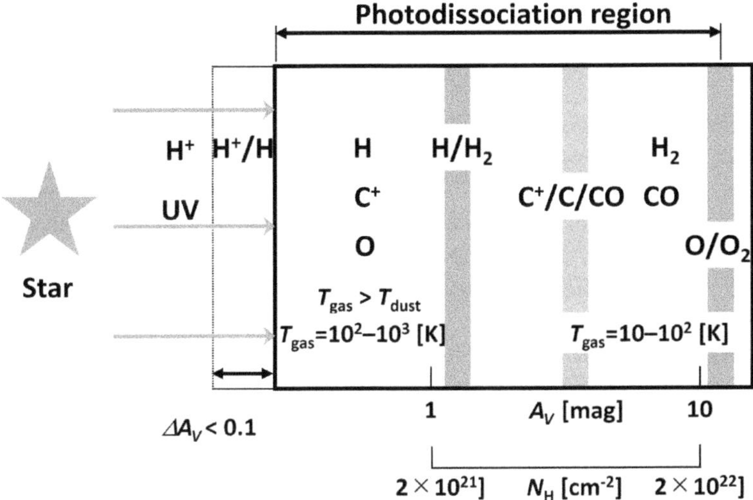

Figure 3.3 A schematic diagram of a photodissociation region (PDR). The PDR is illuminated from the left, extending from the predominantly atomic surface region to the point where O_2 is no longer appreciably photodissociated. The PDR thus includes gas where hydrogen is mainly in the form of H_2 and carbon is mostly CO. Large columns of warm oxygen, carbon, C^+, and CO, along with vibrationally excited H_2, are produced within the PDR.

acronym. This region is characterized by a photodissociation surface, where the hydrogen is half atomic and half molecular. Nearly all atomic and much of the molecular gas in a galaxy can be found in a PDR. Within a PDR, the following photodissociation reaction occurs

$$H_2 + UV \longrightarrow 2H \tag{3.48}$$

This means that the PDR is where the heating mechanisms, chemical composition, and reactions are dominated by far-ultraviolet (FUV) radiation.

3.5.1 SELF-SHIELDING

In PDRs, H_2 is the most common molecule, about 10,000 times more abundant than CO. Most H_2 forms on the surfaces of dust grains. It is important to note that not all FUV photons dissociate H_2. In fact, only 10–15 % of FUV photons will cause dissociation upon absorption. Otherwise, the H_2 remains in a vibrationally excited state. Collisional de-excitation from this state can heat the gas (e.g., Hollenbach & Tielens, 1999). Due to the high density of H_2, their FUV lines become optically thick, creating a sharp transition between HI and H_2. The location of this transition is determined by self-shielding (in high-density, low-radiation regimes) and by dust opacity (in low-density, high-radiation regimes).

In PDRs, C^+ and C are balanced by photoionization and radiative recombination. CO molecules are formed through

1. reactions of neutral radicals like CH and CH_2,
2. sequential reactions

$$C^+ + OH \longrightarrow CO^+ + H, \tag{3.49}$$

$$CO^+ + H_2 \longrightarrow HCO^+ + H, \tag{3.50}$$

$$HCO^+ + e \longrightarrow CO + H. \tag{3.51}$$

CO formation proceeds slowly due to the low abundance of its reactants. Unlike H_2, CO is not significantly self-shielded due to its lower abundance. Thus, CO is primarily found deep within clouds, with its transition determined by dust extinction.

3.5.2 HEATING AND COOLING

Heating and cooling processes in PDRs play a crucial role in the energy balance of the ISM.

Heating mechanisms in PDRs include

1. **Photoelectric heating**

   Dust grains and polycyclic aromatic hydrocarbons (PAHs) are central to photoelectric heating. A PAH is a hydrocarbon made up of only carbon and hydrogen, consisting of multiple aromatic rings. PAHs emit strongly at various mid-infrared (MIR) wavelengths, such as 6.2, 7.7, 8.2, 8.6, 11.2, 11.9, 12.7, 13.5, 14.5, and 16.4 μm. Heating in PDRs is dominated by small grains and PAHs, as they absorb UV radiation more efficiently than H_2 (see Fig. 3.4).

2. **Collisional ee-excitation**

   When H_2 molecules absorb a photon, they transition to a vibrationally excited state. At high densities, these molecules can de-excite through collisions with atomic hydrogen, heating the gas.

3. **Cosmic-ray heating**

   In high-density regions where UV radiation cannot penetrate, cosmic ray ionization becomes an important heating process.

Cooling processes in PDRs include

1. **Atomic line emission**

   Due to the high excitation energy required for hydrogen, cooling in the interstellar gas is mainly through metal lines. FIR forbidden lines, like [CII] 158 μm and [OI] 63 μm and 146 μm, are essential. The [CII] 158μm line often dominates the cooling in the ISM of galaxies.

2. **Molecular line emission**

   Vibro-rotational emission lines from H_2 are efficient cooling mechanisms. UV-excited H_2 decays via electric quadrupole transitions, primarily emitting in the near-infrared (NIR). CO, with its dipole moment, also acts as an effective radiator, with many high-J transitions working efficiently.

3. **Collisional excitation**

   Collisional excitation of atoms, molecules, and ions by electrons, followed by radiative decay, effectively cools the gas.

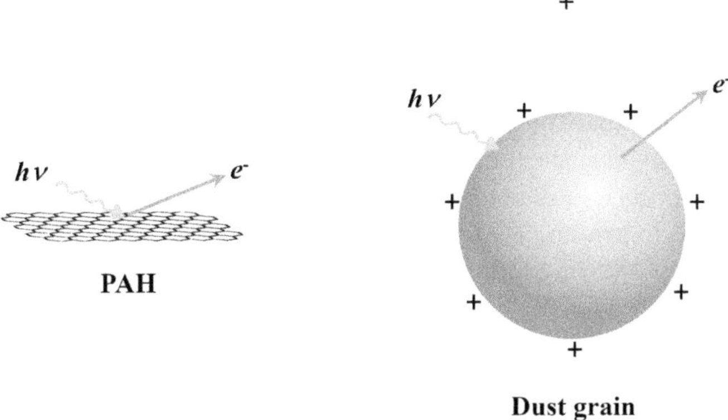

Figure 3.4 The mechanism of photoelectric heating. A far-ultraviolet photon absorbed by a dust grain produces a photoelectron that diffuses on the surface of the grain, losing energy through collisions or escaping. For PAHs, diffusion does not play a role. A simplified expression for the heating efficiency, ε, is provided.

4. **Grain cooling**

   Dust grains absorb optical and UV radiation and re-emit in the mid- to far-infrared (MIR–FIR). Collisions between gas particles and grains, which are generally cooler than the gas, can also lead to net cooling.

The properties of PDRs in galaxies are becoming increasingly important in the study of high-redshift galaxies.

3.6 JEANS INSTABILITY

The interstellar medium (ISM) can be treated as a fluid. To study its dynamics, we start with the basic equations of fluid mechanics in astrophysics

1. The equation of continuity

$$\frac{\partial \rho}{\partial t} + \nabla \cdot (\rho \vec{u}) = 0 \tag{3.52}$$

2. The Euler equation (equation of motion)

$$\rho \left[\frac{\partial \vec{u}}{\partial t} + (\vec{u} \cdot \nabla)\vec{u} \right] = -\nabla p - \rho \nabla \Phi , \tag{3.53}$$

3. The Poisson equation

$$\nabla^2 \Phi = 4\pi G \rho . \tag{3.54}$$

In addition to these three, additional equations, such as the equation of state (relating ρ and p), are required to fully describe fluid dynamics.

Furthermore, we may also consider the energy conservation equation. Let the internal energy per unit mass be denoted by U, then the internal energy per unit volume is ρU. Similarly, the kinetic energy per unit volume is $\rho u^2/2$. If the external force K has a potential Φ_K, then the potential energy per unit volume is $\rho \Phi_K$. The total energy per unit volume is then

$$\mathscr{E} \equiv \frac{\rho u^2}{2} + \rho U + \rho \Phi_K \,. \tag{3.55}$$

Next, we consider the flow of energy and the net work done on the fluid element. The energy influx due to pressure is expressed as $\nabla \cdot (p\vec{u})$. Thus, the energy conservation equation takes the following form

$$\frac{\partial}{\partial t}\left(\frac{\rho u^2}{2} + \rho U + \rho \Phi_K\right) + \nabla \cdot \left[\left(\frac{\rho u^2}{2} + \rho U + \rho + \Phi_K\right)\vec{u} + p\vec{u} + \vec{F}_{\mathrm{rad}}\right] = Q_+ - Q_- \,. \tag{3.56}$$

In this equation, $\vec{F}_{\mathrm{rad}}$, Q_+, and Q_- represent terms not directly related to the fluid element. In the second term on the left-hand side, the expression inside the brackets is called the energy flux. The terms represent, respectively, kinetic energy transport, internal energy transport, gravitational work, and pressure work. The term $\vec{F}_{\mathrm{rad}}$ accounts for energy flux loss due to processes such as radiation or convection, which are not directly tied to fluid variables. On the right-hand side, Q_+ represents heating in the fluid element, while Q_- represents cooling. These terms act as direct sources or sinks for the energy density of the fluid element.

3.6.1 WHAT IS FLUID STABILITY?

Many astrophysical processes, such as star formation, are driven by fluid mechanical instabilities in the ISM. A steady-state solution to the hydrodynamic equations, which we often assume, may not exist or persist due to inherent small perturbations in any system. If such small perturbations grow over time, the steady-state solution is deemed unstable and cannot represent a realistic physical scenario.

Fluid instabilities play a critical role in astrophysical phenomena. We will study some of these instabilities relevant to galaxy formation and evolution. To examine the stability of a fluid system, we conduct a stability analysis, which consists of two main steps

1. Linearize the governing equations around the equilibrium state.
2. Determine whether small perturbations grow over time.

3.6.2 FORMULATION OF THE JEANS INSTABILITY

Gravitational instability, or Jeans instability, is the fundamental reason why matter in the Universe is not distributed uniformly. Stars and galaxies are believed to be the result of perturbations that began to grow due to the Jeans instability. We start with the basic fluid equations that include gravity: the equation of continuity [eq. (3.52)]

$$\frac{\partial \rho}{\partial t} + \nabla \cdot (\rho \vec{u}) = 0 \,,$$

the Euler equation [eq. (3.53)]

$$\rho \left[\frac{\partial \vec{u}}{\partial t} + (\vec{u} \cdot \nabla)\vec{u} \right] = -\nabla p - \rho \nabla \Phi ,$$

and the Poisson equation

$$\nabla^2 \Phi = 4\pi G \rho .$$

First, to find the equilibrium state, we set all time derivatives to zero

$$\nabla p_0 = -\rho_0 \nabla \Phi_0 \tag{3.57}$$

$$\nabla^2 \Phi_0 = 4\pi G \rho_0 \tag{3.58}$$

$$\rho_0 = \text{constant} \tag{3.59}$$

$$\vec{u} = \vec{0} . \tag{3.60}$$

Next, we introduce small perturbations to the equilibrium state

$$\rho = \rho_0 + \rho_1(\vec{x},t) \tag{3.61}$$

$$p = p_0 + p_1(\vec{x},t) \tag{3.62}$$

$$\Phi = \Phi_0 + \Phi_1(\vec{x},t) \tag{3.63}$$

$$\vec{u} = \vec{u}_1(\vec{x},t) , \tag{3.64}$$

where all first-order quantities are assumed to be small.

By substituting these perturbed quantities [eqs. (3.61)–(3.64)] into the governing equations and neglecting second-order terms (linearization), we obtain the following linearized equations

$$\frac{\partial \rho_1}{\partial t} + \nabla \cdot (\rho_0 \vec{u}_1) = 0 \tag{3.65}$$

$$\rho_0 \frac{\partial \vec{u}_1}{\partial t} = -\nabla p_1 - \rho_0 \nabla \Phi_1 = -c_s^2 \nabla \rho_1 - \rho_0 \nabla \Phi_1 \tag{3.66}$$

$$\nabla^2 \Phi_1 = 4\pi G \rho_1 . \tag{3.67}$$

Here, we assume the adiabatic relation between p_1 and ρ_1 as

$$p_1 = \left(\frac{\partial p}{\partial \rho} \right)_{\text{ad}} \rho_1 = c_s^2 \rho_1 . \tag{3.68}$$

This relation holds for adiabatic perturbations, where c_s is the sound speed. Under adiabatic conditions,

$$p \propto \rho^\gamma , \tag{3.69}$$

and

$$c_s^2 = \frac{\gamma p}{\rho} = \frac{\gamma k_B T}{m} . \tag{3.70}$$

Eliminating $\vec{u}_1$ and Φ_1, we arrive at the equation

$$\frac{\partial^2 \rho_1}{\partial t^2} = c_{\mathrm{s}}^2 \nabla^2 \rho_1 + 4\pi G \rho_0 \rho_1 \,. \tag{3.71}$$

This resembles a wave equation. Assuming perturbations of the form $\propto e^{i(\vec{k}\cdot\vec{x}-\omega t)}$, we derive the dispersion relation

$$\omega^2 = k^2 c_{\mathrm{s}}^2 - 4\pi G \rho_0$$
$$k \equiv |\vec{k}| \,. \tag{3.72}$$

We then have the following stability conditions

$$\omega^2 > 0 \qquad \text{Stable}$$
$$\omega^2 < 0 \qquad \text{Unstable (Jeans instability)}$$

If $\omega^2 < 0$, then $\omega = \pm i\lambda$ (imaginary), indicating that the perturbations grow exponentially. This implies that a gravitationally bound structure forms when $\omega^2 < 0$, leading to gravitational instability, or Jeans instability.

3.6.3 JEANS CRITERION

The boundary between stable and unstable states is determined by the Jeans wavenumber

$$k_{\mathrm{J}} \equiv \sqrt{\frac{4\pi G \rho_0}{c_{\mathrm{s}}^2}} \,, \tag{3.73}$$

Waves with a wavenumber smaller than this are subject to instability. The corresponding Jeans wavelength is

$$\lambda_{\mathrm{J}} \equiv \frac{2\pi}{k_{\mathrm{J}}} = c_{\mathrm{s}} \left(\frac{\pi}{G\rho_0} \right)^{\frac{1}{2}} \,, \tag{3.74}$$

Half of the Jeans wavelength is called the Jeans radius, $R_{\mathrm{J}} \equiv \lambda_{\mathrm{J}}/2$. The mass contained within a sphere of radius R_{J} is the Jeans mass

$$\mathcal{M}_{\mathrm{J}} \equiv \frac{4\pi R_{\mathrm{J}}^3 \rho_0}{3} \,. \tag{3.75}$$

The Jeans mass plays a central role in the formation of stars and galaxies.

3.7 STAR FORMATION IN THE CONTEXT OF GALAXIES

Star formation is the fundamental process driving galaxy evolution. While the Jeans instability in the ISM is likely the most important physical process, many other baryonic mechanisms are now understood to play a role in modern star formation theory

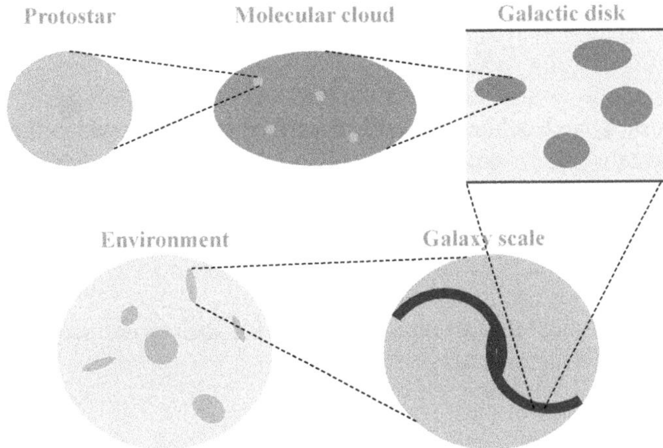

Figure 3.5 Star formation across various spatial scales, from protostellar cores (on the order of a few AU) to the galactic environment (up to Mpc scales).

(e.g., Krumholz, 2015). The spatial scale of star formation itself is typically less than a few parsecs, ultimately reaching the scale of AU (see Fig. 3.5).

Despite extensive research on star formation at these scales, insights from ISM physics have not always been integrated in a well-organized manner. Consequently, star formation processes are often treated as simple threshold mechanisms for gas (e.g., Boissier, 2017).

3.7.1 STAR FORMATION CRITERIA

3.7.1.1 Toomre's Q Parameter

In a seminal paper, Toomre (1964) extensively studied the stability of rotating disks, a concept often connected to the criterion for star formation in galactic disks. For stellar disks, Toomre's Q parameter is defined as

$$Q_* = \frac{\sigma_* \kappa_{\mathrm{epi}}}{3.36 G \Sigma_*} , \tag{3.76}$$

where σ_* and Σ_* are the velocity dispersion and surface density of the stellar disk, and κ_{epi} is the epicyclic frequency (e.g., Binney & Tremaine, 2008).

Similarly, for a gas disk, the Q parameter is

$$Q_{\mathrm{gas}} = \frac{\sigma_{\mathrm{gas}} \kappa_{\mathrm{epi}}}{\pi G \Sigma_{\mathrm{gas}}} , \tag{3.77}$$

where σ_{gas} and Σ_{gas} are the velocity dispersion and surface density of the gas disk, respectively, with all quantities depending on the radius r.

For a disk containing both stars and gas, the combined Q parameter is given by

$$Q \simeq \frac{\sigma_{\text{gas}} \kappa_{\text{epi}}}{\pi G \Sigma_{\text{gas}}} \left(1 + \frac{\sigma_* \Sigma_{\text{gas}}}{\sigma_{\text{gas}} \Sigma_*} \right)^{-1} . \tag{3.78}$$

The condition $Q = 1$ defines the critical surface density at which the galactic disk becomes unstable to self-gravitational collapse, setting a threshold for star formation. For gaseous disks, this critical surface density is

$$\Sigma_{\text{crit}} \equiv \frac{c_s \kappa_{\text{epi}}}{\pi G} = Q \Sigma_{\text{gas}} . \tag{3.79}$$

A related threshold that accounts for the disk's shear is defined by the Q_A parameter, proposed by Hunter et al. (1998)

$$Q_A = \frac{2.5 \sigma_{\text{gas}} A}{\pi G \Sigma_{\text{gas}}} , \tag{3.80}$$

$$A \equiv 0.5 R \frac{\mathrm{d}\Omega(R)}{\mathrm{d}R} . \tag{3.81}$$

This criterion incorporates the effect of shear, caused by the differential rotation of the disk, which tends to suppress gravitational instability.

3.7.1.2 Phase Transition Criterion

Gas in galaxies cools and contracts due to various triggers. As the temperature decreases, the ISM transitions from ionized to neutral, and ultimately to molecular form in dense regions. The most abundant element in the universe, hydrogen, forms molecular clumps in these dense regions.

Schaye (2004) proposed a threshold based on the phase transition of the ISM from hot to cold gas. The critical surface density, above which a cold phase can exist, depends on several factors, including

f_{gas} Gas mass fraction
f_{th} Ratio of thermal to total pressure
Z Metallicity
$\mathcal{N}_{\text{UV}}$ UV photon flux density

It may also depend on the dust-to-metal ratio. The minimum column density for which $Q = 1$ is

$$\begin{aligned}
\log N_{\text{H,crit}} =\ & 20.68 + 0.28 \log f_{\text{gas}} + 0.020 (\log f_{\text{gas}})^2 \\
& - 0.35 \log f_{\text{th}} + 0.030 (\log f_{\text{th}})^2 \\
& - 0.30 \log \frac{Z}{0.1\,[Z_\odot]} - 0.047 \log \left(\frac{Z}{0.1\,[Z_\odot]} \right)^2 \\
& + 0.22 \log \frac{\mathcal{N}_{\text{UV}}}{10^6\,[\text{cm}^{-2}]} + 0.022 \left(\log \frac{\mathcal{N}_{\text{UV}}}{10^6\,[\text{cm}^{-2}]} \right)^2 .
\end{aligned} \tag{3.82}$$

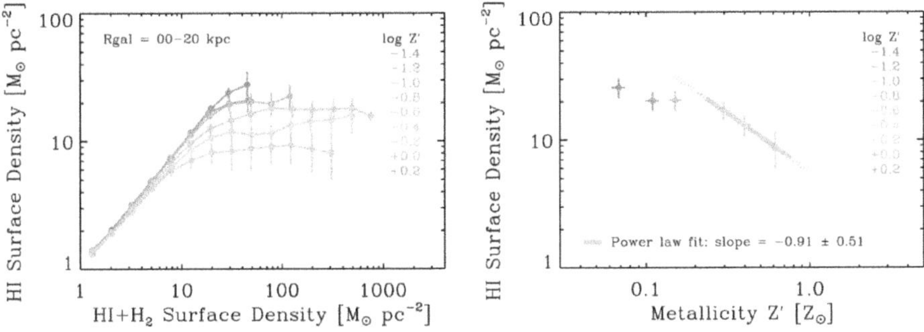

Figure 3.6 HI saturation as a function of total gas mass surface density (left) and metallicity (right). Credit: Schruba, A., Bialy, S., & Sternberg, A. 2018, Astrophysical Journal, 862, 110, Fig. 3, reproduced by permission of AAS.

Schaye (2004) highlights that $N_{\mathrm{H,crit}}$ depends explicitly on f_{th}, since the Q parameter is directly influenced by turbulent velocity dispersion.

To delve deeper into star formation, it is important to consider the transition from cold neutral gas to the molecular phase in the ISM, as star formation typically occurs in molecular clouds. Schruba et al. (2018) studied the metallicity dependence of the saturation surface density of HI, which serves as the threshold for the transition from atomic to molecular hydrogen (see Fig. 3.6).

The threshold depends on the metallicity Z of the ISM, as molecular formation in present-day galaxies occurs primarily on dust grains (e.g., Draine, 2011). Their observations show that for $Z = 0.3$–2, the HI surface density increases almost linearly with decreasing metallicity

$$\log \Sigma_{\mathrm{HI}} = (-0.86 \pm 0.19)\log Z + (0.98 \pm 0.04)\,, \tag{3.83}$$

where Σ_{HI} is measured in $\mathcal{M}_\odot\,\mathrm{pc}^{-2}$, and $Z = 1$ corresponds to solar metallicity. For $Z \lesssim 0.3$, the HI surface density flattens at $\Sigma_{\mathrm{HI}} \simeq 20$–$30\,[\mathcal{M}_\odot\,\mathrm{pc}^{-2}]$. This relationship has been further discussed in Bialy & Sternberg (2016).

3.7.2 KENNICUTT–SCHMIDT LAW

As discussed earlier, it is natural to assume that the efficiency of star formation (SF) would be linked to gas density. Dating back to the 1950s, Schmidt (1959) proposed that the star formation rate (SFR) is proportional to the nth power of gas mass density ρ, leading to what is now known as the Schmidt law of star formation

$$\mathrm{SFR}\,[\mathcal{M}_\odot\,\mathrm{yr}^{-1}] \propto \rho^n\,, \tag{3.84}$$

If $n \simeq 1$, it suggests that the amount of stars formed in a cloud is directly proportional to the gas density. In contrast, $n \simeq 2$ implies that the SFR depends on the square of the gas mass density, hinting at processes like cloud-cloud collisions. The power-law index n provides insights into the underlying mechanisms that trigger star formation.

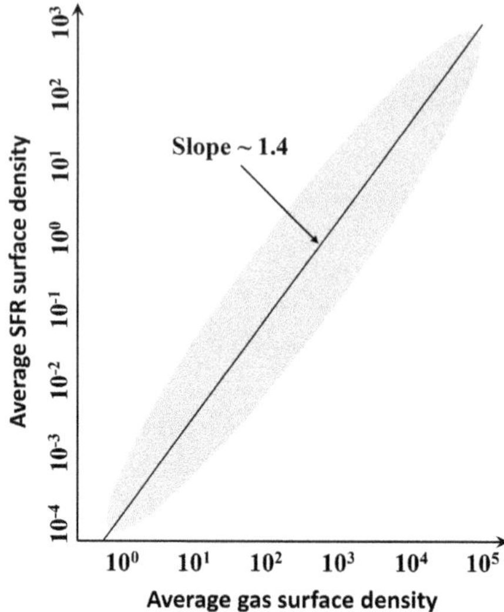

Figure 3.7 A schematic disk-averaged Kennicutt–Schmidt law for total gas surface density. Both star formation rate (SFR) and gas surface density are averaged within the radius of the main star-forming disk.

Schmidt (1959) studied the vertical distribution of HI and stars in the Milky Way, suggesting $n \simeq 2$. However, in those days, measuring gas density or SFR with accuracy was challenging. Later studies by Sanduleak (1969) and Hartwick (1971) examined this relation in the Small Magellanic Cloud (SMC) and M31, finding a surface density relation

$$\Sigma_{\text{SFR}} \propto \Sigma_{\text{HI}}^{N} \tag{3.85}$$

with $N = 1.84 \pm 0.14$ for SMC and $N = 3.50 \pm 0.12$ for M31. If the scale height in the galaxy remains constant, $N = n$.

Subsequent work by Madore et al. (1974) explored this relationship in M33, noting a variation in n between the inner and outer regions of the galaxy. Other studies by Newton (1980), Tosa & Hamajima (1975), and Hamajima & Tosa (1975) extended this analysis to multiple nearby galaxies, finding N values between 1.5 and 2.9. During these investigations, the primary focus was on HI gas as the tracer.

The discovery of the CO emission line by Penzias et al. (1971) revolutionized the study of molecular gas. By the 1980s, observations of CO lines revealed that total hydrogen gas (both atomic HI and molecular H_2) was crucial for understanding SF in galaxies. Thirty years after Schmidt's initial proposal, Kennicutt (1989) significantly advanced this work by demonstrating a log-linear relation between gas mass surface density and SFR. This was further refined in Kennicutt (1998), where improved data quality confirmed an index of $n = 1.4$ over six orders of magnitude in surface density. This relation is now known as the Kennicutt–Schmidt (K-S) law, named after its key contributors.

The K-S law represents the culmination of various processes that govern the transformation of gas into stars. Numerous theoretical models have been proposed to explain the power-law index n, with particular attention given to $n = 1.4$, which lies between the expected values of 1 and 2.

If stars form from gas with volume density ρ_{gas} over a star formation timescale t_{SF} and efficiency ε, the SFR is

$$\text{SFR} = \varepsilon \frac{\rho_{gas}}{t_{SF}} \, . \tag{3.86}$$

If ε is constant and star formation depends solely on the mass of giant molecular clouds (GMCs), the SFR per unit molecular gas mass remains constant, leading to $N = 1$. Alternatively, if the free-fall time of gas, which scales as $1/\sqrt{G\rho}$, is related to the K-S law, we obtain

$$\rho_{SFR} \propto \frac{\rho_{gas}}{(G\rho_{gas})^{-\frac{1}{2}}} \propto \rho_{gas}^{1.5} \, , \tag{3.87}$$

as discussed by Elmegreen (e.g., 1994); Kennicutt (e.g., 1998); Li et al. (e.g., 2006).

Silk (1997) proposed a more general model where

$$\Sigma_{SFR} \propto \frac{\Sigma_{gas}}{\tau_{dyn}} \propto \Sigma_{gas}\Omega_{gas} \tag{3.88}$$

with τ_{dyn} as the local dynamical timescale of the disk and Ω_{gas} as the angular rotation speed.

On the other hand, if SF is driven by cloud-cloud collisions, we expect $N = 2$ (e.g., Tan, 2000). Komugi et al. (2006) provided a comprehensive discussion of cloud-cloud collisions. Considering GMCs moving through a galactic disk with radius R, height h, mass m_{GMC}, and velocity v, the mean free path λ_{GMC} of GMCs can be derived as

$$\lambda_{GMC} = \frac{1}{\sqrt{2}\sigma_{GMC}n_{GMC}} \, , \tag{3.89}$$

where σ_{GMC} is the cross-section and n_{GMC} is the number density of GMCs. The collision timescale t_{cc} is

$$t_{cc} = \frac{\lambda_{GMC}}{v} = \frac{m_{gas}h}{\sqrt{2}\pi a^2 \Sigma_{gas} v} \, . \tag{3.90}$$

Setting $t_{SF} = t_{cc}$, the SFR scales as

$$\Sigma_{SFR} \propto \Sigma_{gas}^2 \, . \tag{3.91}$$

More recent numerical simulations have incorporated turbulence and magnetic fields into star formation models. Krumholz & McKee (2005) proposed that stars form in molecular clouds where self-gravity and supersonic turbulence are balanced.

They found that their results align with the K-S law for gas surface densities $\Sigma_{gas} > 1 \, [\mathcal{M}_\odot \, \text{kpc}^{-2}]$.

Numerous other scenarios have been proposed to explain the K-S law, and many researchers claim to have uncovered the underlying physics.

Before the 2000s, discussions on the K-S law were largely focused on global properties, i.e., surface densities of various quantities averaged over entire galaxies. It is important to emphasize that the scale of a galaxy is significantly larger than that of the local star formation (SF) regions. Therefore, it is essential to begin with the global K-S law when investigating the relationship between local and global SF processes. Naturally, additional insights gained from spatially resolved analyses of galaxies are crucial for bridging the gap between macro- and micro-scale physics in galaxies. This shift led to a growing interest in spatially resolved studies of the K-S law.

Wong & Blitz (2002) analyzed the relationship between HI, H_2, and SFR in seven molecular gas-rich galaxies. Their study demonstrated that H_2 correlates more strongly with SFR than HI, and that the HI gas surface density saturates at $10 \, \mathcal{M}_\odot \, \text{pc}^{-2}$. The K-S law indices they derived were $N_{mol} = 0.7$–2.1 for H_2 and $N_{gas} = 1.2$–2.1 for total gas. Various studies have reported different K-S law indices. For example, Boissier et al. (2003) found $N_{gas} \simeq 2$ from a sample of 16 galaxies. Onodera et al. (2010) examined the K-S law in M33 across different spatial scales (from 80 pc to 1 kpc) and found that the K-S law breaks down at the scale of giant molecular clouds (GMCs). In the nearby spiral galaxy M51, Schuster et al. (2007) reported $N_{gas} = 1.4 \pm 0.6$, while Kennicutt et al. (2007) obtained $N_{gas} = 1.56 \pm 0.04$ for the same galaxy.

Recently, statistical investigations of the spatially resolved K-S law have expanded, thanks to large-scale mapping surveys of molecular gas in galaxies. One notable study was conducted by Bigiel et al. (2008), who assembled a sample of eighteen nearby galaxies, consisting of eleven HI-dominant galaxies and seven galaxies with an H_2-rich central region, alongside FUV (1530 Å), MIR (24 μm), HI, and CO maps. They divided the galaxy images into 750-pc pixels and examined the spatially resolved K-S law.

The most significant result from this study was the distinction between the K-S laws for HI and H_2 gas. They found that HI gas surface density does not strongly correlate with SFR surface density, exhibiting a relationship that cannot be well-approximated by a single power-law. This behavior was attributed to a threshold surface density of $\Sigma_{HI} = 9 \, \mathcal{M}_\odot \, \text{kpc}^{-2}$, beyond which HI gas cannot exist. In stark contrast, molecular gas displayed a strong correlation with SFR, with a power-law index of $N = 1.0 \pm 0.2$.

Bigiel et al. (2008) proposed that the linear relationship in the molecular K-S law is driven not by the internal gas mass density of GMCs, but by the number density of GMCs over a spatial extent of approximately 750 pc. Their findings suggest the following phenomenological scenario

1. A threshold exists for SF to be initiated.
2. HI gas is not strongly linked to SF.

3. SF occurs in regions where the ISM density is sufficiently high for molecules to form.

Despite the importance of the K-S law in galaxy evolution studies, its physical interpretation remains an active area of research. Many studies continue to explore the underlying physics of the K-S law. Therefore, while the K-S law remains heuristic, it holds tremendous significance in our understanding of galaxy evolution.

3.8 DUST

Although dust accounts for only about $\sim 1\ \%$ of the total ISM mass, it plays a significantly important role in the physics of the ISM, as discussed in various contexts. One prominent effect of dust is extinction and attenuation. Dust substantially reduces radiation in the UV-optical range, with the degree of extinction being wavelength-dependent. In general, shorter wavelengths are more efficiently scattered and absorbed. This phenomenon is known as selective extinction, which leads to reddening. Another crucial role of dust is in the formation of molecules, particularly H_2, where the efficiency of molecular formation on dust grains is several orders of magnitude higher than gas-phase reactions. Photoelectric heating and cooling also play essential roles in the thermal and chemical processes of the ISM.

3.8.1 BASIC PROPERTIES OF DUST

3.8.1.1 Extinction Curve and Related Properties

First, we define the extinction A_λ in units of magnitude as

$$A_\lambda \equiv 2.5 \log \left(\frac{S_\lambda^0}{S_\lambda} \right) , \qquad (3.92)$$

where S_λ^0 is the unextinguished flux density. From radiative transfer theory, we have

$$S_\lambda = S_\lambda^0 \, e^{-\tau_\lambda^{\rm d}} , \qquad (3.93)$$

$$\tau_\lambda^{\rm d} \equiv \int_0^x C_{\rm ext}(\lambda) n_{\rm d} {\rm d}x' , \qquad (3.94)$$

where $C_{\rm ext}$ is the extinction cross-section of a dust grain at wavelength λ, and $n_{\rm d}$ is the number density of dust grains. We note that $C_{\rm ext}$ can be decomposed as

$$C_{\rm ext}(\lambda) = \pi a^2 Q_{\rm ext}(\lambda) , \qquad (3.95)$$

where $Q_{\rm ext}(\lambda)$ is the extinction efficiency. Thus, we obtain

$$A_\lambda = 2.5 \log e^{-\tau_\lambda^{\rm d}} = 1.086 \tau_\lambda^{\rm d} . \qquad (3.96)$$

This shows that extinction is equivalent to the dust optical thickness, aside from a numerical factor.

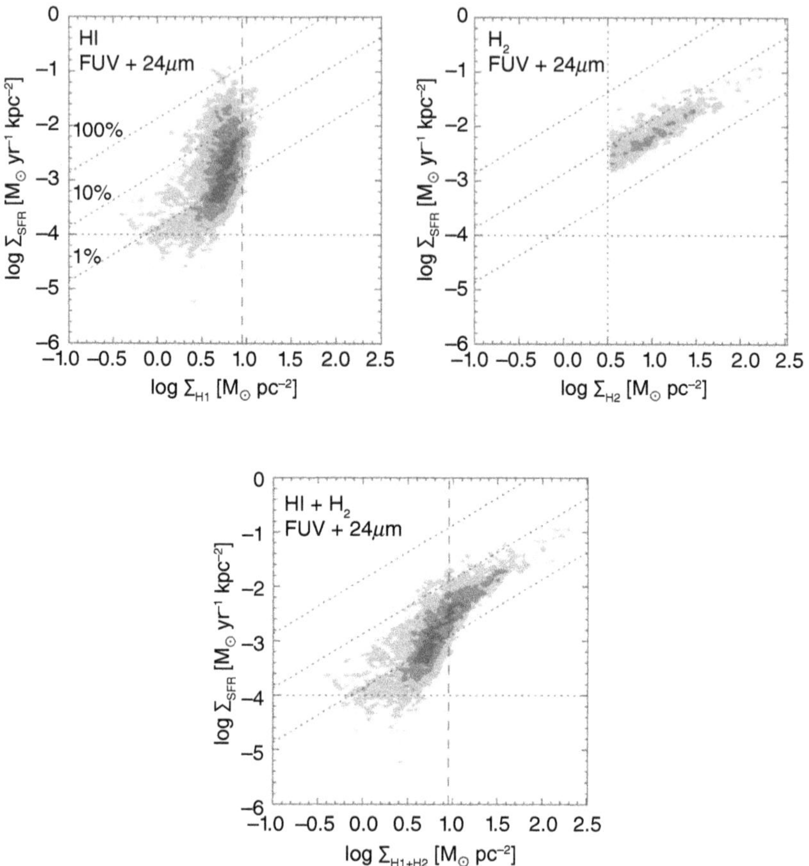

Figure 3.8   An example of the resolved K-S law for Local galaxies. All data from seven spiral galaxies are plotted together. The left panel shows the Σ_{SFR}–Σ_{HI} relation, the middle panel shows the Σ_{SFR}–Σ_{H_2} relation, and the right panel shows the Σ_{SFR}–Σ_{gas} relation. The sensitivity limit of each SF tracer is indicated by a horizontal dotted line. The vertical dashed lines indicate the value at which Σ_{HI} saturates, while the vertical dotted lines represent the sensitivity limit of the CO data. Credit: Bigiel, F., Leroy, A., Walter, F., et al. 2008, Astronomical Journal, 136, 2846, Fig. 8, reproduced by permission of AAS.

Cardelli et al. (1989) constructed an average extinction law based on observations of various lines of sight, covering a wavelength range of $\lambda = 0.125$–3.5 μm. This law applies to both diffuse and dense regions of the ISM (see Fig. 3.9). The extinction law depends on a single parameter

$$R_V \equiv \frac{A_V}{E(B-V)},$$

(3.97)

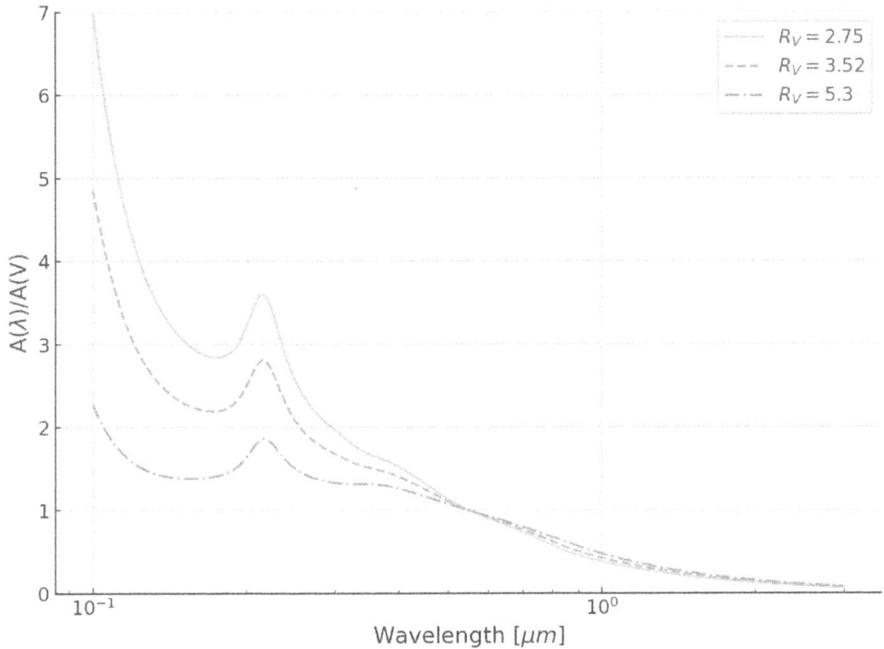

Figure 3.9 Extinction curves from NIR to UV as a function of $R_V = A_V / E(B-V)$.

and is described by

$$\frac{A_\lambda}{A_V} = a(x) + \frac{b(x)}{R_V} . \tag{3.98}$$

Here,

$$E(B-V) \equiv A_B - A_V , \tag{3.99}$$

is referred to as the color excess or reddening. Cardelli et al. (1989) have constructed a formula to describe A_λ as a function of R_V. The behavior of the extinction curve is shown in Fig. 3.9.

As we can see, R_V controls the overall shape of the extinction curve. We can derive information about the grain size distribution from the global slope of the curve. For example, the average extinction in the Milky Way suggests $R_V \simeq 3.0$. Specific features, such as the prominent bump around $\lambda \simeq 2170$ Å are thought to strongly constrain the small fragments of carbonaceous grains, like PAHs (polycyclic aromatic hydrocarbons). The extinction at $\lambda \simeq 5500$ [Å] (V-band) is mainly caused by grains with radii around $a \simeq 0.1$ [mm]. Grains with radii $a < 50$ [Å] dominate extinction at wavelengths $\lambda < 0.3$ [mm].

The extinction formula derived by Cardelli et al. (1989) is primarily designed for the Milky Way, but it is well-known that extinction curves differ significantly

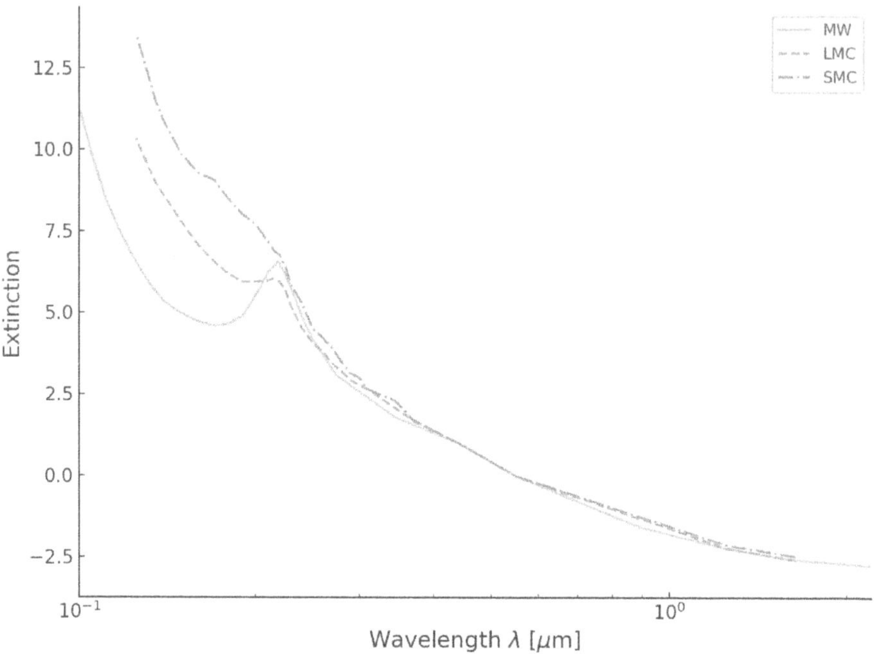

Figure 3.10 Comparison of extinction curves between the Milky Way (MW), Large Magellanic Cloud (LMC), and Small Magellanic Cloud (SMC). All curves are normalized at B-band (4400 Å). Data are taken from Pei (1992).

between galaxies. Figure 3.10 compares the extinction curves of the Milky Way, the Large Magellanic Cloud (LMC), and the Small Magellanic Cloud (SMC).

3.8.1.2 Optical Properties of Dust Grains: Mie Theory

Extinction results from the combined effects of absorption and scattering of radiation. This phenomenon can be theoretically modeled as a scattering problem of radiation by small spherical or ellipsoidal particles. This theory is referred to as Mie theory, named after Gustav Mie, who provided a rigorous solution, although the problem was also studied by others (Debye, 1909; Mie, 1908). Since the theory involves complex analytic calculations, here we present only the useful approximations derived from the final result.

Similar to eq. (3.95), we define the scattering and absorption cross-sections as follows

$$C_{\text{sca}}(\lambda) = \pi a^2 Q_{\text{sca}}(\lambda)\,, \tag{3.100}$$

$$C_{\text{abs}}(\lambda) = \pi a^2 Q_{\text{abs}}(\lambda)\,. \tag{3.101}$$

For ellipsoidal particles, the grain radius a is replaced by the effective radius a_{eff}, defined as

$$a_{\text{eff}} \equiv \left(\frac{3V}{4\pi} \right)^{\frac{1}{3}} , \tag{3.102}$$

where V is the volume of the grain. For simplicity, our discussion is restricted to spherical grains. In Mie theory, a dimensionless size parameter x is introduced

$$x \equiv \frac{2\pi a}{\lambda} , \tag{3.103}$$

along with the complex refractive index of the grain material

$$m_{\text{ref}} = n_{\text{ref}} - i k_{\text{ref}} . \tag{3.104}$$

When $x \ll 1$ (i.e., when the particle is much smaller than the wavelength), the following approximations are obtained

$$Q_{\text{sca}}(\lambda) \simeq \frac{8}{3} x^4 \left| \frac{m_{\text{ref}}^2 - 1}{m_{\text{ref}}^2 + 2} \right|^2 \propto \lambda^{-4} , \tag{3.105}$$

$$Q_{\text{abs}}(\lambda) \simeq 4x \, \text{Im} \left(\frac{m_{\text{ref}}^2 - 1}{m_{\text{ref}}^2 + 2} \right) \propto \lambda^{-1} . \tag{3.106}$$

Equations (3.105) and (3.106) are commonly used in the treatment of dust extinction. As we see, scattering cross-sections $Q_{\text{sca}}(\lambda)$ and absorption cross-sections $Q_{\text{abs}}(\lambda)$ have distinct wavelength dependencies, with scattering more sensitive to shorter wavelengths.

3.8.1.3 Grain Size Distribution

From an observed extinction curve, we can reconstruct the grain size distribution $n(a)$. Mathis et al. (1977) proposed a heuristic model for the grain size distribution of Milky Way dust as

$$n(a) \, da \propto a^{-3.5} \, da , \tag{3.107}$$

which is known as the MRN distribution. In this model, large grains dominate the dust mass, while small grains dominate the surface area.

Although the MRN model successfully reproduces many observations of the Milky Way, modifications are necessary to explain certain detailed features. For example, the observed infrared spectra show emissions in the 3–12 μm bands. These emissions require small grains containing approximately 50 atoms, corresponding to a size of $a \sim 3.5$ Å, i.e., PAHs (polycyclic aromatic hydrocarbons). Therefore, PAHs need to be included at the smallest end of the grain size distribution. Weingartner & Draine (2001) proposed an extended model to account for these small grains. However, the inferred size distribution can vary significantly depending on the assumed optical properties of the grain species (see, e.g., Draine, 2011).

3.8.2 FORMATION OF DUST GRAINS

Since dust consists of metals, its formation occurs in metal-rich environments. Possible dust formation processes include

- The cooling process of hot gas: If the pressure is high, gas condenses into liquid before solidifying. If the pressure is low, gas directly condenses into a solid phase.
- Collision and reaction of atoms and molecules: This process requires that the ingredients of dust be sufficiently dense.

In normal interstellar space, where the hydrogen number density is $n_{\rm H} = 0.1\text{--}1$ [cm^{-3}], the silicon number density is typically $n_{\rm Si} = 10^{-7}\text{--}10^{-6}$ [cm^{-3}]. In such conditions, the growth timescale for dust grains exceeds 10^{10} [yr], which is on the order of the age of the Universe and therefore unrealistic. Consequently, dust formation is limited to the late stages of stellar evolution, where large amounts of mass are ejected. If we denote the stellar mass during the main sequence as $\mathscr{M}_{\rm MS}$, dust formation occurs in the following astrophysical environments

Stars with $\mathscr{M}_{\rm MS} < 8\,[\mathscr{M}_\odot]$
   Dust forms during mass ejection as stars evolve through the red giant branch (RGB), early asymptotic giant branch (E-AGB), and thermal pulse asymptotic giant branch (TP-AGB):

- RGB (first dredge-up): $\dot{\mathscr{M}} \leq 10^{-7}\,[\mathscr{M}_\odot\,{\rm yr}^{-1}]$,
- E-AGB (second dredge-up): $10^{-7} \leq \dot{\mathscr{M}} \leq 10^{-4}\,[\mathscr{M}_\odot\,{\rm yr}^{-1}]$,
- TP-AGB (third dredge-up): $10^{-7} \leq \dot{\mathscr{M}} \leq 10^{-4}\,[\mathscr{M}_\odot\,{\rm yr}^{-1}]$.

Stars with $\mathscr{M}_{\rm MS} > 8\,[\mathscr{M}_\odot]$
   Dust forms during Type II and Type Ib/c supernova explosions, with mass ejection of $\mathscr{M} > 6\,[\mathscr{M}_\odot]$.

Type Ia supernovae in binaries
   Mass ejection during the explosion is limited to $\mathscr{M}_{\rm ejecta} \leq 1.4\,[\mathscr{M}_\odot]$.

Novae in binary systems
   Dust is ejected with a mass of $\sim 10^{-4}\,[\mathscr{M}_\odot\,\text{per eruption}]$.

Carbon-rich Wolf-Rayet (WR) stars
   Mass ejection of $\mathscr{M}_{\rm ejecta} \geq 10^{-5}\,[\mathscr{M}_\odot]$ occurs during their evolution.

Details of stellar evolution and its connection to dust formation will be covered in Chapter 4.

3.8.3 ATTENUATION IN GALAXIES

Since galaxies possess spatial structure, dust, stars, and other radiation sources are intermixed in a complex manner. Consequently, the effective dust extinction differs from the case of a point source, leading to a phenomenon known as attenuation. Understanding attenuation is crucial for interpreting galaxy spectra and requires both observational and theoretical exploration.

Calzetti et al. (1994) introduced an important attenuation curve for local starburst galaxies, commonly known as the Calzetti curve. It is given by

$$
\frac{E(\lambda - V)}{E(B - V)} = \begin{cases} 2.659(-1.857 + 1.040\lambda^{-1}) \\ \qquad\qquad\qquad \text{for } 0.63\,\mu\text{m} \leq \lambda \leq 2.20\,\mu\text{m} \\ 2.659(-2.156 + 1.509\lambda^{-1} - 0.198\lambda^{-2} + 0.011\lambda^{-3}) \\ \qquad\qquad\qquad \text{for } 0.12\,\mu\text{m} \leq \lambda \leq 0.63\,\mu\text{m} \end{cases},
$$

$$(3.108)$$

and is shown in Fig. 3.11. The curve reveals that attenuation depends less strongly on wavelength compared to extinction, primarily due to UV scattering, which redirects UV photons toward the observer, thereby diluting the extinction effect.

3.8.4 DUST EMISSION

The energy absorbed by dust is re-emitted due to energy conservation. The intensity and spectrum of dust emission depend strongly on the dust grain temperature, which is primarily controlled by the energy and intensity of the UV–optical radiation field. Additionally, the dust grain composition plays a role in determining the emission properties. In this subsection, we will formulate dust emission to conclude the discussion.

3.8.4.1 Dust Temperature Distribution

UV and optical photons emitted by stars heat the dust grains. These heated dust grains re-emit the absorbed energy as M/FIR radiation. The temperature of a dust grain is determined by its interaction with the surrounding radiation field. Larger grains establish thermal equilibrium with the radiation field, while smaller grains cannot due to their smaller surface area and heat capacity (Draine & Anderson, 1985; Draine & Li, 2001; Li & Draine, 2001; Takeuchi et al., 2003a, 2005c).

Because of their small heat capacity, small grains are easily heated by absorbed photons and then quickly cool down. This causes their temperature fluctuations to be stochastic rather than in thermal equilibrium. The temperature distribution of such grains must be calculated explicitly to understand their emission.

3.8.4.2 Stochastic Heating

The probability rate that a dust grain absorbs a photon with energy in the range $[E, E + dE]$ and time interval $[t, t + dt]$ is given by

$$
d\mathscr{P}(a, \lambda) = \pi a^2 Q_{\text{abs}}(a, \lambda) u_\lambda \frac{\lambda^3}{h_{\text{p}}^2 c} dE dt,
\tag{3.109}
$$

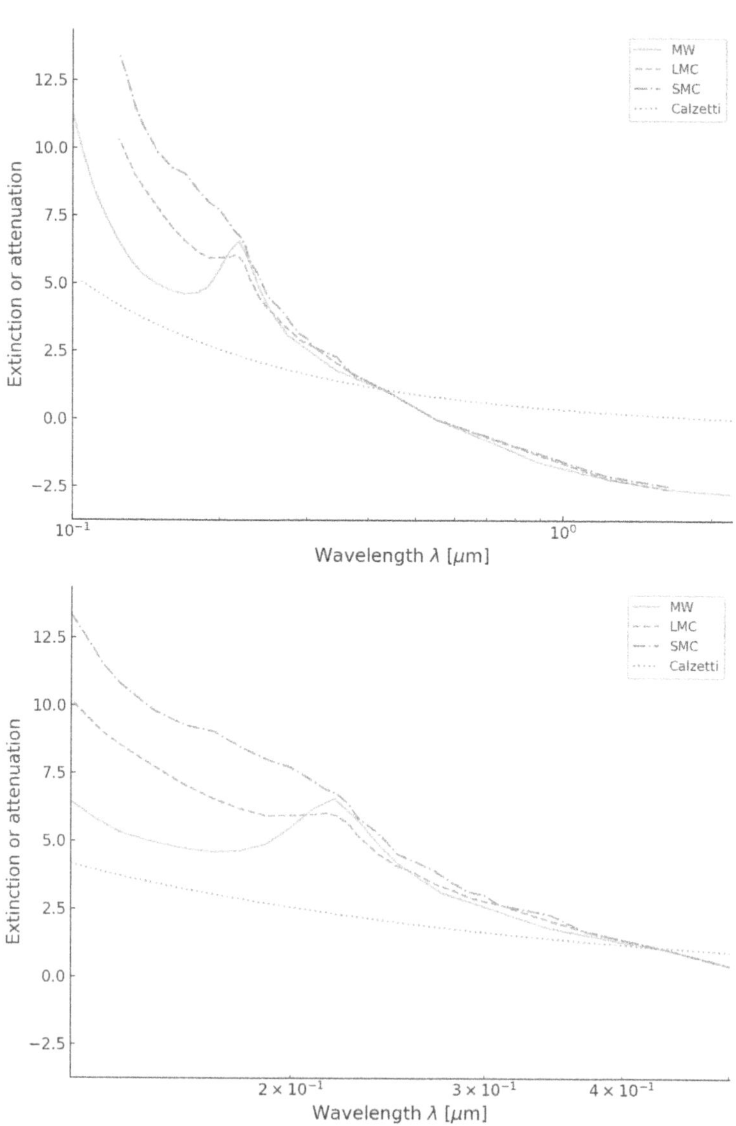

Figure 3.11 The attenuation curve for local starburst galaxies proposed by Calzetti et al. (1994). Here, $R_V = 4.05$ is assumed. Extinction curves of the Milky Way, LMC, and SMC are overplotted for comparison. Top panel shows the global behavior of each extinction/attenuation, and bottom panel is a zoom up at the optical wavelength range. All the curves are normalized at B-band.

where u_λ is the mean energy density of the radiation field at wavelength λ. This can also be interpreted as the probability of absorption if the time interval dt is sufficiently short.

Assuming that all the energy of an absorbed photon is used to heat the dust grain, the temperature change can be written as

$$E(T + \Delta T) = E(T) + \frac{h_P c}{\lambda} . \tag{3.110}$$

The heat capacity of dust grains is often modeled using the Debye model and its variants (Li & Draine, 2001). The heat capacities for silicate and graphite grains are expressed as

$$C_{\mathrm{sil}}(T) = (N_{\mathrm{atom}} - 2)k_B \left[2f_2 \left(\frac{T}{500\,[\mathrm{K}]} \right) + f_3 \left(\frac{T}{1500\,[\mathrm{K}]} \right) \right], \tag{3.111}$$

$$C_{\mathrm{gra}}(T) = (N_C - 2)k_B \left[f_2 \left(\frac{T}{863\,[\mathrm{K}]} \right) + 2f_2 \left(\frac{T}{2504\,[\mathrm{K}]} \right) \right], \tag{3.112}$$

where

$$f_n(a) \equiv n \int_0^1 \frac{y^n\,dy}{\exp\left(\frac{y}{x}\right) - 1} , \tag{3.113}$$

with the subscripts 'sil' and 'gra' representing silicate and graphite grains, respectively.

The instantaneous temperature T_{ins} and energy $E(T_{\mathrm{ins}})$ are related by

$$E(T_{\mathrm{ins}}) = \int_{T_{\mathrm{low}}}^{T_{\mathrm{ins}}} C(T)\,dT , \tag{3.114}$$

where T_{low} is the temperature when a UV photon is absorbed, and T_{ins} is the highest temperature reached by the grain. In interstellar space, T_{low} is often taken to be the temperature of the cosmic microwave background, T_{CMB}.

For PAHs (polycyclic aromatic hydrocarbons), the C–C bond modes are treated similarly to graphite, while the C–H bond modes are added to the heat capacity formula in equation (3.112)

$$E_{\mathrm{PAH}}(T) = E_{\mathrm{gra}} + \frac{H}{C} N_C \sum_{j=1}^{3} \frac{h_P \nu_j}{\exp\left(h_P \nu_j / k_B T\right) - 1} . \tag{3.115}$$

Here, the index j refers to the C–H out-of-plane bending modes ($\nu_1/c = 886$ cm^{-1}), in-plane bending modes ($\nu_2/c = 1161$ cm^{-1}), and stretching modes ($\nu_3/c = 3030$ cm^{-1}) (Draine & Li, 2001). The hydrogen-to-carbon ratio, H/C, is adopted as

$$\frac{H}{C} = \begin{cases} 0.5 & (N_C < 25) \\[2mm] \dfrac{0.5}{\sqrt{N_C/25}} & (25 < N_C < 100) \\[2mm] 0.25 & (N_C > 100) \end{cases} , \tag{3.116}$$

(Li & Draine, 2001).

3.8.4.3   Dust Cooling

The emission of dust grains with radius a can be formulated as

$$4\pi\varepsilon(T,a) = 4\pi \int \frac{2h_{\mathrm{P}}c^2}{\lambda^5} \frac{\pi a^2 Q_{\mathrm{abs}}(\lambda)\,\mathrm{d}\lambda}{\exp\left(\dfrac{h_{\mathrm{P}}c}{\lambda k_{\mathrm{B}}T}\right) - 1} , \tag{3.117}$$

where $\varepsilon(T,a)$ represents the emissivity per unit time and per unit solid angle. The cooling time, t_{cool}, is then given by

$$t_{\mathrm{cool}}(T,a) = \frac{E(T,a)}{4\pi\varepsilon(T,a)} . \tag{3.118}$$

When the dust grain does not absorb energy during the cooling process, the temperature evolution is described by

$$\Delta T(a) = -\frac{1}{t_{\mathrm{cool}}(T+\Delta T,a)} + \frac{1}{t_{\mathrm{cool}}(T,a)} . \tag{3.119}$$

These processes, such as the cooling of dust grains, are often simulated using Monte Carlo methods.

3.8.4.4   Dust emission

Dust emission can be computed from the temperature distribution $\mathrm{d}\mathscr{P}_i(a)/\mathrm{d}T$. The monochromatic luminosity of dust grains for species i is expressed as

$$\mathscr{L}_\lambda^{\,\mathrm{grain},i}(a,\lambda) = 4\pi \int \pi a^2 Q_{\mathrm{abs}}^i(\lambda) B_\lambda(T) \frac{\mathrm{d}\mathscr{P}_i(a)}{\mathrm{d}T}\,\mathrm{d}T , \tag{3.120}$$

where $B_\lambda(T)$ is the Planck function.

Finally, the total monochromatic luminosity of dust at wavelength λ is obtained as

$$\mathscr{L}_\lambda^{\mathrm{dust}} = \sum_i Q_{\mathrm{abs}}^i(\lambda) \int \mathscr{L}_\nu^{\,\mathrm{grain},i}(a,\lambda) n^i(a)\,\mathrm{d}a , \tag{3.121}$$

where $n^i(a)$ represents the grain size distribution function for dust species i.

4 Chemical Evolution of Galaxies

4.1 MAIN SEQUENCE EVOLUTION OF STARS

First, we review stellar evolution, which directly drives the internal evolution of galaxies. During the main sequence, stars undergo stable nuclear fusion, burning hydrogen in a process that represents the most stable phase of stellar life. The subsequent evolution and lifetime of a star strongly depend on its mass during the main sequence phase.

The internal structure of stars must be considered separately for those with $\mathcal{M} < 1.5\,[\mathcal{M}_\odot]$ and those with $\mathcal{M} > 1.5\,[\mathcal{M}_\odot]$, as the primary chain reaction differs: the p–p chain for stars with $\mathcal{M} < 1.5\,[\mathcal{M}_\odot]$ and the CNO cycle for stars with $\mathcal{M} > 1.5\,[\mathcal{M}_\odot]$. This distinction results in significant differences in stellar structure. When the p–p chain dominates in low-mass stars, the stellar core is in radiative equilibrium. In contrast, stars with $\mathcal{M} > 1.5\,[\mathcal{M}_\odot]$ exhibit a convective core, driven by the CNO cycle, which is highly sensitive to temperature gradients.

In high-mass stars ($\mathcal{M} > 1.5\,[\mathcal{M}_\odot]$), the core temperature exceeds $T > 2 \times 10^7\,[\text{K}]$, allowing the CNO cycle to dominate. Since this reaction is extremely sensitive to temperature, it creates a steep temperature gradient, triggering convective instability and leading to a convective core. On the other hand, in low-mass stars ($\mathcal{M} < 1.5\,[\mathcal{M}_\odot]$), the outer layers are relatively cooler. Opacity in low-mass stars is primarily driven by free-free absorption, where $\kappa \propto \rho T^{-3.5}$. This results in high opacity, which induces convection in the outer layers.

Main sequence stars maintain equilibrium, establishing a relationship between luminosity and mass. By employing simple dimensional analysis, we can estimate

$$\mathcal{M} \sim \rho R^3 \tag{4.1}$$

and

$$\frac{p}{R} \sim \frac{G\mathcal{M}\rho}{R^2}. \tag{4.2}$$

Additionally, for stars primarily in radiative equilibrium, the energy transfer equation becomes

$$L \sim \frac{a_{\text{SB}} c T^4 R}{\kappa \rho}. \tag{4.3}$$

Where $T \sim 10^7$ K, the approximate temperature of hydrogen fusion. If gas pressure dominates, the equation of state is

$$p \sim \frac{\rho k_{\text{B}} T}{m_{\text{H}}}, \tag{4.4}$$

DOI: 10.1201/9781003104315-4

"

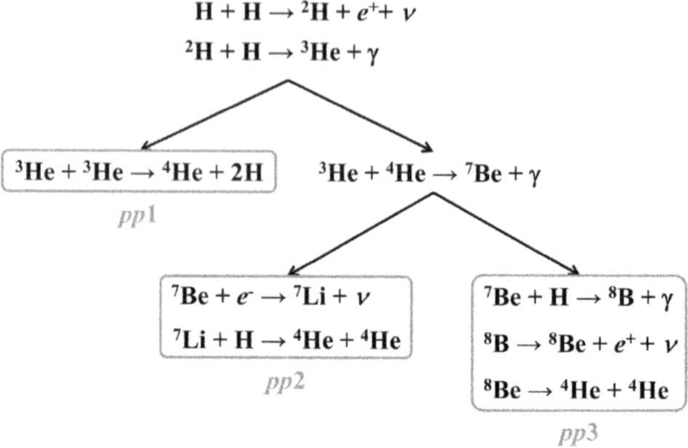

Figure 4.1   The proton–proton (p–p) chain. This is the primary chain reaction that takes place in low-mass stars with $\mathcal{M} < 1.5\,[\mathcal{M}_\odot]$.

and equation (4.2) becomes

$$\frac{k_\mathrm{B}T}{m_\mathrm{H}} \sim \frac{G\mathcal{M}}{R} \,. \tag{4.5}$$

Thus, we find

$$R \sim \mathcal{M} \,, \tag{4.6}$$

$$\rho \sim \mathcal{M}^{-2} \,, \tag{4.7}$$

$$p \sim \mathcal{M}^{-2} \,. \tag{4.8}$$

This shows that as stellar mass increases, the radius becomes larger while the density decreases.

Low-mass stars

Opacity is dominated by free-free absorption, since the density is high. Hence, $\kappa \propto \rho T^{-3.5}$, leading to

$$L \propto \mathcal{M}^5 \,, \tag{4.9}$$

$$T_\mathrm{eff} \propto \mathcal{M}^{3/4} \,. \tag{4.10}$$

High-mass stars

In these stars, low-density results in opacity dominated by Thomson scattering, giving

$$L \propto \mathcal{M}^3 \,, \tag{4.11}$$

$$T_\mathrm{eff} \propto \mathcal{M}^{1/4} \,. \tag{4.12}$$

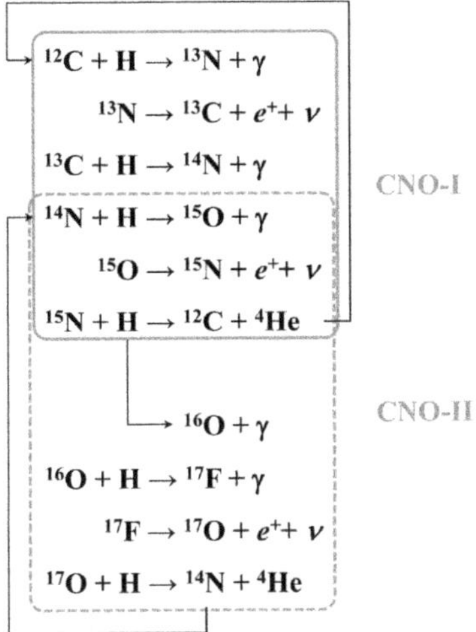

Figure 4.2 The CNO cycle. This is the main reaction occurring in high-mass stars with $\mathcal{M} > 1.5\,[\mathcal{M}_\odot]$.

While this is a rough estimate, it is not far from the result of precise numerical calculations, which yield

$$L \propto \mathcal{M}^{3.5} \tag{4.13}$$

for stars in the mass range $0.5\,[\mathcal{M}_\odot] < \mathcal{M} < 20\,[\mathcal{M}_\odot]$.

At even higher masses, where radiation pressure becomes dominant, the equation of state becomes

$$p = \frac{a_{\mathrm{SB}}T^4}{3}\,, \tag{4.14}$$

which is independent of mass. From dimensional analysis

$$R \sim \mathcal{M}^{1/2}\,, \tag{4.15}$$

$$\rho \sim \mathcal{M}^{1/2}\,. \tag{4.16}$$

And, with constant κ, the relations become

$$L \propto \mathcal{M}\,, \tag{4.17}$$

$$T_{\mathrm{eff}} = \mathrm{const.} \tag{4.18}$$

The corresponding luminosity is known as the Eddington luminosity, L_{Edd}, derived from the balance between radiation pressure and gravity

$$\frac{L_{\mathrm{Edd}}\sigma_{\mathrm{T}}}{4\pi c R^2} = \frac{G\mathcal{M}m_{\mathrm{H}}}{R^2}\,, \tag{4.19}$$

leading to

$$L_{\text{Edd}} = \frac{4\pi c G \mathcal{M} m_{\text{H}}}{\sigma_{\text{T}}} . \tag{4.20}$$

The actual luminosity is typically lower than L_{Edd} due to contributions from gas pressure. Observationally, stars with $\mathcal{M} > 100 \, [\mathcal{M}_\odot]$ are extremely rare because they become unstable due to excessive radiative pressure.

The main sequence lifetime is

$$\tau_{\text{MS}} = \frac{\varepsilon q \mathcal{M} c^2}{L} , \tag{4.21}$$

where $\varepsilon \simeq 7 \times 10^{-3}$ is the energy conversion efficiency, and q represents the fraction of hydrogen that burns. For massive stars, where L_{Edd} is approximately constant

$$\tau_{\text{MS}} \simeq 10^7 \, [\text{yr}] . \tag{4.22}$$

For lower-mass stars

$$\tau_{\text{MS}} = 10^{10} \left(\frac{\mathcal{M}}{\mathcal{M}_\odot} \right)^{-2.5} [\text{yr}] . \tag{4.23}$$

This equation demonstrates that stellar lifetime is highly mass-dependent: the more massive a star, the shorter its lifetime. Stars with $\mathcal{M} < 0.8 \, [\mathcal{M}_\odot]$ have not evolved beyond the main sequence in the 13.8 billion-year age of the Universe.

4.2 POST-MAIN SEQUENCE EVOLUTION

The post-main sequence evolution depends heavily on the star's mass during the main sequence, and stars ultimately evolve into white dwarfs, neutron stars, or black holes. Massive stars explode as supernovae (SNe), playing a critical role in a galaxy's chemical evolution. Here, we outline the later evolutionary stages, including the supernova explosions at the end of a star's life.

4.2.1 VERY LOW-MASS STARS ($\mathcal{M} < 0.45 \, [\mathcal{M}_\odot]$)

The lowest-mass stars eventually develop electron-degenerate cores. The details of electron degeneracy are discussed in Section 4.2.2. For the lowest-Mass Stars ($\mathcal{M} < 0.08 \, [\mathcal{M}_\odot]$), during gravitational contraction, electron degeneracy sets in, halting further contraction. Since temperature cannot rise, hydrogen fusion never ignites, and these objects become brown dwarfs. For stars with $0.08 \, [\mathcal{M}_\odot] < \mathcal{M} < 0.45 \, [\mathcal{M}_\odot]$, after hydrogen fusion ends, the helium core becomes electron-degenerate, and no further nuclear reactions take place. These stars ultimately evolve into helium white dwarfs.

4.2.2 LOW-MASS STARS ($0.45\ [\mathcal{M}_\odot] < \mathcal{M} < 2.0\ [\mathcal{M}_\odot]$)

The upper boundary of this mass range marks the threshold for electron degeneracy. Practically, stars with $\mathcal{M} < 0.8\ [\mathcal{M}_\odot]$ do not evolve into the post-main sequence phase within the current age of the Universe (13.8 Gyr), so we will not discuss them further.

4.2.2.1 Helium Core Formation

Stars begin to deviate from the main sequence when hydrogen in the core is depleted. During hydrogen burning, a helium core gradually forms at the center. Since this core is not very massive, its self-gravity is balanced by the pressure gradient. However, hydrogen surrounding the helium core continues to undergo nuclear fusion. This is referred to as hydrogen shell burning.

As the helium core's mass increases through shell burning, the core's density rises, and its temperature increases uniformly because there is no energy flow. This process further accelerates hydrogen shell burning, producing more energy and causing the outer layers to expand. The surface temperature decreases due to this expansion, leading the star into the red giant branch (RGB) phase.

The evolution into the RGB phase is driven by the increasing core mass and occurs slowly. As the surface temperature decreases, the difference in temperature between the surface and the core becomes significant, triggering strong convection. This convection efficiently transports energy from the core to the surface without altering the star's overall structure. During this phase, RGB stars increase in luminosity without significant expansion. This process continues until helium core burning begins at a temperature exceeding $T > 10^8$ [K]. The mass of the helium core at this point is approximately $0.46\ \mathcal{M}_\odot$.

Another important consequence of the strong convection is the dredge-up of heavy elements to the surface. This event, known as the first dredge-up, is a key episode in the evolution of low-mass stars.

4.2.2.2 Electron Degeneracy

Electron degeneracy plays a crucial role in the evolution of low-mass stars. When electrons become degenerate, the behavior of the gas deviates from that of an ideal gas. Degeneracy sets in at densities of $\rho \simeq 10^4$ [g cm^{-3}] for temperatures between $T \simeq 10^7$ and 10^8 [K]. This critical density is lower at lower temperatures.

In general, gas pressure increases with temperature and density. As density rises, more electrons are found in the same velocity range. However, quantum mechanics limits the number of fermions (such as electrons) that can occupy the same quantum state. This limit increases with higher velocities, larger particle masses, and higher temperatures.

In stellar interiors, composed of nuclei and electrons, electrons—being the lightest particles—are the first to be affected by degeneracy. As the density increases, the electron density predicted by the classical Maxwell-Boltzmann distribution exceeds the maximum allowed by quantum mechanics. Electrons unable to occupy lower velocity states move to higher velocities, increasing the pressure beyond what would be predicted for an ideal gas.

At high densities, degeneracy pressure dominates the total pressure. In this state, the total pressure depends solely on density, not on temperature. This is the condition of degeneracy. In such degenerate regions, heat conductivity is high, and the region approaches an isothermal state.

4.2.2.3 Helium Flash

Once helium core burning begins, the triple-alpha reaction

$$^{4}\text{He} + {}^{4}\text{He} \longleftrightarrow {}^{8}\text{Be}$$
$$^{8}\text{Be} + {}^{4}\text{He} \longleftrightarrow {}^{12}\text{C} + \gamma \tag{4.24}$$

takes place, and the core temperature rises. However, at this stage, the star does not expand in response to the temperature increase. Since degeneracy pressure is nearly independent of temperature, the rise in temperature does not cause a corresponding increase in pressure, leading to a rapid increase in core temperature.

This uncontrolled heating results in the helium flash. When the temperature reaches several times 10^8 K, the electron degeneracy weakens, allowing the gas to expand. This expansion halts the runaway helium flash. Despite releasing energy at a rate 10^{11} times higher than the energy production of the Sun, the expansion absorbs all the energy, preventing any outward release.

4.2.2.4 Horizontal Branch (HB)

After the helium flash, the helium core expands and behaves more like an ideal gas. This stabilizes the helium burning, allowing the star to enter the horizontal branch (HB) phase. During this phase, both helium core burning and hydrogen shell burning occur simultaneously. The core reaches convective equilibrium.

Due to the expansion of the hydrogen shell, the temperature decreases, resulting in a luminosity of approximately $50\,L_\odot$. As helium burning continues, helium in the core is exhausted, and hydrogen and helium burning shells form around a carbon-oxygen core.

The core contracts due to gravitational forces, raising its temperature and igniting helium shell burning. This contraction and heating proceed rapidly, enhancing helium shell burning and driving the outer layers to expand. The surface temperature decreases, creating a significant temperature gradient that promotes the development of a convective zone. During this phase, the star brightens without a significant change in surface temperature, marking the asymptotic giant branch (AGB) phase. A second dredge-up of heavy elements occurs due to convection during the AGB phase.

4.2.3 INTERMEDIATE-MASS STARS ($2\,[\mathcal{M}_\odot] < \mathcal{M} < 8\,[\mathcal{M}_\odot]$)

In intermediate-mass stars, hydrogen burning during the main sequence phase is primarily driven by the CNO cycle. As mentioned earlier, the core is in convective equilibrium, while the outer layers are in radiative equilibrium. Since the core is convective, hydrogen is consumed uniformly across the core, and when hydrogen burning ends, the energy supply ceases, causing the entire star to undergo gravitational contraction.

Electron degeneracy does not occur in the helium core because the core is hotter and less dense than in low-mass stars. As the star contracts, the core temperature increases, eventually igniting hydrogen shell burning, which halts the contraction of the hydrogen-rich outer layers. The pressure gradient alone is insufficient to counteract the core's gravity, leading to the formation of a temperature gradient along with a density gradient.

Due to the temperature gradient, energy is radiatively transported outward, which intensifies the hydrogen shell burning. This process generates more energy than necessary to support the outer layers, causing them to expand. As a result, the surface temperature cools down while the luminosity remains largely unchanged, and the star shifts from left to right on the Hertzsprung–Russell (HR) diagram. The timescale for this evolution is short, making stars in this phase relatively rare, which is why the region on the HR diagram is called the Hertzsprung gap.

Once the star enters the red giant branch (RGB) phase, the outer layers become opaque, and most of the outer envelope becomes convective. Even though hydrogen shell burning increases the energy output, convection efficiently transfers the energy to the surface. Meanwhile, the core continues to contract and heat up, leading to the ignition of helium burning at a temperature of approximately $T \simeq 10^8$ K via the triple-alpha reaction. Since the core is not degenerate in this mass range, no helium flash occurs, and nuclear reactions proceed stably.

The triple-alpha reaction is more sensitive to temperature than the CNO cycle, so the core becomes convective. As the core expands due to helium burning, the temperature in the hydrogen-burning shell decreases, making the shell less active. Consequently, the outer layers contract and the star becomes bluer, creating a phase known as the blue loop. During this phase, the star may cross the instability strip, becoming a Cepheid variable. This blue loop phase is sometimes referred to as the Cepheid loop.

The duration of stable helium burning is about one-fifth of the main sequence lifetime. After the helium core burning ends, the hydrogen shell burning is reactivated, causing the outer layers to expand once again. At this point, the star follows a similar evolutionary path to stars with $0.45\,[\mathcal{M}_\odot] < \mathcal{M} < 2\,[\mathcal{M}_\odot]$. The helium shell ignites at $T \simeq 10^8$ K, driving another expansion of the outer layers and a drop in surface temperature. The outer layers become convective, similar to low-mass stars, and the star enters the asymptotic giant branch (AGB) phase, where the luminosity increases without a significant change in surface temperature. The carbon-oxygen core becomes degenerate after helium burning ceases.

4.2.4   MASSIVE STARS ($\mathcal{M} > 8\ [\mathcal{M}_\odot]$)

In massive stars, the carbon-oxygen core contracts further, allowing for additional nuclear reactions. In stars with masses between 8 $\mathcal{M}_\odot$ and 10 $\mathcal{M}_\odot$, an oxygen-neon-magnesium core forms and eventually becomes degenerate without progressing to neon burning. In even more massive stars, nuclear reactions continue up to the iron-burning stage.

Stars with masses greater than 15 $\mathcal{M}_\odot$ develop massive helium cores by the end of the main sequence. These stars are hot enough to ignite helium burning without significant gravitational contraction, meaning they do not pass through a Hertzsprung gap. Instead, they skip directly to the edge of the blue loop on the HR diagram, bypassing the Hayashi track.

Extremely massive stars experience strong radiative pressure that drives powerful stellar winds, resulting in substantial mass loss. The hotter the star, the stronger the radiative pressure becomes, with wind velocities reaching up to 1500 km s^{-1}. For stars with masses greater than 40 $\mathcal{M}_\odot$, mass loss significantly alters their evolution. These stars become bluer on the HR diagram without entering the RGB phase.

Wolf-Rayet (WR) stars are examples of such massive stars, where strong stellar winds strip away the outer layers, exposing the underlying helium. WR stars are classified into two subclasses: WN and WC. WN stars show higher nitrogen abundance along with helium, while WC stars exhibit excess carbon. The high nitrogen abundance in WN stars can be explained by the CNO cycle, which converts nearly all carbon and oxygen into nitrogen. WC stars are thought to be stars whose carbon layer, formed by helium burning, has been exposed to the strong stellar wind. WR stars maintain high surface temperatures and eventually develop iron cores before exploding as supernovae.

4.3   FINAL STAGE OF STELLAR EVOLUTION

In this section, we discuss the final stages of stellar evolution. Stars with masses greater than 8 $[\mathcal{M}_\odot]$ typically end their lives as supernovae, while stars less massive than this threshold evolves into white dwarfs.

4.3.1   THERMAL PULSE (SHELL FLASH)

As helium is depleted in the core, helium shell burning begins around the core. As helium is consumed, the thickness of this burning layer decreases. Once it becomes sufficiently thin, the helium burning rate runs away, increasing the energy generation rate by over a thousand times. This phenomenon is known as a thermal pulse or shell flash.

Though this process is similar to the helium flash (as it is also a runaway reaction), it is not caused by electron degeneracy. When the temperature of the helium shell increases slightly due to fluctuations, the helium burning rate intensifies. At the same time, radiative energy transfer becomes more efficient because the opacity, which is proportional to $\rho T^{-3.5}$, decreases, leading to a steeper temperature gradient.

However, because helium burning is highly sensitive to temperature, energy production eventually surpasses the radiative loss, causing energy to accumulate in the shell. The shell expands and its density decreases, but due to its thinness, this expansion does not affect the overall stellar structure. To maintain the star's structure, the temperature in the helium shell rises, triggering a runaway reaction. As the runaway continues, the burning shell thickens rapidly, and its outer layers establish convective equilibrium. This enhances energy transfer, leading to a reduction in the energy generation rate.

During each thermal pulse, the core, consisting of carbon, oxygen, neon, and magnesium, grows in mass, and the star becomes more luminous. Although thermal pulses are rare to observe directly, they alter the elemental abundances on the stellar surface, allowing scientists to estimate the number of pulses through spectroscopic observations. Reactions such as

$$\text{He} \longrightarrow \text{C} \longrightarrow \text{O}, \tag{4.25}$$

$$\text{N} \longrightarrow \text{Ne} \longrightarrow \text{Mg} + n \tag{4.26}$$

occur, with the newly formed heavy elements transported to the stellar surface in a process known as the third dredge-up.

4.3.2 STARS WITH $\mathcal{M} < 8\,[\mathcal{M}_\odot]$

For stars within this mass range, helium burning occurs, but the stars do not reach the temperature necessary for carbon burning. After helium burning ceases, thermal pulses repeat, and the degenerate $C + O$ core increases in mass. The star ascends the asymptotic giant branch (AGB), increasing in luminosity until the outer layers are eventually ejected.

Following this mass ejection, the star contracts, and its surface temperature rises to several $\times 10^4$ [K], ionizing the ejected outer layers. This creates a planetary nebula. With the core degenerate and no further nuclear reactions occurring, the star gradually cools and fades into a $C + O$ white dwarf (WD), typically with a mass between $0.5\,[\mathcal{M}_\odot]$ and $1\,[\mathcal{M}_\odot]$.

4.3.3 STARS WITH $8\,[\mathcal{M}_\odot] < \mathcal{M} < 10\,[\mathcal{M}_\odot]$

The evolution of stars in this mass range is somewhat uncertain, as it depends heavily on the mass loss rate during the AGB phase. The $C + O$ core contracts, eventually reaching temperatures of $T \simeq 10^9$ K, igniting carbon burning and forming an $O + Ne + Mg$ core through reactions such as

$$^{12}\text{C} + {}^{12}\text{C} \longrightarrow {}^{24}\text{Mg} \longrightarrow \begin{cases} {}^{24}\text{Mg} + \gamma \\ {}^{20}\text{Ne} + {}^4\text{He} \\ {}^{23}\text{Na} + p \\ {}^{23}\text{Mg} + n \\ {}^{16}\text{O} + 2\,{}^4\text{He} \end{cases} . \tag{4.27}$$

Two possible outcomes exist for stars in this mass range

1. If the outer layers of hydrogen and helium are completely lost and the O + Ne + Mg core is exposed, and the star becomes an O + Ne + Mg white dwarf.
2. If the core mass reaches the Chandrasekhar limit before the outer layers are ejected, electron capture reduces the electron degeneracy pressure, leading to further gravitational contraction and, ultimately, a supernova explosion.

In the second scenario, the core collapses to form a neutron star.

4.3.4   STARS WITH $\mathcal{M} > 10\,[\mathcal{M}_\odot]$

In the most massive stars, nuclear reactions continue uninterrupted, producing increasingly heavier elements up to iron. After carbon burning, neon burning begins at $T \simeq 1.5 \times 10^9$ K

$$^{20}\text{Ne} + \gamma \longrightarrow {}^{16}\text{O} + {}^4\text{He} \tag{4.28}$$

$$^{20}\text{Ne} + {}^4\text{He} \longrightarrow {}^{24}\text{Mg} + \gamma, \tag{4.29}$$

followed by oxygen burning at $T \simeq 2 \times 10^9$ K

$$^{16}\text{O} + {}^{16}\text{O} \longrightarrow \begin{cases} {}^{32}\text{S} + \gamma \\ {}^{31}\text{Si} + n \\ {}^{31}\text{P} + p \\ {}^{28}\text{Si} + {}^4\text{He} \\ {}^{24}\text{Mg} + 2\,{}^4\text{He} \end{cases}, \tag{4.30}$$

and finally, silicon burning at $T > 3 \times 10^9$ [K], proceeding through a process known as photodisintegration rearrangement

$$^{28}\text{Si} + {}^4\text{He} \longrightarrow {}^{32}\text{S} \tag{4.31}$$

$$^{32}\text{S} + {}^4\text{He} \longrightarrow {}^{36}\text{Ar} \tag{4.32}$$

$$^{36}\text{Ar} + {}^4\text{He} \longrightarrow {}^{40}\text{Ca} \tag{4.33}$$

$$^{40}\text{Ca} + {}^4\text{He} \longrightarrow {}^{44}\text{Ti} \tag{4.34}$$

$$^{44}\text{Ti} + {}^4\text{He} \longrightarrow {}^{48}\text{Cr} \tag{4.35}$$

$$^{48}\text{Cr} + {}^4\text{He} \longrightarrow {}^{52}\text{Fe} \tag{4.36}$$

$$^{52}\text{Fe} + {}^4\text{He} \longrightarrow {}^{56}\text{Ni} \tag{4.37}$$

$$^{56}\text{Ni} + {}^4\text{He} \longrightarrow {}^{60}\text{Zn}. \tag{4.38}$$

This sequence typically ends with the production of ^{56}Ni or ^{56}Fe. While it is often said that iron is the most stable element because it has the highest nuclear binding energy per nucleon, the true reason for the halt at iron is the balance between photodisintegration and the alpha process, which makes photodisintegration around iron the most efficient.

This process creates an onion-like structure in massive stars, with heavier elements at the center and nuclear reactions occurring at various layers (see Fig. 4.3).

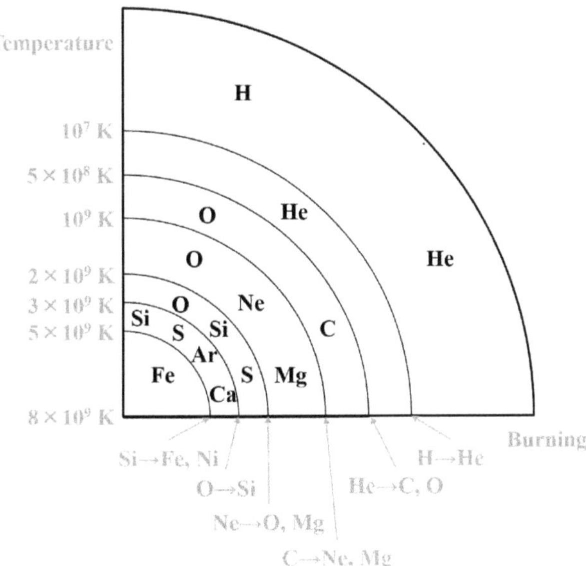

Figure 4.3 Internal structure of massive stars at their final stage of evolution.

Once the core temperature reaches $T \simeq 10^{10}$ [K], iron nuclei undergo photodisintegration

$$^{56}\text{Fe} \longrightarrow 13\,^4\text{He} + 4n - 124.4\,\text{MeV} \tag{4.39}$$

$$^4\text{He} \longrightarrow 2p + 2n - 28.3\,\text{MeV} . \tag{4.40}$$

This collapse occurs within about 0.1 seconds. The outer layers are expelled in a violent supernova explosion, while the core collapses into either a neutron star or a black hole.

4.3.5 SUPERNOVA EXPLOSION

A supernova is a stellar explosion triggered by the recoil of a star's gravitational collapse. Supernovae can be classified into different types, with each category producing distinct elements and exhibiting unique explosion timescales.

4.3.5.1 Classification and Origin of Luminosity

The classical classification of supernovae is based on their observed spectra. Supernovae displaying hydrogen absorption lines in their optical spectra are classified as Type II, while those without hydrogen lines are classified as Type I. Type I supernovae are further subdivided into three subclasses

1. Type Ia: Strong silicon absorption is observed.
2. Type Ib: Weak silicon absorption and strong helium absorption are present.

3. Type Ic: Both silicon and helium absorption features are weak.

This observational classification, however, does not necessarily reflect the underlying physical processes driving the explosion. Based on the mechanism of the explosion, supernovae are classified as core-collapse supernovae and thermonuclear runaway supernovae. Type Ib, Ic, and II are core-collapse supernovae, while Type Ia is classified as a thermonuclear runaway event. We will follow this physical classification.

The luminosity of a supernova originates from two sources

1. Heating caused by the blast wave of the explosion as it propagates through the star.
2. Heating by gamma rays and positrons produced during the radioactive decay of ^{56}Ni to ^{56}Co, and subsequently to ^{56}Fe.

Supernova light curves typically exhibit an initial peak in brightness, followed by a slow decline. The peak is powered by shock heating, while the gradual fading is caused by the radioactive decay of nickel and cobalt. The half-life of ^{56}Ni is approximately five days, and ^{56}Co has a half-life of 77 days, which explains the extended decline in luminosity.

4.3.5.2 Core-Collapse Supernova

A core-collapse supernova occurs when a massive star ($\mathcal{M} > 8 \, [\mathcal{M}_\odot]$) reaches the end of its life. The explosion is triggered by the collapse of the iron core, driven by photodisintegration. The core ultimately forms either a neutron star or a black hole. Detailed models of the explosion mechanism can be found in works such as Burrows & Vartanyan (2021). Core-collapse supernovae primarily produce α elements through nuclear reactions that generate heavy elements by fusing helium nuclei. The elements produced by the α process are shown in Fig. 4.4.

Type Ib and Ic supernovae lack hydrogen features in their spectra, suggesting that the progenitor stars have lost their outer hydrogen and helium layers. Such mass loss is often associated with WR stars. Type Ib progenitors have ejected most of their hydrogen, while Type Ic progenitors have lost both their hydrogen and helium shells.

If the progenitor mass lies between $140 \, \mathcal{M}_\odot$ and $300 \, \mathcal{M}_\odot$, intense radiation fields cause the creation of electron–positron pairs during oxygen core burning. This depletes the star's internal energy, reducing the pressure and causing gravitational contraction. As a result, oxygen burning runs away, leading to a violent explosion that completely disperses the star. This type of core-collapse supernova is known as a pair-instability supernova, and it produces relatively heavy elements such as calcium and iron.

4.3.5.3 Thermonuclear Runaway Supernova

A thermonuclear runaway supernova occurs in a white dwarf (WD) within a binary system. As material (hydrogen and helium) is accreted from a companion star, the WD's mass increases and its surface gravity strengthens. This causes the WD to

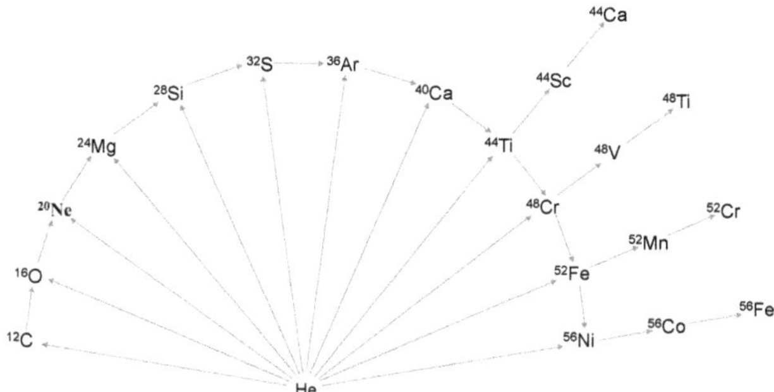

Figure 4.4 Elements produced by the α process. The outer four branches represent decay processes from unstable nuclei produced during the α process sequence.

contract, raising the surface temperature and reigniting nuclear reactions, primarily carbon fusion, which produces carbon and oxygen.

When the core mass approaches the Chandrasekhar limit

$$\mathcal{M}_{\text{Ch}} = \frac{\omega\sqrt{3\pi}}{2}\left(\frac{\hbar c}{G}\right)^{\frac{3}{2}}\frac{1}{(\mu m_{\text{H}})^2} \tag{4.41}$$

(where μ is the average molecular weight per electron, m_{H} is the mass of a hydrogen atom, and $\omega \simeq 2.0182$ is a dimensionless factor from the Lane–Emden equation for $n = 3$), a runaway reaction begins, causing the star to explode. For practical purposes, $\mathcal{M}_{\text{Ch}} \simeq 1.44\,[\mathcal{M}_\odot]$.

A key characteristic of thermonuclear supernovae is their production of large quantities of iron-group elements, releasing as much as $0.8\,\mathcal{M}_\odot$ of iron from a WD with a total mass of $1.44\,\mathcal{M}_\odot$ (typically $0.6\,\mathcal{M}_\odot$).

Since white dwarfs have similar masses (around $1\,\mathcal{M}_\odot$), Type Ia supernovae tend to have relatively uniform luminosities. In fact, the luminosity of a Type Ia supernova can rival that of its host galaxy. This makes them ideal standard candles in cosmology (cf. Appendix A).

The timescale for a thermonuclear runaway supernova is about $\sim 10^9$ years, while core-collapse supernovae have shorter timescales of 10^6–10^7 years. These two types of supernovae contribute different elements to their environments. By studying the abundance ratios of heavy elements, we can trace the chemical evolution of galaxies.

4.4 ORIGIN OF HEAVY ELEMENTS AND YIELD

As discussed earlier, helium and heavier elements are synthesized in stars. At the end of their life cycles, stars produce and release elements heavier than iron into the interstellar medium. The total mass of heavy elements produced by stars, per unit of stellar mass, is called the (stellar) yield.

4.4.1 ORIGIN OF ELEMENTS UP TO IRON

4.4.1.1 Low-Mass Stars ($\mathscr{M} < 8\ \mathscr{M}_\odot$)

Low-mass stars go through the AGB phase and eventually cool down, forming $C+O$ white dwarfs (WDs). The typical mass of a WD is around 0.6 $\mathscr{M}_\odot$ (0.4 $\mathscr{M}_\odot <$ $\mathscr{M}_{WD} < 1.4\ \mathscr{M}_\odot$), meaning these stars return most of their mass to the interstellar medium. During the thermal pulse AGB phase, neutrons are captured by elements such as iron, forming heavier elements like barium (Ba) and strontium (Sr). This neutron capture process is slower than the β-decay of neutrons, leading to the formation of stable nuclei, known as the s-process (slow neutron capture process). The s-process elements, along with carbon synthesized in the star's interior, are transported to the surface via dredge-up mechanisms.

If the WD is part of a binary system, it can explode as a thermonuclear runaway supernova. During this explosion, a significant amount (0.6 $\mathscr{M}_\odot$) of ^{56}Ni is produced, which later decays into ^{56}Fe. Hence, thermonuclear runaway supernovae are major sources of iron, cobalt, and nickel in the universe.

4.4.1.2 High-Mass Stars ($\mathscr{M} > 8\ \mathscr{M}_\odot$)

High-mass stars end their lives in core-collapse supernovae, contributing α elements such as oxygen (O), neon (Ne), magnesium (Mg), silicon (Si), sulfur (S), argon (Ar), calcium (Ca), and titanium (Ti) to the interstellar medium. Additionally, pair-instability supernovae are responsible for producing heavier elements like iron and calcium.

In stars with $\mathscr{M} > 20\ \mathscr{M}_\odot$, some supernovae exhibit explosion energies an order of magnitude greater than typical Type II supernovae (around 10^{52} erg). These are known as hypernovae and are significant sources of elements like zinc (Zn) and cobalt (Co).

4.4.2 ORIGIN OF ELEMENTS HEAVIER THAN IRON

Elements heavier than iron are produced through neutron capture followed by β-decay. Certain elements, such as strontium (Sr), barium (Ba), and lead (Pb), possess a magic number of neutrons (50, 82, and 126), making their nuclei particularly stable and resistant to further neutron capture.

Neutron capture processes occur in environments where a stable supply of neutrons is available. In conditions with low neutron densities, neutron capture occurs at a slower rate than β-decay, known as the s-process. This process typically takes place during the thermal pulse AGB phase and follows a path along stable isotopes.

In contrast, the r-process (rapid neutron capture) occurs on a much faster timescale, lasting only a few seconds. This process produces elements such as gold (Au), platinum (Pt), and uranium (U). During the r-process, neutrons are captured so quickly that the nucleus becomes unstable, undergoing β-decay to form stable isotopes. The neutron capture rate is so rapid that the final mass numbers of stable nuclei differ from those produced by the s-process.

While the r-process was traditionally thought to occur during core-collapse supernovae, challenges to this theory have emerged. For example, supernovae require neutrons to trigger the explosion, leading to questions about whether enough neutrons remain for r-process nucleosynthesis. Additionally, observations of stellar abundance ratios in the Milky Way at low metallicities exhibit a large scatter, implying that r-process elements may have originated from rarer events than core-collapse supernovae.

One compelling alternative site for r-process nucleosynthesis is the kilonova (or macronova), which results from the merger of neutron star binaries. These events are promising candidates because the r-process is expected to occur robustly in such environments. However, kilonovae are much rarer than supernovae, occurring only about once per 100 supernovae.

In 2017, the detection of gravitational waves from the neutron star merger event GW170817 was reported (Abbott et al., 2017a). Just two seconds after the gravitational wave detection, a gamma-ray burst was observed, followed by optical and near-infrared counterparts. This provided the first observational evidence of a kilonova event (Abbott et al., 2017b), supporting the idea that neutron star mergers are a significant site of r-process element production. Nonetheless, the primary origin of r-process elements remains an open question in astrophysics.

4.4.3 ORIGIN OF ELEMENTS: SUMMARY

We summarize the origin of elements produced through nuclear reactions.

H, He, and Li: All of hydrogen (H), most of helium (He), and a small amount of lithium (Li) are formed during Big Bang Nucleosynthesis (BBN). The remaining He and Li are synthesized during hydrogen burning in stars.

C and N: Roughly half of the carbon (C) and nitrogen (N) is produced during helium burning, while the rest is formed in runaway nuclear reactions when stars explode as core-collapse supernovae.

Na and Al: Sodium (Na) and aluminum (Al) are produced during shell burning in massive stars. The yield depends on both the mass of the shell, which correlates with the star's initial mass and its initial metallicity.

Mg: Magnesium (Mg), one of the α-elements, is primarily produced during carbon shell burning. The yield is a function of the star's initial mass, and it is rarely produced in Type Ia supernovae.

O, Si, Ca, and Ti: Oxygen (O) is produced during helium shell burning, while heavier α-elements like silicon (Si), calcium (Ca), and titanium (Ti) are formed during core-collapse supernovae.

Iron-Peaked Elements: Sc, V, Cr, Mn, Fe, Co, Ni, Cu, and Zn: Most iron-peaked elements (except copper (Cu) and zinc (Zn)) are produced by both Type Ia and Type II supernovae. The origin of Cu remains somewhat debated, though Type

Ia supernovae likely contribute minimally. The weak s-process in massive stars is considered an important source of Cu (e.g., Romano & Matteucci, 2007). Cobalt (Co) is produced in Type II supernovae, while zinc (Zn) formation requires hypernovae. The production efficiency of manganese (Mn), cobalt (Co), and copper (Cu) depends on the initial metallicity.

Sr, Y, and Zr: Strontium (Sr), yttrium (Y), and zirconium (Zr) are produced through both the s-process and r-process.

Bi: Bismuth (Bi) is synthesized via the s-process during the asymptotic giant branch (AGB) phase of stellar evolution.

Ba and Pb: Barium (Ba) and lead (Pb) are predominantly (> 80 %) produced by the s-process under solar abundance conditions. However, at very low metallicities, the r-process becomes more significant.

Eu, Pt, Au, Th, U: Under solar abundance conditions, europium (Eu), platinum (Pt), and gold (Au) are predominantly produced by the r-process (over 90%). Thorium (Th) and uranium (U) are entirely (100%) produced by the r-process in solar abundance conditions. Pt, Au, and U, in particular, are produced during kilonovae.

4.5 CHEMICAL EVOLUTION OF GALAXIES

Heavy elements are synthesized within the interiors of stars or during their final evolutionary stages, and subsequently expelled into the interstellar medium (ISM). These elements mix with the ISM, leading to a gradual increase in the ISM's metallicity over time. Through repeated cycles of star formation and death, the initial metallicity of stars evolves as well, resulting in a distribution of stellar metallicities. As discussed earlier, the abundance of each element reflects both the star formation history and the stellar mass distribution. We now formulate the evolution of chemical abundances in galaxies, a process known as chemical evolution.

4.5.1 INITIAL MASS FUNCTION (IMF)

The chemical evolution of galaxies fundamentally depends on the mass distribution of newly formed stars, known as the initial mass function (IMF). More precisely, the IMF, $\Phi(m)$, is defined such that the relative number of stars born within a mass range $(m, m + \mathrm{d}m)$ is given by $\Phi(m)\mathrm{d}m$,

$$\Phi(m) \equiv C \frac{\mathrm{d}n_*}{\mathrm{d}m}, \tag{4.42}$$

where n_* is the number density of newly formed stars, and C is a normalization constant. The IMF plays a crucial role in determining the ratio of massive to less massive stars, and thus it has a profound impact on the chemical evolution of galaxies.

We adopt the following normalization for the IMF

$$\int_{\mathcal{M}_{\mathrm{low}}}^{\mathcal{M}_{\mathrm{up}}} m\Phi(m)\,\mathrm{d}m = 1\,[\mathcal{M}_\odot]\,, \tag{4.43}$$

where typical lower and upper mass bounds are $\mathcal{M}_{\mathrm{low}} = 0.08\,[\mathcal{M}_\odot]$ and $\mathcal{M}_{\mathrm{up}} = 100\,[\mathcal{M}_\odot]$, although different values are occasionally used. For a total stellar mass of $\mathcal{M}_*$, the number and mass of stars in a mass range $(m, m+\mathrm{d}m)$ are given by $\mathcal{M}_*\Phi(m)\mathrm{d}m$ and $\mathcal{M}_*m\Phi(m)\mathrm{d}m$, respectively. It is important to note that a different normalization convention is sometimes used

$$\int_{\mathcal{M}_{\mathrm{low}}}^{\mathcal{M}_{\mathrm{up}}} \Phi_n(m)\,\mathrm{d}m = 1\,, \tag{4.44}$$

where the subscript n denotes normalization by number. In this case,

$$\langle \mathcal{M}_{*,\mathrm{new}} \rangle = \int_{\mathcal{M}_{\mathrm{low}}}^{\mathcal{M}_{\mathrm{up}}} m\Phi_n(m)\,\mathrm{d}m \tag{4.45}$$

represents the average mass of newly born stars.

In a pioneering study, Salpeter (1955) determined the mass function of stars with $m > 1\,[\mathcal{M}_\odot]$, obtaining a power-law function

$$\Phi(m) \propto m^{-\alpha} \tag{4.46}$$

with $\alpha = 2.35$, a form now known as the Salpeter IMF. Subsequent research revealed a shallower slope at the lower-mass end. For instance, Kroupa (2002) proposed the following form

$$\Phi(m) \propto \begin{cases} m^{-0.3} & (0.01\,[\mathcal{M}_\odot] < m < 0.08\,[\mathcal{M}_\odot]) \\ m^{-1.3} & (0.08\,[\mathcal{M}_\odot] < m < 0.5\,[\mathcal{M}_\odot]) \\ m^{-2.3} & (0.5\,[\mathcal{M}_\odot] < m < 1\,[\mathcal{M}_\odot]) \\ m^{-2.7} & (1\,[\mathcal{M}_\odot] < m < 100\,[\mathcal{M}_\odot]) \end{cases}. \tag{4.47}$$

Additionally, Chabrier (2003) proposed a log-normal form for the IMF, which yields a distribution similar to the Kroupa IMF. Figure 4.5 shows a schematic comparison of these IMFs.

4.5.2 CHEMICAL EVOLUTION MODEL: BASIC EQUATIONS

We begin by describing the general framework for chemical evolution in a galaxy. The governing equations are expressed as follows

$$\mathcal{M} = \mathcal{M}_* + \mathcal{M}_{\mathrm{ISM}} \tag{4.48}$$

$$\frac{\mathrm{d}\mathcal{M}}{\mathrm{d}t} = \mathscr{F} - \mathscr{E} \tag{4.49}$$

$$\frac{\mathrm{d}\mathcal{M}_*}{\mathrm{d}t} = \mathrm{SFR}_{\mathrm{tot}} - R \tag{4.50}$$

$$\frac{\mathrm{d}\mathcal{M}_{\mathrm{ISM}}}{\mathrm{d}t} = -\mathrm{SFR}_{\mathrm{tot}} + R + \mathscr{F} - \mathscr{E}\,, \tag{4.51}$$

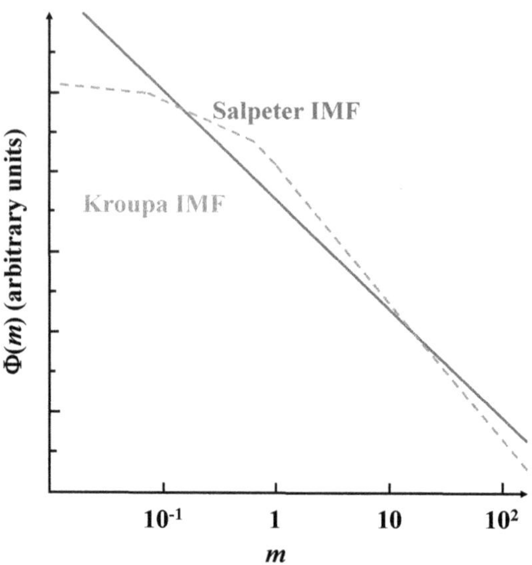

Figure 4.5 Schematic depiction of two representative initial mass functions (IMFs). The solid line represents the Salpeter IMF (Salpeter, 1955), while the dashed line illustrates the Kroupa IMF (Kroupa, 2002).

where

$$\mathcal{M} : \text{total baryonic mass,}$$
$$\mathcal{M}_* : \text{mass in stars,}$$
$$\mathcal{M}_{\text{ISM}} : \text{mass in the ISM,}$$
$$\mathcal{F} : \text{gas infall rate,}$$
$$\mathcal{E} : \text{gas outflow rate,}$$
$$\text{SFR}_{\text{tot}} : \text{star formation rate,}$$
$$R : \text{gas ejection rate from stars.}$$

The total gas ejection rate from stars is given by

$$R(t) = \int_{m_{\text{TO}}}^{\infty} (m - w_m)\, \text{SFR}_{\text{tot}}(t - \tau_m)\Phi(m)\mathrm{d}m \tag{4.52}$$

where m_{TO} is the turnoff mass at time t, representing the lowest mass of stars dying at time t, τ_m is the main sequence lifetime for a star of mass m, and w_m is the remnant mass of a star of mass m. The remnant mass w_m can refer to the final stage of stellar evolution or long-lived stars depending on the context. Numerical results or fitting formulae based on such results are typically used to determine w_m (e.g., Ritter et al., 2018; Yan et al., 2019). The difference $(m - w_m)$ represents the net ejected mass. The term $\text{SFR}_{\text{tot}}(t - \tau_m)\Phi(m)$ denotes the birth rate at $t - \tau_m$, which is equal to the death rate at time t.

The evolution of the metal abundance Z is written as

$$\frac{\mathrm{d}\mathcal{M}_Z}{\mathrm{d}t} = \frac{\mathrm{d}(Z\mathcal{M}_{\text{ISM}})}{\mathrm{d}t}$$
$$= -Z\text{SFR}_{\text{tot}} + R_Z + Y_Z + Z_{\mathscr{F}}\mathscr{F} - Z\mathscr{E}$$
$$\equiv -Z\text{SFR}_{\text{tot}} + E_Z + Z_{\mathscr{F}}\mathscr{F} - Z\mathscr{E} , \tag{4.53}$$

where E_Z is the ejection rate of metals from stars (including main sequence, Wolf-Rayet stars, and supernovae), $Z_{\mathscr{F}}$ is the metallicity of infalling gas, and $\mathcal{M}_Z$ is the mass of metals in the ISM.

The metal ejection rate E_Z is expressed as

$$E_Z = \int_{m_{\text{TO}}}^{\infty} \left[(m - w_m)Z_{t-\tau_m} + mp_{Z_m}\right]\text{SFR}_{\text{tot}}(t - \tau_m)\Phi(m)\mathrm{d}m$$
$$= \int_{m_{\text{TO}}}^{\infty} (m - w_m)Z_{t-\tau_m}\text{SFR}_{\text{tot}}(t - \tau_m)\Phi(m)\mathrm{d}m$$
$$+ \int_{m_{\text{TO}}}^{\infty} mp_{Z_m}\text{SFR}_{\text{tot}}(t - \tau_m)\Phi(m)\mathrm{d}m . \tag{4.54}$$

Here, $p(Z_m)$ represents the metal mass fraction synthesized and ejected by a star of mass m (as discussed in Section 4.4), $(m - w_m)Z_{t-\tau_m}$ is the metal mass locked in a star of mass m at time $t - \tau_m$ and later ejected at time t, and $mp(Z_m)$ is the newly produced metal mass from a star of mass m, formed from gas of metallicity Z.

Thus, we have

$$R_Z = \int_{m_{\text{TO}}}^{\infty} (m - w_m)Z_{t-\tau_m}\text{SFR}_{\text{tot}}(t - \tau_m)\Phi(m)\mathrm{d}m , \tag{4.55}$$

$$Y_Z = \int_{m_{\text{TO}}}^{\infty} mp_{Z_m}\text{SFR}_{\text{tot}}(t - \tau_m)\Phi(m)\mathrm{d}m . \tag{4.56}$$

The key assumption in eqs. (4.53) and (4.54) is the instantaneous mixing approximation, which presumes that the metals are immediately and homogeneously mixed within the ISM. While this is a rough approximation, we will revisit this issue later in more detail.

4.5.3 INSTANTANEOUS RECYCLING APPROXIMATION

To facilitate solving the equations of chemical evolution, it is helpful to introduce two important quantities. The returned mass per unit mass of stars formed is defined as

$$r \equiv \int_{m_{\text{TO}}}^{\infty} (m - w_m)\Phi(m)\mathrm{d}m . \tag{4.57}$$

By definition of the IMF, note that $r < 1$. This returned mass fraction is independent of the star formation rate (SFR), and it holds true for a single generation of stars.

The mass of metals returned to the interstellar medium (ISM) per unit of remaining stellar mass (including stellar remnants) is referred to as the yield

$$y \equiv \frac{1}{1-r} \int_{m_{\mathrm{TO}}}^{\infty} m p_{Z_m} \Phi(m) \mathrm{d}m \,. \tag{4.58}$$

With these quantities, we introduce the instantaneous recycling approximation: massive stars die immediately after their formation, while less massive stars live indefinitely, and the elements they produce are instantly mixed with the ISM. This approximation is valid when the SFR remains relatively constant over timescales of 10^7 years for lighter elements like O, C, N, and Mg (which are formed in core-collapse supernovae) and over 10^8–10^9 years for heavier elements like Fe (originating from Type Ia supernovae).

This approximation simplifies the equations as follows. Assuming the IMF is constant over time ($r = \mathrm{const.}$), the returned mass is now given by

$$R(t) = r\mathrm{SFR}_{\mathrm{tot}}(t) \,. \tag{4.59}$$

The metal ejection rate becomes

$$E_Z(t) = rZ(t)\,\mathrm{SFR}_{\mathrm{tot}}(t) + (1-r)y(t)\,\mathrm{SFR}_{\mathrm{tot}}(t) \,. \tag{4.60}$$

Thus, eq. (4.53) simplifies to

$$\begin{aligned}
\frac{\mathrm{d}\mathcal{M}_Z(t)}{\mathrm{d}t} &= -Z(t)\,\mathrm{SFR}_{\mathrm{tot}}(t) + rZ(t)\,\mathrm{SFR}_{\mathrm{tot}}(t) + (1-r)y(t)\,\mathrm{SFR}_{\mathrm{tot}}(t) + Z_{\mathscr{F}}\mathscr{F}(t) \\
&\quad - Z(t)\mathscr{E}(t) \\
&= (1-r)\left[-Z(t) + y(t)\right]\mathrm{SFR}_{\mathrm{tot}}(t) + Z_{\mathscr{F}}(t)\mathscr{F}(t) - Z(t)\mathscr{E}(t) \,. \tag{4.61}
\end{aligned}$$

For the stellar mass, eq. (4.50) becomes

$$\frac{\mathrm{d}M_*(t)}{\mathrm{d}t} = (1-r)\mathrm{SFR}_{\mathrm{tot}}(t) \,, \tag{4.62}$$

or equivalently

$$\mathcal{M}_*(t) = (1-r)\int_0^t \mathrm{SFR}_{\mathrm{tot}}(t')\,\mathrm{d}t' \,. \tag{4.63}$$

For the gas mass, we have

$$\frac{\mathrm{d}\mathcal{M}_{\mathrm{ISM}}(t)}{\mathrm{d}t} = -(1-r)\mathrm{SFR}_{\mathrm{tot}}(t) + \mathscr{F}(t) - \mathscr{E}(t) \,. \tag{4.64}$$

Applying the Leibniz rule, we get

$$\frac{\mathrm{d}\mathcal{M}_Z(t)}{\mathrm{d}t} = \frac{\mathrm{d}\left[Z(t)\mathcal{M}_{\mathrm{ISM}}(t)\right]}{\mathrm{d}t} = \frac{\mathrm{d}Z(t)}{\mathrm{d}t}\mathcal{M}_{\mathrm{ISM}}(t) + Z(t)\frac{\mathrm{d}\mathcal{M}_{\mathrm{ISM}}(t)}{\mathrm{d}t} \,. \tag{4.65}$$

Thus, from eq. (4.64), the metal mass evolution becomes

$$\frac{dZ(t)}{dt}\mathcal{M}_{\mathrm{ISM}}(t) = (1-r)\left[-Z(t)+y(t)\right]\mathrm{SFR}_{\mathrm{tot}}(t) + Z_{\mathcal{F}}(t)\mathcal{F}(t) - Z(t)\mathcal{E}(t)$$
$$+ (1-r)Z(t)\,\mathrm{SFR}_{\mathrm{tot}}(t) - Z(t)\mathcal{F}(t) + Z(t)\mathcal{E}(t)$$
$$= (1-r)y(t)\,\mathrm{SFR}_{\mathrm{tot}}(t) + \left[Z_{\mathcal{F}}(t) - Z(t)\right]\mathcal{F}(t)\,. \tag{4.66}$$

While the instantaneous recycling approximation is a useful first-order estimate that provides reasonable solutions, it is no longer used for precise modern calculations. Today's computational capabilities allow us to solve the chemical evolution equations numerically.

4.5.4 CLOSED-BOX MODEL

To further simplify the equations, we make additional assumptions. We assume a closed system containing only gas with zero metallicity (though the zero-metallicity assumption is not essential) and no stars. Since $\mathcal{F} = \mathcal{E} = 0$ (i.e., $\mathcal{M} = \mathcal{M}_* + \mathcal{M}_{\mathrm{ISM}} = $ const.), $\mathcal{M}_{\mathrm{ISM}}(t=0) = \mathcal{M}$, and $\mathcal{M}_*(t=0) = 0$, eq. (4.66) simplifies to

$$\frac{dZ(t)}{dt}\mathcal{M}_{\mathrm{ISM}}(t) = (1-r)y(t)\,\mathrm{SFR}_{\mathrm{tot}}(t)\,. \tag{4.67}$$

This can be solved as

$$\frac{1}{\mathcal{M}_{\mathrm{ISM}}(t)}\frac{d\mathcal{M}_{\mathrm{ISM}}(t)}{dZ} = -\frac{1}{y(t)}\,, \tag{4.68}$$

yielding

$$\ln\mathcal{M}_{\mathrm{ISM}}\bigg|_{\mathcal{M}}^{\mathcal{M}_{\mathrm{ISM}}(t)} = -\int_0^{Z(t)}\frac{dZ}{y(t)} \simeq -\frac{Z}{\bar{y}}\,. \tag{4.69}$$

The solution for the metallicity evolution $Z(t)$ is

$$Z(t) = \bar{y}\ln\frac{\mathcal{M}_{\mathrm{ISM}}(0)}{\mathcal{M}_{\mathrm{ISM}}(t)} = \bar{y}\ln\frac{\mathcal{M}}{\mathcal{M}_{\mathrm{ISM}}(t)} \equiv \bar{y}\ln\mathfrak{M}^{-1}(t)\,, \tag{4.70}$$

where we define the gas fraction as $\mathfrak{M}(t) = \mathcal{M}_{\mathrm{ISM}}(t)/\mathcal{M}$.

The stellar metallicity Z_* is given by

$$Z_*\mathcal{M}_* + Z\mathcal{M}_{\mathrm{ISM}} = \int_0^t\int_0^\infty mp_{Z_m}\mathrm{SFR}(t')\Phi(m)\,dm\,dt'$$
$$= \int_0^t(1-r)y(t')\mathrm{SFR}(t')\,dt' = (1-r)\bar{y}\,\overline{\mathrm{SFR}}t\,. \tag{4.71}$$

From eq. (4.63), we have

$$\mathcal{M}_* = (1-r)\overline{\mathrm{SFR}}t\,, \tag{4.72}$$

leading to

$$Z_* \mathcal{M}_* = \bar{y} \mathcal{M}_* - Z \mathcal{M}_{\mathrm{ISM}} \, . \tag{4.73}$$

In the limit where $\mathcal{M}_{\mathrm{ISM}} \ll \mathcal{M}_*$, we find

$$Z_* \simeq \bar{y} \, , \tag{4.74}$$

meaning the stellar metallicity cannot exceed the average yield. This trend is observed in the baryonic mass–metallicity relation in SDSS galaxies (e.g., Tremonti et al., 2004).

4.5.5 NUMERICAL EXAMPLE OF A CLOSED-BOX MODEL WITH SCHMIDT LAW

The metal evolution does not depend on $\mathrm{SFR}_{\mathrm{tot}}(t)$ in the closed-box model, as we have seen in eq. (4.70). However, for actual numerical calculations, we introduce the Schmidt law for the SFR (Section 3.7.2),

$$\mathrm{SFR}_{\mathrm{tot}}(t) \propto \rho^n \, , \tag{4.75}$$

where ρ is gas density, and $n = 1$–2 [eq. (3.84)]. In the context of the chemical evolution theory of galaxies, a variant of eq. (3.84),

$$\mathrm{SFR}_{\mathrm{tot}}(t) = \frac{\mathcal{M}_{\mathrm{ISM}}(t)^n}{\tau_{\mathrm{SF}}} \tag{4.76}$$

is often used. Here, τ_{SF} is the typical timescale of gas depletion by star formation if $n = 1$.[1]

As discussed in Section 3.7.2, $n = 1$ is preferred for molecular gas. For τ_{SF}, we adopt the value inferred from the solar neighborhood: $\tau_{\mathrm{SF}} = 5$ [Gyr]. It is known that $\tau_{\mathrm{SF}} = 1$ [Gyr] represents the star formation history of passive elliptical galaxies, while $\tau_{\mathrm{SF}} = 3$–10 [Gyr] represents typical disk galaxies. For $n = 1$, the SFR becomes an exponential solution

$$\mathrm{SFR}_{\mathrm{tot}}(t) = \frac{1}{\tau_{\mathrm{SF}}} \exp\left(\frac{t}{\tau_{\mathrm{SF}}} \right) \, . \tag{4.77}$$

The metallicity monochromatically increases with the galactic age. On the scale of this figure, the relation looks almost linear. Hence, higher-metallicity stars are gradually formed with time in a galaxy.

4.5.6 THE G-DWARF PROBLEM AND INFALL MODEL

Now we can examine the metallicity distribution function of the stellar population Z_*. The total mass of stars that formed before the gas metallicity reaches a certain value Z is

$$M_*(<Z) = \mathcal{M} - \mathcal{M}_{\mathrm{ISM}}(Z) \, , \tag{4.78}$$

---

[1] Otherwise, the physical meaning is less clear.

which corresponds to the cumulative mass distribution of stars with metallicity $< Z$. Let the present-day galaxy age be T, and the time corresponding to the gas metallicity Z be $t = t(Z)$. If we denote the cumulative distribution of such stars as $F(< Z)$, from eq. (4.70), we have

$$F(< Z) = \frac{\mathcal{M}_*(t)}{\mathcal{M}_*(T)} = \frac{\mathcal{M} - \mathcal{M}_{\mathrm{ISM}}(t)}{\mathcal{M} - \mathcal{M}_{\mathrm{ISM}}(T)}$$

$$= \frac{\mathcal{M}\left[1 - \exp\left(-\dfrac{Z}{\bar{y}}\right)\right]}{\mathcal{M}\left[1 - \mathfrak{M}(T)\right]} = \frac{1 - \exp\left(-\dfrac{Z}{\bar{y}}\right)}{1 - \mathfrak{M}(T)}$$

$$= \frac{1 - \exp\left[-\dfrac{Z(T)}{\bar{y}}\dfrac{Z}{Z(T)}\right]}{1 - \mathfrak{M}(T)} = \frac{1 - \mathfrak{M}^{\frac{Z}{Z(T)}}}{1 - \mathfrak{M}(T)} . \tag{4.79}$$

We often measure metallicity by the logarithm of Z compared to the solar abundance ratio, as

$$\left[\frac{Z}{H}\right] = \log\left(\frac{n_Z}{n_H}\right) - \log\left(\frac{n_Z}{n_H}\right)_\odot , \tag{4.80}$$

where n is the number density, and Z is the metal element of interest. For example, we often use $[\mathrm{Fe/H}]$. Therefore, it is convenient to derive a probability density function of stars with metallicity Z as a function of $\log Z$. This can be obtained by differentiating eq. (4.79) by $\log Z$, as

$$f(\log Z) = \frac{dF}{d\log Z} = \frac{\ln 10}{1 - \exp\left[\dfrac{10^{\log Z(T)}}{\bar{y}}\right]} \frac{10^{\log Z}}{\bar{y}} \exp\left(\frac{10^{\log Z}}{\bar{y}}\right) . \tag{4.81}$$

Figure 4.6 shows the distribution of stars as a function of metallicity[2]. From this figure, we can observe that the number of stars with $[Z] = -1$ ($1/10\, Z_\odot$) is approximately $1/3$ of that of solar abundance stars. However, observationally, there are very few stars with $Z \simeq 1/3\, Z_\odot$, which clearly contradicts the prediction of the closed-box model (e.g., Audouze & Tinsley, 1976; Tinsley, 1980). This discrepancy is known as the G-dwarf problem, named after the observations of G-dwarfs in the solar neighborhood that revealed it. Since the closed-box model successfully reproduces the stellar metallicity distribution of halo and bulge stars, the problem seems to arise in the context of disk formation.

Clearly, a galaxy disk would not be a closed system. Therefore, we expect that the G-dwarf problem can be resolved by considering the infall of gas onto the disk. This process inhibits excessive star formation in the early phase of a galaxy, allowing stars to form gradually after the metallicity increases.

---

[2]This can be obtained by replacing $\log Z$ with $[Z]$ in eq. (4.81).

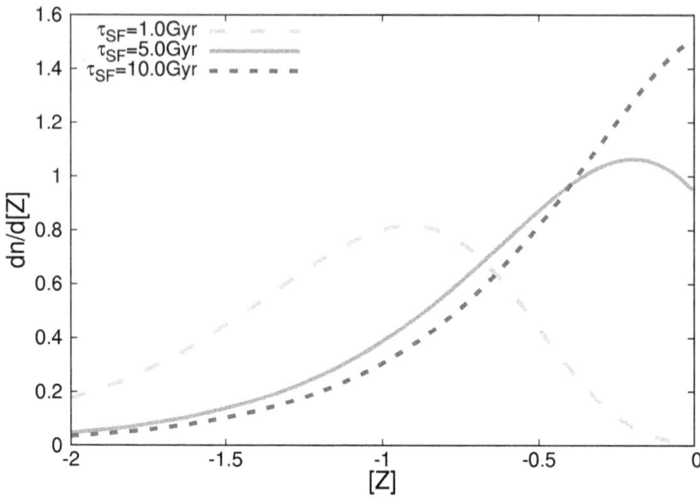

Figure 4.6 The stellar metallicity distribution with different star formation timescales for the closed-box model. Courtesy: Ryosuke S. Asano.

Here, we assume instantaneous mixing and recycling. If we include gas infall, from eq. (4.51), we have

$$\frac{d\mathcal{M}_{\mathrm{ISM}}(t)}{dt} = -(1-r)\mathrm{SFR}_{\mathrm{tot}}(t) + \mathscr{F}(t) \tag{4.82}$$

and the following form is assumed for the infall term (see, e.g., Dekel et al., 2009)

$$\mathscr{F}(t) = \frac{\mathcal{M}}{\tau_{\mathrm{infall}}} \exp\left(-\frac{t}{\tau_{\mathrm{infall}}}\right) \tag{4.83}$$

where τ_{infall} is the infall timescale. As a fiducial model, we assume $\tau_{\mathrm{SF}} = 5$ [Gyr]. For the infall model, the initial baryonic mass is assumed to be zero, i.e.,

$$\mathcal{M}(t=0) = \mathcal{M}_{\mathrm{ISM}}(t=0) = 0. \tag{4.84}$$

Additionally, it is generally assumed that the infalling gas has a pristine abundance, i.e., $Z_{\mathscr{F}} = 0$. Therefore, the equation for metallicity evolution remains the same as in the closed-box model

$$\frac{d\mathcal{M}_Z(t)}{dt} = (1-r)\left[-Z(t) + y(t)\right]\mathrm{SFR}_{\mathrm{tot}}(t). \tag{4.85}$$

An example of the SFR for the infall model is shown in Fig. 4.7. The returned mass is assumed to be $r = 0.4$, and the metallicity evolution is normalized

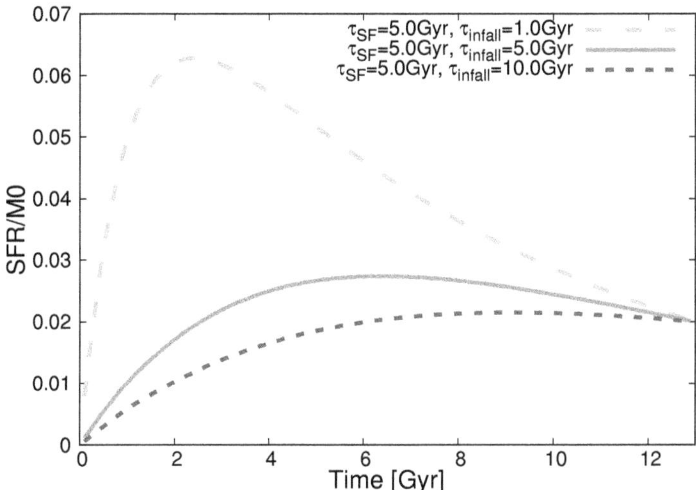

Figure 4.7 The star formation history for the general infall model, calculated using the Schmidt law for $n = 1$ as before. The star formation timescale is $\tau_{SF} = 5$ [Gyr]. Results with $\tau_{infall} = 1, 5$, and 10 [Gyr] are presented. Courtesy: Ryosuke S. Asano.

to reach $Z = 0.02$ at $t = 13.5$ [Gyr]. As shown in Fig. 4.7, a longer infall timescale, or slow infall, effectively suppresses early star formation. The estimated SFR for the Milky Way is $SFR_{tot} \simeq 1$–3 $[\mathcal{M}_\odot \, yr^{-1}]$. Assuming a total baryonic mass of $\mathcal{M}_{MW} \simeq 10^{11}$ $[\mathcal{M}_\odot]$, Fig. 4.7 suggests $SFR_{tot} = 2$ $[\mathcal{M}_\odot \, yr^{-1}]$, which agrees well with observed values. Based on this star formation history (SFH), Fig. 4.8 presents the metallicity evolution history for the infall model. As the infall timescale increases, the metallicity in the early phase of galaxy evolution rises more slowly.

Now, we compare the distribution of stellar metallicity between the closed-box and infall models in Fig. 4.9. In contrast to the closed-box model, the infall model predicts that stars with $Z \simeq 1/10$ $[Z_\odot]$ constitute less than 10% of the total stellar population, consistent with observations. This suggests that star formation proceeds more slowly in the infall model when metallicity is low and accelerates as metallicity increases over time.

Additionally, the requirement for gas infall is supported by observations of deuterium (D) abundance in the ISM. The abundance of D produced during the BBN relative to H is $\simeq 4.5 \times 10^{-5}$, while the present-day ISM shows an abundance of $\simeq 1.6 \times 10^{-5}$. Deuterium is converted into ^{3}He at relatively low temperatures and is absent from stellar ejecta. Therefore, without infall, D should be depleted in the ISM. The significant presence of D indicates the accretion of pristine gas from infall.

4.6 EVOLUTIONARY SYNTHESIS OF GALAXY SPECTRA

We have already discussed that galaxies evolve alongside the evolution of different stellar populations, coupled with the chemical evolution driven by stellar processes.

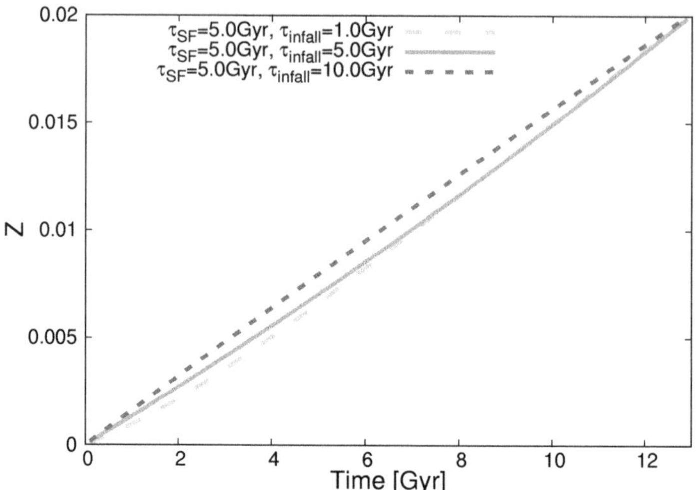

Figure 4.8   The metallicity history for the general infall model. The model settings are the same as in Fig. 4.7. Courtesy: Ryosuke S. Asano.

By combining stellar spectra and evolutionary tracks, we can reproduce the evolution of galaxy spectra from ultraviolet to near-infrared wavelengths, where stellar radiation dominates[3]. In the context of galactic astrophysics, the global shape of this spectrum is often referred to as the spectral energy distribution (SED).

By comparing model SEDs with observed spectra, we can derive important information about the stellar mass, age of the stellar populations, metallicity, and star formation history of a galaxy. In earlier studies, a more empirical approach was developed. This approach used SED templates of star clusters with various ages to reproduce the observed spectra of galaxies by superimposing these templates with different weights (e.g., Bica et al., 1990). The result was evaluated using a goodness-of-fit statistic, treating the weights as free parameters. This method is referred to as population synthesis in its original form. However, since the solution was not uniquely determined, a more physically motivated method was desired, leading to the introduction of evolutionary synthesis based on chemical evolution.

4.6.1   FORMULATION OF SPECTRAL SYNTHESIS

We first define the simple stellar population (SSP)[4]. An SSP represents the spectrum of a star cluster that consists of stars of the same age, as used in classical population synthesis. Each SSP is calculated based on the assumed IMF, libraries of stellar

---

[3]Here we do not discuss dust attenuation and emission.

[4]Often also referred to as the single stellar population.

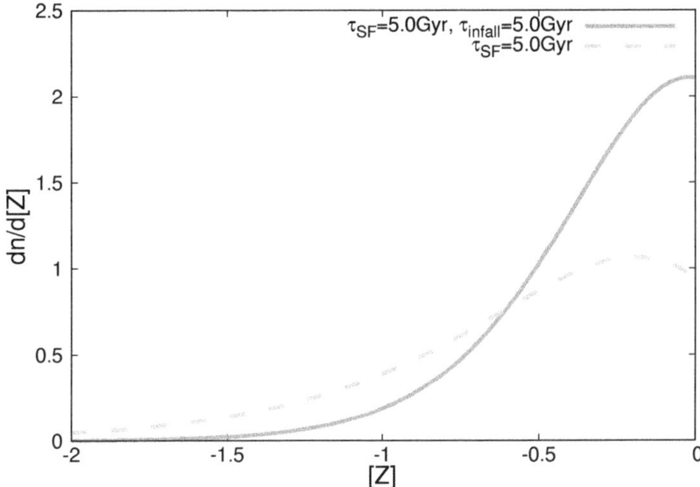

Figure 4.9 Comparison between the stellar metallicity distributions for the closed-box and infall models. The SF and infall timescales are set to $\tau_{SF} = 5$ [Gyr] and $\tau_{infall} = 5$ [Gyr], respectively. Courtesy: Ryosuke S. Asano.

evolutionary tracks, and stellar atmosphere models with various metallicities. An example of SSP spectra is shown in Fig. 4.10, with notable features described below:

1. $t = 10^6$ yr: Spectra of OB stars dominate, with a very blue spectral slope and significant ionizing photon emission ($\lambda < 912$ Å).
2. $t = 10^7$ [yr]: OB stars remain dominant, with continued ionizing photon emission. The near-infrared (NIR) continuum brightens due to the appearance of red supergiants from stars with $\mathcal{M} \simeq 20 \, [\mathcal{M}_\odot]$ (red flash).
3. $t = 10^8$ [yr]: Ionizing photons rapidly disappear due to the short lifetimes of OB stars. Nonionizing UV remains prominent, showing a blue slope, and Balmer absorption lines become significant due to the increasing contribution of A stars. The NIR continuum grows brighter, driven by the AGB population.
4. $t = 4 \times 10^8$ [yr]: Balmer absorption lines peak in prominence, the UV slope flattens (becoming redder), and the NIR continuum weakens as AGB stars evolve into post-AGB stars, which start to emit ionizing photons at $\lambda < 912$ Å.
5. $t = 10^9$ [yr]: The global SED becomes redder, and metal absorption lines appear. The NIR continuum is now dominated by RGB stars, and the 4000 Å break is formed by the Balmer discontinuity at $\lambda = 3800$ Å.
6. $t \gtrsim 4 \times 10^9$ [yr]: The SED becomes even redder, metal absorption lines increase in prominence, and the NIR continuum continues to be dominated by RGB stars. Ionizing photons are still emitted by post-AGB stars.

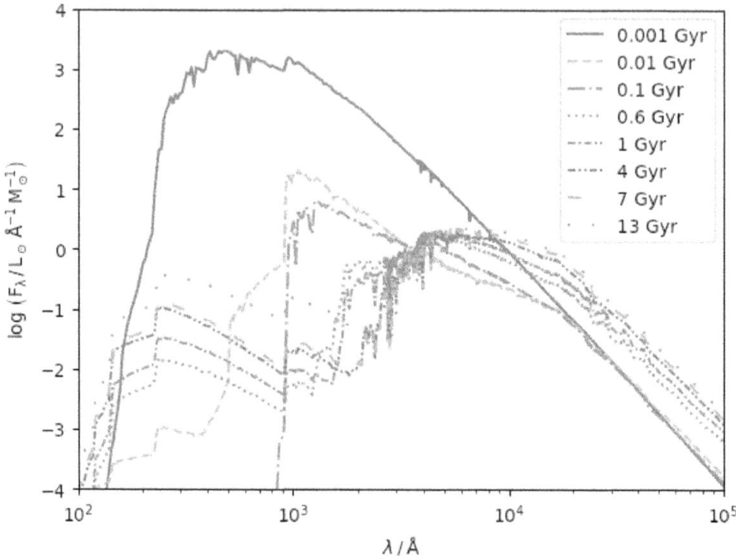

Figure 4.10 Simple stellar population spectra with various ages in units of [Gyr]. Kroupa IMF (Kroupa, 2002) is adopted for the generation of the SED. Courtesy: Aina May So.

By using SSPs, we can model the evolution of a galaxy SED as

$$\mathscr{L}_\lambda(t) = \int_0^t \text{SFR}_{\text{tot}}(t - \tau)\, \text{SSP}_\lambda\,[Z(t - \tau), \tau]\, d\tau \,. \qquad (4.86)$$

where $\text{SSP}_\lambda\,[Z(t - \tau), \tau]$ represents the SSP with age τ and metallicity Z evaluated at time $t - \tau$. This SED model depends on specifying $\text{SFR}_{\text{tot}}(t)$. In practice, the star formation history of galaxies is complex, often involving sequential merging events. Nevertheless, a smooth functional model for $\text{SFR}_{\text{tot}}(t)$ often works well for estimating physical properties from galaxy SEDs (see Chapter 5).

The exponential SFR [eq. (3.84)] is often assumed and is sometimes referred to as the τ model. For small τ, it approximates a burst of star formation, while $\tau \longrightarrow \infty$ approximates constant star formation. This functional form is derived from the chemical evolution model with the Schmidt law [eq. (4.76)] for $n = 1$:

$$\text{SFR}_{\text{tot}}(t) = \frac{\mathscr{M}_{\text{ISM}}}{\tau_{\text{SF}}} \,. \qquad (4.87)$$

Solving for $\text{SFR}_{\text{tot}}(t)$, we obtain

$$\text{SFR}_{\text{tot}}(t) = \frac{1}{\tau_{\text{SF}}} \exp\left(-\frac{t}{\tau_{\text{SF}}}\right) \,. \qquad (4.88)$$

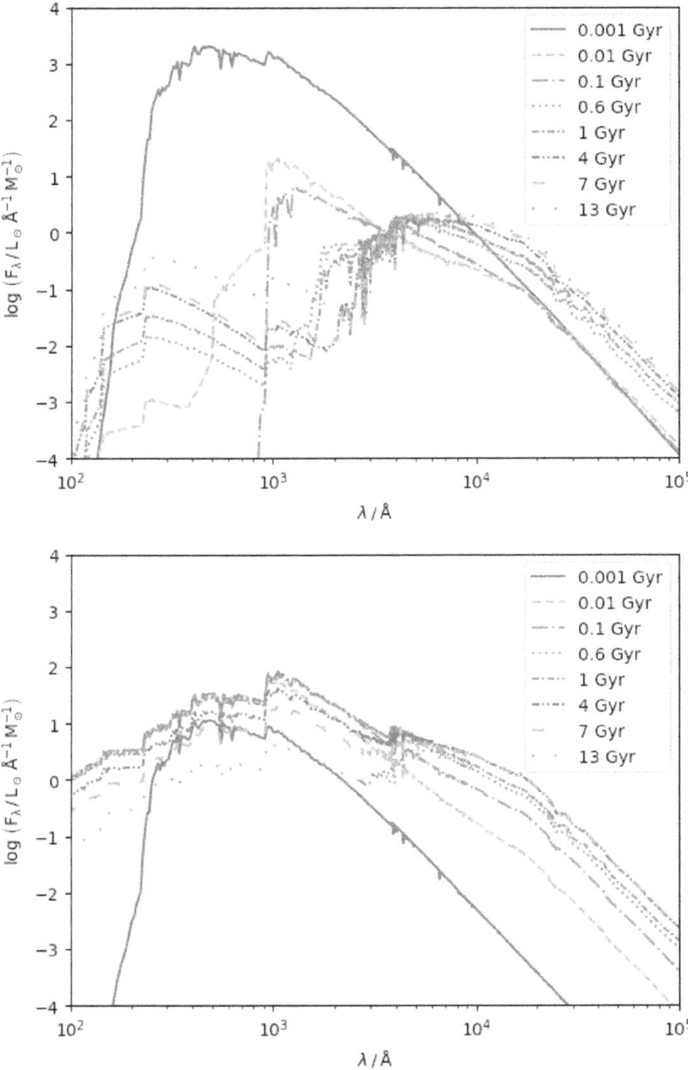

Figure 4.11 The evolution of synthesized SEDs with an instantaneous burst, $\tau_{\mathrm{SF}} = 3$ [Gyr], 7 [Gyr], and ∞ (constant SFR), from top left to bottom right. The age of each SED is indicated with each line in units of [Gyr]. Other parameters are the same as Fig. 4.10 Curtesy: Aina May So.

Figs. 4.11 and 4.12 shows the evolution of the SED for various τ_{SF}. In the instantaneous burst model, the UV continuum undergoes drastic changes, and the red flash is prominent. For $\tau_{\mathrm{SF}} = 3$ [Gyr], the rapid UV evolution becomes milder due to OB star evolution. In the case of constant SFR, the UV continuum remains unchanged while the NIR luminosity increases monotonically.

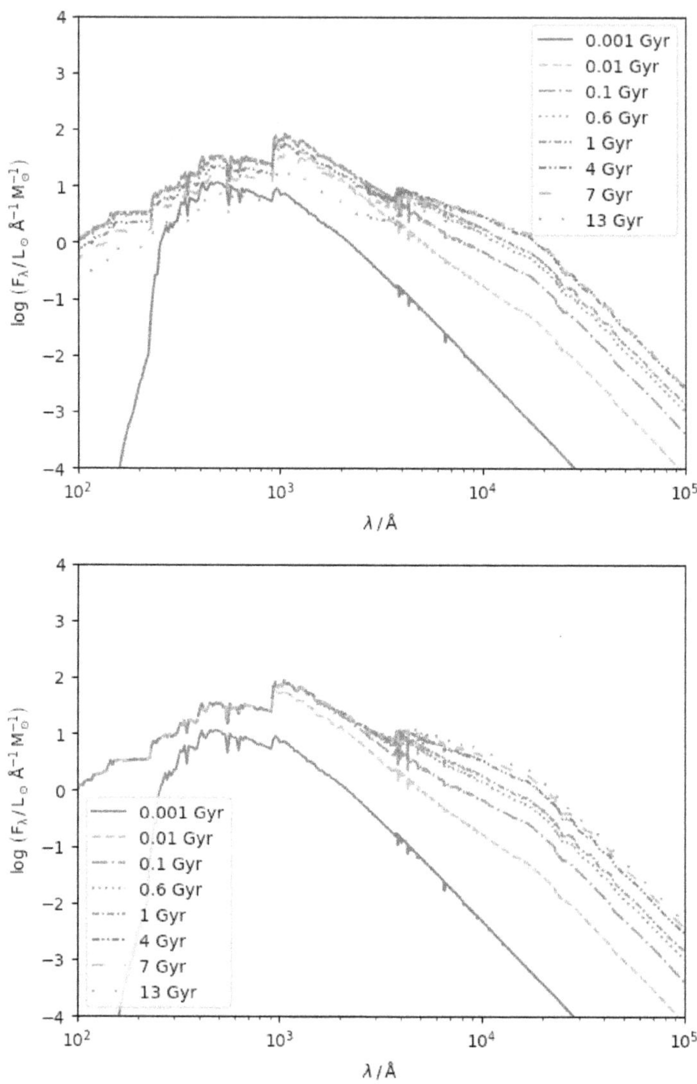

Figure 4.12 Same as Fig. 4.11 but with $\tau_{SF} = 5$ [Gyr] and ∞ (constant SFR) from top to bottom Curtesy: Aina May So.

5 Observational Star Formation Rate Indicator

5.1 PRIMARY STAR FORMATION RATE (SFR) ESTIMATOR

5.1.1 SFR AND OBSERVABLES

The basic equation to derive the SFR at a given time t, assuming no dust attenuation, is

$$
\mathcal{L}_\lambda(t) = \int_0^t \int_{\mathcal{M}_{\rm low}}^{\mathcal{M}_{\rm up}} {\rm SFR}_{\rm tot}(t-\tau)\, \mathscr{F}_{\lambda, Z_{(t-\tau)}}(m,\tau)\Phi(m)\,{\rm d}m\,{\rm d}\tau
$$
$$
= \int_0^t {\rm SFR}_{\rm tot}(t-\tau)\, {\rm SSP}_{\lambda, Z_{(t-\tau)}}(\tau)\,{\rm d}\tau\,, \tag{5.1}
$$

where $\mathcal{M}_{\rm up}$ and $\mathcal{M}_{\rm low}$ are the upper and lower bounds of the considered initial mass function (IMF), $\mathscr{F}_{\lambda, Z_{(t-\tau)}}$ is the spectrum of a star with mass m and metallicity $Z(t-\tau)$, and ${\rm SSP}_{\lambda, Z_{(t-\tau)}}(\tau)$ represents the simple stellar population with metallicity $Z_{(t-\tau)}$ and age τ, calculated based on the assumed IMF $\Phi(m)$. More generally, we should also account for emission from gas components, including nebular emission lines and continuum emission, in eq. (5.1). Many current spectral synthesis models incorporate these effects, but for simplicity, we do not explicitly include them in the spectral synthesis equations here.

5.1.1.1 Simple Recipes

To estimate the SFR from observed data by comparing them with eq. (5.1), the simplest approach is to assume a constant SFR over time. In this case, the SFR is proportional to the intrinsic monochromatic luminosity, and thus we can express it as

$$
{\rm SFR}_{\rm tot} = \frac{\mathcal{L}_\lambda}{\displaystyle\int_0^t {\rm SFR}_{\rm tot}(t-\tau)\, {\rm SSP}_{\lambda, Z_{(t-\tau)}}(\tau)\,{\rm d}\tau}\,. \tag{5.2}
$$

It is important to note that this formula is only practically useful when the luminosity reaches a steady state, i.e., $\mathcal{L}_\lambda(t) = \mathcal{L}_\lambda$.

5.1.1.2 Fitting with Spectral Synthesis Models

Instead of using the limited analytical solution of eq. (5.2), a more refined method involving spectral energy distribution (SED) fitting is commonly employed. In this method, a time sequence of galaxy SEDs is prepared from eq. (5.1), this time explicitly labeled with age. Through various SED fitting techniques to the observed spectra

DOI: 10.1201/9781003104315-5

or photometric measurements, we can determine the optimal solution for $\text{SFR}_{\text{tot}}(t)$. With this approach, $\mathscr{L}_\lambda(t)$ can be computed for any galaxy age t and at any wavelength λ. However, SED fitting techniques must be developed carefully (see, e.g., Boselli, 2011; Conroy, 2013; Walcher et al., 2011).

5.1.1.3 Ultraviolet SFR Estimator

One of the most straightforward applications of the simple recipe is the UV continuum SFR estimator. The UV continuum directly measures star formation in galaxies, as it is dominated by radiation from OB stars. Because of the short lifetimes of OB stars, they serve as ideal tracers of the instantaneous SFR at a given time t. The main challenge in this measurement is correcting for dust attenuation, which can significantly impact the observed UV radiation.

Assuming a Salpeter IMF

$$\Phi(m) \propto m^{-2.35} , \tag{5.3}$$

we obtain

$$\text{SFR}_{\text{tot}}\left[\mathscr{M}_\odot \text{yr}^{-1}\right] = 1.4 \times 10^{-28} \mathscr{L}_\nu \left[\text{erg s}^{-1}\text{Hz}^{-1}\right] . \tag{5.4}$$

This formula assumes that ν corresponds to the frequency range of 1500–2800 Å and that the SFR is constant over $t > 10^8$ [yr] (Kennicutt, 1998). For a Kroupa IMF, the conversion factor is reduced to 8.8×10^{-29} in eq. (5.4) (see, e.g., Kennicutt et al., 2009). For example, in the *GALEX* FUV and NUV bands ($\lambda \simeq 1516$ and 2300 Å), we derive

$$\text{SFR}_{\text{tot}} = 2.70 \times 10^{-10} L_{\text{FUV}} \left[L_\odot\right], \tag{5.5}$$

$$\text{SFR}_{\text{tot}} = 4.14 \times 10^{-10} L_{\text{NUV}} \left[L_\odot\right] \tag{5.6}$$

where L_{FUV} and L_{NUV} are defined by $\nu \mathscr{L}_\nu$ at FUV and NUV, respectively.

5.1.1.4 Hα SFR Estimator

Another important primary SFR estimator is the Hα luminosity. Massive stars with masses greater than $\mathscr{M} \simeq 20\ [\mathscr{M}_\odot]$ produce ionizing photons that ionize surrounding gas. Following this ionization, hydrogen recombination cascades generate emission lines such as Hα (6563 Å) and Hβ (4861 Å). As these lines are located in the optical wavelength range, they are central to SFR estimation (e.g., Kennicutt, 1998).

In an optically thick nebula (Case B, which is typically assumed in astrophysical situations of interest for SFRs, see, e.g., Osterbrock & Ferland 2006), the relation between the Hα luminosity and the ionizing photon rate is expressed as

$$L(\text{H}\alpha) = \frac{\alpha_{\text{H}\alpha}^{\text{eff}}}{\alpha_{\text{B}}} h\nu_{\text{H}\alpha} Q(\text{H}^0) \tag{5.7}$$

where $\alpha_{\mathrm{H}\alpha}^{\mathrm{eff}}$ is the effective recombination coefficient at Hα, and α_{B} is the Case B recombination coefficient. Case B refers to the condition in which the gas is optically thick to all Lyman photons. The effective recombination rate is defined as

$$\alpha_{nn'}^{\mathrm{eff}} \equiv \frac{4\pi j_{nn'}}{n_e^2 h \nu_{nn'}} \, , \tag{5.8}$$

where j is the emission coefficient for the transition from level n to n'. Substituting numerical values and assuming the following conditions

1. Case B recombination,
2. $T_e = 10^4$ [K],
3. Instantaneous SFR,

we obtain

$$L(\mathrm{H}\alpha) = 1.37 \times 10^{-12} Q(\mathrm{H}^0) \, . \tag{5.9}$$

By applying the Salpeter IMF again, we derive the final relation

$$\mathrm{SFR}_{\mathrm{tot}} \, [\mathcal{M}_\odot \mathrm{yr}^{-1}] = 7.9 \times 10^{-42} L(\mathrm{H}\alpha) \, [\mathrm{erg\ s}^{-1}] \, , \tag{5.10}$$

(Kennicutt, 1998). For a Kroupa IMF, the coefficient becomes 5.5×10^{-42}.

5.2 SFR ESTIMATOR BASED ON DUST EMISSION

Infrared observations have revealed the importance of star formation obscured by dust. This has been made possible by facilities such as *IRAS* (1980s), *ISO* (1990s), *Spitzer* (2000s), *AKARI* (2000s), *Herschel* (2010s), ALMA (2010–), *WISE* (2010s), and others. At least half of the star formation at $z = 0$ is hidden within dusty regions, only detectable in the infrared (IR), and this fraction increases with redshift (up to $z = 2$–3; Burgarella et al. 2013; Cucciati et al. 2012; Takeuchi et al. 2005a, as discussed in Chapter 7). Given this, it is natural to use dust emission as an alternative SFR estimator.

5.2.1 SFR ESTIMATOR BASED ON TOTAL IR LUMINOSITY

The temperature of dust grains depends strongly on the distance to the heating source (typically massive stars) and the global configuration of dust and stars (see also, e.g., Silva et al., 1998). To properly estimate the total energy from heating photons absorbed by dust, it is necessary to integrate the dust emission over a wide frequency range that encompasses the entire thermal radiation spectrum. This is why the total dust emission, L_{FIR} (covering 8–1000μm), is commonly used to estimate the SFR.

In practice, it is often assumed that

$$L_{\mathrm{TIR}} = L_{\mathrm{SF}} \tag{5.11}$$

(e.g., Kennicutt, 1998), where L_{SF} represents the luminosity of a galaxy originating from star formation activity. This assumes a continuous burst of star formation with a timescale of 100 Myr. The total IR luminosity is also sometimes denoted as L_{bol} (bolometric luminosity), but for clarity, we use L_{SF} throughout this book. The symbol L_{bol} is reserved for the bolometric stellar emission. In practice, slightly different wavelength ranges are also used, such as 3–1000 μm or 5–1000 μm. While these choices do not significantly affect the results, they can introduce a systematic difference of about 10% (e.g., Takeuchi et al., 2005b).

This integration is, however, not very practical unless the whole M/FIR spectrum is observed, as in the case of some very luminous nearby galaxies. For most galaxies, even in the Local Universe, only a few broadband flux densities are available. Thus, following *IRAS*, many attempts have been made to estimate the total IR luminosity

$$L_{\text{TIR}} = \int_{v_{\text{low}}}^{v_{\text{up}}} \mathscr{L}_v \, dv \,. \tag{5.12}$$

The *IRAS* team proposed a definition of the FIR luminosity, L_{FIR}, based on the *IRAS* bands (central wavelengths: 12, 25, 60, and 100 μm)

$$L_{\text{FIR}} = \int_{v@122\mu m}^{v@42\mu m} \mathscr{L}_v \, dv = 3.29 \times 10^{-22} \left[2.58 \mathscr{L}_v(60) + \mathscr{L}_v(100) \right] [L_\odot], \tag{5.13}$$

where $\mathscr{L}_v$(band) stands for the monochromatic luminosity at an *IRAS* band (Helou et al., 1985) in units of $[\text{erg}\,\text{s}^{-1}\text{Hz}^{-1}]$. As discussed in Takeuchi et al. (2005b), L_{FIR} does not capture all of the IR energy emitted by dust. To address this limitation, several L_{TIR} estimators with additional bands have been developed.

Dale et al. (2001) proposed a correction to L_{FIR}, dependent on the FIR color $x \equiv \log[\mathscr{L}_v(60)/\mathscr{L}_v(100)]$, given by

$$\begin{aligned} L_{\text{TIR}}^{\text{Dale}} &\equiv \int_{v@1.1mm}^{v@3\mu m} \mathscr{L}_v \, dv \\ &= 10^{(a_0 + a_1 x + a_2 x^2 + a_3 x^3 + a_4 x^4)} L_{\text{FIR}} \, [L_\odot], \end{aligned} \tag{5.14}$$

where

$$(a_0, a_1, a_2, a_3, a_4) = (0.2738, -0.0282, 0.7281, 0.6208, 0.9118)$$

at $z = 0$. Dale & Helou (2002) proposed another L_{TIR} estimator that uses additional *IRAS* bands

$$L_{\text{TIR}}^{\text{DH}} \equiv 2.403 v \mathscr{L}_v(25) - 0.2454 v \mathscr{L}_v(60) + 1.6381 v \mathscr{L}_v(100) \, [L_\odot]. \tag{5.15}$$

The IR luminosity between $\lambda = 8$–1000 $[\mu m]$, denoted L_{FIR}, was introduced by Sanders & Mirabel (1996) as

$$\begin{aligned} L_{\text{IR}}^{\text{SM}} &\equiv \int_{v@1mm}^{v@8\mu m} \mathscr{L}_v \, dv \\ &= 4.93 \times 10^{-22} \left[13.48 \mathscr{L}_v(12) + 5.16 \mathscr{L}_v(25) + 2.58 \mathscr{L}_v(60) + \mathscr{L}_v(100) \right] [L_\odot]. \end{aligned} \tag{5.16}$$

Though *IRAS* data are now considered classical, they remain one of the most valuable datasets in astronomy, and these formulae continue to be widely used.

We now summarize the total IR luminosity calibrations from other facilities. Using *Spitzer* bands, the total IR luminosity can be estimated as

$$L_{\text{TIR}}(Spitzer) = 1.559\nu\mathscr{L}_\nu(24) - 0.7686\nu\mathscr{L}_\nu(70) + 1.347\nu\mathscr{L}_\nu(160), \quad (5.17)$$

as described by Boselli (2011) and references therein. Note that $\nu\mathscr{L}_\nu$ is used for *Spitzer* formulae.

For *AKARI*, which includes four Far-Infrared Surveyor (FIS) bands: *N60*, *WIDE-S*, *WIDE-L*, and *N160*, the total IR luminosity can be estimated as

$$L_{\text{TIR}}(AKARI) = L_{AKARI}^{\text{3bands}}$$
$$\equiv \Delta\nu(N60)\mathscr{L}_\nu(65) + \Delta\nu(WIDE\text{-}S)\mathscr{L}_\nu(90) + \Delta\nu(WIDE\text{-}L)\mathscr{L}_\nu(140), \quad (5.18)$$

or

$$L_{\text{TIR}}(AKARI) = L_{AKARI}^{\text{2bands}}$$
$$\equiv \Delta\nu(WIDE\text{-}S)\mathscr{L}_\nu(90) + \Delta\nu(WIDE\text{-}L)\mathscr{L}_\nu(140), \quad (5.19)$$

(Solarz et al., 2016), where

$$\Delta\nu(N60) = 1.58 \times 10^{12}\ [\text{Hz}], \quad (5.20)$$
$$\Delta\nu(WIDE\text{-}S) = 1.47 \times 10^{12}\ [\text{Hz}], \quad (5.21)$$
$$\Delta\nu(WIDE\text{-}L) = 0.831 \times 10^{12}\ [\text{Hz}] \quad (5.22)$$

(Hirashita et al., 2008).

Herschel also made significant contributions in this field. It is important to note that these formulae are only valid for galaxies at $z = 0$. When applying them to galaxies at $z > 0$, a K-correction must first be applied. If the *Herschel* SPIRE luminosity is defined using flux densities from three bands

$$L_{\text{SPIRE}}^{\text{3bands}} = \Delta\nu(250)\mathscr{L}_\nu(250) + \Delta\nu(350)\mathscr{L}_\nu(350) + \Delta\nu(500)\mathscr{L}_\nu(500), \quad (5.23)$$

the total IR luminosity can be expressed as

$$L_{\text{TIR}}(Herschel) = 1.6 \times 10^{-1} \left(L_{\text{SPIRE}}^{\text{3bands}}\right)^{1.16}, \quad (5.24)$$

as described by Galametz et al. (2013).

Then, using any of these total IR luminosities, the dust-based SFR can be estimated by

$$\text{SFR}_{\text{tot}}\ [\mathscr{M}_\odot\text{yr}^{-1}] = 1.74 \times 10^{-10}L_{\text{TIR}}\ [L_\odot], \quad (5.25)$$

assuming eq. (5.11) (Kennicutt, 1998). The conversion factor for the Kroupa IMF is 2.80×10^{-10} (e.g., Eufrasio et al., 2017). We note that, as a substitute for the total

IR luminosities, a monochromatic IR luminosity is often used, but it inevitably has a larger uncertainty and complication due to K-correction.

Thus far, we have discussed how SFR estimators based on dust luminosity can efficiently estimate star formation rates. However, the fundamental assumption that UV photons are completely absorbed by dust is unrealistic, as some amount of UV photon leakage is expected in most galaxies. The scattering of UV photons by dust also plays an important role in shaping the spectral energy distributions (SEDs) of galaxies. Therefore, it is reasonable to introduce a hybrid SFR estimator that combines both UV and dust-based SFR indicators.

5.3 HYBRID SFR ESTIMATOR

We pointed out the potential UV leakage that leads to an underestimation of the SFR by dust-related estimators. On the other hand, the IR estimators are affected by the contribution of older stellar populations and cannot be considered direct SFR indicators. Hybrid SFR estimators allow us to avoid corrections for dust attenuation or non-SF dust emission, providing us with an unbiased estimate of the SFR.

5.3.1 PHYSICAL BASIS OF THE HYBRID SFR ESTIMATOR

First, we outline how to construct such hybrid SFR estimators. By combining L_{FIR} and L_{FUV}, we can derive the total light emitted by young stars. As discussed earlier, the dust emission from galaxies typically consists of two components: emission from star-forming (SF) regions and that from non-SF regions. To accurately estimate the SFR of galaxies, we must account for the contribution of non-SF dust emission, which can lead to an overestimation SFR of a galaxy. The luminosity contribution from SF regions, or the SFR, is expressed as

$$L_{\mathrm{SF}} = L_{\mathrm{UV}}^{\mathrm{obs}} + L_{\mathrm{TIR}}^{\mathrm{SF}} \,. \tag{5.26}$$

Here, $L_{\mathrm{UV}}^{\mathrm{obs}}$ is the *observed* UV luminosity of a galaxy,

$$L_{\mathrm{UV}}^{\mathrm{obs}} = L_{\mathrm{UV}}^{\mathrm{int}} e^{-0.4 A_{\mathrm{UV}}} \,, \tag{5.27}$$

where $L_{\mathrm{UV}}^{\mathrm{int}}$ is the *intrinsic* UV luminosity, and A_{UV} is the extinction at a certain UV band (e.g., $\lambda = 1600$ Å). The second term in eq. (5.26), $L_{\mathrm{TIR}}^{\mathrm{SF}}$, refers to the total IR luminosity from SF regions. "Total" here signifies dust emission in the wavelength range $\lambda = 8\text{--}1000\,\mu$m.

The dust attenuation in the UV is expressed as

$$A_{\mathrm{UV}} = 2.5 \log\left[1 + (1 - \eta)10^{\mathrm{IRX}}\right] \,, \tag{5.28}$$

where IRX refers to the infrared excess, defined as

$$\mathrm{IRX} \equiv \log\left(\frac{L_{\mathrm{TIR}}}{L_{\mathrm{UV}}^{\mathrm{obs}}}\right) \,. \tag{5.29}$$

More generally, UV dust attenuation is often assumed to be a function of IRX

$$A_{UV} = f(IRX) , \qquad (5.30)$$

where a polynomial function is frequently adopted (see, e.g., Boissier, 2013; Buat et al., 2005) for more details.

The assumption that L_{FIR} represents the total dust emission from SF regions holds only for intense, dusty starbursts and does not apply to more typical galaxies. Since dust is also heated by low-mass stars not directly related to current SF activity, we must explicitly introduce this contribution[1]. The actual IR luminosity is expressed as

$$L_{FIR} = L_{TIR}^{SF} + L_{TIR}^{nonSF} , \qquad (5.31)$$

where L_{TIR}^{nonSF} is the contribution from non-SF regions. At $z = 0$, observational and theoretical analyses estimate that the contribution of non-SF stars to dust heating, η, is $\eta = 0.2$–0.4 for normal galaxies and $\eta = 0$ for starburst galaxies (e.g., Bell, 2003; Buat et al., 2011, 2005; Buat et al., 2007; Hirashita et al., 2003; Iglesias-Páramo et al., 2006). Thus, we have

$$L_{SF} = L_{UV}^{obs} + (1 - \eta)L_{TIR} . \qquad (5.32)$$

Meurer et al. (1999) proposed a relation between the UV spectral slope and IRX, which is widely used to estimate the extinction of UV-luminous galaxies because it does not require IR emission data for a galaxy[2]. We discuss this method further in Section 5.4. Overzier et al. (2011) derived a formula using the method of Meurer et al. (1999) to describe the SF luminosity L_{SF} as

$$L_{SF} = L_{UV}^{obs} + 0.6L_{TIR} . \qquad (5.33)$$

This formula is similar to eq. (5.32), and it is interesting to examine the physical reasons for this similarity. Meurer et al. (1999) demonstrated that IRX can be approximated as

$$IRX = \left(10^{0.4A_{UV}} - 1\right) \frac{BC_*}{BC_{IR}} . \qquad (5.34)$$

Here, the bolometric correction $BC_* = L_{bol}/L_{UV}^{int}$ is defined as the ratio between the bolometric luminosity of stars, L_{bol}, and the intrinsic UV luminosity, L_{UV}^{int}, while BC_{IR} is the ratio between the integrated IR luminosity and L_{TIR}. We set $BC_{IR} = 1$ to simplify the expression. Thus,

$$IRX = \left(10^{0.4A_{UV}} - 1\right) BC_* . \qquad (5.35)$$

---

[1]Additionally, we must also account for heating by AGNs, although this is a more complex issue and will not be addressed here.

[2]The original relation was derived for the attenuation of central starbursts in galaxies.

Therefore,

$$
\begin{aligned}
L_{\mathrm{FIR}} &= \left(10^{0.4A_{\mathrm{UV}}} - 1\right) \mathrm{BC}_* L_{\mathrm{UV}}^{\mathrm{obs}} \\
&= \left(10^{0.4A_{\mathrm{UV}}} - 1\right) \mathrm{BC}_* L_{\mathrm{UV}}^{\mathrm{int}} 10^{-0.4A_{\mathrm{UV}}} \\
&= \left(1 - 10^{-0.4A_{\mathrm{UV}}}\right) \mathrm{BC}_* L_{\mathrm{UV}}^{\mathrm{int}} .
\end{aligned}
\tag{5.36}
$$

If no dust is present, the SF luminosity should be the intrinsic UV luminosity. Thus,

$$
\begin{aligned}
L_{\mathrm{SF}} &= L_{\mathrm{UV}}^{\mathrm{int}} \\
&= L_{\mathrm{UV}}^{\mathrm{obs}} + L_{\mathrm{TIR}}^{\mathrm{SF}} \\
&= 10^{-0.4A_{\mathrm{UV}}} L_{\mathrm{UV}}^{\mathrm{int}} + \left(1 - 10^{-0.4A_{\mathrm{UV}}}\right) L_{\mathrm{UV}}^{\mathrm{int}} .
\end{aligned}
\tag{5.37}
$$

Next, we examine L_{TIR}.

$$
\begin{aligned}
L_{\mathrm{TIR}} &= L_{\mathrm{TIR}}^{\mathrm{SF}} + L_{\mathrm{TIR}}^{\mathrm{nonSF}} \\
&= \left(1 - 10^{-0.4A_{\mathrm{UV}}}\right) \mathrm{BC}_* L_{\mathrm{UV}}^{\mathrm{int}} \\
&= \left(1 - 10^{-0.4A_{\mathrm{UV}}}\right) L_{\mathrm{UV}}^{\mathrm{int}} + L_{\mathrm{TIR}}^{\mathrm{nonSF}} .
\end{aligned}
\tag{5.38}
$$

Thus, we obtain

$$
\begin{aligned}
L_{\mathrm{TIR}}^{\mathrm{nonSF}} &= (\mathrm{BC}_* - 1)\left(1 - 10^{-0.4A_{\mathrm{UV}}}\right) L_{\mathrm{UV}}^{\mathrm{int}} \\
&= \frac{\mathrm{BC}_* - 1}{\mathrm{BC}_*} L_{\mathrm{FIR}} \equiv \eta L_{\mathrm{TIR}} ,
\end{aligned}
\tag{5.39}
$$

leading to

$$
L_{\mathrm{TIR}}^{\mathrm{SF}} = (1 - \eta) L_{\mathrm{TIR}} .
\tag{5.40}
$$

Using η, we finally have the same formula as eq. (5.32)

$$
L_{\mathrm{SF}} = L_{\mathrm{UV}}^{\mathrm{obs}} + L_{\mathrm{TIR}}^{\mathrm{SF}} = L_{\mathrm{UV}}^{\mathrm{obs}} + (1 - \eta) L_{\mathrm{FIR}} .
\tag{5.41}
$$

This equation can be interpreted as follows. The bolometric luminosity from stars is rewritten as

$$
L_{\mathrm{bol}} = L_{\mathrm{UV}}^{\mathrm{int}} + (\mathrm{BC}_* - 1) L_{\mathrm{UV}}^{\mathrm{int}} = (1 - \eta) L_{\mathrm{bol}} + \eta L_{\mathrm{bol}} ,
\tag{5.42}
$$

meaning the second term represents the luminosity from stars that do not emit strong UV radiation. Thus, in eq. (5.41), the term including $1 - \eta$ represents the correction for non-SF dust emission, consistent with the interpretation in works such as Buat et al. (2007).

5.3.2　UV/IR RATIO (IRX) AS AN ATTENUATION ESTIMATOR

We discussed the UV attenuation A_{UV} in the previous discussion. Actually, the balance between the absorbed UV and re-emitted IR emissions is evaluated by the IR/UV flux ratio $L_{\mathrm{FUV}}/L_{\mathrm{TIR}}$, and it serves as the most accurate method to correct the observed UV flux. The advantage of this method is that it does not depend on the spatial geometry of the system or the details of the extinction curve (e.g., Buat

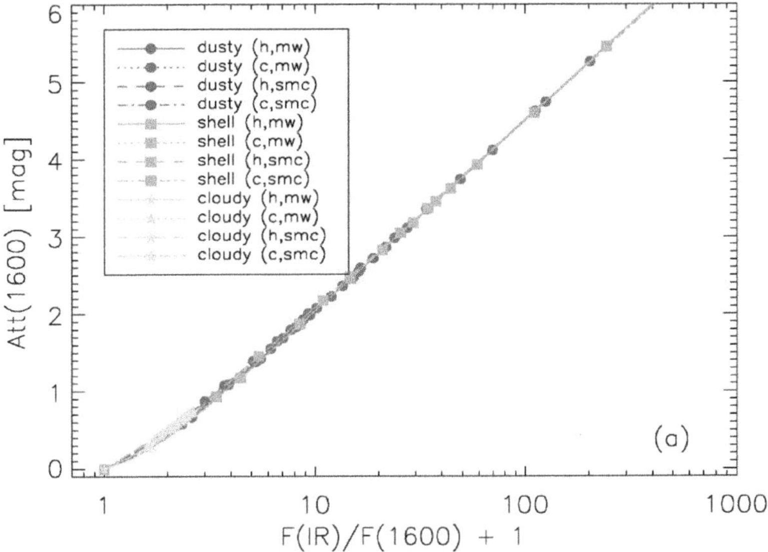

Figure 5.1 Relation between the IR/UV flux ratio (denoted as $F(\mathrm{IR})/F(1600\,\text{Å})$) and A_{1600} for all the parameters of the model proposed by Gordon et al. (2000) and Witt & Gordon (2000) (DIRTY model). The parameters of the DIRTY radiative transfer model curves include three star/gas/dust geometries (cloudy, dusty, shell), two dust grain properties, Milky Way (MW) and the Small Magellanic Cloud (SMC), and two dust distributions (h: homogeneous, c: clumpy). Credit: Gordon, K. D., Clayton, G. C., Witt, A. N., et al. 2000, Astrophysical Journal, 533, 236, Fig. 1, reproduced by permission of AAS.

& Xu, 1996; Gordon et al., 2000; Witt & Gordon, 2000). The robustness of this method is clearly seen in Fig. 5.1. This is because the dust is heated mainly by massive stars in HII regions and the IR does not suffer self-absorption and is radiated almost isotropically. Unless the geometry is in a very extreme configuration (such as dust and massive stars being completely separated), the flux ratio provides a reliable estimate for attenuation.

The following empirical relation is widely used to describe the dust attenuation for star-forming galaxies (Buat et al., 2005). For *GALEX* NUV (2310 Å),

$$
A_{\mathrm{NUV}} = 0.2269 + 0.8998 \left[\log\left(\frac{L_{\mathrm{TIR}}}{L_{\mathrm{NUV}}} \right) \right]
$$

$$
+ 0.4718 \left[\log\left(\frac{L_{\mathrm{TIR}}}{L_{\mathrm{NUV}}} \right) \right]^2 - 0.0495 \left[\log\left(\frac{L_{\mathrm{TIR}}}{L_{\mathrm{NUV}}} \right) \right]^3, \tag{5.43}
$$

and for FUV (1530 Å),

$$
A_{\mathrm{FUV}} = 0.4967 + 1.1960 \left[\log\left(\frac{L_{\mathrm{TIR}}}{L_{\mathrm{FUV}}} \right) \right]
$$

$$
+ 0.3522 \left[\log\left(\frac{L_{\mathrm{TIR}}}{L_{\mathrm{FUV}}} \right) \right]^2 - 0.0333 \left[\log\left(\frac{L_{\mathrm{TIR}}}{L_{\mathrm{FUV}}} \right) \right]^3. \tag{5.44}
$$

Here, $A(\lambda)$ is expressed in magnitudes. For intense starburst galaxies, a different formula better describes the attenuation (Calzetti et al., 2000):

$$A_{1600} = 2.5 \log \left[\frac{1}{0.9} \frac{L_{\mathrm{TIR}}}{1.75 \nu \mathscr{L}_\nu (1600\,\text{Å})} + 1 \right] \tag{5.45}$$

where A_{1600} is the attenuation at 1600 Å.

However, these formulae are not valid for quiescent galaxies in the Local Universe. In these cases, the dust emission from SF regions is not the main contributor, and the cirrus emission from dust heated by the general interstellar radiation field (mainly from intermediate and old stars) dominates the IR luminosity (e.g., Kong et al., 2004). In these cases, the $A(\lambda)$–$L_{\mathrm{TIR}}/L_{\mathrm{FUV}}$ relation depends on the SED of the stellar population. Cortese et al. (2008) proposed an attenuation formula for quiescent populations of galaxies, depending on their SFH, as

$$A_{\mathrm{FUV}} = a_1 + a_2 \left[\log \left(\frac{L_{\mathrm{TIR}}}{L_{\mathrm{FUV}}} \right) \right] + a_3 \left[\log \left(\frac{L_{\mathrm{TIR}}}{L_{\mathrm{FUV}}} \right) \right]^2$$
$$+ a_4 \left[\log \left(\frac{L_{\mathrm{TIR}}}{L_{\mathrm{FUV}}} \right) \right]^3 + a_5 \left[\log \left(\frac{L_{\mathrm{TIR}}}{L_{\mathrm{FUV}}} \right) \right]^4. \tag{5.46}$$

Here, the coefficients a_i depend on the SFH, characterized by the SF timescale τ_{SF} (Fig. 5.2). The dependence of a_i on τ_{SF} is tabulated in Cortese et al. (2008). As shown in Fig. 5.2, formulae for SF or starburst galaxies, such as eqs. (5.44), (5.43), and (5.45), significantly overestimate the actual attenuation in quiescent galaxies.

As we discussed, the balance between the absorbed UV and the re-emitted IR is crucial in understanding UV attenuation. The IRX serves as a valuable tool for estimating dust attenuation in star-forming galaxies. However, different approaches may be required for quiescent galaxies or galaxies with complex star formation histories.

The relationship between the IRX and attenuation in galaxies is influenced by a number of factors, including the star formation history and metallicity of the galaxy. Careful consideration of these factors ensures more accurate SFR estimations. The empirical relations and hybrid models discussed offer a variety of ways to improve the estimation of SFR by accounting for dust attenuation in different environments and galaxy types.

5.4   THE INFRARED EXCESS–ULTRAVIOLET SLOPE (IRX–β) RELATION

As previously discussed, estimating UV attenuation, A_{UV}, is essential for accurately determining the star formation rate (SFR) in galaxies. Although A_{UV} can be estimated using the IRX (infrared excess), infrared observations are generally less efficient and more time-consuming than optical observations, even with the advancements provided by ALMA. Therefore, an alternative method to estimate A_{UV} without relying on IR data would be highly beneficial for studies of galaxy SFRs.

Meurer et al. (1999) discovered that the slope of the UV continuum can serve as a proxy for dust attenuation. This conclusion was based on observations of local

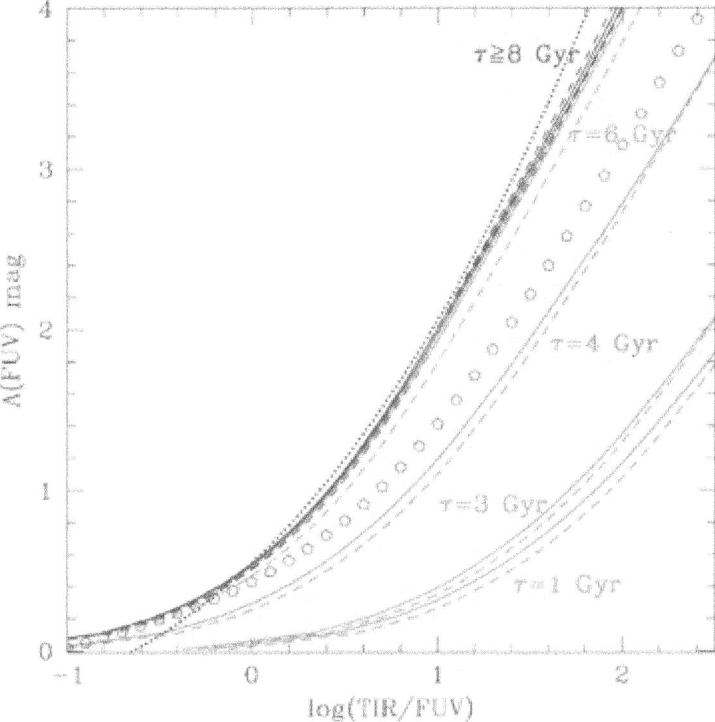

Figure 5.2 The relationship between the TIR/FUV ratio and the FUV attenuation A_{FUV} obtained from a model for different values of τ_{SF} by Cortese et al. (2008). The solid and dashed lines show the relations for stellar metallicities $Z = 2.5Z_\odot$ and $0.02Z_\odot$, respectively. The dotted black line indicates the age-independent relation given by Buat et al. (2005). The empty pentagons show the relation proposed by Kong et al. (2004). Credit: Cortese, L., Boselli, A., Franzetti, P., et al. 2008, Monthly Notices of the Royal Astronomical Society, 386, 1157, Fig. 1.

starburst galaxies made by *IRAS* and *IUE*. Utilizing this finding, they proposed a method for estimating dust attenuation using only the UV spectral slope β, which is defined as

$$\mathscr{L}_\lambda \propto \lambda^\beta, \tag{5.47}$$

in the wavelength range of 1200–2500 Å.

In this section, we revisit the original formulation of the IRX–β relation

$$\mathrm{IRX}^{\mathrm{M}} \equiv \frac{S(\mathrm{FIR})}{S(1600)} = \frac{S(\mathrm{Ly}\alpha) + \int_{912\,\text{Å}}^{\infty} S_{\lambda',0}\left(1 - 10^{-0.4A_{\lambda'}}\right) d\lambda'}{S(1600,0)\,10^{-0.4A_{1600\,\text{Å}}}} \frac{S(\mathrm{FIR})}{S(\mathrm{TIR})}, \tag{5.48}$$

where $S(\lambda) \equiv \nu S_\nu$ at wavelength λ, $S_{\lambda',0}$ is the unattenuated flux density of the emitted spectrum, A_λ represents the attenuation at wavelength λ, and $S(\mathrm{Ly}\alpha)$ is the Lyα

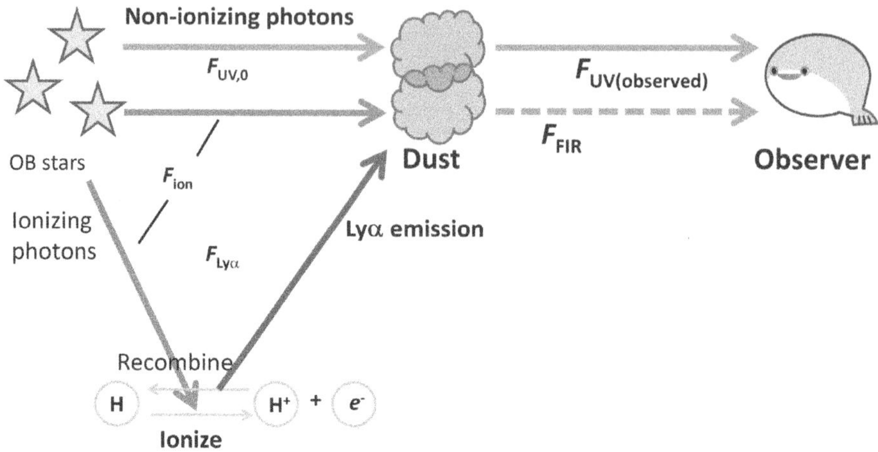

Figure 5.3 The schematic diagram explaining the physics behind the IRX–β relation. It is assumed that ionizing photons do not significantly contribute to dust heating and are not directly absorbed by dust grains. This means that the horizontal arrow from OB stars to the dust cloud (denoted as F_{ion}) can be ignored.

flux. In eq. (5.48), it is assumed that ionizing photons do not contribute significantly to dust heating, meaning they are not directly absorbed by dust grains. This assumption implies that the horizontal arrow from OB stars to a dust cloud (denoted as F_{ion}) in Fig. 5.3 can be ignored. While this is not strictly accurate, it does not significantly affect the outcome of the analysis (see Takeuchi et al., 2012).

The Lyα flux $S(\mathrm{Ly}\alpha)$ is produced by a cascade of photons with wavelengths shorter than $912\,\text{Å}$, which eventually reach the transition from $n : 2 \to 1$. If the ionized gas is in Case B, Lyα photons are resonantly scattered by hydrogen atoms and ultimately absorbed by dust. Thus, the numerator of eq. (5.48) represents the total flux absorbed by dust, which includes two contributions: heating from Lyα and heating from non-ionizing photons. The denominator in eq. (5.48) represents the observed $1600\,\text{Å}$ flux, attenuated by dust.

Therefore

$$S(\mathrm{TIR}) = S(\mathrm{Ly}\alpha) + \int_{912\,\text{Å}}^{\infty} S_{\lambda',0}(1 - 10^{-0.4A_{\lambda'}})\mathrm{d}\lambda' , \tag{5.49}$$

$$S(1600) = S(1600,0)\,10^{-0.4A_{1600\,\text{Å}}}, \tag{5.50}$$

$$\frac{S(\mathrm{FIR})}{S(\mathrm{TIR})} = \frac{\text{Flux of dust emission within } \lambda = 42\text{--}122\,\mu\mathrm{m}}{\text{Total flux of dust emission}}. \tag{5.51}$$

Most of the photons absorbed by dust are emitted by massive stars. Therefore, by assuming that most of the radiation from massive stars is in the UV, we can substitute

A_λ with A_{1600} in eq. (5.48) as follows

$$\mathrm{IRX}^{\mathrm{M}} = (10^{0.4 A_{1600 \text{Å}}} - 1) \frac{S(\mathrm{Ly}\alpha) + \int_{912\,\text{Å}}^{\infty} S_{\lambda',0} d\lambda'}{S(1600,0)} \frac{S(\mathrm{FIR})}{S(\mathrm{TIR})} . \tag{5.52}$$

We further rearrange eq. (5.52) as follows.

$$\mathrm{IRX}^{\mathrm{M}} = (10^{0.4 A_{1600\,\text{Å}}} - 1)B , \tag{5.53}$$

$$B = \frac{\mathrm{BC}(1600)_*}{\mathrm{BC}(\mathrm{FIR})_{\mathrm{dust}}} . \tag{5.54}$$

Here BC stands for the bolometric correction. The subscript asterisk refers to stellar emission, while the subscript dust refers to dust emission. Specifically,

$$\mathrm{BC}(1600)_* = \frac{S(\mathrm{Ly}\alpha) + \int_{912}^{\infty} S_{\lambda',0} d\lambda'}{S(1600,0)} , \tag{5.55}$$

$$\mathrm{BC}(\mathrm{FIR})_{\mathrm{dust}} = \frac{S(\mathrm{TIR})}{S(\mathrm{FIR})} . \tag{5.56}$$

We assume

$$B = \frac{\mathrm{BC}(1600)_*}{\mathrm{BC}(\mathrm{FIR})_{\mathrm{dust}}} = \mathrm{const.} \tag{5.57}$$

On one hand, $\mathrm{BC}(1600)_*$ can be approximated as a constant if we fix the IMF and assume a constant SFR. Adopting the Salpeter IMF with an upper mass limit of 100 $\mathcal{M}_\odot$, we obtain $\mathrm{BC}(1600)_* = 1.66 \pm 0.15$ (Meurer et al., 1999). The uncertainty arises from variations in $\mathrm{BC}(1600)_*$ based on the burst duration. On the other hand, $\mathrm{BC}(\mathrm{FIR})_{\mathrm{dust}}$ is the ratio of total IR flux to FIR flux. From empirical *IRAS* observations (Iglesias-Páramo et al., 2006), Meurer et al. (1999) adopted $\mathrm{BC}(\mathrm{FIR})_{\mathrm{dust}} = 1.4 \pm 0.2$, resulting in $B = 1.19 \pm 0.20$.

Combining these with eq. (5.53), we derive

$$\begin{aligned}
\log \mathrm{IRX}^{\mathrm{M}} &= \log(10^{0.4 A_{1600}} - 1) + \log B \\
&= \log(10^{0.4 A_{1600}} - 1) + 0.076 \pm 0.044 .
\end{aligned} \tag{5.58}$$

Assuming a linear relation between A_{1600} and β, we express it as

$$A_{1600} = b_0 + b_1 \beta , \tag{5.59}$$

Thus, the general relation between IRX and β becomes

$$\log \left[\frac{S(\mathrm{FIR})}{S(\mathrm{FUV})} \right] = \log \left[10^{0.4(b_0 + b_1 \beta)} - 1 \right] + 0.076 . \tag{5.60}$$

By performing a least-square fit to observed data, Meurer et al. (1999) obtained

$$A_{1600} = 4.43 + 1.99\beta \, . \tag{5.61}$$

Combining eqs. (5.61) and (5.60), we arrive at

$$\log\left[\frac{S(\mathrm{FIR})}{S(\mathrm{FUV})}\right] = \log\left[10^{0.4(4.43+1.99\beta)} - 1\right] + 0.076 \, . \tag{5.62}$$

This is the original IRX–β relation proposed by Meurer et al. (1999).

Since eq. (5.62) was derived from *IUE* observations, the small aperture of the instrument introduces a significant underestimation of UV flux (Overzier et al., 2011; Takeuchi et al., 2012). Takeuchi et al. (2012) investigated the aperture effect using *GALEX* and *AKARI* images of the same sample galaxies as the original study. Since *GALEX* data was taken in broadband filters FUV and NUV, we define

$$\beta_{\mathrm{GALEX}} \equiv \frac{\log S_\lambda(\mathrm{FUV}) - \log S_\lambda(\mathrm{NUV})}{\log \lambda_{\mathrm{FUV}} - \log \lambda_{\mathrm{NUV}}} \, . \tag{5.63}$$

We also redefine IRX, as *AKARI* measures the total IR flux, as seen in eqs. (5.18) or (5.19)

$$\mathrm{IRX} \equiv \log\left(\frac{F_{\mathrm{TIR}}}{F_{\mathrm{FUV}}}\right) \, . \tag{5.64}$$

Thanks to this, $\mathrm{BC(FIR)}_{\mathrm{dust}}$ no longer needs consideration. From this analysis, Takeuchi et al. (2012) obtained a corrected formula

$$\log\left[\frac{S(\mathrm{TIR})}{S(\mathrm{FUV})}\right] = \log\left[10^{(0.4(3.06+1.58\beta)} - 1\right] + 0.22 \, . \tag{5.65}$$

This represents the accurate IRX–β relation for local starburst galaxies (see Fig. 5.4). The original relation [eq. (5.62)] remains more applicable for active starbursts at higher redshifts.

At the end of this section, we note that there is considerable variation in the IRX–β relations across different galaxy populations. The original relation was derived for a sample of UV-luminous nuclear starbursts, and while it was found to be tight for that sample, subsequent studies have shown that it does not hold universally for all types of galaxies. For instance, quiescent star-forming (SF) galaxies and IR-luminous galaxies deviate from this relationship. Buat et al. (2005) demonstrated that quiescent SF galaxies tend to lie below the original curve of Meurer et al. (1999), whereas actively IR-luminous galaxies are distributed above the line. It is also important to consider the redshift evolution of the IRX–β relation. While many studies have explored this topic, the nature of the redshift evolution remains debated and controversial (e.g., Cousin et al., 2019; Koprowski et al., 2020).

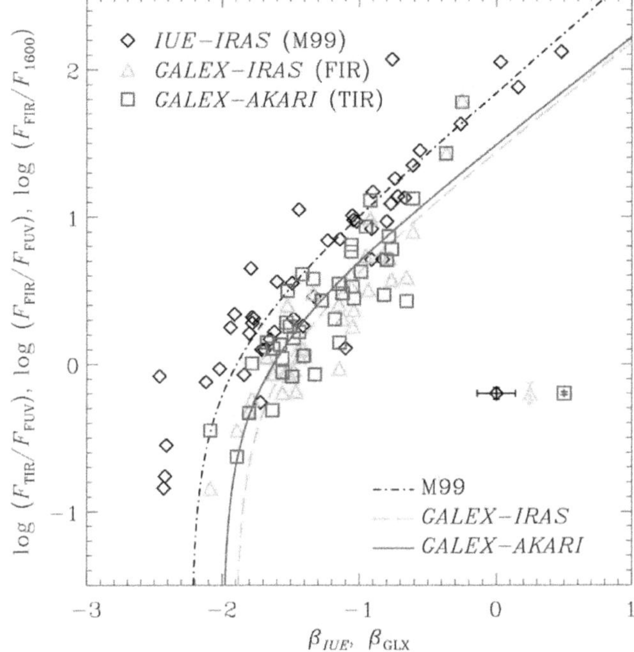

Figure 5.4 The IRX–β relation which was obtained using *GALEX* and *AKARI* diffuse maps. Relations derived from *IUE–IRAS* and *GALEX–IRAS* are also shown. Diamonds, triangles, and squares represent *IUE–IRAS*, *GALEX–IRAS*, and *GALEX–AKARI* values, respectively. The dot-dashed curve represents the original formula from Meurer et al. (1999), the dashed curve shows the *GALEX–IRAS* fit, and the solid curve represents the new formula with *GALEX–AKARI* measurements. The slope β derived from *GALEX* data is denoted as β_{GLX}. Credit: Takeuchi, T. T., Yuan, F.-T., Ikeyama, A., et al. 2012, Astrophysical Journal, 755, 144, Fig. 9, reproduced by permission of AAS.

5.5 SECONDARY SFR ESTIMATOR

5.5.1 OPTICAL FORBIDDEN LINES

Hydrogen recombination lines Hα and Hβ move out of the optical wavelength regime at $z \geq 0.5$ and $z \geq 0.9$, respectively, and then they are not ideal SFR indicators for deep surveys. For such situations, oxygen forbidden lines [OII] λ3727 Å and [OIII] λ5007 Å are discussed as a possible alternative (e.g., Gallagher et al., 1989; Kennicutt, 1992; Moustakas et al., 2006). The basis of this idea is that the oxygen forbidden lines are also emitted from HII regions. However, since the forbidden lines are not directly connected to the number luminosity of ionizing photons, their luminosity strongly depends on the gas metallicity and ionization state. Currently the following formula for the [OII] line is popularly used

$$\mathrm{SFR}_{\mathrm{tot}} = (1.4 \pm 0.4) \times 10^{-41} L([\mathrm{OII}]) \, [\mathrm{erg\, s}^{-1}] , \tag{5.66}$$

(Kennicutt, 1998; Rosa-González et al., 2002).

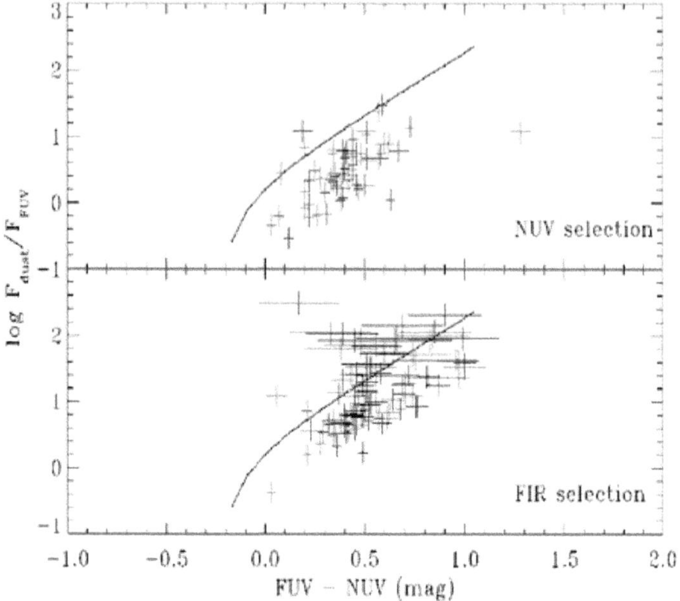

Figure 5.5   The IRX–UV color relation for NUV- and FIR-selected samples. The solid line represents the mean relation expected for starburst galaxies (Kong et al., 2004). Credit: Buat et al. (2005), Fig. 4.

Due to its strong dependence on these quantities, [O III] $\lambda 5007\,\text{Å}$ has been less popular as an SFR estimator. In spite of this potential issue, some attempts are made to make use of this line for SFR estimation. Villa-Vélez et al. (2021) proposed a calibration formula[3]

$$\log\mathrm{SFR}_{\mathrm{tot}} = (0.83 \pm 0.06)\log L([\mathrm{O\,III}])\,[\mathrm{erg\,s^{-1}}] - (34.01 \pm 2.63)\,. \tag{5.67}$$

5.5.2   METAL FINE STRUCTURE LINES

Since the star formation occurs through the fragmentation and contraction of cold, dense molecular gas, the cooling process is tightly related to the SFR. As we have discussed in Chapter 3, metal fine structure lines such as [C II] 158 μm, [O I] 63 μm, and [O III] λ 88 μm are the main pathway of gas cooling (Glover & Clark, 2014). Then, these lines can be used as SFR indicators (Brauher et al., 2008; Hunter et al., 2001).

5.5.2.1   [C II] 158 μ Fine-Structure Line

The [C II] 157.74 μm line has been regarded as a potentially good tracer of the SFR in galaxies (e.g, Boselli et al., 2002; De Looze et al., 2011; Sargsyan et al., 2012; Stacey et al., 1991, 2010). Since the [C II] line is the strongest line in the FIR wavelengths, it is the dominant coolant for neutral atomic gas in the ISM and therefore one of

---

[3]We do not homogenize the style of the formulae because of different fitting schemes.

the brightest emission lines originating from star-forming galaxies (e.g., De Looze et al., 2014, and references therein). In particular, low-metallicity galaxies show exceptionally strong [CII] line emission (e.g., Cormier et al., 2010; Hunter et al., 2001; Israel & Maloney, 2011; Madden et al., 1997; Poglitsch et al., 1995). De Looze et al. (2014) obtained a formula

$$\log \text{SFR}_{\text{tot}} = -(6.99 \pm 0.14) + (1.01 \pm 0.02) \log L([\text{CII}]) \, [L_\odot] \tag{5.68}$$

for Local galaxies. However, this is valid for normal galaxies. The steep decline of the $L(\text{CII})/L_{\text{FIR}}$ ratio found in luminous IR galaxies (LIRGs: $L_{\text{FIR}} > 10^{11} \, L_\odot$) implies that the physical state of the ISM is different for this category of galaxies (e.g., Luhman et al., 2003). To address this issue, De Looze et al. (2014) present a formula for each of the subclass of metal-poor dwarfs, starbursts, LIRGs, and those with AGNs. They also provide the formulae for spatially resolved galaxies.

5.5.2.2 [OI] 63 μ Fine-Structure Line

The [OI] 63 μm line is an efficient coolant in dense and warm photodissociation regions (PDRs, see Chapter 3). However, there is a caveat to regard the [OI] 63 μm line as an SFR estimator. The self-absorption (e.g., Poglitsch et al., 1996) and optical depth effects, and the possible excitation of [OI] 63 μm through shocks (Hollenbach & McKee 1989) make the relation obscured. Further, De Looze et al. (2014) argue that, in the situation that the gas heating is no longer dominated by the photoelectric effect but significantly contributed from other heating mechanisms, e.g., mechanical heating and soft X-ray heating, the origin of the line emission might differ from warm and/or dense PDRs. A possible origin of [OI] 63 μm emission that is different from PDRs can be expected in most metal-poor galaxies characterized by overall low metal content and diminished polycyclic aromatic hydrocarbon (PAH) abundances. Thus, we treat [OI] 63 μm line as a secondary SFR indicator. De Looze et al. (2014) give a formula

$$\log \text{SFR}_{\text{tot}} = -(6.79 \pm 0.22) + (1.00 \pm 0.03) \log L([\text{OI}]) \, [L_\odot] \, . \tag{5.69}$$

5.5.2.3 [OIII] 88 μ Fine Structure Line

The [OIII] 88 μm line originates from diffuse highly ionized regions near young O stars. Ionized gas tracers such as [OIII] 88 μm have become more and more important to explore low-metallicity environments where PDRs occupy only a limited volume of the ISM (e.g., Boselli et al., 2004; Engelbracht et al., 2005, 2008; Galliano et al., 2008; Poglitsch et al., 1996) and low CO abundance (e.g., Israel & Maloney, 2011; Madden et al., 1997; Poglitsch et al., 1995). De Looze et al. (2014) regard the [OIII] 88 μm line as a tracer of ionized gas. They cast a caution that, because of the low critical densities for [OIII] 88 μm excitation with electrons, other lines such as [OIII] 52 μm and optical lines ([OIII] λ5007 Å and Hα) possibly dominate the cooling of ionized gas media for intermediate and high gas densities. Even so, the brightness of the [OIII] 88 μm emission line in metal-poor galaxies hints at an SFR tracer with great potential for the high-z Universe. The formula is

$$\log \mathrm{SFR_{tot}} = -(7.48 \pm 0.42) + (1.12 \pm 0.06) \log L([\mathrm{OIII}]) \, [L_\odot] \, . \tag{5.70}$$

5.5.3 X-RAY LUMINOSITY

Without a luminous AGN, X-ray emission from a normal galaxy is mainly contributed from X-ray binaries, and diffuse radiation from supernova remnants (SNRs) and hot gas is added (e.g., Fabbiano, 2006). X-ray binaries consist of compact objects (such as neutron stars and black holes) and companion stars. Usually, mass is accreted onto the compact object from the companion, and forms an accretion disk. They are classified as high-mass X-ray binaries (HMXB) and low-mass X-ray binaries (LMXB). Typically the companion star in HMXB has a mass of $\geq 10 \, \mathcal{M}_\odot$, and its lifetime is $\sim 10^7$ yr. In contrast, the mass of the LMXB companion is $\leq 1 \, \mathcal{M}_\odot$ and the lifetime is $\sim 10^{10}$ yr. Thus, the HMXB can be used as an SFR estimator (Griffiths & Padovani, 1990).

The advantage of X-rays as an SFR estimator is that usually, the ISM is almost transparent to X-rays above 2 keV. Thus, X-rays can be used to estimate the SFR of galaxies heavily obscured by dust. On the other hand, X-rays are dominated by AGNs even if their luminosities are low, and it is usually difficult to subtract the AGN contribution.

Based on the observations of nearby galaxies by *ASCA* and *BeppoSAX*, the following formulae are obtained

$$\mathrm{SFR_{tot}} = 2.2 \times 10^{-40} L(0.5\text{--}2 \text{ keV}) \, [\mathrm{erg\,s^{-1}}] \tag{5.71}$$

and

$$\mathrm{SFR_{tot}} = 2.0 \times 10^{-40} L(2\text{--}10 \text{ keV}) \, [\mathrm{erg\,s^{-1}}] \tag{5.72}$$

(Ranalli et al., 2003). The uncertainty of soft X-rays is larger due to the internal absorption.

5.5.4 RADIO CONTINUUM

Radio continuum from galaxies is composed of two components: non-thermal synchrotron emission associated with high-energy electrons accelerated in the magnetic field of a galaxy, and thermal Bremsstrahlung (free–free) emission around massive star-forming regions (e.g., Condon, 1992; Murphy et al., 2011). We follow the comprehensive discussion given by Murphy et al. (2011) to derive the radio-based SFR estimators.

At high radio frequencies, the ionizing photon number luminosity is directly proportional to the thermal luminosity $\mathscr{L}_\nu^{\mathrm{T}}$, varying only weakly with electron temperature T_e (Rubin, 1968),

$$Q(\mathrm{H}^0) \, [\mathrm{s^{-1}}] = 6.3 \times 10^{25} \left(\frac{T_e}{10^4 \, [\mathrm{K}]} \right)^{-0.45} \left(\frac{\nu}{[\mathrm{GHz}]} \right)^{0.1} \mathscr{L}_\nu^{\mathrm{T}} \, [\mathrm{erg\,s^{-1}Hz^{-1}}] \tag{5.73}$$

With the Salpeter IMF, as we discussed in Section 5.1.1,

$$\mathrm{SFR}_{\mathrm{tot}} = 1.08 \times 10^{-53} Q(\mathrm{H}_0)[\mathrm{s}^{-1}]. \tag{5.74}$$

Combining eqs. (5.74) and (5.73), we obtain the thermal radio continuum estimator of the SFR as

$$\mathrm{SFR}_{\mathrm{tot}}^{\mathrm{T}}(\nu) = 6.8 \times 10^{-28} \left(\frac{T_e}{10^4\,[\mathrm{K}]}\right)^{-0.45} \left(\frac{\nu}{[\mathrm{GHz}]}\right)^{0.1} \mathscr{L}_\nu^{\mathrm{T}}. \tag{5.75}$$

At lower radio frequencies, the radio continuum is typically dominated by non-thermal synchrotron emission, the total core-collapse supernova rate is obtained by

$$\nu_{\mathrm{SN}} = \int_{8\,\mathscr{M}_\odot}^{\mathscr{M}_{\mathrm{up}}} \Phi(\mathscr{M})\mathrm{d}\mathscr{M}, \tag{5.76}$$

(e.g., Condon, 1992). Then, this is related to the SFR by

$$\mathrm{SFR}_{\mathrm{tot}} = 1.4 \times 10^2\, \nu_{\mathrm{SN}}\, [\mathrm{yr}^{-1}]. \tag{5.77}$$

The calibration based on the Milky Way gives

$$\mathscr{L}_\nu^{\mathrm{NT}}\, [\mathrm{erg\, s}^{-1}\mathrm{Hz}^{-1}] = 1.3 \times 10^{30} \nu_{\mathrm{SN}} \left(\frac{\nu}{[\mathrm{GHz}]}\right)^{-\alpha^{\mathrm{NT}}} \tag{5.78}$$

where typically $\alpha^{\mathrm{NT}} \simeq 0.8$ (Condon & Yin, 1990). By combining eqs. (5.76) and (5.77), we can express the SFR as a function of the non-thermal radio emission as

$$\mathrm{SFR}_{\mathrm{tot}}^{\mathrm{NT}}(\nu) = 1.1 \times 10^{-28} \left(\frac{\nu}{[\mathrm{GHz}]}\right)^{\alpha^{\mathrm{NT}}} \mathscr{L}_\nu^{\mathrm{NT}}. \tag{5.79}$$

Since the observed radio continuum emission consists of both free–free and Synchrotron emission, we should combine eqs. (5.75) and (5.79) to construct a single formula for the SFR from the total radio continuum emission at a given frequency ν. Suppose that the estimated SFRs are the same, i.e., $\mathrm{SFR}_{\mathrm{tot}}^{\mathrm{T}}(\nu) = \mathrm{SFR}_{\mathrm{tot}}^{\mathrm{NT}}(\nu)$. Then, the combined SFR should be the sum of the inverse of each coefficient, which yields

$$\mathrm{SFR}_{\mathrm{tot}}(\nu) = \frac{10^{-27}\mathscr{L}_\nu\,[\mathrm{erg\, s}^{-1}\mathrm{Hz}^{-1}]}{1.47 \left(\dfrac{T_e}{10^4\,[\mathrm{K}]}\right)^{0.45} \left(\dfrac{\nu}{[\mathrm{GHz}]}\right)^{-0.1} + 9.09 \left(\dfrac{\nu}{[\mathrm{GHz}]}\right)^{-\alpha^{\mathrm{NT}}}}. \tag{5.80}$$

These formulae are derived from the theoretical basis of star formation. However, widely used radio SFR estimation is instead based on the tight empirical FIR–radio correlation (Helou et al., 1985). Often the coefficient q_{IR} is defined as

$$q_{\mathrm{IR}} \equiv \log \frac{L_{\mathrm{TIR}}}{3.75 \times 10^{12} \mathscr{L}_\nu(1.4\,\mathrm{GHz})}, \tag{5.81}$$

and if we use $L_{\rm TIR}$ as the total IR luminosity, $q_{\rm IR} = 2.64 \pm 0.26$ dex (Bell, 2003). After considering the contribution of the old stellar population to $L_{\rm TIR}$, Bell (2003) gives the following formula

$$\text{SFR}_{\rm tot} = 5.52 \times 10^{-29}\, \mathscr{L}_\nu(1.4\,\text{GHz})\,[\text{erg s}^{-1}\text{Hz}^{-1}] \tag{5.82}$$

for $\mathscr{L}_\nu(1.4\,\text{GHz}) > 6.4 \times 10^{28}\,[\text{erg s}^{-1}\text{Hz}^{-1}]$ and

$$\text{SFR}_{\rm tot} = 5.52 \times 10^{-29}\,\frac{\mathscr{L}_\nu(1.4\,\text{GHz})\,[\text{erg s}^{-1}\text{Hz}^{-1}]}{0.1 + 0.9\left[\dfrac{\mathscr{L}_\nu(1.4\,\text{GHz})}{6.4 \times 10^{28}}\right]^{0.3}} \tag{5.83}$$

for $3 \times 10^{26}\,[\text{erg s}^{-1}\text{Hz}^{-1}] \leq \mathscr{L}_\nu(1.4\,\text{GHz}) \leq 6.4 \times 10^{28}\,[\text{erg s}^{-1}\text{Hz}^{-1}]$. This is based on the FIR–radio correlation, which was established for integrated FIR and radio continuum of galaxies. Then, we should note that it applies to galaxies as a whole (Murphy et al., 2011), and should be handled with care when a variation of thermal and nonthermal contributions are expected within a galaxy (see, e.g., Calistro Rivera et al., 2017; Smith et al., 2021).

5.6 SPECIFIC SFR AND RELATED QUANTITIES

So far, we have discussed the absolute values of the SFR. However, it is important to note that larger galaxies generally have higher SFRs simply due to their size. Therefore, to facilitate meaningful comparisons, we need a 'normalized' value for the SFR. In this section, we introduce such quantities.

5.6.1 BIRTHRATE PARAMETER AND SPECIFIC SFR

The birthrate parameter, b, is defined as the ratio between the current SFR and the SFR averaged over the entire lifetime of the galaxy (Sandage, 1986; Scalo, 1986). If we denote the age of a galaxy as $t_{\rm G}$, b can be expressed as

$$b \equiv \frac{\text{SFR}_{\rm tot}}{\langle \text{SFR}_{\rm tot} \rangle} = \frac{t_{\rm G}\left[1 - \displaystyle\int_0^{t_{\rm G}} R(t)\,\mathrm{d}t\right]\text{SFR}_{\rm tot}}{\mathscr{M}_*} \tag{5.84}$$

where $R(t)$ is the gas ejection rate (as defined in eq. (4.50)) and $\mathscr{M}_*$ is the total stellar mass in the galaxy (Boselli et al., 2001; Kennicutt et al., 1994). Because a star ejects mass based on its evolutionary stage and the SFR changes over time, R is a time-dependent function. Kennicutt et al. (1994) demonstrated that over 90% of the returned mass was injected by stars formed within the first few Gyr, with more than half of it from the population formed during the first 200 million years. Therefore, b is typically calculated using the instantaneous recycling approximation, with a constant $R = 0.3$ for the Salpeter IMF.

The equivalent width of the Hα emission line, EW(Hα), represents the ratio of the Hα line flux, which corresponds to the youngest stellar population, to the underlying stellar continuum. Thus, EW(Hα) is an observable quantity that is directly related to the birthrate parameter.

The specific SFR (SSFR or sSFR) is another proxy for the birthrate parameter and is frequently used in modern studies (e.g., Brinchmann et al., 2004). It is defined as

$$\mathrm{SSFR}\ [\mathrm{yr}^{-1}] = \frac{\mathrm{SFR_{tot}}}{\mathscr{M}_*} = \frac{b}{t_\mathrm{G}\left(1 - \int R\,dt\right)} \tag{5.85}$$

SSFR represents the SFR per unit stellar mass and provides a way to describe star formation activity independent of size of a galaxy.

5.6.2 STAR FORMATION EFFICIENCY AND GAS CONSUMPTION TIMESCALE

The star formation efficiency (SFE) is defined as the rate at which a galaxy's gas content is converted into stars

$$\varepsilon_{\mathrm{SF}} = \frac{\tau_{\mathrm{SF}}\,\mathrm{SFR_{tot}}}{\Sigma_{\mathrm{gas}}} \tag{5.86}$$

where τ_{SF} is the SF timescale and Σ_{gas} is the gas surface density (Wang & Silk, 1994). The SF timescale depends on the growth rate of gravitational instability in the gas and is therefore related to both the velocity dispersion and surface density of the gas (Wang & Silk, 1994). However, because measuring τ_{SF} is difficult, the SFE is commonly expressed as

$$\mathrm{SFE}\ [\mathrm{yr}^{-1}] = \frac{\mathrm{SFR_{tot}}}{\mathscr{M}_{\mathrm{ISM}}} \tag{5.87}$$

(Young et al., 1996). Since stars are generally formed within molecular clouds, the total ISM mass, $\mathscr{M}_{\mathrm{ISM}}$, is often replaced with the molecular gas mass, $\mathscr{M}_{\mathrm{H_2}}$.

The SFE can also be related to the gas consumption timescale, which accounts for the gas returned to the ISM from stars. This timescale is known as the Roberts time, τ_R (Roberts, 1963), and is defined as

$$\tau_\mathrm{R}\ [\mathrm{yr}] = \frac{\mathscr{M}_{\mathrm{ISM}}}{\left(1 - \int R\,dt\right)\mathrm{SFR_{tot}}} = \frac{1}{\left(1 - \int R\,dt\right)\mathrm{SFE}} \tag{5.88}$$

(Boselli et al., 2001).

6 Clusters, Clustering of Galaxies, and the Large-Scale Structure

6.1 CLUSTERS OF GALAXIES: PHENOMENOLOGY

Rich clusters are the largest gravitationally bound and virialized structures in the Universe. Unlike galaxies, they are not distinct concentrations when observed in the sky, as galaxies within clusters are sparse, and there are many objects both in front of and behind them. Clusters can be observed through

1. galaxies (optical),
2. X-ray gas via Bremsstrahlung (X-ray and radio),
3. the Sunyaev–Zel'dovich effect (radio, submillimeter).

Clusters consist mostly of dark matter, which accounts for 76% of their total mass. Of the remaining mass, 20% is in the form of hot gas (the intracluster medium or ICM), and only 4% consists of the luminous parts of galaxies. Among the direct observables, (2) and (3) relate to the hot ICM, while (1) is obviously connected to galaxies. While we cannot directly observe dark matter, its amount and distribution can be inferred through gravitational lensing. Thus, the majority of a cluster's mass is dark matter, and most of the baryonic matter exists as hot gas, not in the form of stars or galaxies. The fact that most of the baryons remain as hot gas can be explained by the cooling processes of the ICM.

Despite their small contribution to the total mass, clusters of galaxies were originally discovered and defined using optical images. Therefore, we begin with a discussion of clusters as observed in the optical.

6.2 THE MISSING MASS PROBLEM IN CLUSTERS

6.2.1 MASS FROM GALAXY DYNAMICS IN CLUSTERS

Zwicky (1933) was the first to demonstrate the existence of missing mass in clusters of galaxies. He discovered a significant discrepancy between the dynamical mass of a cluster and the total mass obtained by summing the mass-to-light ratio ($\mathcal{M}/L$) of each galaxy. Although some of the difference could be attributed to the presence of hot gas (the ICM), the majority of the dynamical mass could not be explained by the baryonic mass alone. Here, we follow Zwicky's argument.

The observed velocity dispersion of typical cluster galaxies along the line of sight was found to be $\sigma_v \sim 10^3$ [km s^{-1}]. The crossing time for a cluster with radius r is

DOI: 10.1201/9781003104315-6

given by

$$t_{\text{cross}} = \frac{R}{\sigma_v} \simeq 1 \left(\frac{R}{1 h^{-1} \, [\text{Mpc}]} \right) \left(\frac{\sigma_v}{10^3 \, [\text{km s}^{-1}]} \right)^{-1} [\text{Gyr}]. \qquad (6.1)$$

This implies that the inner region of the system ($\lesssim 1 h^{-1}$ [Mpc]) can dynamically relax within a Hubble time ($t_{\text{H}} \simeq 10 h^{-1}$ Gyr), while relaxation is marginal for the outer region (at scales ~ 10 Mpc). The following two assumptions are made

1. The mass distribution follows the galaxy distribution ($\mathcal{M}/L = \text{const.}$),
2. The velocity distribution within the cluster is isotropic.

Assuming virial equilibrium, the typical cluster mass can be expressed as

$$\mathcal{M} \simeq \frac{R\sigma_v^2}{G} \simeq \left(\frac{R}{1 h^{-1} \, [\text{Mpc}]} \right) \left(\frac{\sigma_v}{10^3 \, [\text{km s}^{-1}]} \right)^2 10^{15} h^{-1} \, [\mathcal{M}_\odot]. \qquad (6.2)$$

Using this, we can estimate the mass of the Coma cluster

$$\mathcal{M}_{\text{Coma}} = 1.79 \times 10^{15} h^{-1} \, [\mathcal{M}_\odot], \qquad (6.3)$$

(Merritt, 1987), and for the core of Coma, within a radius of $R \simeq 1 h^{-1}$ [Mpc],

$$\mathcal{M}_{\text{Coma,core}} = 6.1 \times 10^{14} h^{-1} \, [\mathcal{M}_\odot]. \qquad (6.4)$$

Thus, the mass-to-light ratio ($\mathcal{M}/L$) in the core of the Coma cluster is $350 h \, \mathcal{M}_\odot/L_\odot$. In this region, the cluster is dominated by elliptical and lenticular galaxies, which have $\mathcal{M}/L \simeq 10\text{–}20 \mathcal{M}_\odot/L_\odot$. This implies that there is approximately 20 times more mass (both luminous and dark) in the cluster than is present in galaxies alone (Zwicky, 1933, 1937). Smith (1936) used a similar method to estimate the mass of the Virgo cluster and found that it significantly exceeded the value expected from its optical luminosity. These findings provided the first evidence for the existence of dark matter in the Universe.

6.2.2 MASS FUNCTION OF CLUSTERS

The cluster X-ray LF is commonly modeled with a Schechter function as

$$\phi(L_{\text{X}}) \text{d}L_{\text{X}} = \phi_* \left(\frac{L_{\text{X}}}{L_{\text{X}*}} \right)^{-\alpha} \exp \left(-\frac{L_{\text{X}}}{L_{\text{X}*}} \right) \frac{\text{d}L_{\text{X}}}{L_{\text{X}*}}, \qquad (6.5)$$

where α is the faint–end slope, $L_{\text{X}*}$ is the characteristic luminosity, and ϕ_* is the space–density normalization, the same as defined for galaxy LFs [eq. (2.13)]. The estimation of the X-ray LF is performed statistically, again exactly in the same way as discussed in Chapter 2.

The shape of the cluster X-ray LF can be regarded as a direct reflection of the dark halo mass function, since clusters are safely assumed to be in virial equilibrium, as we have seen above. The mass function of dark halos is theoretically derived by Press & Schechter (1974), as

$$n(\mathcal{M}) = \frac{2\alpha}{\sqrt{\pi}} \frac{\bar{\rho}}{\mathcal{M}_*^2} \left(\frac{\mathcal{M}}{\mathcal{M}_*}\right)^{\alpha-2} \exp\left[-\left(\frac{\mathcal{M}}{\mathcal{M}_*}\right)^{2\alpha}\right]. \tag{6.6}$$

This is referred to as the Press–Schechter mass function. We will derive this equation in Chapter 7 as eq. (7.117). As we have mentioned in Chapter 2, the Schechter function for the LF was inspired by the functional shape of the Press–Schechter mass function. For clusters, this is the most obviously seen from the observed X-ray LFs. Because of this advantage, the X-ray LFs play an important role in cosmological studies (e.g., Kitayama, 2014; Kitayama & Suto, 1996; Rosati et al., 2002).

6.3 STATISTICAL METHODS FOR QUANTIFYING GALAXY DISTRIBUTION

6.3.1 THE LARGE-SCALE STRUCTURE IN THE UNIVERSE

It has long been known that the sky distribution of 'nebulae' is not homogeneous, showing visible fluctuations even before they were proven to be extragalactic. In the 19th century, William Herschel and Charles Messier had already noted that nebulae were more frequently found in certain parts of the sky than others, particularly in the constellation Virgo. More quantitatively, Shapley (1933) and Shapley (1937) reported a prominent uneven distribution of galaxies based on photographic plate surveys (Shapley & Ames, 1932).

The Palomar Sky Survey provided groundbreaking data for exploring the clustering of galaxies. Zwicky and collaborators systematically cataloged the positions and magnitudes of thousands of galaxies on these plates, which culminated in the publication of the so-called 'Zwicky Catalog' (Zwicky et al., 1963, 1961). Abell (1958) conducted a systematic survey of rich clusters of galaxies, producing the renowned Abell catalog, which includes thousands of clusters.

The first map of the sky that revealed widespread clustering and superclustering of galaxies came from the Lick survey, carried out by Shane & Wirtanen (1967). This survey provided the foundational data for numerous important analyses of the sky distribution of galaxies. Through such studies, a hierarchy of structures—such as groups, clusters, and superclusters—was identified in the spatial distribution of galaxies. Peebles & Hauser (1974) conclusively demonstrated the superclustering of galaxies by analyzing the power spectrum of cluster distribution. Their work revealed that clusters of galaxies are not randomly distributed but themselves form clusters, introducing the concept of biased galaxy formation (see Section 7.4.8).

These early studies were conducted using 2-dim data on the celestial sphere. The 3-dimensional structure of galaxy distribution was first uncovered by the celebrated Center for Astrophysics (CfA) Galaxy Redshift Survey (e.g., Geller & Huchra, 1989). However, even before the CfA survey, smaller surveys, such as those targeting the Pisces–Perseus supercluster or the Coma–A1367 region, had already revealed rich structures in the distribution of galaxies (e.g., Chincarini et al., 1983a,b;

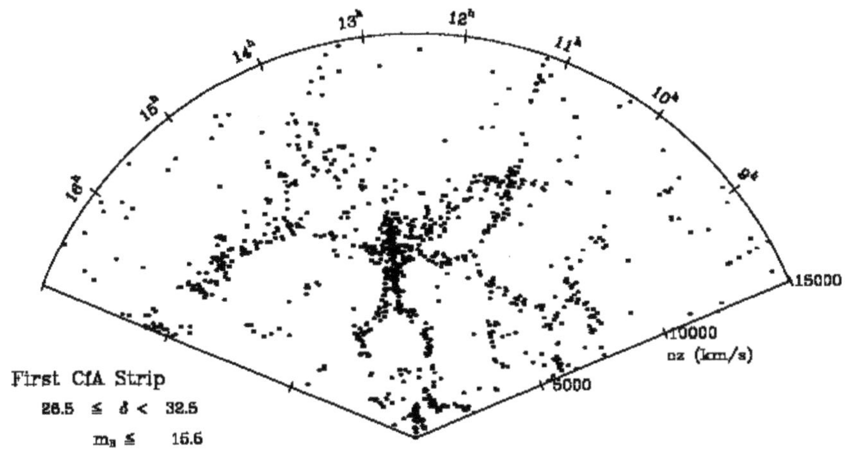

Figure 6.1 A 3-dimensional map from the Center for Astrophysics Redshift Survey slice (de Lapparent et al., 1986). Credit: de Lapparent, V., Geller, M. J., & Huchra, J. P. 1986, Astrophysical Journal, 302, L1, Fig. 1, reproduced by permission of AAS.

Giovanelli & Haynes, 1985; Giovanelli & Haynes, 1989, 1993; Giovanelli et al., 1986). These surveys provided insights into localized cosmic structures but were too limited in scope to represent the entire Universe. The first significant CfA redshift survey, conducted by Huchra et al. (1983), mapped approximately 2,400 galaxies down to $m \leq 14.5$ [mag], using data from the Zwicky catalog. However, this dataset was too sparse to reveal the full structure of the Universe. It was the first CfA-II slice, published by de Lapparent et al. (1986), that truly illuminated the large-scale structure of the cosmos (Fig. 6.1). Their survey presented the foam-like, or "bubbly," nature of the large-scale structure. A comprehensive overview of redshift surveys from this era can be found in Giovanelli & Haynes (1991). Subsequent large-scale redshift surveys, such as the Southern Sky Redshift Survey (e.g., da Costa et al., 1991, 1998), the Two-Degree Field (2dF) Redshift Survey (e.g., Colless et al., 2001; Cross et al., 2001; Peacock et al., 2001), and the Sloan Digital Sky Survey (SDSS) (e.g., Loveday, 2002), have since confirmed the large-scale structure of the Universe.

6.3.2 ENSEMBLE AND VOLUME AVERAGE

In general, the spatial distribution of galaxies can be regarded as a set of points or point cloud data in 3-dimensional space. To quantify the statistical properties of a randomly distributed point cloud generated from a certain stochastic process, various mathematical tools have been developed (e.g., Illian et al., 2008; Stoyan & Stoyan, 1994). Since this is not a specialized cosmology textbook, we will focus on the most frequently used measure for clustering points: moment measures.

First, we introduce an important concept in the statistical analysis of clustering in cosmology. Similar to the approach used in statistical mechanics or statistical mathematics in general, we consider a set (ensemble) of a large number of realizations of the Universe. Thus, even at a fixed spatial position $\vec{x}$, we have a statistical distribution of a certain value $\mathscr{Q}(\vec{x})$. Suppose we have a set of $\mathscr{N}$ realizations of $\mathscr{Q}(\vec{x})$, denoted as $\{Q_1(\vec{x}), Q_2(\vec{x}), \ldots, Q_{\mathscr{N}}(\vec{x})\}$. We can take the ensemble average of $\mathscr{N}$ realizations

$$\langle \mathscr{Q}(\vec{x}) \rangle \equiv \frac{1}{\mathscr{N}} \sum_{i=1}^{\mathscr{N}} Q_i(\vec{x}) \,. \tag{6.7}$$

Note that $\langle \mathscr{Q}(\vec{x}) \rangle$ remains a function of $\vec{x}$.

However, we can observe only one Universe. This means that we cannot perform the operation in eq. (6.7). Instead, we take an average using a different approach. Typically, the ensemble average is replaced by a volume average

$$\overline{\mathscr{Q}(\vec{x})} \equiv \frac{1}{\mathscr{M}} \sum_{\vec{x}_j \in V} \mathscr{Q}(\vec{x}_j) \,, \tag{6.8}$$

where $j = 1, \ldots, \mathscr{M}$, and $\mathscr{M}$ is the number of objects in a certain class within the considered volume V.[1] It is assumed that $V \longrightarrow \infty$ (and accordingly, $\mathscr{M} \longrightarrow \infty$). In this case, the position $\vec{x}_j$ is averaged out, and $\overline{\mathscr{Q}(\vec{x})}$ no longer has spatial position dependence.

In cosmology, the following *ansatz* is adopted for a meaningful discussion on the statistics, known as the Ergodic hypothesis

Ansatz 1. (Ergodic hypothesis)
The ensemble average and volume average are equal, i.e.,

$$\langle \mathscr{Q}(\vec{x}) \rangle = \overline{\mathscr{Q}(\vec{x})} \,. \tag{6.9}$$

From here on, we will not explicitly emphasize this assumption and will simply denote the average as $\langle \mathscr{Q} \rangle$.

6.3.3 2-POINT CORRELATION FUNCTION

Even if the distribution of a given quantity does not follow a Gaussian probability density function (PDF) exactly, its second-order moment typically contains valuable statistical information unless the distribution deviates significantly from Gaussian. This is one reason why the 2-point correlation function is widely used in clustering analysis. Another advantage of the 2-point correlation function is its clear physical interpretation.

---

[1] In this sense, the average can be regarded as a sample average over j.

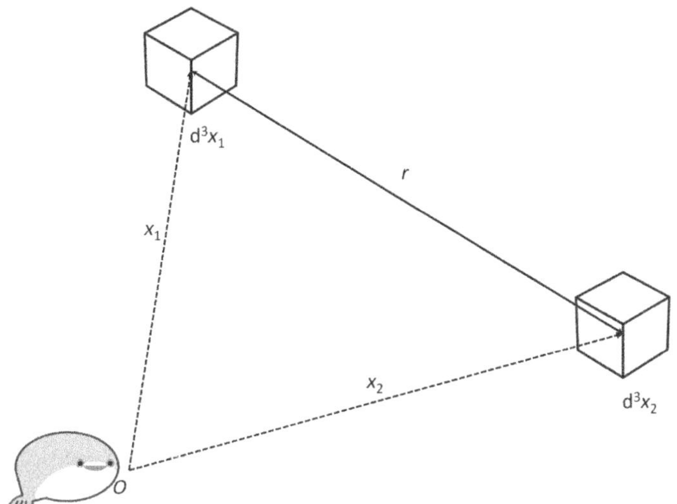

Figure 6.2 Physical meaning of the 2-point correlation function. Density and overdensity are denoted as $n_j = n(\vec{x}_j)$ and $\delta_j = \delta(\vec{x}_j)$ ($j = 1, 2$), respectively.

6.3.3.1 2-Point Correlation Function of Galaxies in Real Space

We can observe galaxy counts $n(\vec{x})$ and compare them to the ensemble average counts $\langle n \rangle$. Consider two infinitesimal volumes d^3x_1 and d^3x_2. The joint probability that a galaxy is found in both volumes is given by $\mathscr{P}(\vec{x}_1, \vec{x}_2)d^3x_1 d^3x_2$ (Fig. 6.2). If there is no clustering, the probability of finding a galaxy in an infinitesimal volume is

$$\mathscr{P}(\vec{x})d^3x = \langle n \rangle d^3x, \tag{6.10}$$

so the joint probability of finding a galaxy in volume d^3x_1 and another galaxy in volume d^3x_2 would be

$$\mathscr{P}(\vec{x}_1, \vec{x}_2)d^3x_1 d^3x_2 = \langle n \rangle^2 d^3x_1 d^3x_2. \tag{6.11}$$

However, if galaxies are clustered, the joint probability will be higher

$$\mathscr{P}(\vec{x}_1, \vec{x}_2)d^3x_1 d^3x_2 = \langle n(\vec{x}_1)n(\vec{x}_2)\rangle d^3x_1 d^3x_2$$
$$= \langle n \rangle^2 \left[1 + \xi(\vec{x}_1, \vec{x}_2)\right] d^3x_1 d^3x_2. \tag{6.12}$$

Thus, the correlation function $\xi(\vec{x}_1, \vec{x}_2)$ is defined as

$$\xi(\vec{x}_1, \vec{x}_2) = \frac{\mathscr{P}(\vec{x}_1, \vec{x}_2)}{\mathscr{P}(\vec{x}_1)\mathscr{P}(\vec{x}_2)} - 1 = \frac{\langle n(\vec{x}_1)n(\vec{x}_2)\rangle}{\langle n \rangle^2} - 1. \tag{6.13}$$

Intuitively, the correlation function is an average measure of the excess probability of finding a galaxy at a specific volume compared to a random, unclustered case.[2]

We define the overdensity (density contrast) of the galaxy number density at $\vec{x}$ as

$$\delta_{\mathrm{gal}}(\vec{x}) \equiv \frac{n(\vec{x}) - \langle n \rangle}{\langle n \rangle} \, . \tag{6.14}$$

This leads to the following relationship

$$\langle \delta_{\mathrm{gal}}(\vec{x}_1)\delta_{\mathrm{gal}}(\vec{x}_2) \rangle = \frac{\langle [n(\vec{x}_1) - \langle n \rangle][n(\vec{x}_2) - \langle n \rangle] \rangle}{\langle n \rangle^2}$$

$$= \frac{\langle n(\vec{x}_1)n(\vec{x}_2) \rangle - \langle n \rangle^2}{\langle n \rangle^2} = \frac{\langle n(\vec{x}_1)n(\vec{x}_2) \rangle}{\langle n \rangle^2} - 1$$

$$= \xi(\vec{x}_1, \vec{x}_2) \, . \tag{6.15}$$

For simplicity, we will drop the subscript "gal" in subsequent equations. Note, however, that the distinction between galaxy distributions as a point cloud and a continuous density field is substantial in a mathematical sense. In this chapter, we consider only point cloud data. We revisit this issue in Chapter 7. Now, we see that the correlation function $\xi(\vec{x}_1, \vec{x}_2)$ can be regarded as a normalized second-order cumulant of $n(\vec{x}_1)$ and $n(\vec{x}_2)$. Alternatively, since δ is a centered statistic[3], $\langle \delta(\vec{x}_1)\delta(\vec{x}_2) \rangle$ itself is a cumulant.

If we handle galaxy data in real space, the correlation function depends only on the difference between $\vec{x}_1$ and $\vec{x}_2$. Thus,

$$\xi(\vec{x}_1, \vec{x}_2) = \xi(\vec{x}_1 - \vec{x}_2) \, . \tag{6.16}$$

Furthermore, due to the translational invariance of the Universe, we obtain the commonly used form of the 2-point correlation function of galaxies

$$\xi(\vec{x}_1, \vec{x}_2) = \xi(|\vec{x}_1 - \vec{x}_2|) \equiv \xi(r) \, . \tag{6.17}$$

6.3.3.2 Observed 2-Point Correlation Function of Galaxies

The functional form of $\xi(r)$ has been a subject of interest since the 1950s. As discussed in Section 6.3.1, statistical functions were initially used to estimate the galaxy correlation function, but with little success as these functions did not provide a good fit for $\xi(r)$. The actual form of $\xi(r)$ was first measured by Totsuji & Kihara (1969) and later by Groth & Peebles (1977). The breakthrough in Totsuji & Kihara (1969) was their assumption that $\xi(r)$ follows a power law. At that time, the researchers' main field was closer to condensed matter physics than astrophysics, and they borrowed the idea of power-law correlations from studies of critical phenomena in phase transitions.

---

[2]If galaxies are anti-correlated, ξ is negative in the range of $-1 \leq \xi \leq 0$.

[3]A statistical quantity from which the mean has already been subtracted.

The estimation of $\xi(r)$ was initially based on angular correlations, i.e., galaxy clustering on the celestial sphere. The power-law correlation maintains its form even when projected onto the sky, and this property was exploited in the analysis (see Fig. 6.3). We will discuss this projection more thoroughly in the next subsection.

Even in the era of precision cosmology, with modern large galaxy surveys, a power law remains a fairly good approximation for the 2-point correlation function of galaxies

$$\xi(r) = \left(\frac{r}{r_0}\right)^{-\gamma}, \tag{6.18}$$

where r_0 is known as the correlation length. For a general population of luminous galaxies, $r_0 = 5.4h^{-1}$ [Mpc] and the exponent is $\gamma \simeq 1.8$. We can similarly define the angular correlation function $\omega(\theta)$ for galaxy clustering on the sky (Peebles, 1980), which has the form

$$\omega(\theta) = \left(\frac{\theta}{\theta_0}\right)^{1-\gamma}, \tag{6.19}$$

where θ_0 depends on the depth of the survey (see, e.g., Groth & Peebles, 1977; Peebles, 1993). The precise values of r_0 and γ vary with galaxy populations, highlighting the concept of galaxy bias, which we introduce in Chapter 7.

6.3.4 POWER SPECTRUM ANALYSIS

We can also analyze fluctuations in Fourier space, where real-space and Fourier analyses provide equivalent information. The Fourier transform of the overdensity $\delta(\vec{x})$ is given by

$$\delta(\vec{x}) = \frac{1}{(2\pi)^3} \int \delta(\vec{k}) e^{i\vec{k}\cdot\vec{x}} d^3 k, \tag{6.20}$$

$$\delta(\vec{k}) = \int \delta(\vec{x}) e^{-i\vec{k}\cdot\vec{x}} d^3 x, \tag{6.21}$$

where $k = 2\pi/x$.

Similar to the 2-point correlation function, we can take the ensemble average of $\langle \delta(\vec{k}_1)\delta^*(\vec{k}_2)\rangle$. From eq. (6.21), we have

$$\langle \delta(\vec{k}_1)\delta^*(\vec{k}_2)\rangle = \langle \delta(\vec{k}_1)\delta(-\vec{k}_2)\rangle$$
$$= \int\int e^{-i(\vec{k}_1\cdot\vec{x}_1+\vec{k}_2\cdot\vec{x}_2)}\xi(|\vec{x}_1-\vec{x}_2|)\,d^3 x_1 d^3 x_2. \tag{6.22}$$

Integrating over $\vec{x}_1$ while keeping $\vec{x}_2$ fixed, we obtain

$$\langle \delta(\vec{k}_1)\delta^*(\vec{k}_2)\rangle = (2\pi)^3 \delta_{\mathrm{D}}^3(\vec{k}_1-\vec{k}_2)\int \xi(x) e^{-i\vec{k}_1\cdot\vec{x}}d^3 x, \tag{6.23}$$

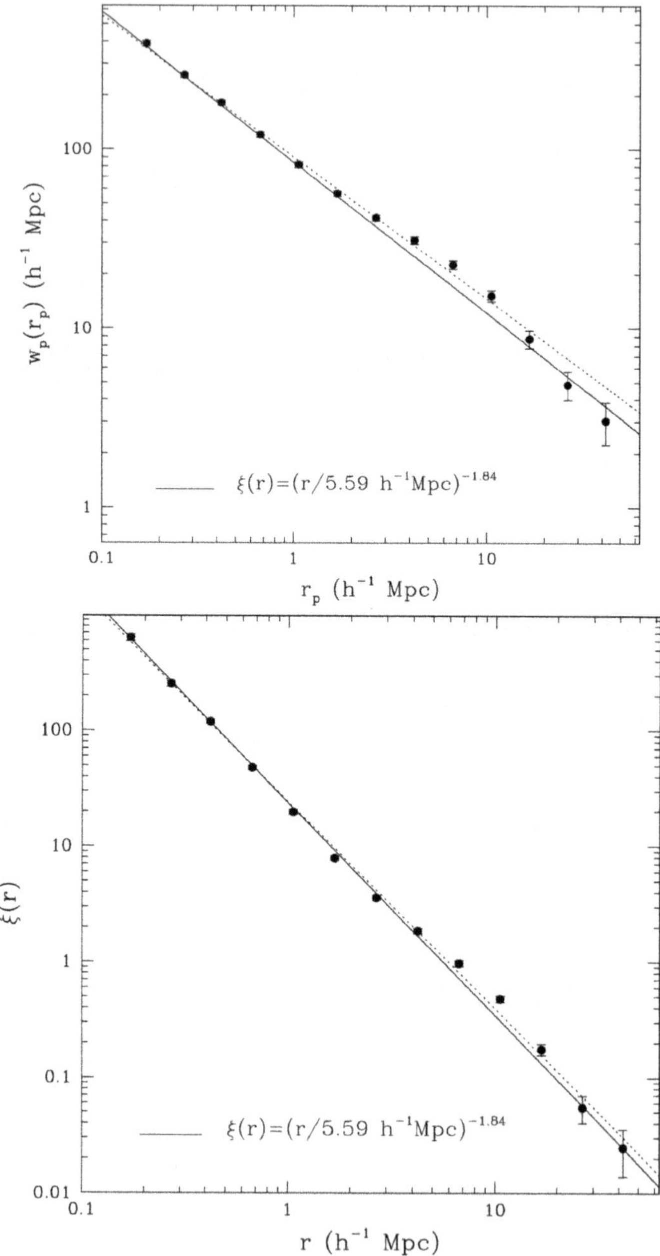

Figure 6.3 Observed 2-point correlation functions. The left panel shows the angular correlation, while the right panel presents the spatial correlation function, both derived from the Sloan Digital Sky Survey data. Both functions are well-fitted by a single power law, with small deviations. Credit: Zehavi, I. et al. 2005, Astrophysical Journal, 630, 1, Figs. 6 and 7, reproduced by permission of the AAS.

where $\vec{x}_1$ has been replaced by $\vec{x}$, and we used the integral formula for Dirac's delta function

$$\delta_{\mathrm{D}}^3(\vec{k}) = \frac{1}{(2\pi)^3} \int e^{-i\vec{k}\cdot\vec{x}} \mathrm{d}^3 x \,. \tag{6.24}$$

Assuming isotropy, we can integrate over angles and obtain

$$\langle \delta(\vec{k})\delta(\vec{k}') \rangle = (2\pi)^3 \delta_{\mathrm{D}}^3(\vec{k}-\vec{k}') \int \frac{\sin(kx)}{kx} \xi(x) x^2 \mathrm{d}x$$
$$\equiv (2\pi)^3 \delta_{\mathrm{D}}^3(\vec{k}-\vec{k}') P(k) \,, \tag{6.25}$$

where we relabelled $\vec{k}_1$ and $\vec{k}_2$ as $\vec{k}$ and $\vec{k}'$. The integral in eq. (6.25) only depends on $|\vec{k}|$ because of the Universe's rotational invariance. This defines the power spectrum.

Thus, the relation between the power spectrum and the 2-point correlation function is

$$P(k) = \int e^{-i\vec{k}\cdot\vec{x}} \xi(x) \mathrm{d}^3 x = 4\pi \int \frac{\sin(kx)}{kx} \xi(x) x^2 \mathrm{d}x \,, \tag{6.26}$$

$$\xi(x) = \frac{1}{(2\pi)^3} \int e^{i\vec{k}\cdot\vec{x}} P(k) \mathrm{d}^3 k = \frac{1}{2\pi^2} \int \frac{\sin(kx)}{kx} P(k) k^2 \mathrm{d}k \,. \tag{6.27}$$

In other words, not only the fluctuations themselves but also their ensemble-averaged quantities, $\xi(x)$ and $P(k)$, are related by Fourier transformation. This is known as the Wiener–Khinchin relation.

6.3.5 HIGHER-ORDER MOMENT MEASURES

A Gaussian random field can be fully characterized by its first and second-order moment measures. For a general random field, however, this is not the case, and infinitely higher-order moments are required for full characterization. Although it is impossible to measure an infinite number of moments, the cosmic matter and galaxy density fields evolve from an almost Gaussian distribution (as reflected in the temperature fluctuations of the Cosmic Microwave Background), so a few low-order moments are sufficient to explore their gravitational evolution.

6.3.5.1 *N*-Point Correlation Functions

As we formulated for the general case, the 2-point correlation function can be extended to infinitely higher-order correlations via the cumulant expansion. The three-point correlation function $\zeta(\vec{x}_1,\vec{x}_2,\vec{x}_3)$ is defined as

$$\zeta(\vec{x}_1,\vec{x}_2,\vec{x}_3) \equiv \langle \delta(\vec{x}_1), \delta(\vec{x}_2), \delta(\vec{x}_3) \rangle_{\mathrm{c}} \,, \tag{6.28}$$

and the four-point function $\eta(\vec{x}_1,\vec{x}_2,\vec{x}_3,\vec{x}_4)$ is defined as

$$\eta(\vec{x}_1,\vec{x}_2,\vec{x}_3,\vec{x}_4) \equiv \langle \delta(\vec{x}_1), \delta(\vec{x}_2), \delta(\vec{x}_3), \delta(\vec{x}_4) \rangle_{\mathrm{c}} \,. \tag{6.29}$$

In general, the N-point correlation function is

$$\xi^N(\vec{r}_1 \ldots \vec{x}_N) = \langle \delta(\vec{x}_1)\ldots\delta(\vec{x}_N)\rangle_c \tag{6.30}$$

(e.g., Bernardeau, 2007; Bernardeau et al., 2002; Peebles, 1980)

6.3.5.2 Polyspectra

Similarly to the power spectrum $P(k)$, the Fourier transformation of N-point correlations provides high-dimensional counterparts of $P(k)$, known as polyspectra. The third-order cumulant is expressed as

$$\zeta(\vec{x}_1,\vec{x}_2,\vec{x}_3) = \langle \delta(\vec{x}_1),\delta(\vec{x}_2),\delta(\vec{x}_3)\rangle_c = (2\pi)^3 \delta_D^3(\vec{k}_1 + \vec{k}_2 + \vec{k}_3)B(\vec{k}_1,\vec{k}_2,\vec{k}_3)\,, \tag{6.31}$$

where $B(\vec{k}_1,\vec{k}_2,\vec{k}_3)$ is the bispectrum. The delta function in eq. (6.31) indicates that the three wavenumber vectors must satisfy $\vec{k}_1 + \vec{k}_2 + \vec{k}_3 = \vec{0}$. Furthermore, due to the isotropy of the Universe, B only depends on $|\vec{k}_1|$, $|\vec{k}_2|$, and $|\vec{k}_3|$. Thus, the degree of freedom for the bispectrum is three.

The fourth-order cumulant gives

$$\eta(\vec{x}_1,\vec{x}_2,\vec{x}_3,\vec{x}_4) = \langle \delta(\vec{x}_1),\delta(\vec{x}_2),\delta(\vec{x}_3),\delta(\vec{x}_4)\rangle_c$$
$$= (2\pi)^4 \delta_D^3(\vec{k}_1 + \vec{k}_2 + \vec{k}_3 + \vec{k}_4)T(\vec{k}_1,\vec{k}_2,\vec{k}_3,\vec{k}_4)\,, \tag{6.32}$$

where $T(\vec{k}_1,\vec{k}_2,\vec{k}_3,\vec{k}_4)$ is the trispectrum. Here, the four wavenumber vectors must satisfy $\vec{k}_1 + \vec{k}_2 + \vec{k}_3 + \vec{k}_4 = \vec{0}$, forming a tetrahedron with $\vec{k}_1,\vec{k}_2,\vec{k}_3$, and $\vec{k}_4$. Again, by the isotropy of the Universe, T is determined solely by the shape of the tetrahedron. The six edges of the tetrahedron uniquely determine the trispectrum T, yielding six degrees of freedom.

In general, the Nth-order polyspectrum is given by

$$\xi^N(\vec{x}_1 \ldots \vec{x}_N) = \langle \delta(\vec{x}_1)\ldots\delta(\vec{x}_N)\rangle_c$$
$$= (2\pi)^N \delta_D^3(\vec{k}_1 + \cdots + \vec{k}_N)P^{(N)}(\vec{k}_1,\ldots,\vec{k}_N)\,. \tag{6.33}$$

The degrees of freedom for the Nth order polyspectrum is $3N - 6$ for $N \geq 3$. A detailed and comprehensive discussion on this topic can be found in Matsubara (1995).

7 Structure and Galaxy Formation in the Universe

7.1 CHARACTERIZATION OF FLUCTUATION

We begin by describing the properties of density fluctuations. To characterize the statistical properties of matter fluctuations in the Universe, we use statistical methods similar to those introduced in Chapter 6. In Chapter 6, we dealt with galaxies in the Universe, or more generally, a point cloud in space.[1] In contrast, here we characterize the continuous stochastic field of matter fluctuation, denoted as $\rho(\vec{x})$. We define the statistical properties of the matter density field here for completeness, and it is important to note that this time, all the quantities refer to a continuous random field.

We start with the density contrast

$$\delta(\vec{x}) = \frac{\rho(\vec{x}) - \bar{\rho}}{\bar{\rho}} \tag{7.1}$$

and its dispersion

$$\sigma^2 = \langle \delta^2 \rangle . \tag{7.2}$$

The Fourier transform of $\delta(\vec{x})$ is given by

$$\delta(\vec{x}) = \frac{1}{(2\pi)^3} \int \delta(\vec{k}) e^{i\vec{k}\cdot\vec{x}} \mathrm{d}^3 k \tag{7.3}$$

$$\delta(\vec{k}) = \int \delta(\vec{x}) e^{-i\vec{k}\cdot\vec{x}} \mathrm{d}^3 x . \tag{7.4}$$

The power spectrum of the density fluctuation is then defined as

$$P(\vec{k}) = \left\langle \left| \delta(\vec{k}) \right|^2 \right\rangle . \tag{7.5}$$

Due to rotational invariance, the power spectrum can be written as

$$P(\vec{k}) = P(k) . \tag{7.6}$$

Higher-order power spectra (such as the bispectrum or trispectrum, corresponding to the 3-point and 4-point correlation functions, respectively) are also defined. In general, an infinite number of moments (or their Fourier counterparts) is required to fully specify the properties of a stochastic field. One particularly important class of

---

[1] Although this distinction is not often emphasized, the exact mathematical treatment of a point set and a continuous field is significantly different (e.g., Daley & Vere-Jones, 2006).

DOI: 10.1201/9781003104315-7

stochastic fields is the Gaussian random field, which has ideal mathematical properties.

A Gaussian random field is a stochastic field whose distribution is described by a Gaussian probability density function (PDF) and whose Fourier phases are uncorrelated, such that

$$f(\delta)\mathrm{d}\delta = \frac{1}{(2\pi\sigma^2)^{\frac{1}{2}}}e^{-\frac{\delta^2}{2\sigma^2}}\mathrm{d}\delta\,, \tag{7.7}$$

$$\langle\phi(\vec{k})\phi(\vec{k}')\rangle = \delta_{\mathrm{d}}(\vec{k}-\vec{k}')\,. \tag{7.8}$$

For Gaussian random fields, all the stochastic properties are uniquely characterized by the power spectrum $P(k)$. The significance of Gaussian random fields lies not only in their mathematical simplicity but also in the fact that the density fluctuations in the Universe, as observed, can be approximated as (nearly) Gaussian.

7.2 INITIAL FLUCTUATION: THE HARRISON–ZEL'DOVICH SPECTRUM

A power-law form

$$P(k) \propto k^n \tag{7.9}$$

for the initial power spectrum has been proposed based on heuristic requirements for structure formation in the Universe. The case with $n = 1$ is specifically called the Harrison–Zel'dovich spectrum (Harrison, 1970; Zeldovich, 1972). This spectrum was originally introduced based on fundamental discussions on the isotropy of the Universe by these authors. Later, the Harrison–Zel'dovich spectrum was understood as a natural consequence of the exponential cosmic expansion that occurred during the very early phase of the Universe, known as inflation.

Inflation refers to an exponential expansion of the Universe that occurred before the Big Bang fireball. The ideas underlying cosmic inflation were developed during the late 1970s and early 1980s to address several problems of Big-Bang cosmology, such as the horizon, smoothness, flatness, and monopole problems (e.g., Albrecht & Steinhardt, 1982; Brout et al., 1978; Guth, 1981; Kazanas, 1980; Linde, 1982, 1983; Sato, 1981; Starobinsky, 1980).

Later, it was realized that inflation also provided a mechanism to generate primordial cosmological perturbations (e.g., Bardeen et al., 1983; Guth & Pi, 1982; Hawking, 1982; Mukhanov, 1985; Mukhanov & Chibisov, 1981, 1982; Starobinsky, 1982). Without going into the detailed physics of generating Gaussian density fluctuations, inflation produces nearly Gaussian fluctuations with a power spectrum

$$P(k) \propto k\,, \tag{7.10}$$

(e.g., Lyth & Liddle, 2009).

7.3 DEFORMATION OF THE HARRISON–ZEL'DOVICH SPECTRUM

As time progresses, the fluctuation power spectrum deforms due to various physical processes in the Universe. The rate at which fluctuations grow is referred to as the 'growth rate.' The growth rate of a density perturbation depends on several factors, such as the epoch (i.e., which component dominates the global expansion dynamics at that time), whether the size of a perturbation is on a superhorizon or subhorizon scale, and the cosmological parameters. Superhorizon and subhorizon refer to fluctuations that are larger or smaller, respectively, than the particle horizon of the Universe at a given cosmic age.

During the radiation-dominated epoch, when the initial power spectrum enters the horizon ($L_H = ct$) (as the horizon expands to the scale of the fluctuation), fluctuations can grow only marginally because cosmic expansion acts to suppress the growth of dark matter perturbations. This phenomenon is known as the Mészáros effect or stagspansion. Since fluctuations larger than the horizon can grow, the power spectrum bends at the horizon scale of that epoch and deviates from a simple power law. The scale at which this bending occurs is

$$k_H(t_{eq}) = a(t_{eq})H(t_{eq}) = 1.028 \times 10^{-2} \left(\frac{\Omega_{M0}h^2}{0.141} \right) [\text{Mpc}^{-1}] . \tag{7.11}$$

It is important to be aware of a significant conceptual challenge when trying to understand superhorizon-scale fluctuations. Fluctuations larger than the horizon must be treated relativistically, but due to the freedom of coordinate transformations in general relativity, the form of these fluctuations cannot be uniquely determined—for example, it is always possible to adopt a coordinate system in which fluctuations vanish entirely. The gauge-invariant formulation is now widely used to address this problem (Bardeen, 1980; Kodama & Sasaki, 1984).

The power spectrum at any given time t can be expressed as the initial power spectrum multiplied by a deformation factor, known as the transfer function of the power spectrum, $T(k)$. In the case of cold dark matter (CDM), since stagspansion is the primary cause of deformation, the transfer function is given by

$$T(k) = \begin{cases} 1 & k \ll k_h(t_{eq}) \\ k^{-2} & k \gg k_h(t_{eq}) \end{cases} . \tag{7.12}$$

Thus, the power spectrum can be written as

$$P(k) \propto kT^2 = \begin{cases} k & k \ll k_h(t_{eq}) \\ k^{-3} & k \gg k_h(t_{eq}) \end{cases} . \tag{7.13}$$

This acts as the effective 'initial condition' for structure formation in the Universe. The corresponding fluctuation is reflected in the CMB temperature fluctuations,[2].

---

[2]The actual temperature fluctuations observed in the CMB are also influenced by various physical processes that occurred during later epochs of the Universe.

7.4 LINEAR THEORY OF THE STRUCTURE FORMATION

Now we move on to the formation of cosmic structures and galaxies. The treatment is divided into two regimes: linear and nonlinear evolution of structures. In the early Universe, the distribution of matter and radiation was almost homogeneous, with tiny fluctuations. These fluctuations are thought to have originated during the inflation period. Inflation models predict that the initial fluctuations were (almost) random and Gaussian.

Since (as far as we know) there is no repulsion in gravitational interaction, tiny fluctuations can attract surrounding matter through gravity and grow monotonically. In the early phase, the amplitude of such fluctuations is still very small compared to the average density. In this regime, second and higher-order terms related to the fluctuation can be neglected in the basic equations, allowing the equations to be linearized. Under this condition, fluctuations with different spatial scales can be treated independently, as there is no interaction between fluctuations of different scales. The Fourier transformation, therefore, plays a central role in the formulation. In contrast, when the amplitude of density fluctuations approaches the average density, we can no longer apply this approximation. In such cases, special treatment is required, which defines the nonlinear regime.

To clarify the relationship between the matter power spectrum and the mass distribution, we introduce the concept of mass contained within a radius R. We denote the average mass within this radius over the Universe as $\mathcal{M}$, then

$$\mathcal{M} = \frac{4\pi R^3}{3}\bar{\rho}\,, \tag{7.14}$$

and

$$\delta\mathcal{M} = \int_{|\vec{x}|\leq R} \bar{\rho}\,\delta(\vec{x})\,\mathrm{d}^3x\,. \tag{7.15}$$

Thus,

$$\frac{\delta\mathcal{M}}{\mathcal{M}} = \frac{3}{4\pi R^3}\int_{|\vec{x}|\leq R}\delta(\vec{x})\,\mathrm{d}^3x \equiv \int W_R(|\vec{x}|)\delta(\vec{x})\,\mathrm{d}^3x\,. \tag{7.16}$$

Here

$$W_R(\vec{x}) = \frac{3}{4\pi R^3}\Theta\left(R - |\vec{x}|\right)\,, \tag{7.17}$$

$$\Theta(x) \equiv \begin{cases} 1 & (x \geq 0)\,, \\ 0 & (x < 0) \end{cases}\,. \tag{7.18}$$

This function $W_R(x)$ is the so-called tophat window function, which is one example of window functions used to isolate a part of the density fluctuation field with scale R. The dispersion of mass fluctuation on scale R is then given by

$$\sigma^2(R) = \left\langle \left(\frac{\delta\mathcal{M}}{\mathcal{M}}\right)^2 \right\rangle = \frac{1}{2\pi^2}\int W^2(kR)P(k)k^2\,\mathrm{d}k\,, \tag{7.19}$$

where

$$W^2(kR) = \int e^{-i\vec{k}\cdot\vec{x}} W_R(|\vec{x}|)\, d^3k \,. \tag{7.20}$$

Equation (7.19) shows that the window function filters out smaller scale fluctuations at $k > R^{-1}$. Thus, $\sigma^2(R)$ represents the amplitude of the power spectrum for scales larger than R. It is often convenient to define the contribution to the mass power spectrum from an infinitesimal logarithmic interval $d\ln k$ as

$$\Delta^2(k) \equiv \frac{k^3}{2\pi^2} P(k) \,. \tag{7.21}$$

For cold dark matter (CDM) with the Harrison–Zel'dovich spectrum, the mass spectrum increases monotonically with k. Since $P(k)$ grows as $\propto a^2$ in the matter-dominated (MD) epoch, fluctuations become nonlinear, starting from smaller scale structures. This is referred to as the bottom-up scenario. In contrast, for hot dark matter (HDM), the velocity of dark matter particles is relativistic, causing their mean free path to become comparable to the size of the horizon. As a result, small-scale fluctuations decay significantly. Fluctuations become nonlinear from larger scales and fragment into smaller-scale structures. This is called the top-down scenario.

We will introduce the theoretical formulation details in the following sections.

7.4.1 FROM PHYSICAL COORDINATES TO COMOVING COORDINATES

On scales much smaller than the horizon scale, or when the motion of matter is much slower than the speed of light, we can safely approximate that matter behaves like a classical fluid. This approximation is known as the "Newtonian fluid approximation." Under this assumption, we have the continuity equation, Euler equation, and Poisson equation as shown in [eqs. (3.52)–(3.54)]

$$\frac{\partial \rho}{\partial t} + \nabla \cdot (\rho \vec{v}) = 0 \,, \tag{7.22}$$

$$\frac{\partial \vec{v}}{\partial t} + (\vec{v} \cdot \nabla)\vec{v} = -\frac{\nabla p}{\rho} - \nabla \phi \,, \tag{7.23}$$

$$\nabla^2 \phi = 4\pi G \rho \,, \tag{7.24}$$

where ρ is the matter density, p is the pressure, $\vec{v}$ is the velocity of matter, and ϕ is the gravitational potential produced by the matter. Equations (7.22)–(7.24) are written in *physical* coordinates.

We introduce comoving coordinates to treat structure formation in the expanding Universe

$$\vec{r} = a(t)\vec{x} \,, \tag{7.25}$$

where $a(t)$ is the scale factor describing cosmic expansion (e.g., Peebles, 1993; Weinberg, 1972). This transformation directly leads to

$$\frac{\partial \vec{r}}{\partial t} = \dot{\vec{r}} = \dot{a}\vec{x} + a\dot{\vec{x}} \equiv H(t)a\vec{x} + a\dot{\vec{x}}$$

$$= H(t)\vec{r} + \vec{u}\,, \tag{7.26}$$

$$H(t) = \frac{\dot{a}(t)}{a(t)}\,, \tag{7.27}$$

where $\vec{u}$ represents the peculiar velocity. Note that $\dot{a}\vec{x} = H(t)\vec{r}$ is due to cosmic expansion and does not represent the actual motion of an object. Therefore, eq. (7.26) can be expressed as

$$\vec{v} = \dot{a}\vec{x} + \vec{u}\,. \tag{7.28}$$

We also obtain the following expressions for the differential operators

$$\left(\frac{\partial}{\partial t}\right)_{r} = \left(\frac{\partial}{\partial t}\right)_{x} - \frac{\dot{a}}{a}\vec{x}\cdot\nabla_{x}\,, \tag{7.29}$$

$$\nabla_{r} = \frac{1}{a}\nabla_{x}\,. \tag{7.30}$$

From here on, we drop the subscript x.

By changing from physical to comoving coordinates, the continuity equation [eq. (7.22)] becomes

$$\frac{\partial \rho}{\partial t} + 3H\rho + \frac{1}{a}\nabla\cdot(\rho\vec{u}) = 0\,, \tag{7.31}$$

and the Euler equation [eq. (7.23)] becomes

$$\frac{\partial \vec{u}}{\partial t} + H\vec{u} + \frac{1}{a}(\vec{u}\cdot\nabla)\vec{u} = -\frac{\nabla p}{a\rho} - \nabla\phi - \ddot{a}\vec{x}\,. \tag{7.32}$$

By defining a new potential

$$\Phi \equiv \nabla\phi + \frac{a\ddot{a}\vec{x}}{2}\,, \tag{7.33}$$

we have

$$\frac{\partial \vec{u}}{\partial t} + H\vec{u} + \frac{1}{a}(\vec{u}\cdot\nabla)\vec{u} = -\frac{\nabla p}{a\rho} - \frac{1}{a}\nabla\Phi\,. \tag{7.34}$$

The Poisson equation [eq. (7.24)] becomes

$$\nabla^{2}\phi = 4\pi G a^{2}\rho\,.$$

From the Friedmann equation [eq. (A.3)], we obtain

$$3a\ddot{a} = (\nabla \vec{x}) = \left(\nabla^2 \frac{x^2}{2}\right) = -4\pi G \bar{\rho} \, .$$

Thus, we have

$$\nabla^2 \Phi = 4\pi G a^2 \left(\rho - \bar{\rho}\right) \, , \tag{7.35}$$

with the solution

$$\Phi(\vec{x}) = -G a^2 \int \mathrm{d}^3\vec{x}' \, \frac{\rho - \bar{\rho}}{|\vec{x}' - \vec{x}|} \, . \tag{7.36}$$

Consider a small fluctuation from the background Universe. The density, peculiar velocity, potential, and pressure are given by

$$\rho_\mathrm{b} = \bar{\rho}(t) \, ,$$
$$\vec{u} = \vec{0} \, ,$$
$$\nabla \Phi = \vec{0} \, ,$$
$$\nabla p = \vec{0} \, .$$

We introduce fluctuations from the homogeneous background as

$$\delta(\vec{x},t) \equiv \frac{\rho - \bar{\rho}}{\bar{\rho}} \, ,$$

[eq. (7.1)] and

$$\delta p(\vec{x},t) \equiv p(\vec{x},t) - \bar{p}(t) \, . \tag{7.37}$$

Then, eq. (7.35) becomes

$$\nabla^2 \Phi = 4\pi G a^2 \bar{\rho} \delta \, . \tag{7.38}$$

The continuity equation [eq. (7.31)] can now be written as

$$\frac{\partial}{\partial t}\left[\bar{\rho}(1+\delta)\right] + 3H\bar{\rho}(1+\delta) + \frac{\bar{\rho}}{a}\nabla \cdot \left[(1+\delta)\vec{u}\right] = 0 \, . \tag{7.39}$$

We should note that, from the mass conservation $\partial(a^3 \bar{\rho})/\partial t = 0$,

$$\frac{\partial \bar{\rho}}{\partial t} + 3H\bar{\rho} = 0 \, . \tag{7.40}$$

From eqs. (7.39) and (7.40), we have

$$\frac{\partial}{\partial t}\left(\bar{\rho}\delta\right) + 3H\bar{\rho}\delta + \frac{\bar{\rho}}{a}\nabla \cdot \left[(1+\delta)\vec{u}\right] = \bar{\rho}\frac{\partial \delta}{\partial t} + \frac{\bar{\rho}}{a}\nabla \cdot \left[(1+\delta)\vec{u}\right] = 0 \, . \tag{7.41}$$

Thus, the continuity equation [eq. (7.31)] becomes

$$\frac{\partial \delta}{\partial t} + \frac{1}{a} \nabla \cdot [(1+\delta)\vec{u}] = 0 \,. \tag{7.42}$$

Similarly, the Euler equation [eq. (7.34)] becomes

$$\frac{\partial \vec{u}}{\partial t} + H\vec{u} + \frac{1}{a}(\vec{u}\cdot\nabla)\vec{u} = -\frac{\nabla \delta p}{a\bar{\rho}(1+\delta)} - \frac{1}{a}\nabla\Phi \,. \tag{7.43}$$

Equations (7.38), (7.42), and (7.43) form the starting point for deriving the solutions that describe the linear growth of fluctuations.

7.4.2 JEANS INSTABILITY IN THE EXPANDING UNIVERSE

Here, we neglect terms that include multiplications of δ, δp, and $\vec{u}$ (linearization). Then, the continuity equation and the Euler equation now read

$$\frac{\partial \delta}{\partial t} + \frac{1}{a} \nabla \cdot \vec{u} = 0 \,, \tag{7.44}$$

$$\frac{\partial \vec{u}}{\partial t} + H\vec{u} + \frac{1}{a}\nabla\Phi + \frac{\nabla \delta p}{a\bar{\rho}} = 0 \,. \tag{7.45}$$

By manipulating eqs. (7.44) and (7.45), we obtain

$$\frac{\partial^2 \delta}{\partial t^2} + 2\frac{\dot{a}}{a^2}(\nabla\cdot\vec{u}) - 4\pi G\bar{\rho}\delta - \frac{\nabla^2 \delta p}{a^2\bar{\rho}}$$

$$= \frac{\partial^2 \delta}{\partial t^2} + 2H\frac{\partial \delta}{\partial t} - \left[4\pi G\bar{\rho}\delta + \frac{\nabla^2 \delta p}{a^2\bar{\rho}}\right] = 0 \,. \tag{7.46}$$

Now, consider the sound speed of the fluid

$$c_s \equiv \sqrt{\left.\frac{\partial p}{\partial \rho}\right|_S} \tag{7.47}$$

where S is the entropy. Assuming that entropy fluctuations are small, we have

$$\delta p = c_s^2 \bar{\rho}\delta \,. \tag{7.48}$$

Since the basic equations are linear, we can deal with the Fourier components for each wavenumber k. Consider

$$\delta(\vec{x}) = \sum_k \delta_k e^{i\vec{k}\cdot\vec{x}} \,. \tag{7.49}$$

Then,

$$\frac{\partial^2 \delta_k}{\partial t^2} + 2H\frac{\partial \delta_k}{\partial t} - \left[4\pi G\bar{\rho} - \frac{c_s^2 k^2}{a^2}\right]\delta_k = 0 \,. \tag{7.50}$$

This is the Jeans instability equation. The third term in the square brackets [ ] controls the evolution of the fluctuation. Consider two cases

1. $[\ ] < 0$

   Fluctuations oscillate and decay. This occurs when c_s is large or when k is large (small scale), because the fluctuation does not contain enough mass to contract gravitationally.

2. $[\ ] > 0$

   Fluctuations grow.

We define k_J so that

$$\frac{c_s^2 k_J^2}{a^2} = 4\pi G \bar{\rho} \,, \tag{7.51}$$

and the Jeans length λ_J is defined as

$$\lambda_J = \frac{2\pi a}{k_J} = c_s \sqrt{\frac{\pi}{G\bar{\rho}}} \,. \tag{7.52}$$

Fluctuations smaller than λ_J decay, while those larger than λ_J can grow due to gravitational instability. The corresponding mass is

$$\mathcal{M}_J \equiv \frac{4\pi}{3} \bar{\rho} \left(\frac{\lambda_J}{2}\right)^3 \tag{7.53}$$

which is referred to as the Jeans mass.

7.4.2.1 Growth of Fluctuation in Various Epochs

Using the Jeans equation, we can analyze the evolution of fluctuations in the early Universe. The evolution of the Universe is governed by the energy density through eqs. (A.2) and (A.3). As discussed in Chapter (A), radiation dominates the energy density in the early Universe (RD epoch), whereas matter dominates later (MD era). Now, consider the evolution of dark matter fluctuations. Dark matter is considered to be pressureless. Therefore, eq. (7.46) or eq. (7.50) becomes (since they are linear, the real or Fourier space treatment does not differ)

$$\frac{\partial^2 \delta}{\partial t^2} + 2H \frac{\partial \delta}{\partial t} - 4\pi G \bar{\rho} \delta = 0 \,. \tag{7.54}$$

In the Einstein-de Sitter universe, $a \propto t^{2/3}$. Then,

$$\frac{\partial^2 \delta}{\partial t^2} + \frac{4}{3t} \frac{\partial \delta}{\partial t} - \frac{2}{3t^2} \delta = 0 \,. \tag{7.55}$$

Assuming a solution of the form

$$\delta = A_+(x) D_+(t) + A_-(x) D_-(t) \,, \tag{7.56}$$

eq. (7.55) yields the general solution

$$\delta = A_+(x) t^{2/3} + A_-(x) t^{-1} \,. \tag{7.57}$$

The second term represents a decaying mode, which can be neglected for larger t. Thus,

$$\delta = A_+(x)t^{2/3} \propto a \, . \tag{7.58}$$

Defining the growth factor $D(t)$ by

$$D(t) \equiv \frac{D_+(t)}{D_+(t_0)} \tag{7.59}$$

where t_0 is the present age of the Universe, eq. (7.58) is also expressed as

$$D(t) \propto a(t) \propto t^{2/3} \, . \tag{7.60}$$

During the radiation-dominated era, dark matter fluctuations are suppressed within the horizon (stagspansion or Mészáros effect), as the expansion rate is too fast.

$$\frac{\partial^2 \delta_M}{\partial t^2} + 2H \frac{\partial \delta_M}{\partial t} - 4\pi G \bar{\rho} \delta_M = 0 \, , \tag{7.61}$$

$$H^2 = \frac{8\pi G}{3} (\bar{\rho}_M + \bar{\rho}_R) \, . \tag{7.62}$$

Here, subscripts M and R denote matter and radiation. We also define

$$\zeta \equiv \frac{\bar{\rho}_M}{\bar{\rho}_R} = \frac{a}{a_{eq}} \tag{7.63}$$

(where a_{eq} is the scale factor at matter-radiation equality). This leads to

$$\frac{\partial^2 \delta_M}{\partial \zeta^2} + \frac{2+3\zeta}{2\zeta(1+\zeta)} \frac{\partial \delta_M}{\partial t} - \frac{3\delta_M}{2\zeta(1+\zeta)} = 0 \, . \tag{7.64}$$

Assuming

$$\frac{\partial^2 \delta_M}{\partial \zeta^2} = 0 \, , \tag{7.65}$$

we find a solution of the form

$$\delta_M \propto 1 + \frac{3}{2}\zeta \, . \tag{7.66}$$

For the MD era ($\zeta \gg 1$),

$$\delta_M \propto a \, , \tag{7.67}$$

and for the RD era ($\zeta \ll 1$),

$$\delta_M \propto \text{constant.} \tag{7.68}$$

(more precisely, $\delta_M \propto \ln a$). Thus, we can quantitatively understand the stagspansion effect.

For baryon fluctuations, gravitational potentials are determined by dark matter

$$\frac{\partial^2 \delta_B}{\partial t^2} + 2H\frac{\partial \delta_B}{\partial t} - 4\pi G\bar{\rho}_M \delta_M = 0 \,, \tag{7.69}$$

where subscript B denotes baryons. Before recombination, baryon fluctuations cannot grow due to the large sound speed $(c/3)$. Thus, baryon fluctuations can only start growing after the recombination epoch ($z \simeq 1100$). If we define $\eta \equiv a/a_{\rm rec}$ (where "rec" stands for the recombination epoch), then

$$\delta_M \propto \eta \,. \tag{7.70}$$

This leads to

$$\eta^{1/2}\frac{\rm d}{{\rm d}\eta}\left(\eta^{3/2}\frac{{\rm d}\delta_B}{{\rm d}\eta}\right) = \frac{3}{2}\delta_M \,. \tag{7.71}$$

The growing solution is

$$\delta_B = \left(1 - \frac{1}{\eta}\right)\delta_M \,. \tag{7.72}$$

This solution implies that

$$\begin{aligned}
a \simeq a_{\rm dec} \Leftrightarrow \eta \simeq 1 \quad &: \ \delta_B \simeq 0 \\
a \gg a_{\rm dec} \Leftrightarrow \eta \gg 1 \quad &: \ \delta_B \simeq \delta_M \,.
\end{aligned} \tag{7.73}$$

This indicates that immediately after decoupling, baryons exhibit no fluctuation, but later ($\eta \gg 1$), baryons develop the same fluctuation as dark matter. This phenomenon is referred to as the "catch-up." By combining this with a fully relativistic treatment of superhorizon fluctuations, we can derive a comprehensive picture of the evolution of the fluctuation power spectrum.

7.4.3 PECULIAR VELOCITY FIELD

Based on the linear theory of structure formation, we can also derive another important quantity: the peculiar velocity field of galaxies. The growth of the gravitational potential is determined by the growth of the density contrast and the Poisson equation [eq. (7.38)]

$$\nabla^2\Phi = 4\pi Ga^2 \bar{\rho}\delta \,.$$

Since $\bar{\rho} \propto a^{-3}$ during the matter-dominated epoch, the time evolution of the potential is given by

$$\Phi(t) \propto \frac{D(t)}{a(t)} \,. \tag{7.74}$$

The Fourier transformation of eq. (7.38) gives

$$\tilde{\Phi}(\vec{k}, t) = -4\pi G a^2 \bar{\rho} \frac{\delta(\vec{k}, t)}{k^2} \,. \tag{7.75}$$

Returning to real space

$$\Phi(\vec{x}, t) = \frac{-4\pi G a^2 \bar{\rho}}{(2\pi)^3} \int e^{i\vec{k}\cdot\vec{x}} \int \delta(\vec{x}', t) e^{-i\vec{k}\cdot\vec{x}'} \mathrm{d}^3 x' \mathrm{d}^3 k$$

$$= -G a^2 \bar{\rho} \int \frac{\delta(\vec{x}', t)}{|\vec{x} - \vec{x}'|} \mathrm{d}^3 x' \,. \tag{7.76}$$

The second line follows from the Green function formula for the Laplacian operator. Note that these equations for the potential are general and not restricted to the linear regime.

Taking the curl of the linearized Euler equation [eq. (7.45)] gives

$$\nabla \times \left[\frac{\partial \vec{u}}{\partial t} + H\vec{u} + \frac{1}{a}\nabla\Phi + \frac{\nabla\delta p}{a\rho} \right] = \frac{\partial}{\partial t}(\nabla \times \vec{u}) + H\nabla \times \vec{u} = 0 \tag{7.77}$$

which leads to

$$\frac{\partial}{\partial t}(\nabla \times \vec{u}) = \frac{\dot{a}}{a}\nabla \times \vec{u} \,, \tag{7.78}$$

meaning that the curl of the velocity field decays as $\propto a^{-1}$ in the linear regime. This implies that the transversal (rotational) component of the velocity field is a decaying mode, decreasing with cosmic expansion. Thus, the growth of the velocity field is essentially potential flow, which can be expressed through the velocity potential $\Psi(\vec{x}, t)$ as

$$\vec{u} = -\frac{\nabla\Psi}{aH} \,. \tag{7.79}$$

Recall that the linearized continuity equation [eq. (7.44)] is

$$\frac{\partial\delta}{\partial t} + \frac{1}{a}\nabla \cdot \vec{u} = 0 \,.$$

For the growing mode, we have

$$\frac{\partial\delta}{\partial t} = \frac{\nabla^2\Psi}{a^2 H} \tag{7.80}$$

which leads to the Poisson equation for the velocity potential

$$\nabla^2\Psi = a^2 H \frac{\partial\delta}{\partial t} = a^2 H^2 f \delta \,, \tag{7.81}$$

where $f(t)$ is the linear growth rate, defined as

$$f(t) \equiv \frac{d \ln D}{d \ln a} \,.$$ (7.82)

By comparing eqs. (7.38) and (7.81), we can relate the gravitational and velocity potentials as

$$\Psi = \frac{H^2 f}{4\pi G \bar{\rho}} \Phi \,.$$ (7.83)

The solution to the velocity Poisson equation [eq. (7.81)] is

$$\Psi(\vec{x}, t) = -\frac{a^2 H^2 f}{4\pi} \int \frac{\delta(\vec{x}', t)}{|\vec{x} - \vec{x}'|} d^3 x' \,.$$ (7.84)

The velocity field is then derived as

$$\vec{u}(\vec{x}, t) = -\frac{\nabla^2 \Phi(\vec{x}, t)}{a^2 H^2} = -\frac{a H f}{4\pi} \int \frac{\vec{x} - \vec{x}'}{|\vec{x} - \vec{x}'|^3} \delta(\vec{x}', t) \, d^3 x' \,.$$ (7.85)

As shown, the linear growth rate [eq. (7.82)] is derived by differentiating the growth factor. In a general case, we have

$$f(t) = -1 - \frac{\Omega_M}{2} + \Omega_\Lambda + \left[\int_0^1 \frac{dy}{\left(\Omega_M y^{-1} + \Omega_\Lambda y^2 + \Omega_K \right)^{\frac{3}{2}}} \right]^{-1} \,.$$ (7.86)

In some cases, this integration can be performed analytically

1. Einstein–de Sitter Universe $\Omega_{M0} = 1$, $\Omega_{\Lambda 0} = 0$

$$f = 1 \,.$$ (7.87)

2. Flat Λ-dominated Universe $\Omega_{M0} + \Omega_{\Lambda 0} = 1$

$$f = \Omega_M \frac{F(4/3, 5/6, 11/6; \Omega_\Lambda)}{F(1/3, 5/6, 11/6; \Omega_\Lambda)}$$ (7.88)

where $F(\alpha, \beta, \gamma; z)$ is the hypergeometric function.

Simple approximations such as

$$f(\Omega_M) \simeq \Omega_M^{0.6}$$ (7.89)

(e.g., Peebles, 1980, 1993) and

$$f(\Omega_M, \Omega_\Lambda) = \Omega_M^{\frac{4}{7}} + \frac{\Omega_\Lambda}{70} \left(1 + \frac{\Omega_M}{2} \right)$$ (7.90)

(e.g., Carroll et al., 1992; Lahav et al., 1991; Lightman & Schechter, 1990) are often used[3].

---

[3] Note that cosmological parameters without a suffix 0 are time-dependent.

7.4.4 BARYON ACOUSTIC OSCILLATION (BAO)

Before decoupling, radiation pressure from photons resists the gravitational compression of the baryon fluid into potential wells, setting up acoustic oscillations in the fluid. A typical example of such a nontrivial phenomenon is the baryon acoustic oscillation (BAO), generated in the matter-radiation fluid of the early Universe (Peebles & Yu, 1970; Sunyaev & Zeldovich, 1970).

Consider a point-like initial perturbation in the primordial matter-photon plasma. In this plasma, matter and photons are coupled into a single fluid. Since the photons are extremely hot and numerous, the fluid possesses tremendous pressure for its density. The pressure seeks to equalize itself with its surroundings, resulting in an expanding spherical sound wave. The sound speed at this early epoch is given by

$$c_{\mathrm{s}} = \frac{c}{\sqrt{3\left(1 + \dfrac{3\rho_{\mathrm{B}}}{4\rho_{\mathrm{R}}}\right)}}, \tag{7.91}$$

where c is the speed of light, and the subscripts B and R refer to baryons and radiation, respectively. Equation 7.91 shows that the sound speed before the matter-radiation decoupling can be around 57% of the speed of light in the early epoch. The sound horizon (final radius) r_{s} is obtained by

$$\begin{aligned} r_{\mathrm{s}} &= \int_0^{t_{\mathrm{dec}}} (1+z)c_{\mathrm{s}}\,\mathrm{d}t \\ &= \int_{z_{\mathrm{dec}}}^{\infty} \frac{c_{\mathrm{s}}}{H_0\sqrt{\Omega_{\mathrm{R},0}(1+z)^4 + \Omega_{\mathrm{M},0}(1+z)^3 + \Omega_{\Lambda,0}}}\,\mathrm{d}z, \end{aligned} \tag{7.92}$$

where the subscript "dec" stands for the time of the decoupling between photons and matter. We observe the superposition of many acoustic waves imprinted on the large-scale structure emerging from the primordial fluctuations. Though BAO is a baryonic phenomenon by definition, the density enhancement structure of baryons and dark matter becomes indistinguishable at the final state (e.g., Eisenstein et al., 2007; Ma & Bertschinger, 1995).

The length scale of the BAO remains constant in comoving coordinates. Therefore, in principle, we can detect the signal in the galaxy 2-point correlation function. However, since the BAO scale is vast compared to typical galaxy scales, the 2-point correlation signal is usually weak. This requires a large and densely sampled galaxy catalog to measure the large-scale signal effectively. The advent of the SDSS, one of the largest optical photometric and spectroscopic surveys covering one-third of the sky, made it realistic to detect the BAO signal in the 2-point correlation function. Eisenstein et al. (2005) first detected the signal around 150 Mpc in the 2-point correlation function. Even with modern surveys, BAO analysis remains feasible only with extensive datasets such as those from SDSS and, more recently, eBOSS, a dedicated survey for BAO studies (e.g., Merz et al., 2021).

The shorter the wavelength of the potential fluctuation, the faster the fluid oscillates, leading to different phases at last scattering, depending on the wavelength.

Since compressed regions (maxima) correspond to hot regions and expanding regions (minima) to cold regions, there will be a harmonic series of peaks associated with the acoustic oscillations. The BAO can be leveraged for various cosmological studies, and developing more flexible and accessible methods for detecting and quantifying the BAO signal would significantly contribute to advancing our understanding of the cosmos.

7.4.5 PRESS–SCHECHTER FORMALISM

7.4.5.1 Spherical Collapse in the Expanding Universe

In comoving coordinates, a sphere centered on a local overdensity shrinks in time as the Hubble expansion is retarded by the overdensity. At some point, the expansion of the sphere stops (turn-around), and the sphere starts to collapse. Consider the evolution of a spherical overdense region as a simple model of the nonlinear evolution of fluctuations (Gunn & Gott, 1972; Tomita, 1969). Let $R(t)$ be the radius of a shell, and $\mathcal{M}$ the mass contained inside the shell. Then, the equation of motion of the shell is

$$\frac{\mathrm{d}^2 R}{\mathrm{d}t^2} = -\frac{G\mathcal{M}}{R^2} \tag{7.93}$$

which gives

$$\left(\frac{\mathrm{d}R}{\mathrm{d}t}\right)^2 = \frac{2G\mathcal{M}}{R} + 2E . \tag{7.94}$$

As in classical mechanics, if the "binding energy" $E < 0$, the shell motion is bounded. In this case, the solution is given parametrically as

$$R = C^2 \left(1 - \cos\theta\right) , \tag{7.95}$$

$$t = \frac{C^3}{\sqrt{G\mathcal{M}}} \left(\theta - \sin\theta\right) , \tag{7.96}$$

where C is an integration constant, corresponding to the size of the shell. This curve is called a "cycloid." If the energy $E > 0$, the motion is unbounded, corresponding to the dynamics of voids (e.g., Sheth & van de Weygaert, 2004).

Densities of the overdense and average regions are

$$\rho = \mathcal{M} \left(\frac{4\pi R^3}{3}\right)^{-1} , \tag{7.97}$$

$$\bar{\rho} = \frac{1}{6\pi G t^2} , \tag{7.98}$$

respectively. Thus, we obtain the density contrast

$$\delta(t) = \frac{9G\mathcal{M}t^2}{2R^3} - 1 = \frac{9}{2}\frac{(\theta - \sin\theta)^2}{(1 - \cos\theta)^3} . \tag{7.99}$$

The motion can be characterized as follows:

1. $\theta = \pi$

   The turn-around point from expansion to contraction:

$$R_{\text{turn}} = 2C^2, \quad t_{\text{turn}} = \frac{\pi C^3}{\sqrt{G\mathcal{M}}} \,. \tag{7.100}$$

2. $\theta = 2\pi$

   Collapse ($R = 0$):

$$t_{\text{coll}} = \frac{2\pi C^3}{\sqrt{G\mathcal{M}}} \,. \tag{7.101}$$

In reality, $R = 0$ is never established; rather, the overdense region becomes an object with radius R_{vir} via mechanisms like violent relaxation. For a mass $\mathcal{M}$, from the conservation of energy and the virial theorem, we have

$$\frac{G\mathcal{M}^2}{R_{\text{turn}}} = \frac{1}{2}\frac{G\mathcal{M}^2}{R_{\text{coll}}} \tag{7.102}$$

which gives

$$R_{\text{coll}} = \frac{R_{\text{turn}}}{2} \,. \tag{7.103}$$

In this case, the density contrast of a collapsed object is

$$\delta(t_{\text{coll}}) = \frac{\mathcal{M}}{\left(\dfrac{4\pi R_{\text{vir}}^2}{3}\right)\bar{\rho}(t_{\text{coll}})} - 1 = 18\pi^2 - 1 \simeq 177 \,. \tag{7.104}$$

7.4.5.2 Connection to the linear theory

At an early phase, the growth should agree with the linear growth. Expanding with θ, we get

$$\delta(t) = \frac{3}{20}\theta^2 + \mathcal{O}(\theta^4) \,,$$

$$t = \frac{C^3}{6\sqrt{G\mathcal{M}}}\theta^3 + \mathcal{O}(\theta^5) \,, \tag{7.105}$$

which gives

$$\delta \propto t^{2/3} \,.$$

If we define this overdensity extrapolated from the linear theory, δ_{L},

$$\delta_{\text{L}}(t) = \frac{3}{20}\left(\frac{6\sqrt{G\mathcal{M}}}{C^3}t\right)^{\frac{2}{3}} \,. \tag{7.106}$$

Since both the nonlinear δ of spherical collapse and linear δ_L are monotonic functions of t, we can estimate the value of δ from δ_L if we have a relation between δ and δ_L, as

$$\delta_L(t_{\text{coll}}) = \frac{3(12\pi)^{\frac{2}{3}}}{20} \simeq 1.69 \,. \tag{7.107}$$

This means that a region with $\delta = 1.69$ is actually a collapsed object.

The discussion here was based on the Einstein–de Sitter Universe, but for the flat Λ-dominated Universe, the correction is small:

$$\delta(t_{\text{coll}}) \simeq 18\pi^2 + 82x_{\text{coll}} - 39x_{\text{coll}}^2 - 1 \,, \tag{7.108}$$

$$\delta_L(t_{\text{coll}}) \simeq \frac{3}{20}(12\pi)^{\frac{2}{3}} \left[1 + 0.0123\log\Omega_M(t_{\text{coll}})\right] \,, \tag{7.109}$$

where

$$x_{\text{coll}} \equiv \Omega_M(t_{\text{coll}}) - 1 \tag{7.110}$$

(Bryan & Norman, 1998; Kitayama & Suto, 1996).

7.4.5.3   Original Press–Schechter Formalism

Press & Schechter (1974) proposed an idea to connect the linear growth solution of density fluctuation and the extrapolation to the nonlinear regime through a spherical collapse model. This gives an analytic model of dark halo formation.

Let the number density of objects whose mass is between $\mathscr{M}$ and $\mathscr{M} + d\mathscr{M}$ be $n(\mathscr{M})d\mathscr{M}$. Then, this $n(\mathscr{M})$ is called the mass function. The Press–Schechter (PS) formalism gives an analytic solution of $n(\mathscr{M})$.

$$\mathscr{M} = \frac{4\pi R^3}{3}\bar{\rho} \,. \tag{7.111}$$

The smoothed (averaged) overdensity in a sphere whose radius R corresponds to a mass $\mathscr{M}$ is called a fluctuation $\delta_{\mathscr{M}}$ of mass scale $\mathscr{M}$.

As mentioned above, the original matter fluctuation δ is Gaussian. In general, if a stochastic field is Gaussian, the Gaussianity is preserved by smoothing. Thus, the smoothed fluctuation $\delta_{\mathscr{M}}$ is also Gaussian.

$$f(\delta_{\mathscr{M}})d\delta_{\mathscr{M}} = \frac{1}{\sqrt{2\pi\sigma(\mathscr{M})^2}} \exp\left[-\frac{\delta_{\mathscr{M}}^2}{2\sigma(\mathscr{M})^2}\right] d\delta_{\mathscr{M}} \,, \tag{7.112}$$

where $\sigma(\mathscr{M})^2$ represents the variance of $\delta_{\mathscr{M}}$. Naturally, the variance $\sigma(\mathscr{M})^2$ becomes smaller than the original σ^2 due to smoothing. At a certain point, if the linear (extrapolated) $\delta_{\mathscr{M}}$ exceeds the threshold value δ_c, a collapsed object with mass $\mathscr{M}$ is formed. We set $\delta_c = \delta_{\text{coll}} = 1.69$ as derived from the spherical collapse model. Recent theoretical studies, however, adopt more complex forms for δ_c to reflect more realistic physical conditions.

The spatial fraction of regions with $\delta > \delta_c$ is

$$F(\delta_{\mathcal{M}} > \delta_c)\mathrm{d}\delta_{\mathcal{M}} = \frac{1}{\sqrt{2\pi\sigma(\mathcal{M})^2}}\int_{\delta_c}^{\infty} e^{-\frac{\delta_{\mathcal{M}}^2}{2\sigma(\mathcal{M})^2}}\mathrm{d}\delta_{\mathcal{M}} = \frac{1}{\sqrt{2\pi}}\int_{\frac{\delta_c}{\sigma(\mathcal{M})}}^{\infty} e^{-\frac{x^2}{2}}\mathrm{d}x\,.$$

$$(7.113)$$

The amount of matter involved in an object with a mass larger than $\mathcal{M}$ per unit volume is $\bar{\rho}F(>\delta_c)(\mathcal{M})$. Hence,

$$\bar{\rho}F(\delta_c)(\mathcal{M}+\mathrm{d}\mathcal{M}) - \bar{\rho}F(\delta_c)(\mathcal{M}) = \bar{\rho}\frac{\mathrm{d}F(\delta_c)}{\mathrm{d}\mathcal{M}}\mathrm{d}\mathcal{M} = n(\mathcal{M})\mathcal{M}\mathrm{d}\mathcal{M}\,. \qquad (7.114)$$

The discussion above ignored the possibility that a once-collapsed object would be involved in a larger object (cloud-in-cloud problem). Also, regions with $\delta < 0$ will never collapse into objects (i.e., $F(\delta_c) \to 1/2$ as $\sigma(\mathcal{M}) \to \infty$). Thus, we multiply by a factor of 2 to correct for these issues:

$$n(\mathcal{M})\mathcal{M}\mathrm{d}\mathcal{M} = 2\bar{\rho}\left|\frac{\mathrm{d}F(\delta_c)}{\mathrm{d}\mathcal{M}}\right|_{\mathcal{M}}\mathrm{d}\mathcal{M}\,. \qquad (7.115)$$

We finally obtain the formula:

$$n(\mathcal{M}) = \sqrt{\frac{2}{\pi}}\,\frac{\bar{\rho}}{\mathcal{M}^2}\,\frac{\delta_c}{\sigma(\mathcal{M})}\left|\frac{\mathrm{d}\ln\sigma(\mathcal{M})}{\mathrm{d}\ln\mathcal{M}}\right|\exp\left[-\frac{\delta_c^2}{2\sigma(\mathcal{M})^2}\right]\,. \qquad (7.116)$$

Since the power spectrum is approximated by a power law such as $P(k) \propto k^n$, or equivalently $\sigma \propto \mathcal{M}^{-\alpha}$ $(\alpha = (n+3)/6)$, the final form becomes

$$n(\mathcal{M}) = \frac{2\alpha}{\sqrt{\pi}}\,\frac{\bar{\rho}}{\mathcal{M}_*^2}\left(\frac{\mathcal{M}}{\mathcal{M}_*}\right)^{\alpha-2}\exp\left[-\left(\frac{\mathcal{M}}{\mathcal{M}_*}\right)^{2\alpha}\right]\,, \qquad (7.117)$$

referred to as the Press–Schechter (PS) mass function. The Schechter function (Schechter, 1976a), often used as an approximation of the galaxy luminosity function, was inspired by the PS mass function [eq. (7.117)].

However, the original formulation by Press & Schechter (1974) did not provide a physical reasoning for the factor of 2. Furthermore, the cloud-in-cloud problem is not easily addressed by their approach. A mathematically consistent derivation was developed in later studies, which we now discuss in the following section.

7.4.6 EXCURSION SET APPROACH FOR THE MASS FUNCTION

7.4.6.1 Excursion set Approach

The current mainstream of the mass function formulation is to derive the PS mass function by modeling the excursion set of the density field. Consider a varying smoothing scale R (or equivalently the mass scale $\mathcal{M}$, as discussed above) for a given Gaussian random field. Here, we define the sharp k-space window function (tophat) as

$$\tilde{W}(\vec{k};k_{\mathrm{c}}) = \begin{cases} 1 & |\vec{k}| \leq k_{\mathrm{c}} \\ 0 & |\vec{k}| > k_{\mathrm{c}} \end{cases} \tag{7.118}$$

where k_{c} is the wavenumber corresponding to the smoothing scale R, i.e., $k_{\mathrm{c}} = 1/R$. The smoothed density field is then obtained as

$$\begin{aligned} \delta_{k_{\mathrm{c}}}(\vec{x},t) &= \int \delta_{\vec{k}}(t)\tilde{W}(\vec{k};k_{\mathrm{c}})e^{-i\vec{k}\cdot\vec{x}}\,\mathrm{d}^3 k \\ &= \int_{|k|\leq k_{\mathrm{c}}} \delta_{\vec{k}}(t)e^{-i\vec{k}\cdot\vec{x}}\,\mathrm{d}^3 k. \end{aligned} \tag{7.119}$$

Here, we consider a one-dimensional stochastic process as a function of k_{c}.

$$\langle \Delta\delta_{k_{\mathrm{c}}}^2 \rangle \equiv \langle [\delta_{k_{\mathrm{c}}+\Delta k_{\mathrm{c}}}(\vec{x},t) - \delta_{k_{\mathrm{c}}}(\vec{x},t)]^2 \rangle. \tag{7.120}$$

We can straightforwardly obtain

$$\langle \Delta\delta_{k_{\mathrm{c}}}^2 \rangle = \sigma_{k_{\mathrm{c}}+\Delta k_{\mathrm{c}}}^2 - \sigma_{k_{\mathrm{c}}}^2 \equiv \Delta\sigma_{k_{\mathrm{c}}}^2. \tag{7.121}$$

For a sharp-k tophat window, this is a Markov process, i.e., no correlation between steps. Actually, this condition can be relaxed to a more general case, but here we present only the core of the argument. For a comprehensive treatment, see, e.g., Maggiore & Riotto (2010a,b,c). Note that $\delta_{k_{\mathrm{c}}} \to 0$ as $\sigma_{k_{\mathrm{c}}}^2 \to 0$.

Consider the mass scale $\mathcal{M}_1(k_{\mathrm{c}1})$. According to the original Press–Schechter formalism, the fraction of trajectories with $\delta > \delta_{\mathrm{c}}$ at $k_{\mathrm{c}1}$ equals the fraction of mass in collapsed objects with $\mathcal{M} > \mathcal{M}_1$ (Fig. 7.1). Thus, trajectory B in Fig. 7.1 is not a part of a collapsed object with $\mathcal{M} > \mathcal{M}_1$. However, as discussed earlier, this object is involved in a collapsed object with $\mathcal{M} > \mathcal{M}_3 > \mathcal{M}_1$, which turns out to be a contradiction. The original formulation does not account for all the Markovian trajectories (such as trajectory B) but only those which, once reaching δ_{c}, stay above that threshold. However, according to the properties of a Markovian random walk, trajectory B is just as probable as trajectory B′. Each trajectory that the original formalism accounts for has a corresponding trajectory that mirrors it after the first upcrossing, such as B′. Thus, the Press–Schechter formalism should be modified so that the probability that $\delta > \delta_{\mathrm{c}}$ is twice the fraction of mass contained in halos with mass greater than $\mathcal{M}$. This correction resolves the issue with the factor of two introduced in the original derivation.

7.4.6.2 Extended Press–Schechter (EPS) Formalism

To derive the Press–Schechter mass function from the extended formalism, consider the fraction of trajectories with their first upcrossing after $k_{\mathrm{c}1}$ (as trajectory A in Fig. 7.1). According to the modified *Ansatz*, these trajectories correspond to mass

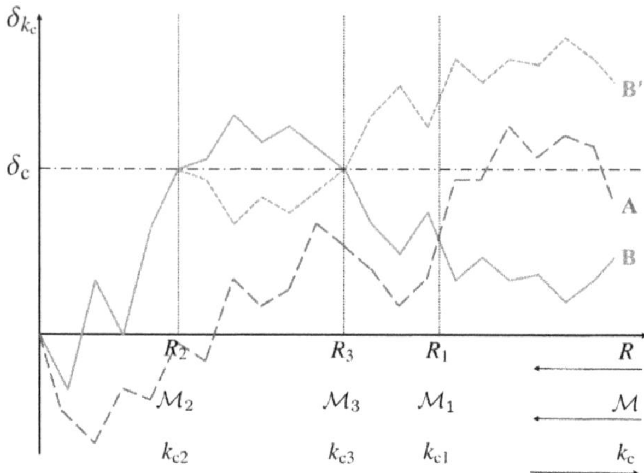

Figure 7.1　A schematic description of some sample paths of the stochastic process of the smoothed density field. The abscissa corresponds either to the smoothing scale R, smoothed mass scale $\mathcal{M}$, or $k_c = 1/R$. The ordinate is the smoothed density enhancement with respect to k_c (or R or $\mathcal{M}$).

elements in collapsed objects of mass $\mathcal{M} < \mathcal{M}_1$. The probability of the first upcrossing is expressed as

$$f_{\mathrm{FU}}(\delta_{k_c}) = \frac{1}{\sqrt{2\pi\sigma_{k_c}^2}}\exp\left(-\frac{\delta_{k_c}^2}{2\sigma_{k_c}^2}\right) - \frac{1}{\sqrt{2\pi\sigma_{k_c}^2}}\exp\left[-\frac{(2\delta_c - \delta_{k_c})^2}{2\sigma_{k_c}^2}\right]. \quad (7.122)$$

This solution is concisely explained by Chandrasekhar (1943).

To obtain this fraction, we need the fraction of trajectories with $\delta_{k_{c1}} < \delta_c$, but we must also subtract the fraction of trajectories as B. Since $\mathscr{P}(\mathrm{B}) = \mathscr{P}(\mathrm{B}')$ (where $\mathscr{P}(A)$ stands for the probability of event A), we can write this fraction as

$$F(< k_{c1}) = \int_{-\infty}^{\delta_c} f_{\mathrm{FU}}(\delta_{k_{c1}})\,\mathrm{d}\delta$$

$$= \frac{1}{\sqrt{2\pi\sigma_{k_{c1}}^2}}\int_{-\infty}^{\delta_c}\left\{\exp\left(-\frac{\delta_{k_{c1}}^2}{2\sigma_{k_{c1}}^2}\right) - \exp\left[-\frac{(2\delta_c - \delta_{k_{c1}})^2}{2\sigma_{k_{c1}}^2}\right]\right\}\,\mathrm{d}\delta.$$

$$(7.123)$$

This fraction is the fraction of masses included in halos with $k > k_{c1}$, i.e., $\mathcal{M} < \mathcal{M}_1$. Since the trajectory is a random walk, every trajectory will cross the barrier as $M \to 0$ as $k_c \to \infty$, and then this fraction will also converge to zero. What we want here is the fraction of masses included in halos with $\mathcal{M} > \mathcal{M}_1$, $F(> \mathcal{M})$ should be equal to $1 - F(< M(k_{c1}))$. Thus this factor of two has been correctly taken into account. After

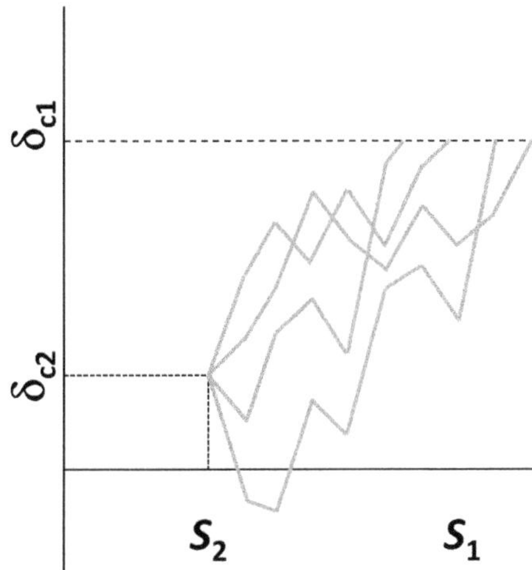

Figure 7.2   Sample trajectories which cross a barrier δ_{c2} at $S_2 = \sigma_2^2$, and then cross a second barrier $\delta_{c1} > \delta_{c2}$ at various $S_1 > S_2$.

changing variables from k_c to $\mathcal{M}$, we finally obtain the PS mass function formula again,

$$
\begin{aligned}
n(\mathcal{M}) &= \frac{\bar{\rho}}{\mathcal{M}} \frac{\partial F(> \mathcal{M})}{\partial M} \\
&= \sqrt{\frac{2}{\pi}} \frac{\bar{\rho}}{\mathcal{M}^2} \frac{\delta_c}{\sigma(\mathcal{M})} \left| \frac{d\ln\sigma(\mathcal{M})}{d\ln\mathcal{M}} \right| \exp\left[-\frac{\delta_c^2}{2\sigma(\mathcal{M})^2} \right].
\end{aligned}
\tag{7.124}
$$

This new derivation is referred to as the extended Press–Schechter (EPS) formalism.

7.4.7   MERGING IN THE EPS FORMALISM

The EPS formalism is not only mathematically consistent but also can be flexibly applied to various related problems. As a very important application, we calculate the properties of progenitors of a given halo based on the EPS approach (Lacey & Cole, 1993). To derive a merger probability per unit time for a halo of a given mass $\mathcal{M}$ at a time t, we consider a subset of trajectories that upcross a barrier δ_{c2} at $S_2 = \sigma_2^2$, and then upcross a second barrier $\delta_{c1} > \delta_{c2}$ at various $S_1 > S_2$. The conditional probability that one of these trajectories first upcrosses a barrier δ_{c1} in the interval $[S_1, S_1 + dS_1]$ can be obtained as

$$
f_{S_1}(S_1, \delta_{c1} | S_2, \delta_{c2}) dS_1 = \frac{\delta_{c1} - \delta_{c2}}{(2\pi)^{\frac{1}{2}} (S_1 - S_2)^{\frac{3}{2}}} \exp\left[-\frac{(\delta_{c1} - \delta_{c2})^2}{2(S_1 - S_2)} \right] dS_1.
\tag{7.125}
$$

Equation (7.125) can be used directly to derive the mass distribution of halos at time t_1 which will form a halo with mass $\mathcal{M}_2$ at time t_2. We evaluate the conditional probability that a trajectory with a first upcrossing of δ_{c1} at S_1 will upcross δ_{c2} at $[S_2, S_2 + dS_2]$,

$$
\begin{aligned}
& f_{S_2}(S_2, \delta_{c2} | S_1, \delta_{c1}) dS_2 \\
&= \frac{f_{S_1}(S_1, \delta_{c1} | S_2, \delta_{c2}) dS_1 f_{S_2}(S_2, \delta_{c2}) dS_2}{f_{S_1}(S_1, \delta_{c1}) dS_1} \\
&= \frac{1}{(2\pi)^{\frac{1}{2}}} \left[\frac{S_1}{S_2(S_1 - S_2)} \right]^{\frac{3}{2}} \frac{\delta_{c2}(\delta_{c1} - \delta_{c2})}{\delta_{c1}} \exp\left[-\frac{(\delta_{c2}S_1 - \delta_{c1}S_2)^2}{2 S_1 S_2 (S_1 - S_2)} \right] dS_2 .
\end{aligned}
\tag{7.126}
$$

Hence, by taking a limit of $\delta_{c2} \to \delta_{c1} = \delta$, we can make a formula of the mean transition rate

$$
\frac{d^2 \mathscr{P}}{dS_2 d\delta}(S_1 \to S_2 | \delta) dS_2 d\delta = \frac{1}{(2\pi)^{\frac{1}{2}}} \left[\frac{S_1}{S_2(S_1 - S_2)} \right]^{\frac{3}{2}} \exp\left[-\frac{\delta^2(S_1 - S_2)}{2 S_1 S_2} \right] dS_2 d\delta .
\tag{7.127}
$$

This represents the probability that a halo mass $\mathscr{M}_1$ will merge with another halo with a mass $\Delta\mathscr{M} = \mathscr{M}_2 - \mathscr{M}_1$, where $\mathscr{M}_2$ and $\mathscr{M}_1$ correspond to S_2 and S_1,

$$
\begin{aligned}
& \frac{d^2 \mathscr{P}}{d\ln\Delta\mathscr{M} dt}(\mathscr{M}_1 \to \mathscr{M}_2 | t) = 2\sigma(\mathscr{M}_2) \left| \frac{d\sigma(\mathscr{M}_2)}{d\mathscr{M}_2} \right| \Delta\mathscr{M} \left| \frac{d\delta}{dt} \right| \frac{d^2 \mathscr{P}}{dS_2 d\delta}(S_1 \to S_2 | \delta) \\
&= \sqrt{\frac{2}{\pi}} \frac{1}{t} \frac{\Delta\mathscr{M}}{\mathscr{M}_2} \frac{\delta_c}{\sigma_2} \left| \frac{d\ln\delta_c}{d\ln t} \right| \left| \frac{d\ln\sigma_2}{d\ln\mathscr{M}_2} \right| \left(\frac{\sigma_1^2}{\sigma_1^2 - \sigma_2^2} \right)^{\frac{3}{2}} \exp\left[-\frac{\delta_c(t)^2}{2} \left(\frac{1}{\sigma_2^2} - \frac{1}{\sigma_1^2} \right) \right]
\end{aligned}
\tag{7.128}
$$

where $\sigma_1 \equiv \sigma(\mathscr{M}_1)$ and $\sigma_2 \equiv \sigma(\mathscr{M}_2)$, respectively. Lacey & Cole (1993) also discussed the survival time of halos.

7.4.8 BIAS

With the development of galaxy surveys, an important and interesting fact was discovered: clusters of galaxies are more strongly clustered than galaxies. This is referred to as 'bias'.

Soon after this was recognized, a simple theoretical explanation was proposed. The basic idea is the so-called 'peak model' of the formation of gravitationally bound objects. Here we introduce the theoretical model which turned out to be very convincing and flexibly extensive. Basic assumption is that, when fluctuations smoothed with a certain scale R, $\delta(R)$, exceeds a threshold δ_c, they start to grow nonlinearly and form gravitationally bound objects (dark halos).

Consider the following quantities of a density fluctuation field: $\delta(\vec{x})$ is the density fluctuation field with a zero mean and dispersion σ^2, $\xi(r)$ is the correlation function of the density field δ, and $\xi_{>\nu}(r)$ is the correlation function of density peaks which lie above a threshold $v\sigma$. Here $\xi_{>v}(r)$ is defined as a fractional probability $\delta(\vec{x}_2) > v\sigma$ given that $\delta(\vec{x}_1) > v\sigma$, where $r = |\vec{x}_2 - \vec{x}_1|$.

If δ is a Gaussian random field, the probability F_1 such that $\delta > v\sigma$ at $\vec{x}_1$ is expressed as

$$F_1 = \int_{v\sigma}^{\infty} f(y)\mathrm{d}y = \frac{1}{\sqrt{2\pi\sigma^2}} \int_{v\sigma}^{\infty} e^{-\frac{y^2}{2\sigma^2}} \mathrm{d}y \tag{7.129}$$

Then, the probability that both δ_1 and δ_2 lie above the threshold $v\sigma$, F_2, is

$$F_2 = \int_{v\sigma}^{\infty} \int_{v\sigma}^{\infty} f(y_1, y_2)\mathrm{d}y_1 \mathrm{d}y_2$$

$$= \frac{1}{2\pi[\xi(0)^2 - \xi(r)^2]^{\frac{1}{2}}}$$

$$\times \int_{v\sigma}^{\infty} \int_{v\sigma}^{\infty} \exp\left\{-\frac{1}{2[\xi(0)^2 - \xi(r)^2]} (y_1, y_2) \begin{bmatrix} \xi(0) & \xi(r) \\ -\xi(r) & \xi(0) \end{bmatrix} \begin{pmatrix} y_1 \\ y_2 \end{pmatrix}\right\} \mathrm{d}y_1 \mathrm{d}y_2$$

$$= \frac{1}{2\pi[\xi(0)^2 - \xi(r)^2]^{\frac{1}{2}}} \int_{v\sigma}^{\infty} \int_{v\sigma}^{\infty} \exp\left\{-\frac{\xi(0)^2 y_1^2 + \xi(0)^2 y_2^2 - 2\xi(r)y_1 y_2}{2[\xi(0)^2 - \xi(r)^2]}\right\} \mathrm{d}y_1 \mathrm{d}y_2 \tag{7.130}$$

Using F_1 and F_2, the correlation of the high-peak regions is defined as

$$1 + \xi_{>v}(r) = \frac{F_2}{F_1^2} \tag{7.131}$$

To calculate this quantity, we should perform some arithmetics. Define $y = \sigma\eta$, and $\mathrm{d}y = \sigma \mathrm{d}\eta$, then we have

$$F_1 = \frac{1}{\sqrt{2\pi\sigma^2}} \int_{v\sigma}^{\infty} e^{-\frac{y^2}{2\sigma^2}} \mathrm{d}y = \frac{1}{\sqrt{2\pi}} \int_{v}^{\infty} e^{-\frac{\eta^2}{2}} \mathrm{d}\eta = \frac{1}{\sqrt{\pi}} \mathrm{erfc}\left(\frac{v}{\sqrt{2}}\right) \tag{7.132}$$

Also, since $\xi(0) = \sigma^2, y_1 = \sigma\eta_1$, and $y_2 = \sigma\eta_2$, we obtain

$$F_2 = \frac{1}{2\pi\left[1 - \frac{\xi(r)^2}{\xi(0)^2}\right]^{\frac{1}{2}}} \int_{v}^{\infty} \int_{v}^{\infty} \exp\left\{-\frac{\left[\eta_1 - \frac{\xi(r)}{\xi(0)}\eta_2\right]^2}{2\left[1 - \frac{\xi(r)^2}{\xi(0)^2}\right]}\right\} \exp\left(-\frac{\eta_2^2}{2}\right) \mathrm{d}\eta_1 \mathrm{d}\eta_2$$

$$= \frac{1}{\sqrt{2\pi}} \int_{v}^{\infty} \mathrm{erfc}\left(-\frac{v - \frac{\xi(r)}{\xi(0)}\eta}{\sqrt{2}\left[1 - \frac{\xi(r)^2}{\xi(0)^2}\right]^{\frac{1}{2}}}\right) \exp\left(-\frac{\eta_2^2}{2}\right) \mathrm{d}\eta \tag{7.133}$$

By using eqs. (7.132) and (7.133), we get

$$\frac{F_2}{F_1^2} = \frac{\sqrt{2} \displaystyle\int_v^\infty \mathrm{erfc}\left(-\frac{v - \frac{\xi(r)}{\xi(0)}\eta}{\sqrt{2}\left[1 - \frac{\xi(r)^2}{\xi(0)^2}\right]^{\frac{1}{2}}}\right) \exp\left(-\frac{\eta^2}{2}\right) d\eta}{\left[\mathrm{erfc}\left(\frac{v}{\sqrt{2}}\right)\right]^2} \tag{7.134}$$

For a limiting case $\xi \ll 1$, eq/ (7.134) is approximated as

$$\xi_{>v}(r) \simeq \frac{1}{\left(e^{\frac{v^2}{2}} \displaystyle\int_v^\infty e^{-\frac{\eta^2}{2}} d\eta\right)^2} \frac{\xi(r)}{\sigma^2} \tag{7.135}$$

and for the other extreme, $\xi \gg 1$,

$$\xi_{>v}(r) \simeq \frac{v^2}{\sigma^2}\xi(r) \tag{7.136}$$

Equation (7.136) is the high-peak bias formula derived by Kaiser (1984). From eq. (7.136), the larger δ_c is, the more strongly the density enhancement localizes, i.e., the fluctuation becomes stronger. This is called the halo bias.

However, in addition to halo bias, another source of bias must be considered. Baryonic gas cools and falls into the gravitational potential well of a halo, contracting to form galaxies. Galaxies are more localized within halos. This is referred to as galaxy bias (or star formation bias, according to Kaiser 1987). Galaxy bias depends on the physics of galaxy formation (and evolution), varying with different populations of galaxies (e.g., red and blue galaxies, luminous and less luminous, massive and less massive, optical and FIR). In essence, galaxy bias is directly tied to the physics of galaxy formation.

These characteristics are clearly reflected in the luminosity and correlation functions of galaxies. For example, bluer galaxies are less clustered than the general population, while redder ones are more strongly clustered. Moreover, more luminous galaxies show stronger clustering (e.g., Pollo et al., 2006; Zehavi et al., 2005). Thus, galaxy bias is intricately linked to the process of galaxy formation.

7.5 GALAXY FORMATION

In this section, we delve into the physics of galaxy formation. Galaxy formation, in essence, describes the transition from dark halos to galaxies. Dark matter is collisionless, interacting only gravitationally. In contrast, baryonic physics involves hydrodynamics and electromagnetic interactions, which are essential for gas cooling and star formation. Star formation is further tied to chemical evolution, dust formation, feedback processes, and black hole (AGN) formation.

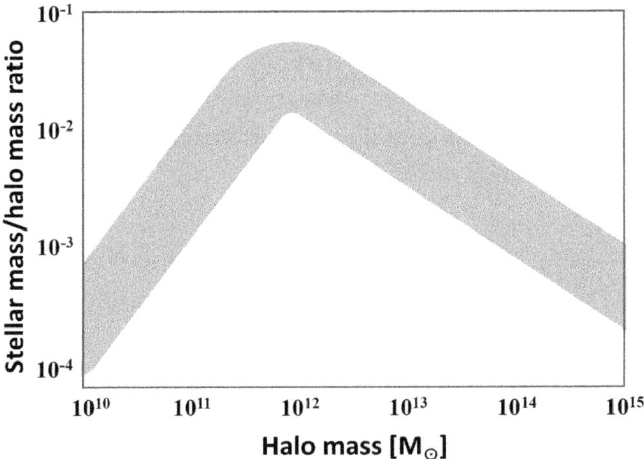

Figure 7.3 Observed stellar mass–halo mass relation. Typical uncertainty between literature is expressed by the thickness of the curve.

The significant difference between the shapes of the halo mass function and the galaxy luminosity function exemplifies the distinction between dark matter and baryonic physics (e.g., Somerville & Primack, 1999). Recently, attention has shifted to the relationship between dark matter and baryonic mass functions (e.g., Behroozi et al., 2019), as illustrated in Fig. 7.3. It is worth noting that neither the lower- nor higher-mass ends are efficient at forming stars. Though this remains a subject of active debate, we will introduce the essence of the issue here.

7.5.1 COOLING OF GAS CLOUDS

Dark halos of cold dark matter (CDM) form from smaller masses and grow over time via merging and mass accretion to create larger objects (see Chapter 7). This is the hierarchical structure formation scenario. From an astronomical perspective, it was proposed by Searle & Zinn (1978). When halos collapse or merge, shock waves are generated in the baryonic matter, heating the gas. As discussed in Chapter 3, hot gas cannot form stars. Thus, the gas must cool in order for galaxies to form. This question is also linked to another puzzle: Why do galaxies typically have masses $\mathcal{M} < 10^{12}$ $[\mathcal{M}_\odot]$, while there are halos with masses as large as 10^{15} $[\mathcal{M}_\odot]$ (e.g., clusters of galaxies)?

To address these questions, we consider two critical timescales for gas clouds. The gas temperature is typically of the order of the virial temperature

$$k_{\rm B} T_{\rm gas} = k_{\rm B} T_{\rm vir} \simeq \frac{G \mathcal{M} m_p}{r} , \tag{7.137}$$

where m_p is the proton mass. For example, galaxies typically have $T_{\rm vir} \sim 10^4$–10^5 [K], while clusters exhibit $T_{\rm vir} > 10^7$ [K] ($\Leftrightarrow$ 1 [keV]). The cooling time

t_{cool} of a baryonic gas clump can be estimated using the cooling function. Let n be the gas number density, and we have

$$t_{\text{cool}} \simeq \frac{k_{\text{B}} T}{n \Lambda_{\text{cool}}} \, , \tag{7.138}$$

where Λ_{cool} is the cooling function, as discussed in Section 3.1. The dynamical time t_{dyn} of an object is estimated from the free-fall time:

$$t_{\text{dyn}} \simeq \frac{1}{\sqrt{G\rho}} \, . \tag{7.139}$$

Rees & Ostriker (1977) proposed a fundamental condition for galaxy formation:

1. $t_{\text{cool}} \ll t_{\text{dyn}}$
   Baryons fall into the halo center on a timescale of t_{dyn} before being influenced by pressure from shock heating.
2. $t_{\text{cool}} \gg t_{\text{dyn}}$
   Baryons are supported by pressure and dissipate their energy quasistatically.

In other words, if condition 1 is satisfied, the gas cloud can collapse to form stars, whereas if condition 2 holds, the gas cannot cool quickly and instead reaches an equilibrium state. This criterion is clearly depicted on the number density n–temperature T diagram (Fig. 7.4). The parameter range for galaxies is located within the envelope

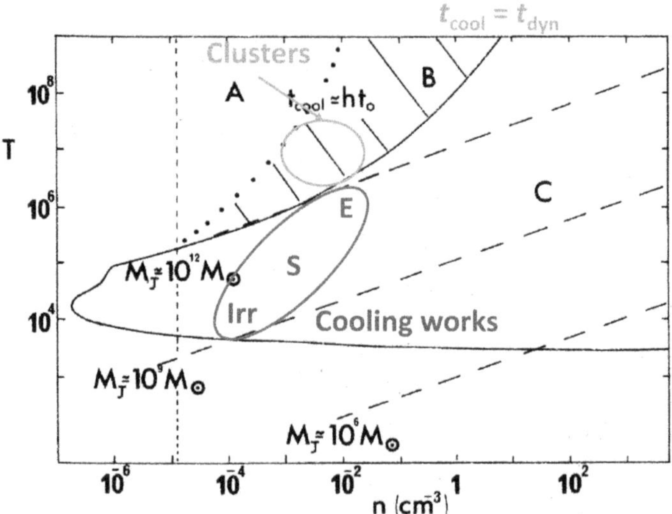

Figure 7.4 A density–temperature diagram of a gas. The solid contour denotes the curve $t_{\text{dyn}} = t_{\text{cool}}$. The gas is assumed to consist of H and He only. The region occupied by galaxies is indicated by an ellipse with symbols E, S, and Irr, while the region for clusters lies outside the solid curve. Lines of constant Jeans mass $\mathcal{M}_{\text{J}}$ have a slope of $1/3$ on this plane, indicated by dashed lines. Credit: adopted from Rees, M. J. & Ostriker, J. P. 1977, Monthly Notices of the Royal Astronomical Society, 179, 541, Fig. 1.

of the curve defined by $t_{\text{cool}} = t_{\text{dyn}}$, indicating that galaxies can cool rapidly enough to form stars. In contrast, the parameter range for clusters of galaxies lies outside the curve, signifying that gas in clusters cools too slowly to form stars. Thus, the differing physical conditions between galaxies and clusters are explained by comparing cooling and dynamical timescales. This also explains the high-mass discrepancy between stellar and dark matter mass functions. This discussion provides a well-established explanation for the absence of galaxies with $\mathcal{M} \simeq 10^{15}\,[\mathcal{M}_\odot]$. However, this is essentially a one-zone argument. Recent studies have revealed that gas can cool efficiently in the center of galaxy clusters when the density profile is considered properly. Massive galaxies at the centers of clusters are often surrounded by gas with a radiative cooling time short enough to enable cooling flows (Fabian, 1994). X-ray observations indicate a significant radiative loss and cooling rates, as given by

$$\dot{\mathcal{M}} = \frac{2}{5}\frac{\mu m_p L}{k_{\text{B}} T}\,, \tag{7.140}$$

where the luminosity L is primarily emitted in the X-ray. This suggests that stellar mass in cluster cores should be growing significantly if the cooled gas forms stars. Although some observations do show such star formation, they do not match the predictions from eq. (7.140). Therefore, we must consider other mechanisms to halt this process. While still debated, AGN feedback is increasingly regarded as a plausible mechanism.

7.5.2 STELLAR FEEDBACK

7.5.2.1 Kinetic galactic wind

Thus far, we have discussed how star formation is quenched in the high-mass end of galaxies. As seen in Fig. 7.3, we must also suppress the number of lower-mass galaxies. This is supported by their high mass-to-light ratios (e.g., Mateo, 1998, and references therein). Additionally, several other peculiarities are observed. For example, there is a systematic difference in surface brightness between giant and dwarf galaxies (e.g., Roberts & Haynes, 1994). The stellar mass–metallicity relation also differs between giant and dwarf galaxies (Lee et al., 2006). To explain these characteristics, Dekel & Silk (1986) proposed the dwarf galactic wind scenario. It is important to note that the concept of galactic wind was discussed much earlier (e.g., Larson, 1974). The notable aspect of Dekel & Silk (1986)'s work was their explicit discussion of kinetic galactic winds.

The energy budget for a Type II supernova is as follows:

- Gravitational energy $\sim 10^{53}$ erg
- Explosion energy $\sim 10^{51}$ erg
- Neutrino energy $\sim 10^{53}$ erg.

Thus, $\sim 10^{51}$ erg is released as kinetic energy in the ejecta, more than 100 times the energy of the integrated luminosity of a supernova. The supernova remnants (SNRs) expand due to the injected energy, and overlapping SNRs ultimately produce galactic winds. Dwarf galaxies, with their shallow gravitational potentials, are especially

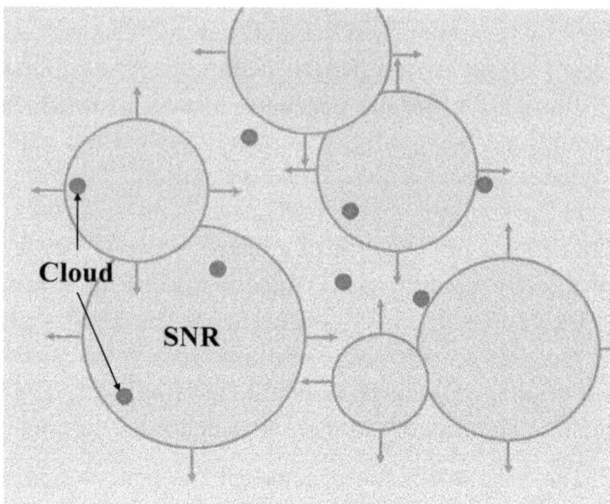

Figure 7.5 Schematic depiction of supernova feedback caused by sequential supernova explosions.

susceptible to various internal and external dynamical effects. Once star formation begins in a dwarf galaxy, subsequent supernovae inject kinetic energy, heating the interstellar medium (ISM) and eventually blowing the gas out of the galaxy. The supernova wind also disrupts molecular clouds through shock waves (Fig. 7.5).

7.5.2.2 Properties of dwarf galaxies in the context of galactic wind

The condition for kinetic galactic wind can be evaluated by comparing the energy injected from supernovae (SNe) and the galaxy's gravitational potential. Since supernovae originate from massive stars, the timescale is short (10^{6-7} yr). Thus, the supernova rate can be safely assumed to be proportional to the star formation rate (SFR). The *accumulated* input energy from SNe into the interstellar medium (ISM) is proportional to the number of supernovae that have occurred in a galaxy, which in turn is proportional to the total stellar mass formed. The binding energy of baryons within the dark halo's potential well is given by

$$E_{\text{bind}} \propto \frac{(b\mathcal{M})\mathcal{M}}{R} \propto \frac{\mathcal{M}_*}{s}\frac{\mathcal{M}}{R}\,, \tag{7.141}$$

where
$\mathcal{M}$ dynamical mass
$\mathcal{M}_*$ stellar mass
R radius of dark halo
b baryonic mass fraction relative to $\mathcal{M}$
s stellar mass fraction relative to the total baryonic mass.
Assuming a constant mass density throughout the galaxy, we have

$$E_{\text{bind}} \propto \left(\frac{\mathcal{M}_*}{s}\right)^{\frac{5}{3}}. \tag{7.142}$$

As discussed above, the total accumulated energy from SNe is

$$E_{\rm SN} \propto \mathscr{M}_* .$$

(7.143)

The critical condition for the energy from SNe to exceed the binding energy of the ISM, leading to gas expulsion from the galaxy, is evaluated as

$$E_{\rm SN} \propto \mathscr{M}_* \propto \left(\frac{\mathscr{M}_*}{s} \right)^{\frac{5}{3}} ,$$

(7.144)

which leads to

$$s \propto \mathscr{M}_*^{\frac{2}{5}} .$$

(7.145)

From these relations, the surface brightness I is expressed as

$$I \propto \frac{\mathscr{M}_*}{R^2} \propto \mathscr{M}_*^{\frac{3}{5}} .$$

(7.146)

Thus, as $\mathscr{M}_*$ becomes smaller, both surface brightness and luminosity decrease.

As discussed in Chapter 4, galaxies with higher gas mass fractions tend to have lower metallicity. This implies that low-luminosity galaxies are more metal-poor. When the stellar mass fraction s is small, we can approximate the metallicity using eq. (4.70) as

$$\begin{aligned}
Z(t) &= \bar{y}\ln\frac{\mathscr{M}_{\rm ISM}(0)}{\mathscr{M}_{\rm ISM}(t)} = \bar{y}\ln\frac{b\mathscr{M}}{\mathscr{M}_{\rm ISM}(t)} \\
&= \bar{y}\ln\frac{\mathscr{M}_* + \mathscr{M}_{\rm ISM}}{\mathscr{M}_{\rm ISM}(t)} \\
&\simeq \bar{y}\frac{\mathscr{M}_*}{\mathscr{M}_{\rm ISM}} \propto s .
\end{aligned}$$

(7.147)

Thus,

$$Z \propto s \propto \mathscr{M}_*^{\frac{2}{5}} \propto L^{\frac{2}{5}} ,$$

(7.148)

which successfully reproduces the observed relations.

Section II

Statistical Approach

8 Basics of Statistics

8.1 PROBABILITY AND STATISTICS

We first define the basic concepts of classical statistical inference. Let us consider the probability density function (often abbreviated as PDF) $f(X)$. The distribution function (DF) $F(x)$ is defined as

$$F(x) \equiv \int_{-\infty}^{x} f(x)\,\mathrm{d}x\,. \tag{8.1}$$

Here we should note that the terminology "distribution function" is used for the PDF in many fields of natural science including physics. Also often a DF is called "cumulative distribution function (CDF)". Though in most cases the risk of confusion is not high, we should keep this issue in ind.

More rigorously, we can first define the DF $F(x)$ as a certain probability measure, which then implies that the PDF $f(x)$ is defined as the Radon–Nikodym derivative of $F(x)$, provided it exists. For detailed discussions on measure theory and real analysis, see, e.g., Rudin (1987). Here, note that in the field of probability theory and mathematical statistics, a stochastic variable is denoted by a capital letter, such as X. When we refer to a specific realization of the variable, we use a lowercase letter, e.g., x_i. We often use the notation

$$f(x)\,\mathrm{d}x = \mathrm{d}F(x)\,, \tag{8.2}$$

which is more precise from a measure-theoretic perspective. The subscript i represents the i-th realization.

8.1.1 MEAN, DISPERSION, AND MOMENTS

We now define the sample mean as

$$\bar{X} \equiv \frac{1}{N}\sum_{i=1}^{N} x_i\,, \tag{8.3}$$

where N is the sample size. Next, we introduce the expectation value of a certain statistical property $Q(X)$ under the PDF $f(X)$ as

$$\mathbb{E}[Q(X)] \equiv \int Q(x)f(x)\,\mathrm{d}x = \int Q(x)\,\mathrm{d}F(x)\,. \tag{8.4}$$

We also define the population mean μ as

$$\mu \equiv \int x f(x)\,\mathrm{d}x = \int x\,\mathrm{d}F(x) = \mathbb{E}[X]\,. \tag{8.5}$$

DOI: 10.1201/9781003104315-8

Another commonly used measure for describing the probability distribution or frequency distribution is the dispersion

$$\mathbb{E}[(x-\mu)^2] \equiv \int (x-\mu)^2 \, dF(x) \, . \tag{8.6}$$

In general, we can define the n-th moment around $x = \mu'$ of the PDF $f(X)$ as

$$\mu_n \equiv \int (x-\mu')^n \, dF(x) \, . \tag{8.7}$$

In particular, we define the n-th central moment as

$$\mu_n \equiv \int (x-\mu)^n \, dF(x) \, . \tag{8.8}$$

8.1.2   LIMIT THEOREMS

In this *Thesis*, we rely on two important limit theorems. Other limit theorems and related topics are discussed extensively in e.g., Gnedenko & Korolev (1996).

Theorem 1. The law of large numbers *Let $X_1,\ldots,X_N$ be independent random variables, each having the same mean value*

$$\mathbb{E}[X_1] = \cdots = \mathbb{E}[X_N] = m \, , \tag{8.9}$$

and suppose there exists $^{\exists}\sigma^2 > 0$ such that

$$\mathbb{E}\left[(X_i - m_i)^2\right] \leq \sigma^2 \quad \text{for} \quad i = 1,\ldots,N \, . \tag{8.10}$$

Then for the random variable

$$\Xi(X) \equiv \frac{1}{N}\sum_{i=1}^{N} X_i \, , \tag{8.11}$$

we have

$$\mathbb{E}\left[\Xi(X)\right] = m \, , \tag{8.12}$$

as $N \longrightarrow \infty$.

The proof of this famous theorem can be found in many probability theory textbooks, so we omit it here (see e.g., Shiryaev, 1996).

Theorem 2. Multivariate central limit theorem
Let a_n^k $(n = 1,\ldots,N)$ be a set of sample values of a vector random variable A^k, which has a mean

$$\mathbb{E}_A\left[A^k\right] = m^k \, , \tag{8.13}$$

($m^k < \infty$) and a dispersion

$$\mathbb{E}_A\left[(A^k - m^k)(A^l - m^l)\right] = \sigma^{kl} \tag{8.14}$$

for ($k = 1, \ldots, K$). We define Z^k as a vector random variable whose sample value z^k is

$$z^k \equiv \sqrt{N}\left\{\frac{1}{N}\sum_{n=1}^{N} a_n^k - m^k\right\}. \tag{8.15}$$

Then, as $N \longrightarrow \infty$, Z^k becomes a Gaussian vector random variable with

$$\mathbb{E}_Z\left[Z^k\right] = 0, \tag{8.16}$$

and

$$\mathbb{E}_Z\left[Z^k Z^l\right] = \sigma^{kl}. \tag{8.17}$$

The proof of this theorem can also be found in many textbooks (e.g., Shiryaev 1996).

8.1.3 ESTIMATION AND HYPOTHESIS TESTING

Consider the situation where we have a set of observational data. Statistical inference is a mathematical method used to construct a probability distribution model and quantitatively assess that model in relation to the data. There are two broad categories in statistical inference: one is statistical estimation, and the other is hypothesis testing. Suppose we have a family of probability density functions (pdfs) with certain parameters. Now, the observations are random variables, and any function of the observations will also be a random variable. A function of the observations alone is called a statistic. If we use a statistic to estimate a parameter θ, it may at times differ significantly from the true value of θ. We must, therefore, be satisfied with formulating a rule that yields good results "in the long run" or "on average," or that has "a high probability of success"—phrases that reflect the fundamental reality that our method of estimation generates a distribution of estimates, which we assess based on the properties of that distribution.

To clarify our ideas, it helps to distinguish between the method or rule of estimation, which we call an *estimator*, and the value it produces in a particular case or realization, the *estimate*. Our objective is not merely to find estimates, but rather to find good estimators. We present the criteria for judging good estimators as follows:

Consistent estimators It is desirable for an estimator to have increasing precision as more data is gathered. If the variance of the sampling distribution of an estimator decreases with increasing sample size N, it is important that the central value should converge to θ. An estimator t_N, computed from a sample of N

values, is called *a consistent estimator* of θ if, for any positive ε and η, there exists some M such that the probability that

$$|t_N - \theta| < \varepsilon \tag{8.18}$$

is greater than $1 - \eta$ for all $N > M$.

Unbiased estimators Consider the sampling distribution of an estimator $t(X)$. Among the class of consistent estimators, we may prefer those where θ equals the central value, not only for large samples but for all sample sizes. If we require that for all N and θ, the mean value of t is θ, i.e.,

$$\mathbb{E}[t(X)] = \theta \,, \tag{8.19}$$

then we call $t(X)$ *an unbiased estimator* of θ. We also define the bias of the estimator as

$$\mathrm{Bias}(\theta) \equiv \mathbb{E}[t(X)] - \theta \,. \tag{8.20}$$

Minimum Variance It is natural to use the sampling variance of an estimator as a criterion for its acceptability. The Cramér–Rao inequality (see, e.g., Stuart et al., 1999) provides the fundamental lower bound for the variance of an estimator, $\mathbb{V}[t(X)]$, given by

$$\mathbb{V}[t(X)] \geq -\frac{1}{\mathbb{E}\left[\dfrac{\partial^2 \ln \mathscr{L}}{\partial \theta^2}\right]}, \tag{8.21}$$

where $\mathscr{L}$ is the likelihood function, which we will introduce in the next subsection.

Instead of seeking the best estimator for an unknown parameter, we often face the problem of deciding whether a predesignated value is acceptable in light of the observations. This is known as hypothesis testing. In a sense, the problem of testing is more fundamental than that of estimation. We compare the observed differences between two samples with what might be expected under the hypothesis that there is no true difference, only random sampling variation. If this hypothesis is not supported, we may proceed to the second step of estimating the magnitude of the difference. If the distribution underlying the observations is assumed to take a certain form and the hypothesis concerns only the value of one or both of its parameters, the hypothesis is called *parametric*. On the other hand, if no parameter values are specified in the hypothesis, it is referred to as *nonparametric*.

8.1.4 MAXIMUM LIKELIHOOD ESTIMATION

We limit our discussion to samples of N independent observations drawn from the same pdf $f(X)$. The joint probability of the observations, regarded as a function of a single unknown parameter θ, is called the likelihood function of the sample and is written as

$$\mathscr{L}(x;\theta) = \prod_{i=1}^{N} f(x_i;\theta) \, . \tag{8.22}$$

The principle of maximum likelihood states that, when faced with a choice of hypotheses, we choose the one that maximizes $\mathscr{L}$, i.e.,

$$\mathscr{L}(X;\hat{\theta}) = \max_{\theta \in \Theta} \mathscr{L}(X;\theta) \, . \tag{8.23}$$

To find the best parameter estimate $\hat{\theta}$, we typically solve the likelihood equation

$$\frac{\partial \ln \mathscr{L}}{\partial \theta} = \sum_{i=1}^{N} \frac{\partial \ln p(x_i;\theta)}{\partial \theta} = 0 \, . \tag{8.24}$$

The maximum likelihood estimator is consistent and, although not generally unbiased, it is asymptotically unbiased. Additionally, the maximum likelihood estimator is asymptotically normally distributed, with variance determined by the right-hand side of inequality (8.21). Because of these ideal properties, the maximum likelihood estimator is regarded as the best available method. The mathematical structure of maximum likelihood estimation is well-understood in the context of information theory. We explore this aspect further in the following sections.

8.2   INFORMATION THEORY AND STATISTICS

8.2.1   INFORMATION ENTROPY

Here, we consider 'information entropy'. The concept of information is closely related to probability. If we can expect an event to occur with certainty, we say the probability of the event is $p \sim 1$. On the contrary, if the event is very unlikely, $p \sim 0$. The self-information is defined as

$$I(p) = -\ln p \, , \tag{8.25}$$

where $\ln p = \log_e p$.

Next, we discuss the case of a (discrete) probability distribution $\{f_i\}_{i=1,\cdots,N}$. The expectation value of $I(f_i)$ is

$$S \equiv \mathbb{E}[I] = -\sum_{i=1}^{N} f_i \ln f_i \, . \tag{8.26}$$

This is called information entropy. A precise meaning of S is discussed in works such as Watanabe (1969).

Along the same lines, when we have two different probability distributions $\{f_i\}_{i=1,\cdots,N}$ and $\{g_i\}_{i=1,\cdots,N}$, we can construct the following quantity

$$D_{\mathrm{KL}}(f,g) \equiv -\sum_{i=1}^{N} f_i \ln \frac{f_i}{g_i} \, . \tag{8.27}$$

This is called Kullback–Leibler information, or more comprehensively, the relative entropy between $\{f_i\}_{i=1,\cdots,N}$ and $\{g_i\}_{i=1,\cdots,N}$ (Kullback & Leibler, 1951).

To understand the meaning of $D_{\mathrm{KL}}(f,g)$, let us define $p_i = p_i(\theta_1,\cdots,\theta_K)$ ($i = 1,\cdots,N$), where $(\theta_1,\cdots,\theta_K)$ are the parameters on which the model depends. Then, let

$$\begin{cases} f_i = p_i(\theta_1^0,\cdots,\theta_K^0)\,, \\[2mm] g_i = p_i(\theta_1^0 + \mathrm{d}\theta_1,\cdots,\theta_K^0 + \mathrm{d}\theta_K)\,. \end{cases} \tag{8.28}$$

Then $\ln g_i$ can be written as

$$\ln g_i \simeq \ln f_i + \sum_{k=1}^{K} \left(\frac{\partial \ln p_i}{\partial \theta_k}\right)_{\theta=\theta^0} \mathrm{d}\theta_k + \frac{1}{2}\sum_{k=1}^{K}\sum_{l=1}^{K}\left(\frac{\partial^2 \ln p_i}{\partial \theta_k \partial \theta_l}\right)_{\theta=\theta^0}\mathrm{d}\theta_k\mathrm{d}\theta_l\,, \tag{8.29}$$

where the subscript $\theta = \theta^0$ means $\theta_i = \theta_i^0$ for $i = 1,\cdots,N$.

Therefore, we obtain

$$\sum_{i=1}^{N} f_i \ln g_i \simeq \sum_{i=1}^{N} f_i \ln f_i + \sum_{i=1}^{N}\sum_{k=1}^{K} f_i \frac{1}{f_i}\left(\frac{\partial p_i}{\partial \theta_k}\right)_{\theta=\theta^0}\mathrm{d}\theta_k$$
$$+ \frac{1}{2}\sum_{i=1}^{N}\sum_{k=1}^{K}\sum_{l=1}^{K} f_i \left(\frac{\partial^2 \ln p_i}{\partial \theta_k \partial \theta_l}\right)_{\theta=\theta^0}\mathrm{d}\theta_k\mathrm{d}\theta_l\,. \tag{8.30}$$

Since $\sum_{i=1}^{N} p_i(\theta) = 1$, the second term on the right-hand side of equation (8.30) becomes

$$\sum_{i=1}^{N}\sum_{k=1}^{K}\left(\frac{\partial p_i}{\partial \theta_k}\right)_{\theta=\theta^0}\mathrm{d}\theta_k = \sum_{k=1}^{K}\left\{\frac{\partial}{\partial \theta_k}\sum_{i=1}^{N} p_i\right\}_{\theta=\theta^0}\mathrm{d}\theta_k = 0\,. \tag{8.31}$$

Now, let us consider the third term. We have

$$\frac{\partial^2 \ln p_i}{\partial \theta_k \partial \theta_l} = \frac{\partial}{\partial \theta_l}\left(\frac{1}{p_i}\frac{\partial p_i}{\partial \theta_k}\right)$$
$$= \frac{1}{p_i}\frac{\partial^2 p_i}{\partial \theta_k \partial \theta_l} - \frac{1}{p_i^2}\frac{\partial p_i}{\partial \theta_k}\frac{\partial p_i}{\partial \theta_l}$$
$$= \frac{1}{p_i}\frac{\partial^2 p_i}{\partial \theta_k \partial \theta_l} - \frac{\partial \ln p_i}{\partial \theta_k}\frac{\partial \ln p_i}{\partial \theta_l}\,, \tag{8.32}$$

and therefore

$$\sum_{i=1}^{N}\sum_{k=1}^{K}\sum_{l=1}^{K} f_i \left(\frac{\partial^2 \ln p_i}{\partial \theta_k \partial \theta_l}\right)_{\theta=\theta^0}\mathrm{d}\theta_k\mathrm{d}\theta_l$$
$$= \sum_{i=1}^{N}\sum_{k=1}^{K}\sum_{l=1}^{K}\left(\frac{\partial^2 p_i}{\partial \theta_k \partial \theta_l}\right)_{\theta=\theta^0} - f_i\left(\frac{\partial \ln p_i}{\partial \theta_k}\frac{\partial \ln p_i}{\partial \theta_l}\right)_{\theta=\theta^0}\mathrm{d}\theta_k\mathrm{d}\theta_l$$
$$= -\sum_{i=1}^{N}\sum_{k=1}^{K}\sum_{l=1}^{K} f_i\left(\frac{\partial \ln p_i}{\partial \theta_k}\frac{\partial \ln p_i}{\partial \theta_l}\right)_{\theta=\theta^0}\mathrm{d}\theta_k\mathrm{d}\theta_l\,. \tag{8.33}$$

We used eq. (8.31) in the second step of the above calculation.

Making the substitution, we have

$$D_{KL}(f,g) = \frac{1}{2} \sum_{i=1}^{N} \sum_{k=1}^{K} \sum_{l=1}^{K} f_i \left(\frac{\partial \ln p_i}{\partial \theta_k} \frac{\partial \ln p_i}{\partial \theta_l} \right)_{\theta=\theta^0} d\theta_k d\theta_l$$

$$\equiv \frac{1}{2} \sum_{k=1}^{K} \sum_{l=1}^{K} I(\theta^0)_{kl} d\theta_k d\theta_l , \tag{8.34}$$

where

$$I_{kl} = \sum_{i=1}^{N} p_i \left(\frac{\partial \ln p_i}{\partial \theta_k} \frac{\partial \ln p_i}{\partial \theta_l} \right) . \tag{8.35}$$

The matrix I_{kl} is called Fisher's information matrix. If we regard I_{kl} as a "metric" of the parameter space of θ_k and θ_l, we can treat $D_{KL}(f,g)$ as the distance between the two probability distributions $\{f_i\}_{i=1,\cdots,N}$ and $\{g_i\}_{i=1,\cdots,N}$, just as we do in differential geometry.

8.2.2 MAXIMUM LIKELIHOOD ESTIMATION IN THE CONTEXT OF INFORMATION THEORY

Here, we will discuss the maximum likelihood method in the context of relative entropy $D_{KL}(f,g)$. Maximum likelihood is the method used to estimate the optimal model for *a given set of data*. Let $x_1,\cdots,x_N$ represent a realization of a random variable X, which follows an unknown probability distribution $f(X)$. We define a model probability distribution $g(x; \theta_1,\cdots,\theta_K)$, which depends on K parameters $\{\theta_k\}_{k=1,\cdots,K}$. The likelihood function $\mathscr{L}$ is then defined by

$$\mathscr{L}(\theta_1,\cdots,\theta_K|x_1,\cdots,x_n) = \prod_{n=1}^{N} g(x_n; \theta) , \tag{8.36}$$

and the logarithmic likelihood is

$$\ln \mathscr{L} \equiv \sum_{n=1}^{N} \ln g(x_n; \theta) . \tag{8.37}$$

Performing maximum likelihood estimation means finding the parameter set $\theta = \theta_1,\cdots,\theta_K$ that defines the most suitable model $g(x; \theta)$ for the unknown *true* probability distribution $f(X)$. If we know the true $f(X)$, using the relative entropy from eq. (8.27), we have

$$D_{KL}(f,g) = \sum_{n=1}^{N} f(x_n) \ln \frac{f(x_n)}{g(x_n; \theta)}$$

$$= \mathbb{E} \left[\ln \frac{f(X)}{g(X; \theta)} \right]$$

$$= \mathbb{E}[\ln f(X)] - \mathbb{E}[\ln g(X; \theta)] . \tag{8.38}$$

However, since we do not know $f(X)$, we cannot obtain $\mathbb{E}[\,\cdot\,]$ directly. Therefore, we use the law of large numbers. If the size of the dataset N is sufficiently large, we have

$$\frac{1}{N}\ln\mathscr{L} = \frac{1}{N}\sum_{i=1}^{N}\ln g(x_n;\theta) \simeq \mathbb{E}[\ln g(X;\theta)] \ . \tag{8.39}$$

This leads to

$$D_{\mathrm{KL}}(f,g) = \mathbb{E}[\ln f(X)] - \frac{1}{N}\sum_{i=1}^{N}\ln g(x_n;\theta) \ . \tag{8.40}$$

The last term can be obtained without knowing the true $p(X)$. Since $\mathbb{E}[\ln f(X)]$ is independent of $\theta = \theta_1,\cdots,\theta_K$, what needs to be maximized is $\ln\mathscr{L} = \sum_{i=1}^{N}\ln g(x_n;\theta)$.
  Consequently, we obtain the likelihood equation

$$\frac{\partial}{\partial\theta_k}\ln\mathscr{L}\bigg|_{\theta=\hat{\theta}} = \frac{\partial}{\partial\theta_k}\left\{\sum_{n=1}^{N}\ln g(x_n;\theta)\right\}_{\theta=\hat{\theta}} = 0 \tag{8.41}$$

for $k = 1,\cdots,K$, where we define $\hat{\theta}$ to satisfy

$$\max_{\theta\in\Theta}(\ln\mathscr{L}) = \ln\mathscr{L}(\hat{\theta}|x_1,\cdots,x_n) \ . \tag{8.42}$$

Here, Θ denotes the family of parameters θ. The solution $\hat{\theta}$ of eq. (8.41) is called the maximum likelihood estimator.

8.2.3 ROLE OF THE INFORMATION MATRIX

The discussions made thus far cover the basics of maximum likelihood estimation among models of the same type, i.e., models with the same assumed form and the same number of parameters. To include the differences between assumed models and compare their goodness, we evaluate the error of the logarithmic likelihood equation [eq. (8.41)]

$$\Delta(\hat{\theta}) \equiv \mathbb{E}[\ln g(X;\theta)]|_{\theta=\hat{\theta}} - \frac{1}{N}\sum_{n=1}^{N}\ln g(x_n;\theta)|_{\theta=\hat{\theta}} \ . \tag{8.43}$$

We evaluate the first term on the right-hand side of eq. (8.43).
  Let $\theta^0 = \theta_1^0,\cdots,\theta_K^0 \in \Theta$ be the estimator that maximizes $\mathbb{E}[\ln g(X;\theta)]$, i.e.,

$$\frac{\partial}{\partial\theta_k}\mathbb{E}[\ln g(X;\theta)]\bigg|_{\theta=\theta^0} = 0 \tag{8.44}$$

for $k = 1, \cdots, K$. Then we have

$$
\mathbb{E}\left[\ln g(X; \theta)\right]|_{\theta=\hat{\theta}} \simeq \mathbb{E}\left[\ln g(X; \theta)\right]|_{\theta=\theta^0} + \sum_{k=1}^{K} \frac{\partial}{\partial \theta_k} \mathbb{E}\left[\ln g(X; \theta)\right]\Big|_{\theta=\theta^0} (\hat{\theta}_k - \theta_k^0)
$$
$$
+ \frac{1}{2} \sum_{k=1}^{K} \sum_{l=1}^{K} (\hat{\theta}_k - \theta_k^0) \left\{ \frac{\partial^2}{\partial \theta_k \partial \theta_l} \mathbb{E}\left[\ln g(X; \theta)\right] \right\}_{\theta=\theta^0} (\hat{\theta}_l - \theta_l^0) .
$$
$$(8.45)$$

The second term vanishes due to eq. (8.44), and it follows that

$$
\mathbb{E}\left[\ln g(X; \theta)\right]|_{\theta=\hat{\theta}} \simeq \mathbb{E}\left[\ln g(X; \theta)\right]|_{\theta=\theta^0}
$$
$$
+ \frac{1}{2} \sum_{k=1}^{K} \sum_{l=1}^{K} (\hat{\theta}_k - \theta_k^0) \left\{ \frac{\partial^2}{\partial \theta_k \partial \theta_l} \mathbb{E}\left[\ln g(X; \theta)\right] \right\}_{\theta=\theta^0} (\hat{\theta}_l - \theta_l^0) .
$$
$$(8.46)$$

Next, for the second term of eq. (8.43), we have

$$
\sum_{n=1}^{N} \ln g(x_n; \theta)\Big|_{\theta=\theta^0} \simeq \sum_{n=1}^{N} \ln g(x_n; \theta)\Big|_{\theta=\hat{\theta}} + \sum_{k=1}^{K} \frac{\partial}{\partial \theta_k} \sum_{n=1}^{N} \ln g(x_n; \theta)\Big|_{\theta=\hat{\theta}} (\theta_k^0 - \hat{\theta}_k)
$$
$$
+ \frac{1}{2} \sum_{k=1}^{K} \sum_{l=1}^{K} (\theta_k^0 - \hat{\theta}_k) \left\{ \frac{\partial^2}{\partial \theta_k \partial \theta_l} \sum_{n=1}^{N} \ln g(x_n; \theta) \right\}_{\theta=\hat{\theta}} (\theta_l^0 - \hat{\theta}_l) .
$$
$$(8.47)$$

From eq. (8.41), we obtain

$$
\sum_{n=1}^{N} \ln g(x_n; \theta)\Big|_{\theta=\hat{\theta}} \simeq \sum_{n=1}^{N} \ln g(x_n; \theta)\Big|_{\theta=\theta^0}
$$
$$
- \frac{1}{2} \sum_{k=1}^{K} \sum_{l=1}^{K} (\hat{\theta}_k - \theta_k^0) \left\{ \frac{\partial^2}{\partial \theta_k \partial \theta_l} \sum_{n=1}^{N} \ln g(x_n; \theta) \right\}_{\theta=\hat{\theta}} (\hat{\theta}_l - \theta_l^0) .
$$
$$(8.48)$$

Now, applying theorem 1, we get

$$
\frac{1}{N} \sum_{n=1}^{N} \ln g(x_n; \theta)\Big|_{\theta=\theta^0} \simeq \mathbb{E}\left[\ln g(X; \theta)\right]|_{\theta=\theta^0}
$$
$$(8.49)$$

and

$$
\frac{1}{N} \sum_{n=1}^{N} \left\{ \frac{\partial^2}{\partial \theta_k \partial \theta_l} \ln g(x_n; \theta) \right\}_{\theta=\hat{\theta}} \simeq \frac{1}{N} \sum_{n=1}^{N} \left\{ \frac{\partial^2}{\partial \theta_k \partial \theta_l} \ln g(x_n; \theta) \right\}_{\theta=\theta^0}
$$
$$
\simeq \mathbb{E}\left[\frac{\partial^2}{\partial \theta_k \partial \theta_l} \ln g(X; \theta) \right]\Big|_{\theta=\theta^0} .
$$
$$(8.50)$$

The first step of eq. (8.50) holds to the first order of $(\hat{\theta}_k - \theta_k^0)$.

Finally, using eqs. (8.33), (8.45), (8.48), (8.49), and (8.50), we obtain the important result

$$
\Delta(\theta^0) \simeq \sum_{k=1}^{N}\sum_{l=1}^{N}(\hat{\theta}_k - \theta_k^0)\mathbb{E}\left[\frac{\partial^2}{\partial\theta_k\partial\theta_l}\ln g(X;\theta)\right]_{\theta=\theta^0}(\hat{\theta}_l - \theta_l^0)
$$

$$
\simeq -\sum_{k=1}^{N}\sum_{l=1}^{N}(\hat{\theta}_k - \theta_k^0)I_{kl}(\hat{\theta}_l - \theta_l^0) . \tag{8.51}
$$

Again, I_{kl} is the information matrix. Thus, we can evaluate the error of the likelihood function using I_{kl}.

8.3   AKAIKE'S INFORMATION CRITERION

In this section, we discuss Akaike's information criterion (AIC: Akaike, 1974a). To the first order of θ, the likelihood equation (8.41) is expressed as

$$
\sum_{n=1}^{N}\left\{\frac{\partial}{\partial\theta_k}\ln g(x_n;\theta)\right\}_{\theta=\theta^0} + \sum_{n=1}^{N}\sum_{l=1}^{K}\left\{\frac{\partial^2}{\partial\theta_k\partial\theta_l}\ln g(x_n;\theta)\right\}_{\theta=\theta^0}(\hat{\theta}_l - \theta_l^0) = 0 .
$$

$$\tag{8.52}$$

To evaluate this equation, we use the multivariate central limit theorem. By the definition of θ^0 (eq. 8.44), we have

$$
\mathbb{E}\left[\frac{\partial}{\partial\theta_k}\ln g(X;\theta)\right]\bigg|_{\theta=\theta^0} = \left\{\frac{\partial}{\partial\theta_k}\mathbb{E}[\ln g(X;\theta)]\right\}_{\theta=\theta^0} \equiv \mathcal{M}^k = 0 . \tag{8.53}
$$

Therefore, with eq. (8.35),

$$
\mathbb{E}\left[\left(\frac{\partial}{\partial\theta_k}\ln g(X;\theta) - \mathcal{M}^k\right)\left(\frac{\partial}{\partial\theta_l}\ln g(X;\theta) - \mathcal{M}^l\right)\right]\bigg|_{\theta=\theta^0}
$$

$$
= \mathbb{E}\left[\left(\frac{\partial}{\partial\theta_k}\ln g(X;\theta)\right)\left(\frac{\partial}{\partial\theta_l}\ln g(X;\theta)\right)\right]\bigg|_{\theta=\theta^0} = I_{kl} . \tag{8.54}
$$

Using the result of theorem 2, we have a vector random variable Z^k, which is Gaussian with:

$$
\mathbb{E}_Z[Z^k] = 0 , \; \mathbb{E}_Z[Z^kZ^l] = I_{kl} , \tag{8.55}
$$

and whose sample value z^k is

$$
z^k = \frac{1}{\sqrt{N}}\sum_{n=1}^{N}\left[\frac{\partial}{\partial\theta_k}\ln g(x_n;\theta)\right]_{\theta=\theta^0} . \tag{8.56}
$$

We then apply Theorem 1 to the second term of eq. (8.52), and it follows that

$$\frac{1}{N}\sum_{n=1}^{N}\left[\frac{\partial^2}{\partial\theta_k\partial\theta_l}\ln g(x_n;\theta)\right]_{\theta=\theta^0} \simeq \mathbb{E}\left[\frac{\partial^2}{\partial\theta_k\partial\theta_l}\ln g(X;\theta)\right]_{\theta=\theta^0} = -I_{kl}\ . \quad (8.57)$$

Substituting eqs. (8.56) and (8.57) into eq. (8.52) gives

$$z^k = \sqrt{N}\sum_{l=1}^{K}I_{kl}(\hat{\theta}_l - \theta_l^0)\ . \quad (8.58)$$

If I_{kl} is non-singular[1], the inverse matrix I_{kl}^{-1} is defined as

$$\sum_{k=1}^{K}I_{uk}I_{kv}^{-1} = \delta_{uv} \quad (8.59)$$

where δ_{uv} is the Kronecker delta, and it leads to

$$(\hat{\theta}_k - \theta_k^0) = \frac{1}{\sqrt{N}}\sum_{l=1}^{K}I_{kl}^{-1}z^l\ . \quad (8.60)$$

Hence, we can write

$$\Delta(\hat{\theta}) = -\sum_{k=1}^{K}\sum_{l=1}^{K}(\hat{\theta}_k - \theta_k^0)I_{kl}(\hat{\theta}_l - \theta_l^0) = -\frac{1}{N}\sum_{u=1}^{K}\sum_{v=1}^{K}I_{uv}^{-1}z^uz^v\ . \quad (8.61)$$

Now, using theorem 2 again, we obtain

$$\Delta(\hat{\theta}) \simeq -\frac{1}{N}\sum_{u=1}^{K}\sum_{v=1}^{K}I_{uv}^{-1}\mathbb{E}_Z[Z^uZ^v]$$

$$= -\frac{1}{N}\sum_{u=1}^{K}\sum_{v=1}^{K}I_{uv}^{-1}I_{uv} - \frac{K}{N}\ . \quad (8.62)$$

Thus, we reach the final important result

$$\mathbb{E}[\ln g(X;\theta)]|_{\theta=\hat{\theta}} \simeq \frac{1}{N}\sum_{n=1}^{N}\ln g(x_n;\theta)\bigg|_{\theta=\hat{\theta}} - \frac{K}{N}$$

$$= \frac{1}{N}\left(\ln\mathscr{L}(\hat{\theta}|x_1,\cdots,x_n) - K\right)\ . \quad (8.63)$$

Since, as eq. (8.38) shows, the distance between the true probability distribution $f(X)$ and model $g(X;\theta)$ is evaluated by $\mathbb{E}[\ln g(X;\theta)]$, we have to choose $\hat{\theta} \in \Theta$ to maximize the quantity $(\ln\mathscr{L} - K)$ for a given K. Though a larger K provides a larger $\ln\mathscr{L}$, it also reduces $(\ln\mathscr{L} - K)$ at the same time. Therefore, the optimal number of parameters K has to be chosen so as to maximize $(\ln\mathscr{L}(\hat{\theta}) - K)$. To make this more similar to the likelihood ratio statistic (e.g., Rao, 1965), the value $-2(\ln\mathscr{L}(\hat{\theta}) - K)$ has been considered. This is Akaike's information criterion (AIC). We now see that the optimal model is the one that minimizes AIC.

---

[1] We can always expect this by reparametrization.

9 Expectation–Maximization (EM) Algorithm

9.1 INTRODUCTION

9.1.1 PROBLEM OF INCOMPLETE DATA

Data are often incomplete. One may agree that this tendency is especially noticeable in astronomy. In such cases, the same methods as for complete data cannot be applied straightforwardly for statistical analysis.

We then first consider a familiar example: data on the heights of adult men and women. Suppose that we model each with a single normal distribution. If the gender of all height data are known, it is easy to determine the ratio of men and women, as well as the mean and variance for each gender. But what if some data are incomplete, with the gender unknown? This kind of issue is known as a typical problem of incomplete data. A simple way to deal with it is to treat the incomplete data as invalid and exclude them, essentially discarding it. Many people actually do this without much thought, and if the amount of incomplete data is small, this method may still yield reasonably accurate results. However, if there are a large amount of incomplete data, this approach becomes untenable. Furthermore, if data collection is costly, discarding incomplete data wastes valuable information. In extreme cases, when the gender of all data points is unknown, this method is simply not possible.

If an incomplete data point happens to be a rather small value within the overall distribution, it is likely to belong to the female group. Conversely, if it is a significantly large value, it is more likely to belong to the male group. But what if the value lies around the overall average, making it unclear whether it belongs to men or women? Indeed, this data point must belong to one of the two, but rather than forcibly assigning it to either gender, we consider the probability that the incomplete data belongs to either the male or female group. In other words, we count incomplete data as, for instance, 0.5 men and 0.5 women, and use it to estimate the distribution of both groups. Naturally, since not all incomplete data lies around the middle of the two distributions, it may be better to assume different proportions when counting.

Now assume some initial estimates for the mean and variance of the male and female distributions. There are various ways to make these initial assumptions, but they remain provisional distributions for the time being. Based on this, we assign fractional counts to the incomplete data for each gender. If there are multiple incomplete data points, we repeat this for each one. As a result, we might have somethinglike

DOI: 10.1201/9781003104315-9

53.3 men and 46.7 women in the dataset. From this weighted dataset, we calculate the updated mean and variance. This step updates the assumed distributions for men and women. If the distributions are updated, the fractional counts of incomplete data will also change. We then recalculate the distributions and repeat this process. Summarizing this procedure, once we have the initial distributions for men and women, we follow these steps:

1. Using the distributions for men and women, calculate the probability (posterior probability) that each incomplete data point belongs to each gender.
2. Use the incomplete data, weighted by these fractional counts, to recalculate the distributions for men and women.

Then, this two-fold process is repeated.

There are various methods for determining the initial distributions. The method outlined here is based on an intuitive approach. While it may seem reasonable, will the solution converge? Even if it does, will the estimated distributions approach the true distributions? Or more precisely, what properties will the estimated distributions have? These questions were theoretically addressed in the seminal paper by Dempster et al. (1977). In this paper, they proved that the above estimation method converges to the maximum likelihood estimate of the distribution. The procedure outlined above can be generalized further. This is known as the Expectation–Maximization (EM) algorithm.

9.1.2 SLOPE OF THE RESOLVED KENNICUTT–SCHMIDT LAW

We move on to a more specific astronomical problem. In general, Astronomical data often involves significant uncertainty due to the method of observations and their dependence on various unknown physical conditions. Consequently, even a simple linear regression problem can be far from straightforward. In this chapter, we introduce an example of applying the EM algorithm to regression problems in astronomy, specifically focusing on the problem of estimating the slope of the spatially resolved Kennicutt–Schmidt law.

9.1.2.1 Kennicutt–Schmidt law

Gas in galaxies cools and condenses by various mechanisms. As the temperature drops, the gas transitions from ionized to neutral phase, and eventually forming molecules. Since hydrogen is the most abundant element in the Universe, molecular clouds, which are clusters of these molecules, are primarily composed of hydrogen molecules. These hydrogen molecules are not only continuously formed but are also dissociated back into atoms by ultraviolet radiation from their surroundings. The boundary of a molecular cloud is defined where the formation and dissociation of hydrogen molecules reach equilibrium. In other words, ultraviolet radiation cannot penetrate inside this boundary. Hence, the interior of the molecular cloud can cool further, eventually forming extremely dense molecular cloud cores. Stars are born within these cores.

As introduced in Section 3.7.2, the power-law relationship between star formation rate and gas density

$$\mathrm{SFR} \propto \rho^n \tag{9.1}$$

is called the Schmidt law (Schmidt, 1959), while the relationship between the surface density of the star formation rate and the gas surface density

$$\Sigma_{\mathrm{SFR}} \propto \Sigma_{\mathrm{gas}}^N \tag{9.2}$$

is referred to as the Kennicutt–Schmidt (K-S) law (Kennicutt, 1989). Nowadays, as discussed in Sec. 3.7.2, the spatially resolved K-S law gathers attention to reveal its underlying physics (e.g., Bigiel et al., 2008).

9.1.2.2 Slope Estimation of the Kennicutt–Schmidt Law from Spatially Resolved Galaxy Dataset

Since a log-linear structure is expected for the relation between the gas surface density and SFR density, we adopt the following equation as a proper statistical model.

$$\log\left(\Sigma_{\mathrm{SFR}}\right) = N\log\left(\Sigma_{\mathrm{gas}}\right) + A \tag{9.3}$$

where A is an intercept and N is a slope of the K-S law. We stress that this formula plays the most fundamental role in the studies of the K-S law. However, as pointed out by many previous studies, the determination of the slope of the K-S law is tricky (e.g., Bigiel et al., 2008; Momose et al., 2010, 2013; Rahmani et al., 2016)). On the reduced galaxy images, noise dominates the measured quantities at the most faint regions. In general, the traditional least-square method tends to estimate the slope shallower than the true slope under the existence of noise, i.e., the more noisy the data is, the flatter the estimated slope becomes (Hoel, 1954). This issue has been considered by statisticians and astronomers and some specially-designed methods are known (e.g., Isobe et al., 1990), but it is difficult to claim which one yields the "true" estimate, because actually different methods generally estimate different statistical quantities.

Even worse, especially for the K-S law studies, the molecular gas surface density map is rather strongly affected by the noise due to the difficulty in the baseline determination. We often see that there is a clear linear structure on the plot, overlapped with another data sequence that is independent of the gas surface density on $\Sigma_{\mathrm{H_2}}$. In such a case, even if the true relation is (log-)linear, the slope estimation can be easily biased (Fig. 9.1). Various methods have been proposed to deal with this problem, some arestatistically-based, some are heuristic and/or ad-hoc. However, if

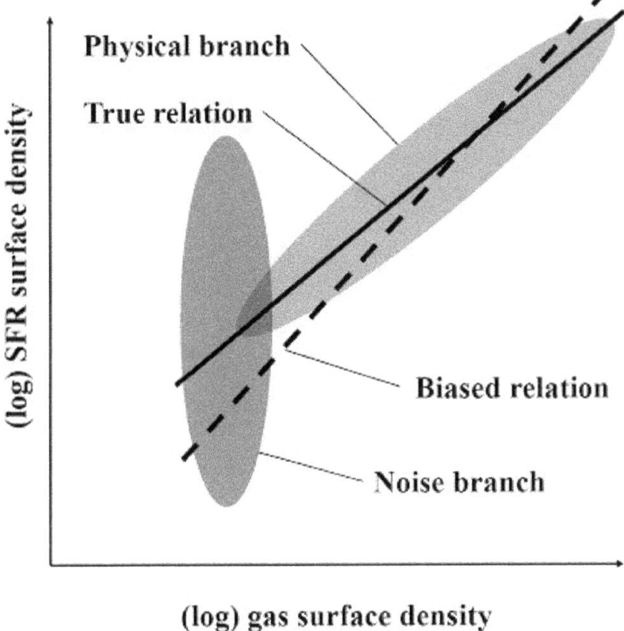

Figure 9.1 Schematic description of the relation between the molecular gas mass density and star formation rate density. Though the physically meaningful sequence of the data point cloud (physical branch) and the overlapping point cloud due to noise are observed.

there are some distinct components in the point data cloud produced from different mechanisms, as our data, it is very difficult to specify the true relation and estimate its slope, and there might be no consensus on which method is the most appropriate.

Here in this work, to address this issue, we estimated parameters with a Gaussian mixture model with the expectation–maximization (EM) algorithm. The Gaussian mixture model assumes that data are drawn from a sum of multiple Gaussian distributions, and is often used to deal with multiply-peaked histograms or data distributions has many peaks in higher dimension, as our K-S law. The EM algorithm judges which Gaussian distribution a data point belongs to at each step, and determines the solution that maximizes the likelihood.

9.2 EM ALGORITHM WITH GAUSSIAN MIXTURE MODEL

9.2.1 GENERAL FORMULATION

Here we introduce a general parameter estimation problem of a Gaussian mixture model with EM algorithm. The Gaussian mixture model fits a D-dimensional Gaussian distribution to each component of the data cloud. We denote a Gaussian distribution by $\mathcal{N}(\vec{x}|\vec{\mu}, \Sigma)$ with parameters, the mean $\vec{\mu} \in \mathbb{R}^D$ and the $(D \times D)$ variance–covariance matrix Σ. Then, we suppose that the data are drawn from a Gaussian mixture distribution

$$p\left(\vec{x}\,|\,\vec{\pi},\vec{\mu},\Sigma\right) = \sum_{k=1}^{K} \pi_k \mathcal{N}\left(\vec{x}\,|\,\vec{\mu}_k,\Sigma_k\right) , \tag{9.4}$$

where π_k is the mixing coefficient. This d-dimensional probability density function describes the situation when M data $\{x_1,\dots,x_M\}$ are divided into K clusters. In this case the log-likelihood function $\ln\mathcal{L}$ is read as

$$\ln\mathcal{L}\left(\vec{x}_1,\dots,\vec{x}_M\,|\,\vec{\pi},\vec{\mu},\Sigma\right) = \ln \prod_{m=1}^{M}\left[\sum_{k=1}^{K}\pi_k\mathcal{N}\left(\vec{x}_m\,|\,\vec{\mu}_k,\Sigma_k\right)\right]$$

$$= \sum_{m=1}^{M}\ln\left[\sum_{k=1}^{K}\pi_k\mathcal{N}\left(\vec{x}_m\,|\,\vec{\mu}_k,\Sigma_k\right)\right] . \tag{9.5}$$

We estimate the parameters by maximizing the log-likelihood function eq. (9.5).

To maximize $\mathcal{L}\left(\vec{x}_1,\dots,\vec{x}_m\,|\,\vec{\pi},\vec{\mu},\Sigma\right)$, the expectation-maximization (EM) algorithm is applied as follows.

1. Set initial values of $\vec{\pi}_{\text{init}}$, $\vec{\mu}_{\text{init}}$, and Σ_{init}.
2. Calculate the log-likelihood function $\mathcal{L}\left(\vec{x}_1,\dots,\vec{x}_m\,|\,\vec{\pi}_{\text{init}},\vec{\mu}_{\text{init}},\Sigma\right)_{\text{init}}$
3. Calculate a contribution function γ_{nk} (E-step)

$$\gamma(z_{nk}) = \frac{\pi_k\mathcal{N}\left(\vec{x}_m\,|\,\vec{\mu}_k,\Sigma_k\right)}{\displaystyle\sum_{j=1}^{k}\pi_j\mathcal{N}\left(\vec{x}_m\,|\,\vec{\mu}_j,\Sigma_j\right)} , \tag{9.6}$$

where z_{nk} is a latent variable.

4. Calculate maximum likelihood solutions of $\vec{\pi}$, $\vec{\mu}$, and Σ, which we denote as $\vec{\pi}_k^*$, $\vec{\mu}_k^*$, and Σ_k^* (M-step),

$$\vec{\mu}_k^* = \frac{1}{N_k}\sum_{m=1}^{N}\gamma(z_{nk})\vec{x}_m , \tag{9.7}$$

$$\Sigma_k^* = \frac{1}{N_k}\sum_{m=1}^{N}\gamma(z_{nk})\left(\vec{x}_m - \vec{\mu}_k\right)\left(\vec{x}_m - \vec{\mu}_k\right)^{\top} , \tag{9.8}$$

$$\pi_k^* = \frac{N_k}{N} = \frac{1}{N}\sum_{m=1}^{N}\gamma(z_{nk}) . \tag{9.9}$$

Derivation of eqs. (9.7)–(9.9) is shown in Sec. 9.2.2.

5. Repeat Step 3 and 4 until the difference of a log-likelihood function satisfies a convergence condition, i.e., for a pre-set number ε,

$$\mathcal{L}_{\text{new}} \equiv \mathcal{L}\left(\vec{x}_1,\dots,\vec{x}_m\,|\,\vec{\pi}^*,\vec{\mu}^*,\Sigma^*\right) \tag{9.10}$$

$$\mathcal{L}_{\text{new}} - \mathcal{L}_{\text{old}} < \varepsilon . \tag{9.11}$$

Here $\top$ stands for the transverse of a matrix.

9.2.2 DERIVATION OF THE LIKELIHOOD SOLUTIONS IN THE EM ALGORITHM

Here we derive the maximum likelihood solutions for $\vec{\mu}_k$, Σ_k, and π_k in eqs. (9.7)–(9.9). We start from $\vec{\mu}_k$.

$$
\begin{aligned}
\frac{\partial \ln \mathscr{L}}{\partial \vec{\mu}_k} &= \frac{\partial}{\partial \vec{\mu}_k} \sum_{m=1}^{N} \ln \left[\sum_{j=1}^{K} \pi_j \mathscr{N}\left(\vec{x}_m \middle| \vec{\mu}_j, \Sigma_j\right) \right] \\
&= \sum_{m=1}^{N} \frac{\pi_k \dfrac{\partial}{\partial \vec{\mu}_k} \mathscr{N}\left(\vec{x}_m \middle| \vec{\mu}_k, \Sigma_k\right)}{\displaystyle\sum_{j=1}^{K} \pi_j \mathscr{N}\left(\vec{x}_m \middle| \vec{\mu}_j, \Sigma_j\right)} \\
&= \sum_{m=1}^{N} \frac{\pi_k \mathscr{N}\left(\vec{x}_m \middle| \vec{\mu}_k, \Sigma_k\right) \Sigma_k^{-1}\left(\vec{x}_m - \vec{\mu}_k\right)}{\displaystyle\sum_{j=1}^{K} \pi_j \mathscr{N}\left(\vec{x}_m \middle| \vec{\mu}_j, \Sigma_j\right)} \\
&= \Sigma_k^{-1} \sum_{m=1}^{N} \frac{\pi_k \mathscr{N}\left(\vec{x}_m \middle| \vec{\mu}_k, \Sigma_k\right)}{\displaystyle\sum_{j=1}^{K} \pi_j \mathscr{N}\left(\vec{x}_m \middle| \vec{\mu}_j, \Sigma_j\right)} \left(\vec{x}_m - \vec{\mu}_k\right) \\
&= \Sigma_k^{-1} \sum_{m=1}^{N} \gamma(z_{nk})\left(\vec{x}_m - \vec{\mu}_k\right) .
\end{aligned}
\tag{9.12}
$$

To proceed from the second line to the third line, it follows the formula

$$
\begin{aligned}
\frac{\partial \mathscr{N}\left(\vec{\mu}, \Sigma\right)}{\partial \vec{\mu}} &= \mathscr{N}\left(\vec{\mu}, \Sigma\right) \frac{\partial}{\partial \vec{\mu}} \ln \mathscr{N}\left(\vec{\mu}, \Sigma\right) \\
&= \mathscr{N}\left(\vec{\mu}, \Sigma\right) \Sigma^{-1}\left(\vec{x} - \vec{\mu}\right).
\end{aligned}
$$

Equating eq. (9.12) to 0, we have

$$
\sum_{m=1}^{N} \gamma(z_{nk})\left(\vec{x}_m - \vec{\mu}_k\right) = 0 ,
\tag{9.13}
$$

since $\Sigma_k^{-1} \neq 0$. It leads to

$$
\sum_{m=1}^{N} \gamma(z_{nk}) \vec{x}_m = \vec{\mu}_k \sum_{m=1}^{N} \gamma(z_{nk}) ,
\tag{9.14}
$$

then we obtain the maximum-likelihood solution for $\vec{\mu}_k$.

$$\vec{\mu}_k = \frac{\sum_{m=1}^{N} \gamma(z_{nk})\vec{x}_m}{\sum_{m=1}^{N} \gamma(z_{nk})}$$

$$= \frac{1}{N_k} \sum_{m=1}^{N} \gamma(z_{nk})\vec{x}_m \,, \tag{9.15}$$

namely, eq. (9.7).

Similarly, we can solve for Σ_k.

$$\frac{\partial \ln \mathscr{L}}{\partial \Sigma_k} = \frac{\partial}{\partial \Sigma_k} \sum_{m=1}^{N} \ln \left[\sum_{j=1}^{K} \pi_j \mathscr{N}\left(\vec{x}_m | \vec{\mu}_j, \Sigma_j\right) \right]$$

$$= \sum_{m=1}^{N} \frac{\pi_k \frac{\partial}{\partial \Sigma_k} \mathscr{N}\left(\vec{x}_m | \vec{\mu}_k, \Sigma_k\right)}{\sum_{j=1}^{K} \pi_j \mathscr{N}\left(\vec{x}_m | \vec{\mu}_j, \Sigma_j\right)}$$

$$= \sum_{m=1}^{N} \frac{\pi_k \mathscr{N}\left(\vec{x}_m | \vec{\mu}_k, \Sigma_k\right)}{\sum_{j=1}^{K} \pi_j \mathscr{N}\left(\vec{x}_m | \vec{\mu}_j, \Sigma_j\right)} \left\{ -\frac{1}{2}\Sigma_k^{-1} + \frac{1}{2}\left[\Sigma_k^{-1}(\vec{x}_m - \vec{\mu}_k)(\vec{x}_m - \vec{\mu}_k)^{\top}\Sigma_k^{-1}\right] \right\}$$

$$= \sum_{m=1}^{N} \gamma(z_{nk}) \left\{ -\frac{1}{2}\Sigma_k^{-1} + \frac{1}{2}\left[\Sigma_k^{-1}(\vec{x}_m - \vec{\mu}_k)(\vec{x}_m - \vec{\mu}_k)^{\top}\Sigma_k^{-1}\right] \right\} \,. \tag{9.16}$$

This time we used the following formula.

$$\ln \mathscr{N}(\mu, \Sigma) = \frac{D}{2}\ln(2\pi) - \frac{\ln(\det \Sigma)}{2} - \frac{1}{2}(\vec{x} - \vec{\mu})^{\top}\Sigma^{-1}(\vec{x} - \vec{\mu})$$

$$= \frac{D}{2}\ln(2\pi) - \frac{\ln(\det \Sigma)}{2} - \frac{1}{2}\mathrm{tr}\left[\Sigma^{-1}(\vec{x} - \vec{\mu})(\vec{x} - \vec{\mu})^{\top}\right] \,. \tag{9.17}$$

Then we have

$$\frac{\partial \ln \mathscr{N}(\mu, \Sigma)}{\partial \Sigma} = -\frac{1}{2}\frac{\partial \ln(\det \Sigma)}{\partial \Sigma} - \frac{1}{2}\frac{\partial}{\partial \Sigma}\mathrm{tr}\left[\Sigma^{-1}(\vec{x} - \vec{\mu})(\vec{x} - \vec{\mu})^{\top}\right]$$

$$= -\frac{1}{2}\Sigma^{-1} - \frac{1}{2}\left[\Sigma^{-1}(\vec{x} - \vec{\mu})(\vec{x} - \vec{\mu})^{\top}\Sigma^{-1}\right] \,,$$

where the second line follows because Σ is symmetric. Equating eq. (9.16) to 0, we have

$$\sum_{m=1}^{N} \gamma(z_{nk}) \left\{ -\frac{1}{2}\Sigma_k^{-1} + \frac{1}{2}\left[\Sigma_k^{-1}(\vec{x}_m - \vec{\mu}_k)(\vec{x}_m - \vec{\mu}_k)^{\top}\Sigma_k^{-1}\right] \right\}$$

$$= -\frac{1}{2}\Sigma^{-1} \sum_{m=1}^{N} \gamma(z_{nk}) \left[1 - (\vec{x}_m - \vec{\mu}_k)(\vec{x}_m - \vec{\mu}_k)^{\top}\Sigma_k^{-1} \right] = 0 \,, \tag{9.18}$$

Where I is the identity matrix. Thus

$$\sum_{m=1}^{N} \gamma(z_{nk}) = \sum_{m=1}^{N} \gamma(z_{nk}) \left[(\vec{x}_m - \vec{\mu}_k)(\vec{x}_m - \vec{\mu}_k)^{\top} \Sigma_k^{-1} \right] , \tag{9.19}$$

then we obtain the maximum-likelihood solution for Σ_k.

$$\Sigma_k = \frac{\sum_{m=1}^{N} \gamma(z_{nk})(\vec{x}_m - \vec{\mu}_k)(\vec{x}_m - \vec{\mu}_k)^{\top}}{\sum_{m=1}^{N} \gamma(z_{nk})}$$

$$= \frac{1}{N_k} \sum_{m=1}^{N} \gamma(z_{nk})(\vec{x}_m - \vec{\mu}_k)(\vec{x}_m - \vec{\mu}_k)^{\top} , \tag{9.20}$$

namely, eq. (9.8).

Finally, we derive the maximum likelihood solution for pi_k As for π_k, we have a constraint

$$\sum_{k=1}^{K} \pi_k = 1 . \tag{9.21}$$

Then, in this case, we introduce a Lagrange multiplier λ_L and maximize

$$\ln \mathcal{M} \equiv \ln \mathcal{L} + \lambda_L \left(\sum_{k=1}^{K} \pi_k - 1 \right) . \tag{9.22}$$

We have

$$\frac{\partial \ln \mathcal{M}}{\partial \pi_k} = \frac{\partial}{\partial \pi_k} \sum_{m=1}^{N} \ln \left[\sum_{j=1}^{K} \pi_j \mathcal{N}(\vec{x}_m | \vec{\mu}_j, \Sigma_j) \right] + \lambda_L \left(\sum_{k=1}^{K} \pi_k - 1 \right)$$

$$= \sum_{m=1}^{N} \frac{\mathcal{N}(\vec{x}_m | \vec{\mu}_k, \Sigma_k)}{\sum_{j=1}^{K} \pi_j \mathcal{N}(\vec{x}_m | \vec{\mu}_j, \Sigma_j)} + \lambda_L$$

$$= \frac{1}{\pi_k} \sum_{m=1}^{N} \gamma(z_{nk}) + \lambda_L$$

$$= \frac{N_k}{\pi_k} + \lambda_L = 0 . \tag{9.23}$$

Thus

9.3 DATA

9.3.1 CO DATA

We used the dataset prepared by the COMING project. COMING is a project approved as a Legacy Project of the NRO from 2014–2017. With the 45 m radio telescope of NRO, COMING performed a mapping of $J = 1$–0 emission of ^{12}CO, ^{13}CO, and $C^{18}O$ for 147 nearby galaxies. This is the largest sample size of targets in molecular gas imaging observations to date (Sorai et al., 2019). This large number was achieved thanks to the two important technical improvements. A multi-beam receiver FOREST (FOur beam REceiver System on the 45-m Telescope: Minamidani et al. 2016) was developed to be equipped with NRO 45-m telescope. Also, a pointing strategy referred to as "on the fly" (OTF) observation was applied for the COMING project. This is the method to move the telescope to a target object continuously to obtain the image (Sawada et al., 2008). By these improvements, the observing time was so drastically reduced to perform such a large survey successfully. Details of the COMING project are presented in the project overview paper by Sorai et al. (2019). Data of COMING are publicly available[1].

The spatial resolution of COMING maps is $17''$ and the velocity resolution is $10\,\mathrm{km\,s^{-1}}$. For all the sample galaxies, an area within a region covering 70 % of D_{25} is observed. The radius D_{25} has been mainly used in optical astronomy, defined as a radius at which the B-band surface brightness per $1\,\square''$ is 25 mag. Namely, D_{25} represents the size of a galaxy determined by stars, and COMING covers most of the area within D_{25}.

We use the public integrated intensity map and integrated error map of COMING. The error map takes into account the error propagation of the spectrum and systematic uncertainty in the determination of the baseline (Sage, 1993). To estimate molecular gas mass surface density, Σ_{H2}, we use $^{12}\mathrm{CO}(J = 1 - 0)$ integrated intensities obtained by COMING.

9.3.2 DATA FOR SFR

Another important observational piece for the studies on the K-S law is the estimation of the SFR. A part of the energy emitted from OB stars is absorbed by dust and re-emitted at MIR–FIR via energy conservation. Then, it is natural to consider the combination of the "leaking" UV and re-emitted FIR luminosities and use it as an unbiased SFR estimator (5). This type of the hybrid estimator can be generally expressed as

$$\mathrm{SFR} = C(\mathrm{UV})L_{\mathrm{UV}} + C(\mathrm{IR})L_{\mathrm{IR}}\,, \qquad (9.24)$$

where symbolically L_{UV} and L_{IR} stand for a luminosity at UV and IR, $C(\mathrm{UV})$ and $C(\mathrm{IR})$ mean the conversion coefficient corresponding to UV and IR, respectively (e.g., Boquien et al., 2016; Hao et al., 2011; Hirashita et al., 2003; Kennicutt & Evans, 2012; Madau & Dickinson, 2014; Takeuchi et al., 2010). We adopted a formula specially prepared for spatially resolved galaxy data by Casasola et al. (2017).

9.3.2.1 FUV data

We used the non-ionizing UV as the indicator of the unobscured SFR of galaxies. As UV data, we used the data obtained by the all-sky surveys conducted by a UV astronomical satellite *GALEX* (*Galaxy Evolution Explorer*: Martin et al. 2005). *GALEX*

---

[1] URL: `https://astro3.sci.hokudai.ac.jp/ radio/coming/`.

observed more than 83 % of the whole sky in its imaging surveys at far-UV (FUV) and near-UV (NUV) as well as some spectroscopic observations. In this study, we made use of the flux density at *GALEX* FUV band, since there is almost no contamination from white dwarfs which are not directly related to the recent SF activity. The FUV band has a wavelength range of 1344–1786 Å, with an effective wavelength of 1538 Å. The spatial resolution evaluated by the FWHM of the point spread function of *GALEX* is $4\rlap{.}{''}2$ (Morrissey et al., 2007).

The units of original *GALEX* images are counts per pixel per second (CPS). We converted the CPS into AB magnitude as

$$m_{AB} = m_0 - 2.5\log\left(f_{UV}\,[\text{CPS}]\right) \tag{9.25}$$

(Morrissey et al., 2007). Here $m_0 = 18.82$ at FUV. Then, the flux density is obtained as

$$S_\nu\,[\text{Jy}] = 10^{0.4(-m_{AB}+A(\text{FUV})+8.9)}\,, \tag{9.26}$$

where $A(\text{FUV})$ is the extinction at FUV band (Chap. 3). The pixel size of *GALEX* FUV is $1\rlap{.}{''}5$, then the solid angle of a pixel corresponds to 5.2884×10^{-11} sr. Thus, the FUV intensity of *GALEX* image is obtained as

$$I_{FUV}\,[\text{MJy}\,\text{sr}^{-1}] = \frac{10^{-6}S_\nu\,[\text{Jy}]}{5.2884 \times 10^{-11}\,[\text{sr}]} = 1.891 \times 10^4 S_\nu\,. \tag{9.27}$$

As for the Galactic extinction correction, we used the V-band Galactic extinction $A(V)$ provided by Schlafly & Finkbeiner (2011). Then, by adopting the extinction curve of the Milky Way of Cardelli et al. (1989), we converted $A(V)$ into $A(\text{FUV})$. We applied the extinction correction uniformly over each galaxy and ignored the possible variation of the Galactic extinction within the image.

9.3.2.2 22-μm data

As for the estimation of the SFR hidden by dust, we used the IR emission from galaxies. Since the dust emission of a galaxy is radiated in a wide range of wavelengths from MIR to submit, it would be ideal to observe a galaxy at a whole range of these wavelengths and integrate over the IR SED. However, because of the technical limitation of IR satellites which must be cooled to a very low temperature, the number of galaxies with well-sampled IR observations is very limited. Instead, it is common to convert a monochromatic MIR or FIR luminosity to the total IR luminosity to estimate the SFR (e.g., Calzetti et al., 2007, 2010; Takeuchi et al., 2005b; Yuan et al., 2012, 2011). We stress that there is another merit to use particularly the MIR, that the angular resolution is much better than FIR or submm images. Thus, we used the 22-μm images obtained by *Wide-field Infrared Survey Explorer* (*WISE*: Wright et al. 2010) in this study. *WISE* performed an all-sky survey at four MIR bands, 3.4, 4.6, 12, and 22 μm. We retrieved the *WISE* data from NASA/IPAC Infrared Science Archive (IRSA). We convert the units of original raw *WISE* data, Digital Numbers (DN), to Jy at *WISE* 22 μm band as

$$F\,[\text{Jy}] = 5.2269 \times 10^{-5}\,F\,[\text{DN}] \tag{9.28}$$

(Cutri et al., 2011).

As well as the extinction correction, we should also deal with the diffuse background emission from Galactic dust at MIR in *WISE* data (more physically it should be referred to as the foreground in this case). Similar to the case of extinction, we also do not consider the internal variation of the background. The *WISE* MIR data are not background-subtracted, we estimated the background intensity by the sigma clipping and subtracted from the whole image for each sample.

For this, we first constructed the histogram of the intensity per pixel on each image and confirmed that the noise distribution is well-approximated by a Gaussian for *WISE* images. Then, we first fitted a Gaussian from the original histogram and estimated its mean and dispersion. These are overestimated because the signal from a target galaxy is included. We then trim out the data above $m\sigma$ (we set $m = 5$) and refit a Gaussian to the rest of the data. We iterate this procedure until the estimated μ and σ converge. This procedure is called "sigma clipping", and is applied at various wavelength observations (cf. Chap. 16).

After these procedures, the final sample size is 103 in this analysis.

9.4 RESULT AND DISCUSSION

For our purpose here, we set the dimension $D = 2$ and the number of clusters as $K = 2$, and applied the above method to the spatially resolved K-S law data. The slope of the log-linear correlation was uniquely determined as one of the parameters of the Gaussian (a component of the covariance matrix), without ambiguity.

It can also separate the contribution of the background noise and remove its harmful influence on the slope estimation. Further, since it fits a 2-dimensional Gaussian components, the parameter estimation and hypothesis test can be performed simultaneously. All these features are ideal for this type of problem.

We present the flow of the analysis in Fig. 9.2. We regard the data point cloud as a realization from a sum of multiple 2-dimensional Gaussians, and the parameters of each Gaussian is determined. First, we do not expect any correlation between the gas axis and SFR axis for the data point cloud stems from the noise (hereafter noise branch). Then we can safely assume that the noise branch is well approximated by a Gaussian distribution with a zero covariance (e.g., the covariance matrix is diagonal). In contrast, the physically meaningful branch (physical branch) is well approximated by a Gaussian distribution with a nonzero covariance.

$$N = \sum_{k=1}^{K} N_k = -\lambda \sum_{k=1}^{K} \pi_k = -\lambda_{\mathrm{L}} \, . \tag{9.29}$$

Putting altogether, we obtain

$$\pi_k = -\frac{N_k}{\lambda_{\mathrm{L}}} = \frac{N_k}{N} \tag{9.30}$$

corresponding to eq. (9.9).

Then, the slope of the K-S relation is obtained from the component of Σ_k corresponding to the physical branch, denoted as Σ_{phys}. We only use Σ_{phys} in the following. Given the maximum likelihood estimate of Σ_{phy}, $\widehat{\Sigma}_{\mathrm{phys}}$, and let $X \equiv \log \Sigma_{\mathrm{H}_2}$,

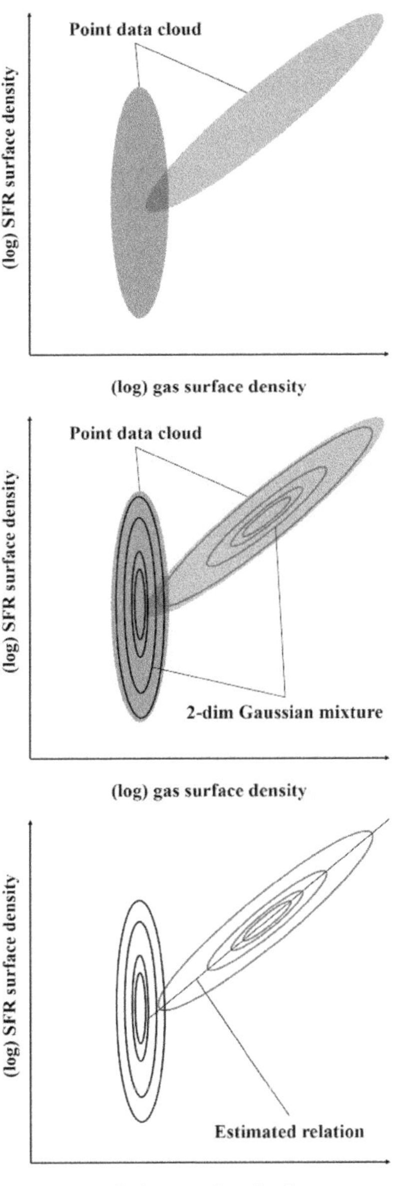

Figure 9.2 Schematic description of how the expectation-maximization algorithm works for the estimation of Gaussian mixture model parameters.

$Y \equiv \log \Sigma_{\mathrm{SFR}}$, we have

$$\widehat{\Sigma}_{\mathrm{phy}} = \begin{pmatrix} \hat{\sigma}_X^2 & \hat{\sigma}_{XY} \\ \hat{\sigma}_{XY} & \hat{\sigma}_Y^2 \end{pmatrix} = \begin{pmatrix} \hat{\sigma}_X^2 & \rho \hat{\sigma}_X \hat{\sigma}_Y \\ \rho \hat{\sigma}_X \hat{\sigma}_Y & \hat{\sigma}_Y^2 \end{pmatrix}, \tag{9.31}$$

where $\hat{\sigma}$s are the maximum likelihood estimates for the variance and covariance for these components and ρ is Pearson's product-moment correlation coefficient.

Then, recalling eq. (9.3), the best-fit line for the physical branch of the molecular K-S law is obtained as

$$Y = NX + A = \text{sgn}(\rho)\frac{\hat{\sigma}_Y}{\hat{\sigma}_X}(X - \hat{\mu}_X) + \hat{\mu}_Y \,, \tag{9.32}$$

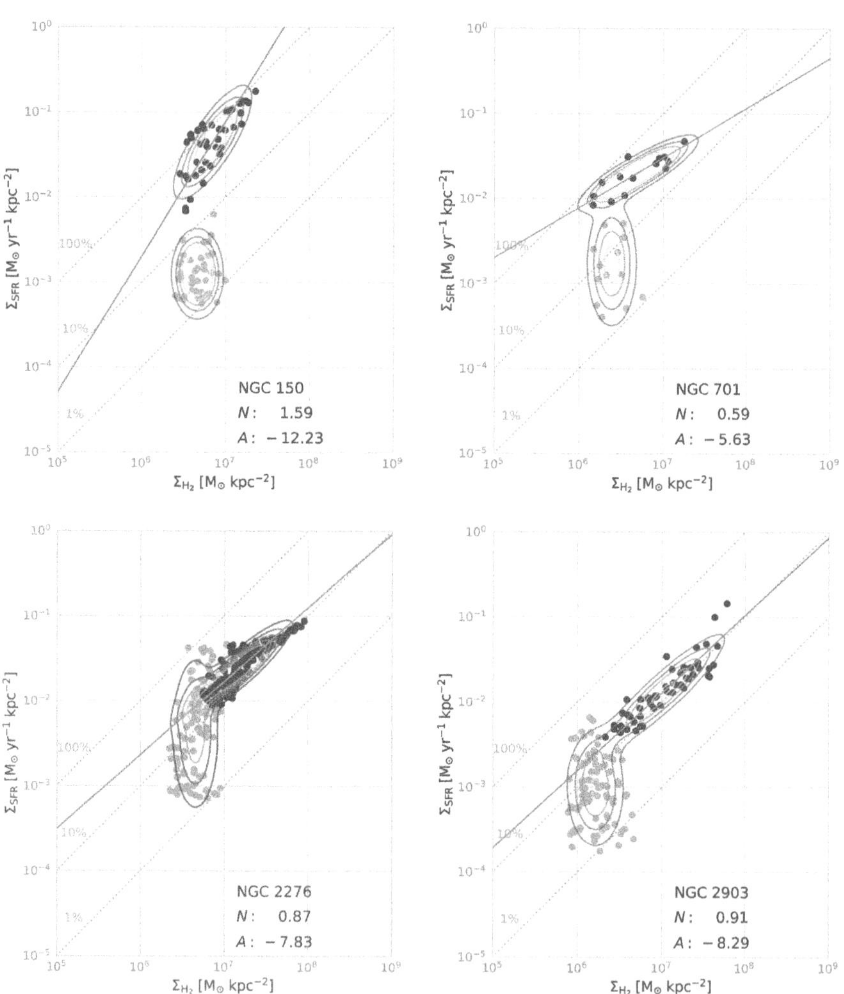

Figure 9.3 Examples of the parameter estimation of the spatially resolved K-S law by a 2-dim Gaussian mixture model. Black symbols represent the data judged as belonging to the physical branch, while gray symbols are judged as noise. Black and gray contours show the 2-dim Gaussian distributions representing physical and noise branches, respectively. Straight lines show the regression line determined from the EM algorithm.

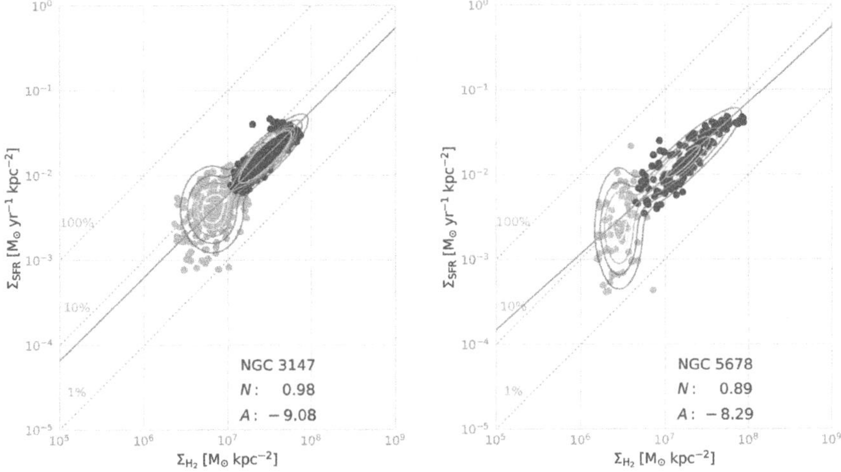

Figure 9.3   (Continued).

where $\mathrm{sgn}(\rho)$ stands for the sign of ρ, $\hat{\mu}_X$ and $\hat{\mu}_Y$ are the estimates for the average of X and Y, respectively.

Figure 9.3 shows some examples of the parameter estimation of the spatially resolved K-S law from the COMING sample. Though the slopes vary from one galaxy to another, and the standard deviation is ~ 0.5. We can also estimate the average slope by stacking all the data points in the physical branch of each galaxy. We performed the 2-dim Gaussian fit to the data point cloud to estimate the K-S law parameters and obtained $N = 1.15$. This is slightly steeper than unity but still consistent with $N = 1.01$ obtained by Bigiel et al. (2008). This result clearly indicates that Σ_{SFR} is almost proportional to Σ_{H_2}, since the sample size is 103, much larger than previous studies. This also provides various interesting implications, but further detailed discussions will be presented elsewhere (Takeuchi et al. 2024, in preparation).

10 Copula and Luminosity and Mass Functions of Galaxies

10.1 COPULA AS A TOOL TO CONSTRUCT A MULTIVARIATE LUMINOSITY/MASS FUNCTION

A luminosity function (LF) of galaxies is one of the fundamental tools to describe and explore the distribution of luminous matter in the Universe. Up to now, studies on the LFs have been rather restricted to a univariate one, i.e., LFs based on a single selection wavelength band. However, such a situation is drastically changing in the era of large and/or deep Legacy surveys. Indeed, a vast number of recent studies are multiband-oriented: they require data from various wavelengths of the ultraviolet (UV) to the infrared (IR) and radio bands. A bivariate LF (BLF) would be a very convenient tool in such studies. To date, it is often defined and used in a confused manner, without careful consideration of complicated selection effects in both bands. This confusion might be partially because of the intrinsically complicated nature of multiband surveys, but also because of the lack of proper recipes to describe a BLF. Then, the situation will be remedied if we have a proper analytic BLF model.

However, it is not a trivial task to determine the corresponding bivariate function from its marginal distributions, if the distribution is not multivariate Gaussian. In fact, there exist infinitely many distributions with the same marginals because the correlation structure is not specified. For such purposes, a general method to construct a bivariate distribution function with pre-defined marginal distributions and correlation coefficient is desired.

In econometrics and mathematical finance, such a function has been commonly used to analyze two covariate random variables. This is called "copula". Especially in a bivariate context, copulae are useful to define nonparametric measures of dependence for pairs of random variables (e.g., Trivedi & Zimmer, 2005). In astrophysics, however, it is only recently that copulae attract researchers' attention and are not very widely known yet (still only a handful of astrophysical applications: Benabed et al., 2009; Jiang et al., 2009; Koen, 2009; Scherrer et al., 2010). Hence the usefulness and limitations of copulae are still not well understood in the astrophysical community.

In this chapter, we first introduce a relatively rigorous definition of a copula. Then, we choose two specific copulae, the Farlie–Gumbel–Morgenstern (FGM) copula and the Gaussian copula to adopt for the construction of a model BLF. Both of them have an ideal property that they are explicitly related to the linear correlation coefficient. Though, as we show in the following, the linear correlation coefficient is not a perfect measure of the dependence of two quantities, this is the most familiar and thus fundamental statistical tool for physical scientists. We focus on the far-infrared (FIR)–far-ultraviolet (FUV) BLF as a concrete example, and discuss its properties and some applications. We restrict our discussion to LFs, but the

DOI: 10.1201/9781003104315-10

theoretical framework for the analysis of the mass function (MF) is essentially the same, and we do not distinguish the two functions.

10.1.1 COPULA

As we discussed in the Introduction, there is an infinite degree of freedom to choose a dependence structure of two variables with a given marginal distribution. However, very often we need a systematic procedure to construct a bivariate distribution function (DF) of two variables[1]. Copulae have a very desirable property from this point of view. In short, copulae are functions that join multivariate DFs to their one-dimensional marginal DFs. However, this statement does not serve as a definition. We first introduce its abstract framework, and move on to a more concrete form which is suitable for the aim of this work (and of many other physical studies).

Before defining the copula, we prepare some mathematical concepts in the following.

Definition 1. *Let S_1 and S_2 be nonempty subsets of $\bar{\mathbb{R}}$ [a union of real number $\mathbb{R}$ and $(-\infty, \infty)$]. Let H be a real function with two arguments (referred to as bivariate or 2-place) such that Dom $H = S_1 \times S_2$. Let $B = [x_1, x_2] \times [y_1, y_2]$ be a rectangle all of whose vertices are in Dom H. Then, the H-volume of B is defined by*

$$V_H(B) \equiv H(x_2, y_2) - H(x_2, y_1) - H(x_1, y_2) + H(x_1, y_1) . \tag{10.1}$$

Definition 2. *A bivariate real function H is 2-increasing if $V_H \geq 0$ for all rectangles B whose vertices lie in Dom H.*

Definition 3. *Suppose S_1 has a least element a_1 and S_2 has a least element a_2. then, a function $H \colon S_1 \times S_2 \to \mathbb{R}$ is grounded if $H(x, a_2) = 0$ and $H(a_1, y) = 0$ for all (x, y) in $S_1 \times S_2$.*

With these concepts, we now define a two-dimensional copula.

Definition 4. *A two-dimensional copula (or shortly 2-copula) is a function C with the following properties:*

1. *Dom $C = [0, 1] \times [0, 1]$;*
2. *C is grounded and 2-increasing;*
3. *For every u and v in $[0,1]$,*

$$C(u, 1) = u \ and \ C(1, v) = v . \tag{10.2}$$

It may be useful to show there are upper and lower limits for the ranges of copulae, which is given by the following theorem.

---

[1] In this work, we use a term DF for a cumulative distribution function (CDF). We use a term probability density function (PDF) to avoid confusion with the term used in physics "distribution function" which stands for a Radon-Nikodym derivative of a DF.

Theorem 3. *Let C be a copula. Then, for every (u,v) in Dom C,*

$$\max(u+v-1,0) \leq C(u,v) \leq \min(u,v). \tag{10.3}$$

Often notations $W(u,v) \equiv \max(u+v-1)$ and $M(u,v) \equiv \min(u,v)$ are used. The former and the latter are referred to as Fréchet-Hoeffding lower bound and Fréchet-Hoeffding upper bound, respectively.

The Fréchet-Hoeffding lower and upper bounds are presented in Fig. 10.1.

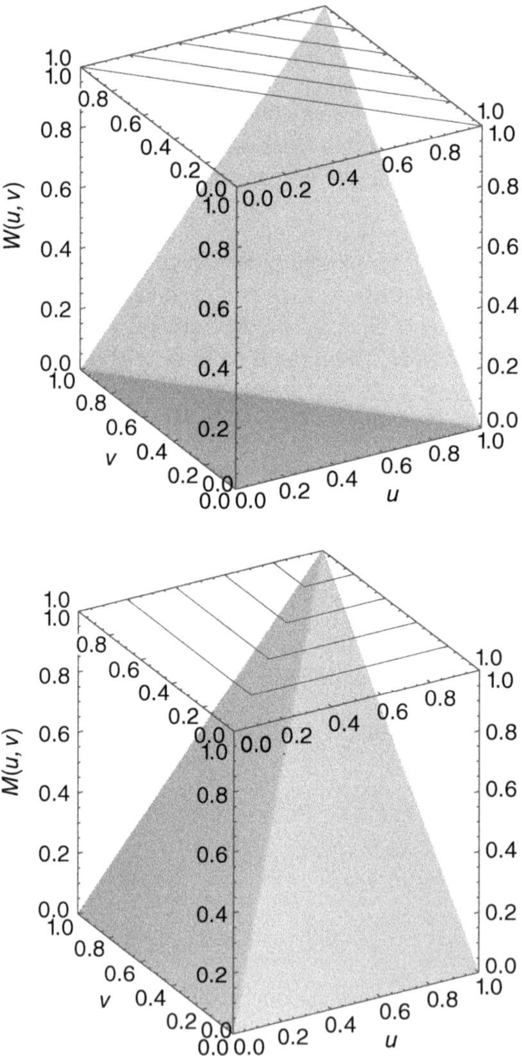

Figure 10.1   The Fréchet-Hoeffding lower and upper bounds (Takeuchi, 2010). Left and right panels show the lower and upper bounds, respectively. The top panels represent the contours of $W(u,v)$ and $M(u,v)$. Credit: Takeuchi, T. T., 2010, Monthly Notices of the Royal Astronomical Society, 406, 1830, Fig. 1.

Any bivariate function which suffices the above conditions can be a copula. Then, there is infinite degrees of freedom for a set of copulae. Using a copula C, we can construct a bivariate DF G with two margins F_1 and F_2 as

$$G(x_1, x_2) = C[F(x_1), F_2(x_2)] . \tag{10.4}$$

However, one may have a natural question: is any bivariate DF written in the above form? This is guaranteed by Sklar's theorem (Sklar, 1959).

Theorem 4. (Sklar's theorem) *Let G be a joint distribution function with margins F_1 and F_2. Then, there exists a copula C such that for all x_1, x_2 in $\bar{\mathbb{R}}$,*

$$G(x_1, x_2) = C[F_1(x_1), F_2(x_2)] . \tag{10.5}$$

If F_1 and F_2 are continuous, then C is unique: otherwise, C is uniquely determined on Range $F_1 \times$ Range F_2.

A comprehensive proof of Sklar's theorem is found in e.g., Nelsen (2006). This theorem gives a basis that *any* bivariate DF with given margins is expressed with a form of eq. (10.5). Then, finally, the somewhat abstract definition of a copula turned out to be really useful for our aim, i.e., to construct a bivariate DF when its marginals are known in some way.

Up to now, we discuss only bivariate DFs and their copulae. It is straightforward to introduce multivariate DFs as a natural extension of the formulation presented here.

10.1.2 COPULAE AND DEPENDENCE MEASURES BETWEEN TWO VARIABLES

The most important statistical aspect of bivariate DFs is their dependence properties between variables. Since the dependence can never given by the marginals of a DF, this is the most nontrivial information which a bivariate DF provides. Since any bivariate DFs are described by eq. (10.5), all the information on the dependence is carried by their copulae.

The most familiar measure of dependence among physical scientists (and others) may be the correlation coefficients, especially Pearson's product-moment correlation coefficient ρ. The bivariate PDF of x_1 and x_2, $g(x_1, x_2)$, is written as

$$g(x_1, x_2) = \frac{\partial^2 C[F_1(x_1), F_2(x_2)]}{\partial x_1 \partial x_2} f_2(x_1) f_2(x_2)$$
$$\equiv c[F_1(x_1), F_2(x_2)] f_1(x_1) f_2(x_2) \tag{10.6}$$

where $f_1(x_1)$ and $f_2(x_2)$ are PDF of $F_1(x_1)$ and $F_2(x_2)$, respectively. Then the correlation coefficient is expressed as

$$\rho = \frac{\int (x_1 - \bar{x_1})(x_2 - \bar{x_2}) g(x_1, x_2) dx_1 dx_2}{\sqrt{\int (x_1 - \bar{x_1})^2 f_1(x_1) dx_1 \int (x_2 - \bar{x_2})^2 f_2(x_2) dx_2}}$$
$$= \frac{\int (x_1 - \bar{x_1})(x_2 - \bar{x_2}) c[(F_1(x_1), F_2(x_2)] f_1(x_1) f_2(x_2) dx_1 dx_2}{\sqrt{\int (x_1 - \bar{x_1})^2 f_1(x_1) dx_1 \int (x_2 - \bar{x_2})^2 f_2(x_2) dx_2}} . \tag{10.7}$$

We should note that ρ can measure only *a linear dependence* of two variables. However, in general, the dependence of two variables would be not linear at all, and it cannot be a sufficient measure of dependence. Further, more fundamentally, eq. (eq:corr) depends not only on the dependence of two variables (copula part) but also its marginals $f_1(x_1), f_2(x_2)$, i.e., the linear correlation coefficient ρ does not measure the dependence purely. In such a situation, a more flexible and genuine measure of dependence, e.g., Spearman's ρ_S or Kendall's τ would be appropriate. Spearman's rank correlation is a nonparametric version of Pearson's correlation using a rank of data. The population version of Spearman's ρ_S is expressed by copula as

$$\rho_S = 12 \int_0^1 \int_0^1 u_1 u_2 dC(u_1, u_2) - 3$$
$$= 12 \int_0^1 \int_0^1 C(u_1, u_2) du_1 du_2 - 3 . \tag{10.8}$$

The definition of Kendall's tau is more complicated. We define a concept of concordance as follows: when we have pairs of data (x_{1i}, x_{2i}) and (x_{1j}, x_{2j}), they are said to be *concordant* if $x_{1i} > x_{1j}$ and $x_{2i} > x_{2j}$ or $x_{1i} < x_{1j}$ and $x_{2i} < x_{2j}$ (i.e., $(x_{1i} - x_{1j})(x_{2i} - x_{2j}) > 0$), and otherwise *discordant*. Let $\{x_{1i}, x_{2i}\}_{i=1,\dots,n}$ denote a random sample of n observations. There are $_nC_2$ pairs (x_{1i}, x_{2i}) and (x_{1j}, x_{2j}) of observations in the sample, and each pair is concordant or discordant. Let n_c denote the number of concordant pairs, and n_d the number of discordant ones. Then, Kendall's τ for the sample, t, is defined as

$$t = \frac{n_c - n_d}{n_c + n_d} . \tag{10.9}$$

The population version of τ is also expressed in a simple form in terms of copula as

$$\tau = 4 \int_0^1 \int_0^1 C(u_1, u_2) dC(u_1, u_2) - 1$$
$$= 4 \int_0^1 \int_0^1 C(u_1, u_2) c(u_1, u_2) du_1 du_2 - 1 . \tag{10.10}$$

Note that both Equations (10.8) and (10.10) are independent of the distributions F_1, F_2, nor G but depends only on the dependence structure described by a copula, unlike the linear correlation coefficient. This is a direct consequence that these estimators are nonparametric, i.e., distribution-free. These are the reasons why the two dependence measures are almost always used in the context of copulae in literature.

10.1.3 ESTIMATORS OF THE NONPARAMETRIC DEPENDENCE MEASURES ρ_S AND τ

Here we present the estimators of the nonparametric measure of dependence introduced in Section 10.1.2. More detailed derivation and properties of these nonparametric measures of dependence are found in e.g., Hettmansperger (1984) and Hollander & Wolfe (1999).

Let $\{R_i\}_{i=1,\dots,n}$ and $\{S_i\}_{i=1,\dots,n}$ be the ranks of $\{x_{1i}\}_{i=1,\dots,n}$ and $\{x_{2i}\}_{i=1,\dots,n}$, respectively. If we denote the estimator of ρ_S for a sample as r_S,

$$r_S = \frac{\sum_{i=1}^{n}\left(R_i - \frac{n+1}{2}\right)\left(S_i - \frac{n+1}{2}\right)}{\sqrt{\sum_{i=1}^{n}\left(R_i - \frac{n+1}{2}\right)\sum_{i=1}^{n}\left(S_i - \frac{n+1}{2}\right)}} \ . \tag{10.11}$$

This is also expressed as

$$r_S = \frac{12\sum_{i=1}^{n}\left(R_i - \frac{n+1}{2}\right)\left(S_i - \frac{n+1}{2}\right)}{n(n^2 - 1)} \ . \tag{10.12}$$

This form is an exact sample counterpart of eq. (10.8). If we define $d_i \equiv S_i - R_i$, eq. (10.12) reduces to the following simpler form

$$r_S = 1 - \frac{6\sum_{i=1}^{n} d_i^2}{n(n^2 - 1)} \ . \tag{10.13}$$

The variance of r_S in the large-sample limit is given by

$$\mathbb{V}[r_S] = \frac{1}{n - 1} \ . \tag{10.14}$$

The most basic form of Kendall's τ for a sample has been already shown in Section 10.1.2 [eq. (10.9)]. It is also expressed as

$$t = \frac{2(n_c - n_d)}{n(n - 1)}$$
$$= 1 - \frac{4}{n(n - 1)} n_d \tag{10.15}$$

since $n_c + n_d = n(n - 1)/2$. The variance of t is given by

$$\mathbb{V}[t] = \frac{2(2n + 5)}{9n(n - 1)} \tag{10.16}$$

(see Valz & McLeod, 1990, for a very concise derivation).

10.1.4 FARLIE–GUMBEL–MORGENSTERN (FGM) COPULA

As seen in the discussion above, the usefulness of the linear correlation coefficient is quite limited, and distribution-free measures of dependence are more appropriate for general joint DFs with non-Gaussian marginals. However, even if it is true, since physicists may cling to the most familiar linear correlation coefficient ρ, a copula which has an explicit dependence on ρ would be convenient. We introduce two special types of copulae with this ideal property.

If the correlation between two variables is weak, a systematic method has been proposed by Morgenstern (1956) and Gumbel (1960) for specific functional forms and later generalized to arbitrary functions by Farlie (1960). This is known as the Farlie–Gumbel–Morgenstern (FGM) distributions after the inventors' names. The studies of the FGM family seem not so tightly connected to that of copulae, but later we will see that the FGM distributions are expressed in terms of a copula (called the FGM copula).

The correlation structure of the FGM distributions was studied by Schucany et al. (1978). Let $F_1(x_1)$ and $F_2(x_2)$ be the (cumulative) distribution functions (DFs) of stochastic variables x and y, respectively, and let $f_1(x_1)$ and $f_2(x_2)$ be their probability density functions (PDFs). The bivariate FGM system of distributions $G(x_1,x_2)$ is written as

$$G(x_1,x_2) = F_1(x_1)F_2(x_2)\left\{1 + \kappa\left[1 - F_1(x_1)\right]\left[1 - F_2(x_2)\right]\right\} \tag{10.17}$$

where $G(x_1,x_2)$ is the joint DF of x_1 and x_2. Here κ is a parameter related to the correlation (see below), and to make $G(x_1,x_2)$ have an appropriate property as a bivariate DF, $|\kappa| \leq 1$ is required (for proof, see Cambanis, 1977). Its PDF can be obtained by a direct differentiation of $G(x_1,x_2)$ as

$$
\begin{aligned}
g(x_1,x_2) &\equiv \left.\frac{\partial^2 H}{\partial x_1 \partial x_2}\right|_{x_1,x_2} \\
&= \left.\frac{\partial^2}{\partial x_1 \partial x_2}\left\{F_1(x_1)F_2(x_2)\left[1 + \kappa\left(1 - F_1(x_1)\right)\left(1 - F_2(x_2)\right)\right]\right\}\right|_{x_1,x_2} \\
&= f_1(x_1)f_2(x_2)\left\{1 + \kappa\left[2F_1(x_1) - 1\right]\left[2F_2(x_2) - 1\right]\right\} .
\end{aligned}
\tag{10.18}
$$

From eq. (10.18), we straightforwardly obtain its covariance function $\mathbb{Cov}(x_1,x_2)$ as

$$\mathbb{Cov}(x_1,x_2) = \kappa \int x_1\left[2F_1(x_1) - 1\right]f_1(x_1)\mathrm{d}x_1 \int x_2\left[2F_2(x_2) - 1\right]f_2(x_2)\mathrm{d}x_2 . \tag{10.19}$$

Then we have a correlation function of two stochastic variables x_1 and x_2, $\rho(x_1,x_2)$ [eq. (10.7)] as follows

$$
\begin{aligned}
\rho(x_1,x_2) &= \frac{\mathbb{Cov}(x_1,x_2)}{\sigma_1 \sigma_2} \\
&= \frac{\kappa}{\sigma_1 \sigma_2} \int x_1\left[2F_1(x_1) - 1\right]f_1(x_1)\mathrm{d}x_1 \int x_2\left[2F_2(x_2) - 1\right]f_2(x_2)\mathrm{d}x_2 .
\end{aligned}
\tag{10.20}
$$

where σ_1 and σ_2 are the standard deviations of x_1 and x_2 with respect to $f_1(x_1)$ and $f_2(x_2)$. It is straightforwardly confirmed that $g(x_1,x_2)$ really has the marginals $f_1(x_1)$ and $f_2(x_2)$, by a direct integration with respect to x_1 or x_2. It is also clear that if we want a bivariate PDF with a prescribed correlation coefficient ρ, we can determine the parameter κ from Equations (10.7) and (10.19).

Here, consider the case that that both $f_1(x_1)$ and $f_2(x_2)$ are the Gamma distributions, i.e.,

$$f_j(x_j) = \frac{x_j{}^{a-1}e^{-x_j/b}}{b^a\Gamma(a)} \tag{10.21}$$

where $\Gamma(x)$ is the gamma function ($j = 1, 2$). In this case, after some algebra, ρ can be written analytically as

$$\rho(x_1, x_2) = \frac{\kappa}{\sqrt{ab}} \left[\frac{2^{-2(a-1)}}{B(a,a)} \right] \left[\frac{2^{-2(b-1)}}{B(b,b)} \right] \tag{10.22}$$

where

$$B(a,b) \equiv \frac{\Gamma(a)\Gamma(b)}{\Gamma(a+b)} . \tag{10.23}$$

This result was obtained by D'Este (1981). Especially when $b = 1$, this corresponds to a bivariate extension of the Schechter function (Schechter, 1976b), and we expect some astrophysical applications.

As we mentioned, the correlation of the FGM distributions is restricted to be weak: indeed, the correlation coefficient cannot exceed $1/3$. Here we prove this. For all (absolutely continuous) $F(x)$,

$$\left\{ \int x \left[2F(x) - 1 \right] f(x) \mathrm{d}x \right\}^2 = \left\{ \int (x - \bar{x}) \left[2F(x) - 1 \right] f(x) \mathrm{d}x \right\}^2$$

$$\leq \int (x - \bar{x})^2 f(x) \mathrm{d}x \int \left[2F(x) - 1 \right]^2 f(x) \mathrm{d}x$$

$$= \frac{\sigma^2}{3} , \tag{10.24}$$

where $\bar{x}$ is the average of x. The second line follows from the Cauchy–Schwarz inequality. From Equations (10.19) and (10.7), and the condition $|\kappa| \leq 1$, we obtain $|\rho| \leq 1/3$.

The copula of the FGM family of distributions is expressed as

$$C^{\mathrm{FGM}}(u_1, u_2; \kappa) = u_1 u_2 + \kappa u_1 u_2 (1 - u_1)(1 - u_2) \tag{10.25}$$

with $-1 < \kappa < 1$. The differential form of the FGM copula follows from Equations (10.6) and (10.25)

$$c^{\mathrm{FGM}}(u_1, u_2; \kappa) = 1 + \kappa(1 - 2u_1)(1 - 2u_2) . \tag{10.26}$$

10.1.5   AN EXTENSION OF THE FGM SYSTEM BY JOHNSON & KOTZ (1977)

Although the FGM system of distributions provides us with a convenient way for the statistical model construction, its usefulness is restricted by the limitation of the correlation strength described above. To overcome this drawback, many attempts have been made to extend the FGM distributions (see, e.g., Kotz et al., 2000; Stuart & Ord, 1994). Among them, Johnson & Kotz (1977) introduced the following iterated generalization of eq. (10.17):

$$G(x_1, x_2) = \sum_{j=0}^{k} \kappa_j \left[F_1(x_1) F_2(x_2) \right]^{\lfloor j/2 \rfloor + 1} \left\{ \left[1 - F_1(x_1) \right] \left[1 - F_2(x_2) \right] \right\}^{\lfloor (j+1)/2 \rfloor} , \tag{10.27}$$

where the symbol in the exponent $[j/2]$ means the maximum natural number which does not exceed $j/2$. We set $\kappa_0 = 1$. Huang & Kotz (1984) examined the dependence structure of eq. (10.27) especially for the case of $k = 2$, and showed that the correlation can be stronger for this extension. In the case of the one-iteration family ($k = 2$), we have the DF as

$$
\begin{aligned}
G(x_1,x_2) &= F_1(x_1)F_2(x_2) \\
&\times \left\{1 + \kappa_1\left[1 - F_1(x_1)\right]\left[1 - F_2(x_2)\right] + \kappa_2 F_1(x_1)F_2(x_2)\left[1 - F_1(x_1)\right]\left[1 - F_2(x_2)\right]\right\}.
\end{aligned}
\tag{10.28}
$$

The corresponding PDF is

$$
\begin{aligned}
g(x_1,x_2) &\\
&= f_1(x_1)f_2(x_2) \\
&\times \left\{1 + \kappa_1\left[2F_1(x_1) - 1\right]\left[2F_2(x_2) - 1\right] + \kappa_2 F_1(x_1)F_2(x_2)\left[3F_1(x_1) - 2\right]\left[3F_2(x_2) - 2\right]\right\}.
\end{aligned}
\tag{10.29}
$$

Then, just the same as in the case of the original FGM distribution ($k = 1$), we obtain the covariance

$$
\begin{aligned}
\mathbb{Cov}(x_1,x_2) &= \iint (x_1 - \bar{x}_1)(x_2 - \bar{x}_2)f_1(x_1)f_2(x_2) \\
&\quad \times \left\{1 + \kappa_1\left[2F_1(x_1) - 1\right]\left[2F_2(x_2) - 1\right]\right. \\
&\qquad \left. + \kappa_2 F_1(x_1)F_2(x_2)\left[3F_1(x_1) - 2\right]\left[3F_2(x_2) - 2\right]\right\} dx_1 dx_2 \\
&= \kappa_1 \int x_1 f_1(x_1)\left[2F_1(x_1) - 1\right]dx_1 \int x_2 f_2(x_2)\left[2F_2(x_2) - 1\right]dx_2 \\
&\quad + \kappa_2 \int x_1 f_1(x_1)F_1(x_1)\left[3F_1(x_1) - 2\right]dx_1 \\
&\qquad \times \int x_2 f_2(x_2)F_2(x_2)\left[3F_2(x_2) - 2\right]dx_2 \\
&\equiv A_1\kappa_1 + A_2\kappa_2.
\end{aligned}
\tag{10.30}
$$

Huang & Kotz (1984) obtained the parameter space for κ_1 and κ_2 as

$$
|\kappa_1| \leq 1,
\tag{10.31}
$$

$$
\kappa_1 + \kappa_2 \geq -1,
\tag{10.32}
$$

$$
\kappa_2 \leq \frac{3 - \kappa_1 + \sqrt{3(1 - \kappa_1)(3 + \kappa_1)}}{2}.
\tag{10.33}
$$

They showed that, for a positive correlation,

$$
\rho \leq \frac{\kappa_1}{3} + \frac{31\kappa_2}{240}.
\tag{10.34}
$$

Under these conditions, we have $\rho \leq 0.5072$, which is considerably better than $1/3$. The BLF constructed with the first-order iterated FGM copula is expressed as

$$
\begin{aligned}
\phi^{(2)}(L_1,L_2) = \phi_1^{(1)}(L_1)\phi_2^{(1)}(L_2) \\
\times \left\{ 1 + \kappa_1 \left[2\Phi_1^{(1)}(L_1) - 1 \right] \left[2\Phi_2^{(1)}(L_2) - 1 \right] \right. \\
\left. + \kappa_2 \phi_1^{(1)}(L_1)\phi_2^{(1)}(L_2) \left[3\Phi_1^{(1)}(L_1) - 2 \right] \left[3\Phi_2^{(1)}(L_2) - 2 \right] \right\} .
\end{aligned}
\tag{10.35}
$$

Clearly, the dependence between the two luminosities is stronger than the original FGM-based BLF. However, now it is not intuitive nor straightforward to relate these two parameters of dependence κ_1 and κ_2 to the linear correlation coefficient.

10.1.6 GAUSSIAN COPULA

As seen in Section 10.1.4, though the FGM distribution has one of the most "natural" example of bivariate DF which has an explicit ρ-dependence, the limitation of the correlation coefficient of the FGM family hampers a flexible application of this DF, though there are many attempts to extend the range of application (see Appendix 10.1.5). Then, the second natural candidate may be a copula related to a bivariate Gaussian DF. The Gaussian copula has also an explicit dependence on a linear correlation coefficient by its construction.

Let

$$
\psi_1(x) = \frac{1}{\sqrt{2\pi}} \exp\left(-\frac{x^2}{2} \right) ,
\tag{10.36}
$$

$$
\Psi_1 = \int_{-\infty}^{x} \Psi(x')dx' ,
\tag{10.37}
$$

$$
\psi_2(x_1,x_2;\rho) = \frac{1}{\sqrt{(2\pi)^2(1-\rho^2)}} \exp\left[-\frac{x_1^2 + x_2^2 - 2\rho x_1 x_2}{2(1-\rho^2)} \right] ,
\tag{10.38}
$$

and

$$
\Psi_2(x_1,x_2;\rho) = \int_{-\infty}^{x_1} \int_{-\infty}^{x_2} \psi_i(x_1',x_2')dx_1' dx_2' .
\tag{10.39}
$$

By using the covariance matrix Σ

$$
\Sigma \equiv \begin{pmatrix} 1 & \rho \\ \rho & 1 \end{pmatrix} ,
\tag{10.40}
$$

Equation (10.38) is simplified as

$$
\psi_2(x_1,x_2;\rho) = \frac{1}{\sqrt{(2\pi)^2 \det \Sigma}} \exp\left(-\frac{1}{2}\vec{x}^\top \Sigma^{-1} \vec{x} \right) ,
\tag{10.41}
$$

where $\vec{x} \equiv (x_1,x_2)^\top$ and superscript T stands for the transpose of a matrix or vector.

Then, we define a Gaussian copula $C^{\mathrm{G}}(u_1, u_2; \rho)$ as

$$C^{\mathrm{G}}(u_1, u_2; \rho) = \Psi_2 \left[\Psi_1^{-1}(u_1), \Psi_1^{-1}(u_2); \rho \right] . \tag{10.42}$$

The density of C^{G}, c^{G}, is obtained as

$$
\begin{aligned}
c^{\mathrm{G}}(u_1, u_2; \rho) &= \frac{\partial^2 C^{\mathrm{G}}(u_1, u_2; \rho)}{\partial u_1 \partial u_2} = \frac{\partial^2 \Psi_2 \left[\Psi_1^{-1}(u_1), \Psi_1^{-1}(u_2); \rho \right]}{\partial u_1 \partial u_2} \\
&= \frac{\psi_2(x_1, x_2; \rho)}{\psi_1(x_1) \psi_1(x_2)} \\
&= \frac{\dfrac{1}{\sqrt{(2\pi)^2 \det \Sigma}} \exp\left(-\dfrac{1}{2} \vec{x}^\top \Sigma^{-1} \vec{x} \right)}{\dfrac{1}{\sqrt{2\pi}} \exp\left(-\dfrac{x_1^2}{2} \right) \dfrac{1}{\sqrt{2\pi}} \exp\left(-\dfrac{x_2^2}{2} \right)} \\
&= \frac{1}{\sqrt{\det \Sigma}} \exp\left[-\frac{1}{2} \left(\vec{x}^\top \Sigma^{-1} \vec{x} - \vec{x}^\top \mathbb{I} \vec{x} \right) \right] \\
&= \frac{1}{\sqrt{\det \Sigma}} \exp\left\{ -\frac{1}{2} \left[\vec{x}^\top \left(\Sigma^{-1} - \mathbb{I} \right) \vec{x} \right] \right\} \\
&= \frac{1}{\sqrt{\det \Sigma}} \exp\left\{ -\frac{1}{2} \left[\vec{\psi}^\top \left(\Sigma^{-1} - \mathbb{I} \right) \vec{\psi} \right] \right\} , \tag{10.43}
\end{aligned}
$$

where $\vec{\psi} \equiv \left[\Psi^{-1}(u_1), \Psi^{-1}(u_2) \right]^\top$ and $\mathbb{I}$ stands for the identity matrix. The second line follows from eq. (10.6).

10.1.7 APPLICATION TO CONSTRUCT THE BIVARIATE LUMINOSITY FUNCTION (BLF) OF GALAXIES

10.1.7.1 Construction of the BLF

We define the luminosity at a certain wavelength band by $L \equiv \nu L_\nu$ (ν is the corresponding frequency). Then the luminosity function is defined as a number density of galaxies whose luminosity lies between a logarithmic interval $[\log L, \log L + \mathrm{d}\log L]$:

$$\phi^{(1)}(L) \equiv \frac{\mathrm{d}n}{\mathrm{d}\log L} , \tag{10.44}$$

where we denote $\log x \equiv \log_{10} x$ and $\ln x \equiv \log_e x$. For mathematical simplicity, we define the LF as *being normalized*, i.e.,

$$\int \phi^{(1)}(L) \mathrm{d}\log L = 1 . \tag{10.45}$$

Hence, this corresponds to a probability density function (PDF), a commonly used terminology in the field of mathematical statistics. We also define the cumulative LF as

$$\Phi^{(1)}(L) \equiv \int_{\log L_{\min}}^{\log L} \phi^{(1)}(L') \mathrm{d}\log L' , \tag{10.46}$$

where L_{min} is the minimum luminosity of galaxies considered. This corresponds to the DF.

If we denote univariate LFs as $\phi_1^{(1)}(L_1)$ and $\phi_2^{(1)}(L_2)$, then the bivariate PDF $\phi^{(2)}(L_1, L_2)$ is described by a differential copula $c(u_1, u_2)$ as

$$\phi^{(2)}(L_1, L_2) \equiv c\left[\phi_1^{(1)}(L_1), \phi_2^{(1)}(L_2)\right] . \tag{10.47}$$

For the FGM copula, the BLF leads from eq. (10.26)

$$\phi^{(2)}(L_1, L_2; \kappa) \equiv \left\{ 1 + \kappa \left[2\Phi_1^{(1)}(L_1) - 1 \right] \left[2\Phi_2^{(1)}(L_2) - 1 \right] \right\} \phi_1^{(1)}(L_1)\phi_2^{(1)}(L_2) . \tag{10.48}$$

The parameter κ is proportional to the correlation coefficient ρ between $\log L_1$ and $\log L_2$. For the Gaussian copula, the BLF is obtained as

$$\phi^{(2)}(L_1, L_2; \rho) = \frac{1}{\sqrt{\det \Sigma}} \exp\left\{ -\frac{1}{2}\left[\vec{\psi}^\top \left(\Sigma^{-1} - \mathbb{1} \right) \vec{\psi} \right] \right\} \phi_1^{(1)}(L_1)\phi_2^{(1)}(L_2) , \tag{10.49}$$

where

$$\vec{\psi} = \left[\Psi^{-1}\left(\Phi_1^{(1)}(L_1) \right), \Psi^{-1}\left(\Phi_2^{(1)}(L_2) \right) \right]^\top \tag{10.50}$$

and Σ is again defined by eq. (10.40).

10.1.7.2   The FIR-FUV BLF

Here, to make our model BLF astrophysically realistic, we construct the FUV–FIR BLF by the copula method. For the IR, we use the analytic form for the LF proposed by Saunders et al. (1990a) which is defined as

$$\phi_1^{(1)}(L) = \phi_* \left(\frac{L}{L_*} \right)^{1-\alpha} \exp\left\{ -\frac{1}{2\sigma^2}\left[\log\left(1 + \frac{L}{L_*} \right) \right]^2 \right\} . \tag{10.51}$$

We adopt the parameters estimated by Takeuchi et al. (2003b) which are obtained from the *IRAS* PSCz galaxies (Saunders & et al., 2000). For the UV, we adopt the Schechter function (Schechter, 1976b, see Chapter 2).

$$\phi_2^{(1)}(L) = (\ln 10)\, \phi_* \left(\frac{L}{L_*} \right)^{1-\alpha} \exp\left[-\left(\frac{L}{L_*} \right) \right] , \tag{10.52}$$

We use the parameters presented by Wyder & et al. (2005) for *GALEX* FUV ($\lambda_{\mathrm{eff}} = 1530$ Å): $(\alpha, L_*, \phi_*) = (1.21, 1.81 \times 10^9 h^{-2}\, [L_\odot], 1.35 \times 10^{-2} h^3\, [\mathrm{Mpc}^{-3}])$. For simplicity, we neglect the K-correction.

10.1.7.3   Result

We show the constructed BLFs from the FGM and Gaussian copulae in Figs. 10.2 and 10.3, respectively. The FGM-based BLF cannot have a linear correlation coefficient larger than $\simeq 0.3$ as explained above, while the Gaussian-based BLF has

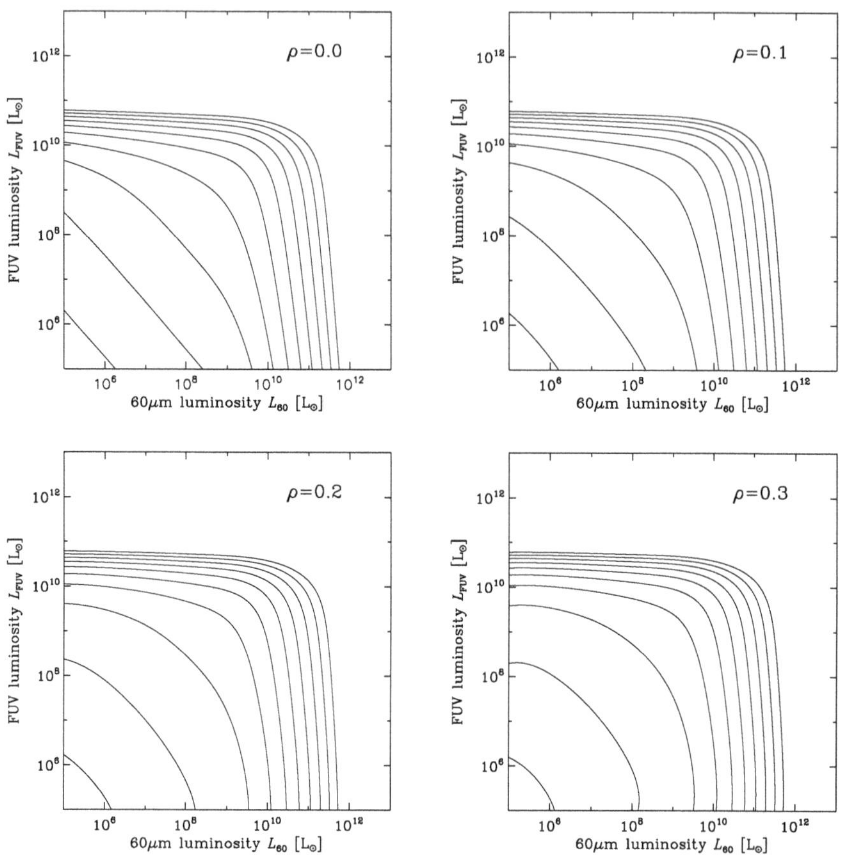

Figure 10.2 The bivariate luminosity functions (BLFs) constructed with the Farlie–Gumbel–Morgenstern (FGM) copula (Takeuchi & Kono, 2020). From top-left to bottom-right, the linear correlation coefficient $\rho = 0.0, 0.1, 0.2$, and 0.3, corresponding to $\kappa = 0.0, 0.33, 0.67$ and 1.0, respectively. The contour intervals are logarithmic. Credit: Takeuchi, T. T., 2010, Monthly Notices of the Royal Astronomical Society, 406, 1830, Fig. 2.

much higher linear correlation. We note that both copulae allow negative correlations, though we do not show them in this article.

First, even if the linear correlation coefficients are the same, the detailed structures of the BLFs with the FGM and Gaussian copulae are different (see the case of $\rho = 0.0$–0.3). For the Gaussian-based BLFs, we see a decline at the faint end, while we do not have such structure in the FGM-based BLFs (see the closed contours in Fig. 10.3). This structure is introduced by the Gaussian copula, and from the physical point of view, it might not be strongly desired. The FGM-based BLF has a more ideal shape.

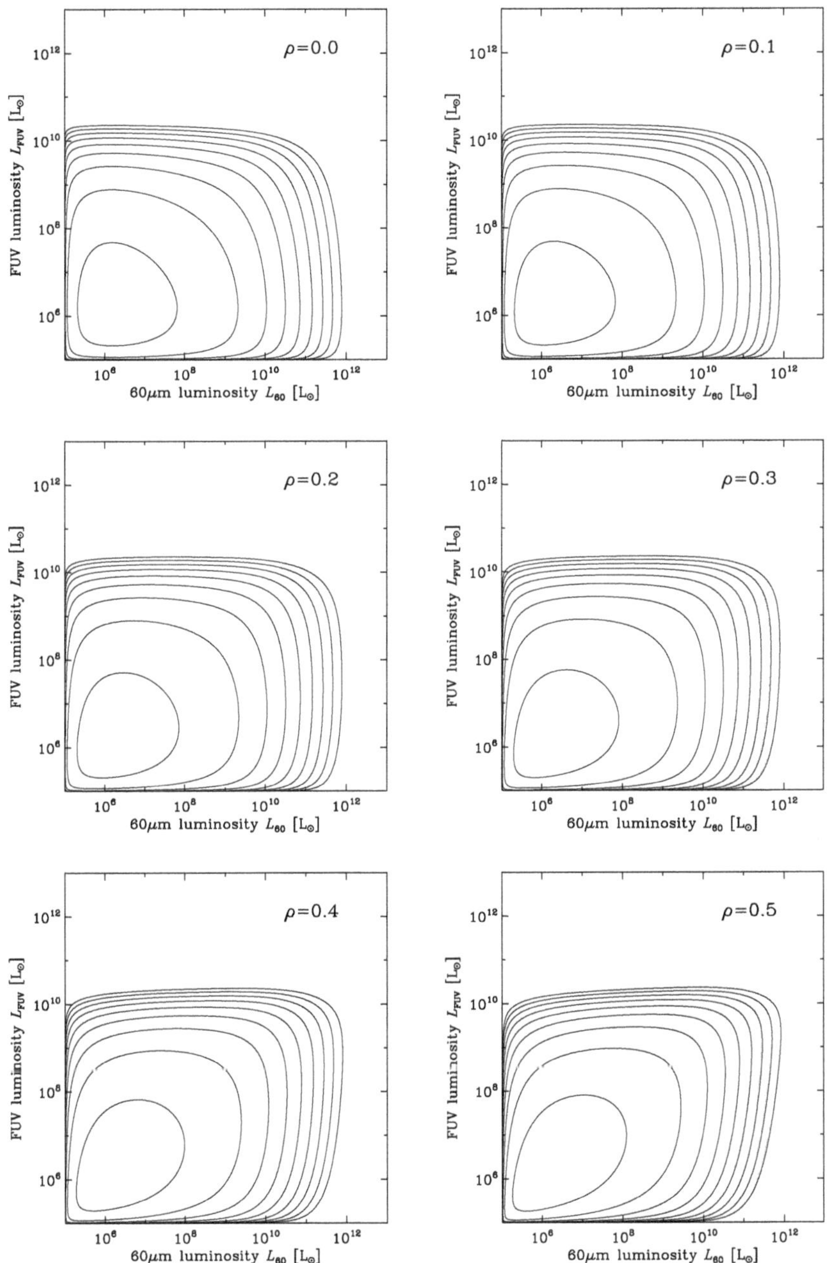

Figure 10.3 The analytical BLF constructed with the Gaussian copula (Takeuchi, 2010). The linear correlation coefficient ρ varies from 0.0 to 0.9 from top-left to bottom-right. The same as Fig. 10.2, the contour intervals are logarithmic. Credit: Takeuchi, T. T., 2010, Monthly Notices of the Royal Astronomical Society, 406, 1830, Fig. 3.

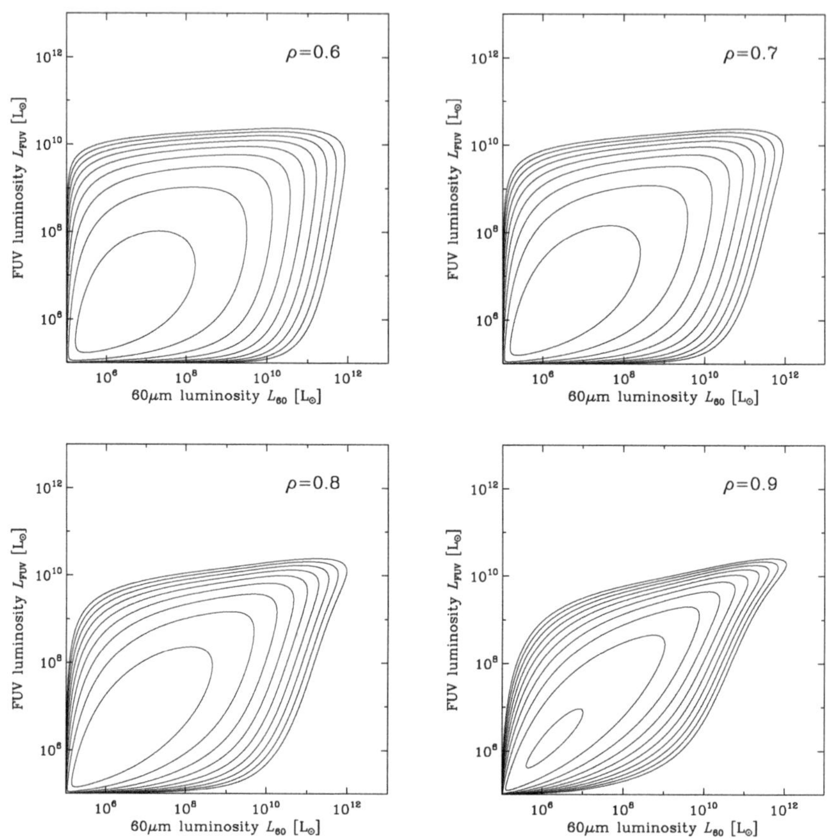

Figure 10.3 Continued.

Second, since the univariate LF shapes are different at FIR and FUV, the ridge of the BLF is not a straight line but clearly nonlinear. This feature is more clearly visible in higher correlation cases in Fig. 10.3 but always exists for the whole range of ρ. This trend is indeed found in the $L_{\rm FIR}$–$L_{\rm FUV}$ diagram (Martin & et al., 2005). Physically this is understood that galaxies with high SFRs are more extinguished by dust (e.g. Buat & et al., 2007a; Buat et al., 2007b).

10.1.7.4   Discussion: Flux Selection Effect in Multiband Surveys

Since we have an explicit form of a BLF, we can discuss the flux selection effect formally. For simplicity, we consider the bivariate case (i.e. sample selected at two bands), but it will be straightforward to extend the formulation to a multiwavelength case with more bands (or more generally, selected with any physical properties). The flux selection is described in terms of luminosity as putting a lower bound L^{lim} on a luminosity–luminosity $(L_1–L_2)$ plane. The lower bound luminosity L^{lim} is defined by the flux (density) detection limit S^{lim} as a function of redshift. In usual surveys, a certain wavelength band is chosen as the primary selection band, like B-band, Ks-band, $60\,\mu$m-selected, etc. The schematic description of a survey is presented in Fig. 10.4.

If we select a sample of objects (in our case galaxies) at band 1, the objects with $L_1 < L_1^{\text{lim}}(z)$ would not be included in the sample at a certain redshift z. Then, the detected sources should have $L_1 > L_1^{\text{lim}}(z)$ and $L_2 > L_2^{\text{lim}}(z)$. Hence, on the $L_1–L_2$ plane, the 2-dim distribution of the detected objects is expressed as

$$\Sigma^{\text{det}}(L_1,L_2) \equiv \int_0^z \frac{d^2V}{dz'd\Omega}\phi^{(2)}(L_1,L_2)\,\Theta\left(L_1^{\text{lim}}(z')\right)\Theta\left(L_2^{\text{lim}}(z')\right)dz', \qquad (10.53)$$

where Ω is a solid angle, and Θ is the Heaviside step function defined as

$$\Theta(a) = \begin{cases} 0 & L < a \\ 1 & L \geq a \end{cases}. \qquad (10.54)$$

The quantity Σ^{det} is proportional to the surface number density of objects detected in both bands on the $L_1–L_2$ plane. We start from a primary selection at band 1, then we would have objects detected at band 1 but not detected at band 2. In such a case we only have upper limits for these objects. The 2-dim distribution of the upper limits at band 2 is similarly formulated as

$$\Sigma^{\text{UL2}}(L_1,L_2) \equiv \int_0^z \frac{d^2V}{dz'd\Omega}\phi^{(2)}(L_1,L_2)\,\Theta\left(L_1^{\text{lim}}(z')\right)\left[1-\Theta\left(L_2^{\text{lim}}(z')\right)\right]dz'.$$
$$(10.55)$$

The superscript UL2 stands for "upper limit at band 2". In statistical terminology, the upper limit case, i.e. we know there is an object but we do only have the upper (or lower) limits of a certain quantity, is referred to as "censored". Though we can define the distribution $\Sigma^{\text{UL2}}(L_1,L_2)$ by eq. 10.55, since the sample objects belonging to this category appear only as upper limits on the plot, a special statistical treatment, referred to as the survival analysis, is required to estimate $\Sigma^{\text{UL2}}(L_1,L_2)$ from the data. Since we select objects at band 1, we do not have upper limits at band 1, because we do not know if there would be an object below the limit. This case is called "truncated" in statistics.

If we select objects at band 2, we can formulate the 2-dim distribution of detected objects and upper limits exactly in the same way as the band 1 selected sample. For the objects detected at both bands, the 2-dim distribution is expressed by eq. (10.53). The objects detected at band 2 but not detected at band 1 are expressed as

$$\Sigma^{\text{UL1}}(L_1,L_2) \equiv \int_0^z \frac{d^2V}{dz'd\Omega}\phi^{(2)}(L_1,L_2)\left[1-\Theta\left(L_1^{\text{lim}}(z')\right)\right]\Theta\left(L_2^{\text{lim}}(z')\right)dz'.$$
$$(10.56)$$

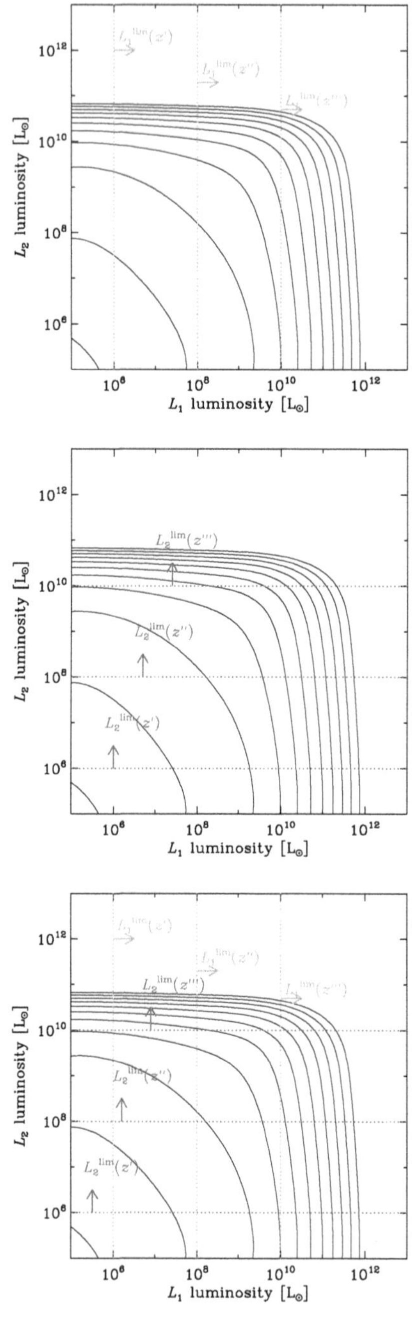

Figure 10.4 Schematic description of the selection effect in a multiband survey (Takeuchi, 2010). Left: selection at band 1, center: selection at band 2, and right: bivariate selection at band 1 and 2. Credit: Takeuchi, T. T., 2010, Monthly Notices of the Royal Astronomical Society, 406, 1830, Fig. 4.

If we can model $L_1^{\lim}(z)$ and $L_2^{\lim}(z)$ precisely including the K-correction, evolutionary effect, etc., we can use the observed bivariate luminosity distribution to estimate the correlation coefficient, or more generally the dependence structure of two luminosities through eqs. (10.53)–(10.55). We can deal with these cases in a unified manner with techniques developed in survival analysis. This will be discussed in the future (Takeuchi et al., in preparation).

10.2 VINE COPULA FOR A CONSTRUCTION OF MULTIVARIATE DISTRIBUTION

Now, studies of galaxy evolution are facing the time for drastic change by multi-band large surveys. Indeed, all of the modern large surveys are performed at many wavelength bands. Connecting the LFs (MFs) obtained at different wavelengths is expected to provide us with new insight into the fundamental physics to drive galaxy evolution (e.g. Caplar et al., 2018; Dutta et al., 2020; Mashian et al., 2016; Vallini et al., 2016, and references therein). This was the motivation to introduce copula to galactic astrophysics and cosmology (Takeuchi, 2010). However, application of the copula was restricted to the bivariate problems, because the copula method was not easy to extend to higher dimensions ($d > 2$), except some special cases like Gaussian. Further, even if we find such a multivariate analytic function in some very fortunate cases, it would probably be very inflexible and impractical, for example, to a realistic statistical estimation in galaxy surveys. Actually, however, a method to improve the copula method and resolve the difficulty of multivariate extension was introduced just some years before Takeuchi (2010). This is based on the decomposition of a general multivariate distribution: a multivariate probability density function can be factorized into a bivariate copulae and univariate density functions. Since we have a rich theoretical method of bivariate copulae, this means that we can extend our methodology to any higher dimension problems (e.g. Aas et al., 2009, and references therein). However, since this decomposition is not unique, we need to sort it out to have a systematic procedure. For this purpose, we introduce the concept of vine copula, invented in the field of graphical modeling (Bedford & Cooke, 2002). Since the work of Aas et al. (2009), vine copulae have been applied to vastly wide range of fields: financial risk management (e.g. Aas, 2016; Allen et al., 2017; Nagler et al., 2019; Sriboonchitta et al., 2014.), insurance (e.g. Mejdoub & Ben Arab, 2018; Peters et al., 2014; Shi & Yang, 2018), weather forecast and engineering (e.g. Alidoost et al., 2019; Kloubert, 2020; Torabi et al., 2020), multivariate time series analysis (e.g. Acar et al., 2019; Almeida et al., 2016; Jäger & Morales Nápoles, 2017), spatio-temporal analysis (e.g. Callau Poduje & Haberlandt, 2018; Gräler, 2014; Gräler & Pebesma, 2011), and technological applications (e.g. Khuntia et al., 2019; Xu et al., 2017a,b), among many others.

10.2.1 VINE COPULA

Here we introduce a vine copula as a systematic method to factorize a multivariate PDF as above. A vine is a concept originally introduced in the field of graphical modeling (Bedford & Cooke, 2002).

10.2.1.1 Factorization of a PDF

As we mentioned in 14.1, this is based on the decomposition of a general multivariate distribution. Let $f(x_1,\ldots,x_d)$ be a joint PDF of a set of d-dimensional vector stochastic variable $\vec{X} = (X_1,\ldots,X_d)$. First, recall the formula of conditional probability

$$f(A,B) = f(B|A)f(A), \tag{10.57}$$

where A and B are events. If we apply this formula to the above PDF, we have

$$
\begin{aligned}
f(x_1,\ldots,x_d) &= f_{2\ldots d|1}(x_2,\ldots,x_d|x_1)f_1(x_1) \\
&= f_{3\ldots d|12}(x_3,\ldots,x_d|x_1,x_2)f_{2|1}(x_2|x_1)f_1(x_1) \\
&\ \ \vdots \\
&= f_{d|123\ldots d-1}(x_d|x_1,\ldots,x_{d-1})\cdots f_{2|1}(x_2|x_1)f_1(x_1)
\end{aligned}
\tag{10.58}
$$

which is known as the chain rule. We start from this well-known mathematical formula. By using a bivariate copula density

$$f_{12}(x_1,x_2) = c\left[F_1(x_1),F_2(x_2)\right]f_1(x_1)f_2(x_2), \tag{10.59}$$

the conditional probability can be expressed as

$$
\begin{aligned}
f_{2|1}(x_2|x_1) &= \frac{f_{12}(x_1,x_2)}{f_1(x_1)} \\
&= c_{12}\left[F_1(x_1),F_2(x_2)\right]f_2(x_2).
\end{aligned}
\tag{10.60}
$$

Similarly, for $f_{23|1}$,

$$f_{23|1}(x_2,x_3|x_1) = c_{23|1}\left[F_{2|1}(x_2|x_1),F_{3|1}(x_3|x_1)\right]f_{2|1}(x_2|x_1)f_{3|1}(x_3|x_1). \tag{10.61}$$

Since

$$\frac{f_{23|1}(x_2,x_3|x_1)}{f_{2|1}(x_2|x_1)} = f_{3|12}(x_3|x_1,x_2), \tag{10.62}$$

we obtain

$$
\begin{aligned}
f_{3|12}(x_3|x_1,x_2) &= c_{23|1}\left[F_{2|1}(x_2|x_1),F_{3|1}(x_3|x_1)\right]f_{3|1}(x_3|x_1) \\
&= c_{23|1}\left[F_{2|1}(x_2|x_1),F_{3|1}(x_3|x_1)\right]c_{13}\left[F_1(x_1),F_3(x_1)\right]f_3(x_3).
\end{aligned}
\tag{10.63}
$$

We can generalize the formula. Set $\vec{v} \equiv (v_1,\ldots,v_k) = x_{i_1},\ldots,x_{i_k}$. If we define $\vec{v}_{-j} \equiv (v_1,\ldots,v_{j-1},v_{j+1},\ldots,v_k)$, the following formula holds

$$f(x|\vec{v}) = c_{x\vec{v}_j|\vec{v}_{-j}}\left[F(x)|\vec{v}_{-j}),F_{v_j}(v_j|\vec{v}_{-j})\right]f(x|\vec{v}_{-j}). \tag{10.64}$$

This is a purely mathematical, direct result of the formula of conditional probability.

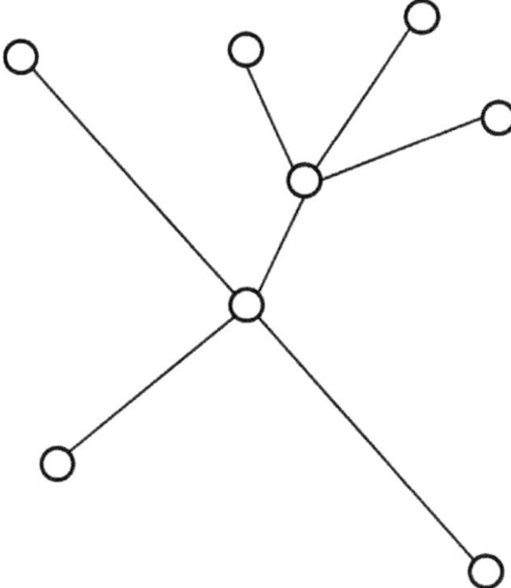

Figure 10.5   An example of a tree graph (Takeuchi & Kono, 2020) (Takeuchi, 2010). Credit: Takeuchi, T. T. & Kono, K. T., 2020, Monthly Notices of the Royal Astronomical Society, 498, 4365, Fig. 1.

We should note that such a decomposition is *not unique* if we consider a permutation of the labels of variables, and when d is large, the number of representations increases dramatically. Hence, we need a systematic procedure to choose which pair combinations should be used to describe the dependence. For this purpose, we introduce the concept of a vine. Since it was invented in the field of graphical modeling (Bedford & Cooke, 2002), it is convenient to use diagrams referred to as graphs.

10.2.1.2   Vine

In order to define it, we have to introduce some graph-theoretical terminologies. We start from the definition of a tree in graph theory.

Definition 5. (tree)
Consider a set of d nodes. When a graph T is connected and has no cycles, T is a tree.

An example of a tree is presented in Fig. 10.5. Based on the concept of tree graph, we define a vine.

Definition 6. (vine)
A vine $\mathcal{V}$ on d elements $\{1,2,\ldots,d\}$ is a set of trees T_i $(i = 1,\ldots,d-1)$, which satisfies the following conditions:

1. *T_1 is a connected tree that have $\{1,2,\ldots,d\}$ as a set of nodes and E_1 as a set of edges,*
2. *For $i = 2,\ldots,d-1$, T_i is a tree that have E_{i-1} as a set of nodes and E_i as a set of edges.*

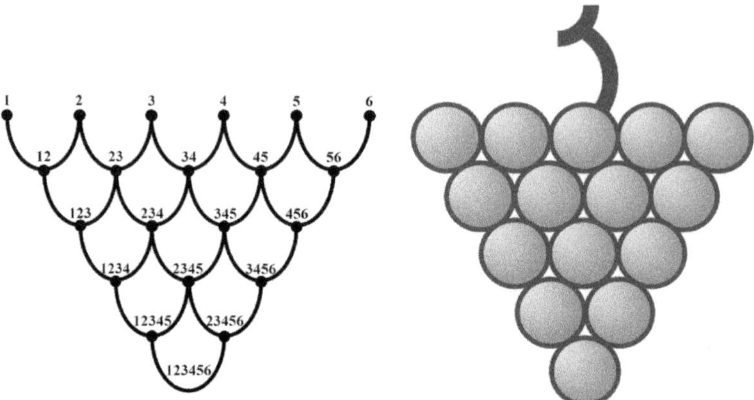

Figure 10.6 A diagrammatic presentation of a concept of vines (Takeuchi & Kono, 2020). As an example, we show the case with six variables. This is called a vine because it looks like a grape (right panel). Credit: Takeuchi, T. T. & Kono, K. T., 2020, Monthly Notices of the Royal Astronomical Society, 498, 4365, Fig. 2.

Definition 7. (regular vine)
If two nodes in tree T_{i+1} are joined by an edge, the corresponding edges in tree i share a node. This is referred to as the proximity condition.

Following discussions will be restricted to regular vines without any loss of generality, since the class of regular vines is still so large that it can treat most of the practical cases. The structure of a regular vine is schematically described in Fig. 10.6. The term "vine" is named after the fact that its botryoidal structure looks similar to a cluster of grapes in its appearance (see Fig. 10.6: e.g. Chapter 1 of Kurowicka & Joe 2011). We note that the tree structure is not strictly necessary for applying the pair-copula methodology, but it helps with identifying the different pair-copula decompositions (Aas et al., 2009).

10.2.1.3 Frequently Used Vines

In practice, two subclasses of vines are frequently used. They are so called "D-vine" and "C-vine", introduced as follows.

Definition 8. D (drawable)-vine
Any joint PDF $f(x_1,\ldots,x_d)$ can be written down by D-vine as follows.

$$f(x_1,\cdots,x_d) = \prod_{j=1}^{d-1}\prod_{i=1}^{d-j} c_{i,(i+j)|(i+1),\ldots,(i+j-1)}$$

$$\times \left[F(x_i|x_i,\ldots,x_{i+j-1}), F(x_{i+j}|x_{i+1},\ldots,x_{i+j-1}) \right]$$

$$\times \prod_{k=1}^{d} f_k(x_k)\,, \tag{10.65}$$

where index j identifies the trees, and i runs over each tree (Bedford & Cooke, 2001).

The definition of D-vine means that for any tree T_i, the number of edges connected to each node never exceeds 2.

To have a concrete idea, we present examples for $d = 3, 4$, and 5.

$$
\begin{aligned}
f(x_1, x_2, x_3) = \\
c_{13|2}\left[F(x_1|x_2)F(x_3|x_2)\right] \\
\times c_{12}\left[F_1(x_1), F_2(x_2)\right] c_{23}\left[F_2(x_2), F_3(x_3)\right] \\
\times f_1(x_1)f_2(x_2)f_3(x_3),
\end{aligned}
\tag{10.66}
$$

$$
\begin{aligned}
f(x_1, x_2, x_3, x_4) = \\
c_{14|23}\left[F(x_1|x_2, x_3)F(x_4|x_2, x_3)\right] \\
\times c_{13|2}\left[F(x_1|x_2), F_3(x_3|x_2)\right] c_{24|3}\left[F(x_2|x_3), F(x_4|x_3)\right] \\
\times c_{12}\left[F_1(x_1), F_2(x_2)\right] c_{23}\left[F_2(x_2), F_3(x_3)\right] c_{34}\left[F_3(x_3), F_4(x_4)\right] \\
\times f_1(x_1)f_2(x_2)f_3(x_3)f_4(x_4),
\end{aligned}
\tag{10.67}
$$

$$
\begin{aligned}
f(x_1, x_2, x_3, x_4, x_5) = \\
c_{15|234}\left[F(x_1|x_2, x_3, x_4)F(x_5|x_2, x_3, x_4)\right] \\
\times c_{14|23}\left[F(x_1|x_2, x_3)F(x_4|x_2, x_3)\right] c_{25|34}\left[F(x_2|x_3, x_4)F(x_5|x_3, x_4)\right] \\
\times c_{13|2}\left[F(x_1|x_2), F_3(x_3|x_2)\right] c_{24|3}\left[F(x_2|x_3), F(x_4|x_3)\right] c_{35|4}\left[F(x_3|x_4), F(x_5|x_4)\right] \\
\times c_{12}\left[F_1(x_1), F_2(x_2)\right] c_{23}\left[F_2(x_2), F_3(x_3)\right] c_{34}\left[F_3(x_3), F_4(x_4)\right] c_{45}\left[F_4(x_4), F_5(x_5)\right] \\
\times f_1(x_1)f_2(x_2)f_3(x_3)f_4(x_4)f_5(x_5).
\end{aligned}
\tag{10.68}
$$

A diagrammatic representation of a D-vine with five variables is shown in Fig. 10.7. This describes the dependence structure of the D-vine well. In this case, it has four layers of the tree structure, labeled as $T_i (i = 1, \ldots, 4)$. Each edge is associated with a pair of copula.

Definition 9. C (canonical)-vine

Any joint PDF $f(x_1, \ldots, x_d)$ can be written down by C-vine as follows.

$$
\begin{aligned}
&f(x_1, \cdots, x_d) \\
&= \prod_{j=1}^{d-1} \prod_{i=1}^{d-j} c_{j,(j+i)|1,\ldots,(j-1)}\left[F(x_j|x_1, \ldots, x_{j-1}), F(j_i|x_1, \ldots, x_{j-1})\right] \prod_{k=1}^{d} f_k(x_k).
\end{aligned}
\tag{10.69}
$$

Each tree T_j has a unique node connected to $d - j$ edges.

When we know a particular variable is a key that governs the interaction in the dataset, the C-vine has a great advantage. We can decide on its "pivot" variable at

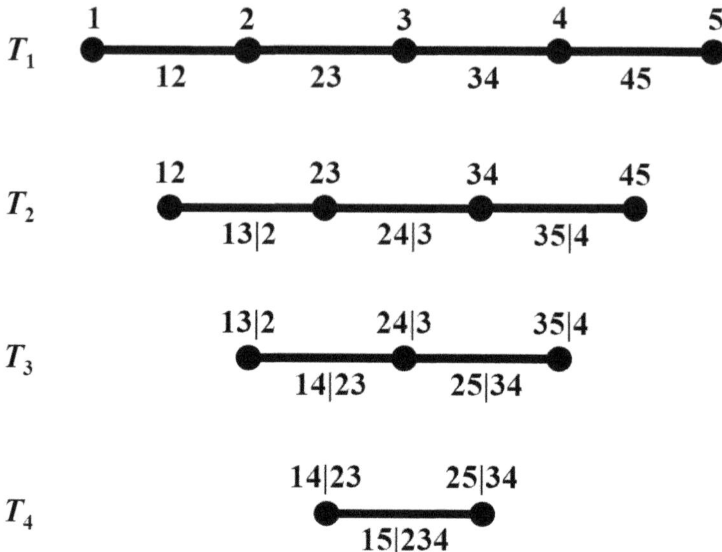

Figure 10.7 A diagrammatic representation of a D-vine with five variables. It contains four trees and ten edges (Takeuchi & Kono, 2020). Each edge is associated with a pair copula. Credit: Takeuchi, T. T. & Kono, K. T., 2020, Monthly Notices of the Royal Astronomical Society, 498, 4365, Fig. 3.

the root of the C-vine. We show examples for $d = 3, 4$, and 5.

$$f(x_1, x_2, x_3) =$$
$$c_{23|1} \left[F(x_2|x_3) F(x_3|x_1) \right]$$
$$\times c_{12} \left[F_1(x_1), F_2(x_2) \right] c_{13} \left[F_1(x_1), F_3(x_3) \right]$$
$$\times f_1(x_1) f_2(x_2) f_3(x_3) \,, \tag{10.70}$$

$$f(x_1, x_2, x_3, x_4) =$$
$$c_{34|12} \left[F(x_3|x_1, x_2) F(x_4|x_1, x_2) \right]$$
$$\times c_{23|1} \left[F(x_2|x_1), F(x_3|x_1) \right] c_{24|1} \left[F(x_2|x_1), F(x_4|x_1) \right]$$
$$\times c_{12} \left[F_1(x_1), F_2(x_2) \right] c_{13} \left[F_1(x_1), F_3(x_3) \right] c_{14} \left[F_1(x_1), F_4(x_4) \right]$$
$$\times f_1(x_1) f_2(x_2) f_3(x_3) f_4(x_4) \,, \tag{10.71}$$

$$f(x_1, x_2, x_3, x_4, x_5) =$$
$$c_{45|123} \left[F(x_4|x_1, x_2, x_3) F(x_5|x_1, x_2, x_3) \right]$$
$$\times c_{34|12} \left[F(x_3|x_1, x_2) F(x_4|x_1, x_2) \right] c_{35|12} \left[F(x_3|x_1, x_2) F(x_5|x_1, x_2) \right]$$
$$\times c_{23|1} \left[F(x_2|x_1), F(x_3|x_1) \right] c_{24|1} \left[F(x_2|x_1), F(x_4|x_1) \right] c_{25|1} \left[F(x_2|x_1), F(x_5|x_1) \right]$$
$$\times c_{12} \left[F_1(x_1), F_2(x_2) \right] c_{13} \left[F_1(x_1), F_3(x_3) \right] c_{14} \left[F_1(x_1), F_4(x_4) \right] c_{15} \left[F_1(x_1), F_5(x_5) \right]$$
$$\times f_1(x_1) f_2(x_2) f_3(x_3) f_4(x_4) f_5(x_5) \,. \tag{10.72}$$

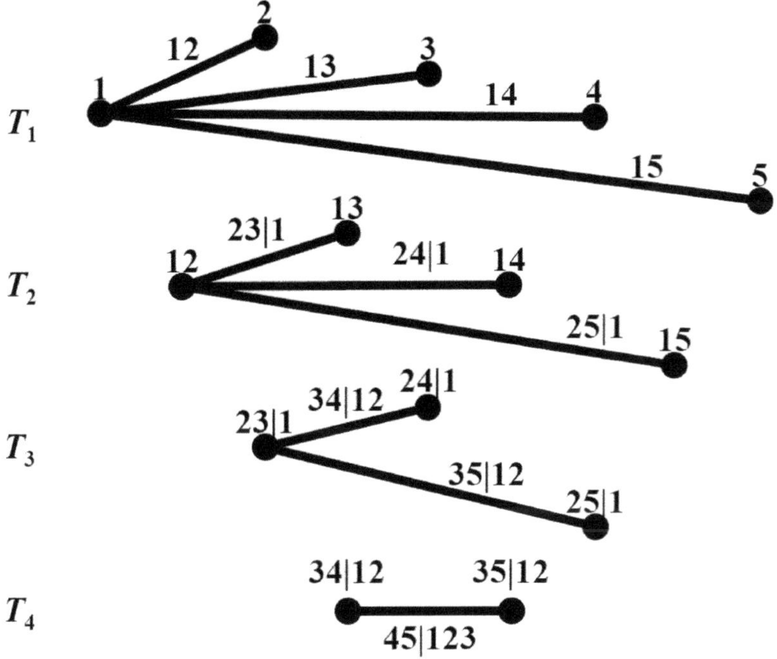

Figure 10.8   Same as Fig. 10.7 but for a C-vine with five variables (Takeuchi & Kono, 2020). Credit: Takeuchi, T. T. & Kono, K. T., 2020, Monthly Notices of the Royal Astronomical Society, 498, 4365, Fig. 4.

For the case of $d = 3$, the general expression for vine structure is expressed as eq. (10.66) or (10.70). It is valid for both D- and C-vines. There are six ways with a permutation between x_1, x_2, and x_3, but only three of them yield different pair-copula decompositions. Further, each of the three corresponds both to D- and C-vine. Namely, both D- and C-vines cover the whole possible structures of pair copula decompositions for $d = 3$. For $d = 4$, there are 24 regular vine decompositions, of which 12 are D-vine and 12 are C-vine ones. There is no overlap between any of the D- and C-vines. Further, there are no other regular vine decompositions. This guarantees the wide range of applicability of D- and C-vines for the pair-copula decomposition. For $d = 5$, there are 60 different D-vines and 60 different C-vines. Between any of these 60 D-vines and 60 C-vines, there is no overlap. However, unlike $d \leq 4$, there are 120 more regular vines that are not D- nor C-vines. Hence in total, there are 240 different possible pair-copula decompositions. In general, for any d, there are $d!/2$ D-vines and the same number of C-vines. Detailed explanations on these results are presented in Aas et al. (2009).

10.2.2   LIKELIHOOD FOR VINES

Practically, we can safely restrict the pair-copula decompositions to D- and C-vines. When we have n data sample $\{\vec{x}_m\} = \{(x_{m,1}, x_{m,2}, \ldots, x_{m,d})\}$ $(m = 1, \ldots, n)$, the log likelihood of a D-vine for the parameter estimation is

$$\ln \mathscr{L}(\vec{\theta}; \vec{x}_m, m = 1, \ldots, n)$$

$$= \sum_{m=1}^{n} \ln f(\vec{\theta}; \vec{x}_m)$$

$$= \sum_{m=1}^{n} \sum_{j=1}^{d-1} \sum_{i=1}^{d-j} \ln c[\vec{\theta}(\text{copula})]_{i,(i+j)|(i+1),\ldots,(i+j-1)} \tag{10.73}$$

$$\times \left[F(x_i | x_i, \ldots, x_{i+j-1}), F(x_{i+j} | x_{i+1}, \ldots, x_{x_{i+j-1}}) \right]$$

$$+ \sum_{k=1}^{d} \ln f_k[\vec{\theta}(\text{marginal}); x_k] \,, \tag{10.74}$$

where we denote the parameter vectors for copulae and marginals in a symbolic way for saving space. Similarly, the log-likelihood of a C-vine is

$$\ln \mathscr{L}(\vec{\theta}; \vec{x}_m, m = 1, \ldots, n)$$

$$= \sum_{m=1}^{n} \sum_{j=1}^{d-1} \sum_{i=1}^{d-j} \ln c[\vec{\theta}(\text{copula})]_{j,(j+i)|1,\ldots,(j-1)} \tag{10.75}$$

$$\times \left[F(x_i | x_i, \ldots, x_{i+j-1}), F(x_{i+j} | x_{i+1}, \ldots, x_{x_{i+j-1}}) \right]$$

$$+ \sum_{k=1}^{d} \ln f_k[\vec{\theta}(\text{marginal}); x_k] \,. \tag{10.76}$$

We should maximize the log-likelihood eq. (10.73) or (10.75) to estimate the parameter set $\vec{\theta} = [\vec{\theta}(\text{copula}), \vec{\theta}(\text{marginal})]$. In principle, both $\vec{\theta}(\text{copula})$ and $\vec{\theta}(\text{marginal})$ can be estimated simultaneously. In research fields such as economics, however, the likelihood estimation of the marginals is often found to be difficult or even implausible. Hence, they do not use the exact form of the likelihood but instead maximize the so-called pseudo-likelihood (Aas et al., 2009). In contrast, in astrophysics, we have a rich field of research on the estimation of the marginals, e.g. the LF or MF of galaxies at a certain observed wavelength (e.g. Johnston, 2011; Takeuchi et al., 2000). Then, we can simply use the estimated marginals and plug in them for the log-likelihood. Namely, we can omit the estimation step for the marginals when we try to estimate copula parameters in a different sense from other research fields.

In this step, we can obtain the error of each parameter and the goodness of fit indicator(s) of the model, since we have the likelihood ellipsoid and maximum likelihood value.

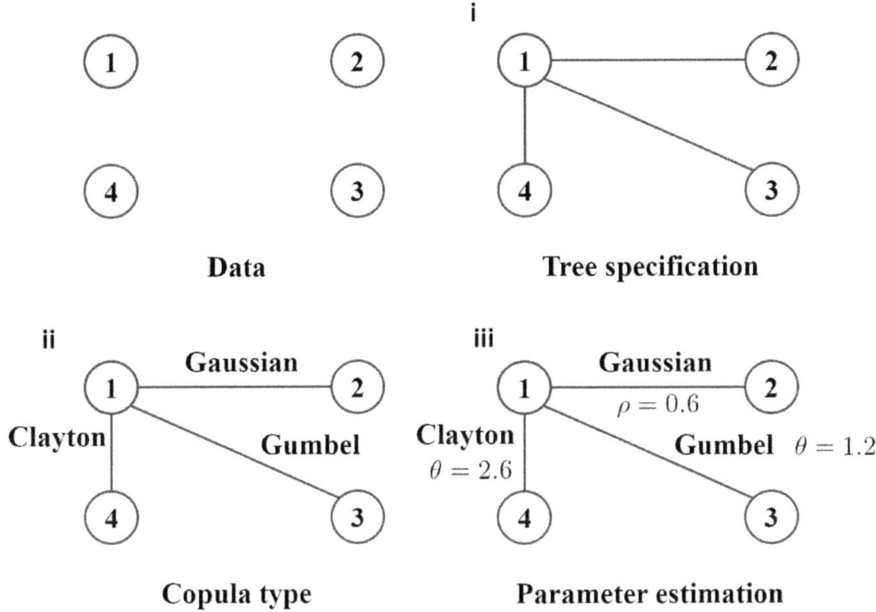

Figure 10.9 A schematic example for the description of the estimation with vines (Takeuchi & Kono, 2020). Gaussian, Clayton, and Gumbel stand for three popular copula types often used in practice. Each of them has some specifying parameters, and they are estimated by usual statistical procedure in Step iii. Credit: Takeuchi, T. T. & Kono, K. T., 2020, Monthly Notices of the Royal Astronomical Society, 498, 4365, Fig. 5.

10.2.3　MODEL SELECTION WITH VINES

The likelihood estimation we discussed above is only one of the steps of the full estimation problem. Schematically, the model to be specified has the following structure as

$$\text{Model} = \text{structure (trees)} + \text{copula families} + \text{copula parameters}. \qquad (10.77)$$

Namely, for the estimation procedures we should consider

1. selection of a specific decomposition,
2. choice of pair-copula types,
3. estimation of the copula parameters.

This is schematically described in Fig. 10.9. As for the copula types since we have an infinitely large degree of freedom for the choice of copulae, the model selection is fundamentally important. Often we have to choose an optimal model from a large choice of candidate models with different numbers of parameters. In such a case, usual goodness-of-fit method does not work, since obviously more parameters give a better fit. For such a case, a model selection procedure should be used instead.

The most popular tool for model selection is the information criterion (e.g. Takeuchi, 2000, and references therein: see Chapter 8). One of the first and most widely used information criteria is the Akaike Information Criterion (AIC: Akaike 1974b),

$$\mathrm{AIC}(q) = -2[\ln \mathscr{L}(\hat{\vec{\theta}}) - q] \tag{10.78}$$

where $\hat{\vec{\theta}}$ stands for the parameter that maximizes the likelihood, and q is the number of parameters. The model that gives the smallest AIC is selected. This is a natural extension of the classical maximum likelihood estimation, corrected for the bias introduced by the parameter estimation step (Akaike, 1974b). Some other information criteria are also used. Among them, the Bayesian Information Criterion (BIC)

$$\mathrm{BIC}(q) = -2 \left[\ln \mathscr{L}(\hat{\vec{\theta}}) - \frac{q}{2} \ln n \right] \tag{10.79}$$

(n: sample size) is also often used (Schwarz, 1978). This model selection is performed in the step of copula selection. Each copula is determined by the evaluation of such information criterion among all possible copula types, as well as the classical goodness-of-fit indicators like χ^2-statistic.

As we saw in Section 10.2.1, the tree determination and subsequent specification of copulae would be computationally heavy because of the factor $d!/2$. Thanks to the present-day development of software, we can treat a problem with a dimension up to $d \sim 500$. Some software packages are available for this problem (e.g. Brechmann & Schepsmeier, 2013). The construction and estimation of a multivariate PDF based on the vine copula is completed by this step.

10.2.4 APPLICATION TO CONSTRUCT A MULTIVARIATE MASS FUNCTION (MF) OF GALAXIES

The relationship between stars, atomic gas, and molecular gas mass is one of the most important issues in galaxy evolution and studied very extensively. For this aim, a multivariate MF is obviously a fundamental tool (see e.g. an elaborate work of Rodríguez-Puebla et al., 2020). Here we present one astrophysically interesting example of multivariate mass function, a three-variate mass function of stellar, atomic gas, and molecular gas mass. The formal procedure is exactly the same for the case of luminosity functions.

10.2.5 MULTIVARIATE MF

First, we prepare some notations for the multivariate MF. A mass function of galaxies is defined as a number density of galaxies whose mass lies between a logarithmic interval $[\log M, \log M + \mathrm{d}\log M]$:

$$\phi^{(1)}(M) \equiv \frac{\mathrm{d}n}{\mathrm{d}\log M}. \tag{10.80}$$

For mathematical simplicity, we define the as *being normalized*, i.e.,

$$\int \phi^{(1)}(M)\mathrm{d}\log M = 1. \tag{10.81}$$

Table 10.1

Parameters for the gas mass function

MF	α	ϕ_* [Mpc^{-3}dex^{-1}]	M_* [$M_\odot$]	Reference
Atomic gas	-1.25 ± 0.02	$(4.5 \pm 0.2) \times 10^{-3}$	8.7×10^9	Jones et al. (2018)
Molecular gas	-1.30 ± 0.16	$(5.9 \pm 2.8) \times 10^{-3}$	9.4×10^9	Kereš et al. (2003)

Hence, this corresponds to a PDF. We also define the cumulative MF as

$$\Phi^{(1)}(M) \equiv \int_{\log M_{\min}}^{\log M} \phi^{(1)}(M')\mathrm{d}\log M' , \tag{10.82}$$

where $M_{\min}$ is the minimum mass of galaxies considered. This corresponds to the DF.

If we denote univariate MFs as $\phi_k^{(1)}(M_k)$ $(k = 1,\ldots,d)$, the joint multivariate PDF $\phi^{(d)}(M_1,\ldots,L_d)$ is described by a differential copula $c(u_1,\ldots,u_k)$ as

$$\phi^{(d)}(M_1,\ldots,M_d) \equiv c\left[\Phi_1^{(1)}(M_1),\ldots,\Phi_d^{(1)}(M_d)\right] \phi_1^{(1)}(M_1) \cdots \phi_d^{(1)}(M_d) . \tag{10.83}$$

For this analysis, we made use of the R package `VineCopula`[2] (Aas et al., 2009). It provides statistical inference of C- and D-vine copulae. This package enables us to construct copula density and tree structures from multivariate data. The optimal combination of copulae and their parameters are chosen through AIC, BIC and maximum likelihood estimation with `RVineStructureSelect` function.

10.2.6 THE STELLAR–ATOMIC GAS–MOLECULAR GAS MULTIVARIATE MF

As marginals of the multivariate MF, we should determine the univariate MFs for stellar mass, atomic gas mass, and molecular gas mass, respectively. The univariate MF for atomic gas mass and molecular gas mass are known to be well described by the Schechter function (Schechter, 1976b, see Chapter 2).

$$\phi^{(1)}(M) = (\ln 10)\, \phi_* \left(\frac{M}{M_*}\right)^{1-\alpha} \exp\left[-\left(\frac{M}{M_*}\right)\right] , \tag{10.84}$$

For the atomic gas mass, we took the parameters from Jones et al. (2018) but in a normalized form with eq. (10.81), i.e., we did not use the normalization factor $\phi_{\mathrm{HI}*}$. Similarly, we took the Schechter function parameters for the molecular gas from Kereš et al. (2003). These parameters are summarized in Table 10.1.

---

[2]https://github.com/tnagler/VineCopula.

Table 10.2

Parameters for the stellar mass function (D'Souza et al., 2015).

α_1	ϕ_{1*} $[\mathrm{Mpc}^{-3}\mathrm{dex}^{-1}]$	M_{1*} $[M_\odot]$	α_2	ϕ_{2*} $[\mathrm{Mpc}^{-3}\mathrm{dex}^{-1}]$	M_{2*} $[M_\odot]$
1.082	6.0×10^{-2}	4.1×10^{10}	1.120	2.5×10^{-3}	9.9×10^{10}

Recent studies revealed that the stellar MF is, however, better described by a double Schechter function (e.g. D'Souza et al., 2015)

$$\phi^{(1)}(M) = \frac{\phi_{1*}}{M_{1*}} \left(\frac{M}{M_{1*}}\right)^{-\alpha_1} \exp\left[-\left(\frac{M}{M_{1*}}\right)\right] + \frac{\phi_{2*}}{M_{2*}} \left(\frac{M}{M_{2*}}\right)^{-\alpha_2} \exp\left[-\left(\frac{M}{M_{2*}}\right)\right].$$

(10.85)

We should note that D'Souza et al. (2015) defined the double Schechter function (eq. (10.85)) for a linear mass interval $\mathrm{d}M$, *not* $\mathrm{d}\log M$. The parameters are shown in Table 10.2.

10.2.7 DATA

In this work, we used a subsample of the combined dataset compiled by Calette et al. (2018). Their original sample consists of Golden, Silver, and Bronze Categories both for $\mathrm{H\,I}$ and H_2. Full details of the original sample are found in the Appendix of Calette et al. (2018). We briefly describe the dataset used here.

10.2.7.1 The Compiled Galaxy Sample with $\mathrm{H\,I}$ Information

We used the following datasets for $\mathrm{H\,I}$ information.
Golden Category

- *GALEX* Arecibo SDSS Survey (GASS; Catinella et al. 2013): an optically-selected a subsample of 760 galaxies more massive than $10^{10}\,M_\odot$ taken from a parent SDSS DR6 sample volume-limited in the redshift range $0.025 < z < 0.05$ and cross-matched with the ALFALFA and *GALEX* surveys.
- Field galaxies from the Herschel Reference Survey (HRS; Boselli et al. 2014a,b; Boselli et al. 2010): a K-band volume-limited ($15 \le D\,[\mathrm{Mpc}] \le 25$) sample of 323 galaxies complete to $K_s = -12$ and -8.7 mag for late-type galaxies and early-type galaxies, respectively.
- Field early type galaxies from the ATLAS^{3D} $\mathrm{H\,I}$ sample (Serra et al., 2012): a sample of 166 local early-type galaxies observed in detail with integral field unities (IFUs; Cappellari et al. 2011). The distance range of the sample is between 10 and 47 Mpc; the sample includes 39 galaxies from the Virgo Cluster, but for the Golden category, the early-type galaxies in the Virgo cluster core was excluded by Calette et al. (2018).

Bronze Category

- Analysis of the interstellar Medium of Isolated GAlaxies (AMIGA; Lisenfeld et al. 2011): a redshift-limited sample ($1500 \leq v_{\text{rec}}[\text{km s}^{-1}] \leq 5000$) consisting of 273 isolated galaxies with reported multi-band imaging and CO data.

10.2.7.2 The Compiled Galaxy Sample with CO (H_2) Information

We used the following datasets for H_2 information.
Golden Category

- Field galaxies from the Herschel Reference Survey (HRS): the same sample described above (excluding Virgo Cluster core), with 155 galaxies with available CO information (101 detections and 54 non-detections).
- CO Legacy Legacy Database for GASS (COLD GASS; Saintonge et al. 2011): a program aimed at observing CO(1–0) line fluxes with the IRAM 30 m telescope for galaxies from the GASS survey described above. From the CO fluxes, the total CO luminosities (and hence the H_2 masses) were calculated for 349 galaxies.
- Field early type galaxies from the ATLAS^{3D} H_2 sample (Young et al., 2011): the same sample described above (excluding the Virgo Cluster core) but with observations in CO using the IRAM 30 m Radio Telescope. The sample amounts for 243 early type galaxies with CO observations.

Bronze Category

- Analysis of the interstellar Medium of Isolated GAlaxies (AMIGA; Lisenfeld et al. 2011: the same sample described above. The authors carried out their own observations of CO(J: 1–0) with the IRAM 30 m or the 14 m FCRAO telescopes for 189 galaxies; 87 more were compiled from the literature.

10.2.8 RESULT OF THE VINE COPULA ANALYSIS

We present the constructed M_*–M_{HI}–M_{H_2} MF with a 3-dim vine copula. The goodness-of-fit analysis of the obtained 3-dim MF to the dataset gave the p-value $p = 0.41$ which suggests that the fit is appropriate. The M_*–M_{HI}–M_{H_2} MF is described by a C-vine copula. The structure is presented in Fig. 10.10. The structure of the estimated C-vine is labeled as (3,1), (3,2), and (2,1| 3) in Fig. 10.10. Here, 1 corresponds to M_*, 2 to M_{HI}, and 3 to M_{H_2}, respectively.

The relation between M_* and M_{H_2} was well described by the BB8 copula

$$C(u_1, u_2; \theta, \delta) = \frac{1}{\delta} \left\{ 1 - \left[1 - \frac{1}{1 - (1-\delta)^{\theta}} \left(1 - (1-\delta u_1)^{\theta} \right) \left(1 - (1-\delta u_2)^{\theta} \right) \right]^{\frac{1}{\theta}} \right\},$$

$$(\theta \geq 1, \ \delta \in (0,1]), \tag{10.86}$$

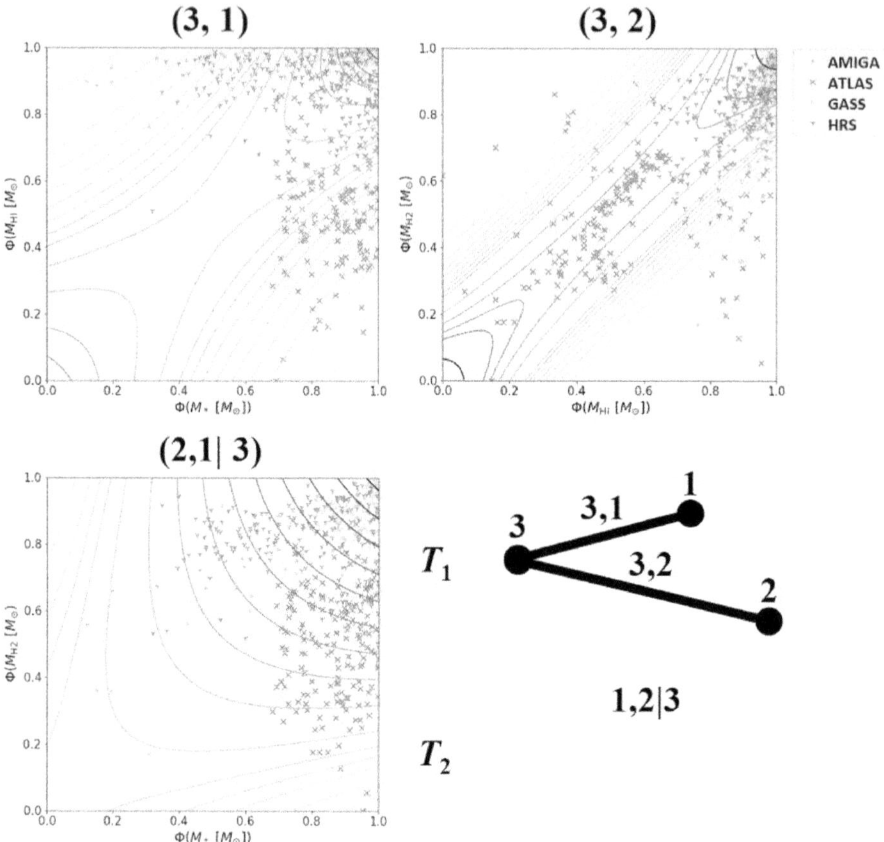

Figure 10.10 Vine copulae which are estimated from the M_*-M_{HI}-M_{H_2} data (Takeuchi & Kono, 2020). Top-left: the estimated copula for M_{H_2}–M_* relation, Top-right: the estimated copula for M_{H_2}–M_{HI} relation, Bottom-left: the estimated conditional copula for M_*–M_{HI} relation with M_{H_2} given, and Bottom-right: tree structure of vine copulae. See ebook for the color version of this figure. Credit: Takeuchi, T. T. & Kono, K. T., 2020, Monthly Notices of the Royal Astronomical Society, 498, 4365, Fig. 6.

with parameters $\theta = 6.00 \pm 1.76$ and $\delta = 0.49 \pm 0.10$. Kendall's τ of this pair is 0.35, reflecting the broad distribution with a loose correlation.

In contrast, the relation between M_{HI} and M_{H_2} was found to be reproduced by the Frank copula,

$$C(u_1, u_2; \delta) = -\frac{1}{\delta} \log \left[1 - \frac{\left(e^{-\delta u_1} - 1\right)\left(e^{-\delta u_2} - 1\right)}{e^{\delta} - 1} \right], \quad (\delta \in \mathbb{R} \setminus \{0\}), \quad (10.87)$$

with a parameter $\delta = 8.80 \pm 0.43$. Kendall's τ of this pair is 0.63. The scatter plot of the data on the M_{HI}–M_{H_2} plane, and the corresponding bivariate PDF are presented in Fig. 10.12. On this plane, We see a moderately strong dependence between these

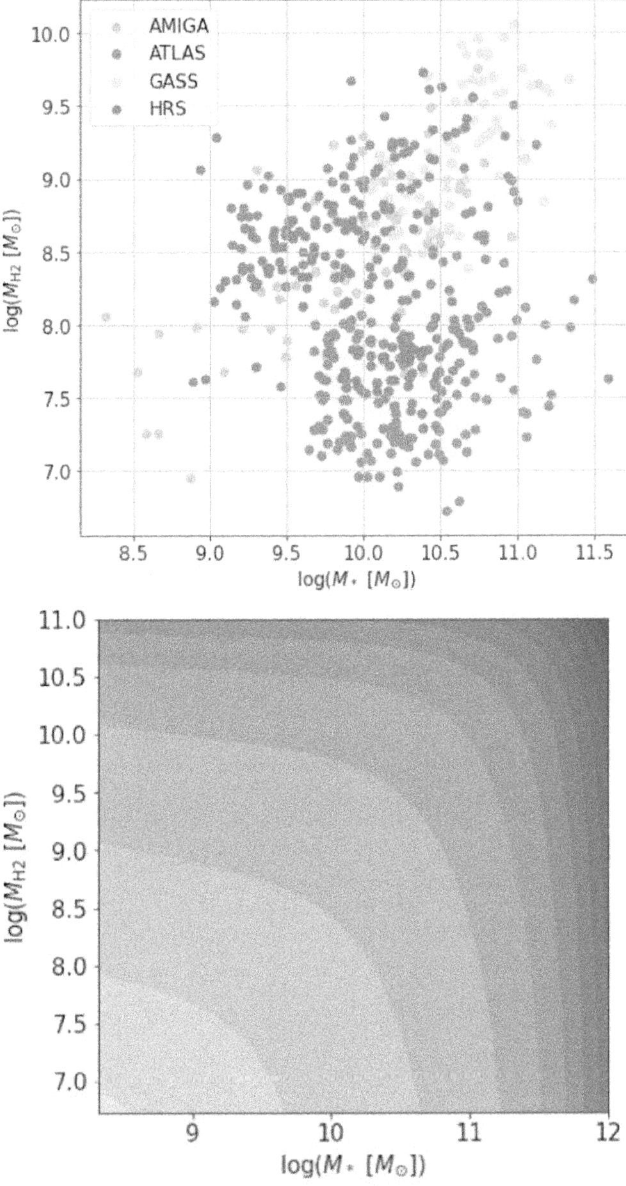

Figure 10.11 Relation between M_* and M_{H_2} (Takeuchi & Kono, 2020). Left: the data distribution on the M_*–M_{H_2} plane. Right: estimated bivariate PDF. The selected copula is the BB8 copula with a parameter of $\theta = 6.00 \pm 1.76$ and $\delta = 0.49 \pm 0.10$. See ebook for the color version of this figure. Credit: Takeuchi, T. T. & Kono, K. T., 2020, Monthly Notices of the Royal Astronomical Society, 498, 4365, Fig. 7.

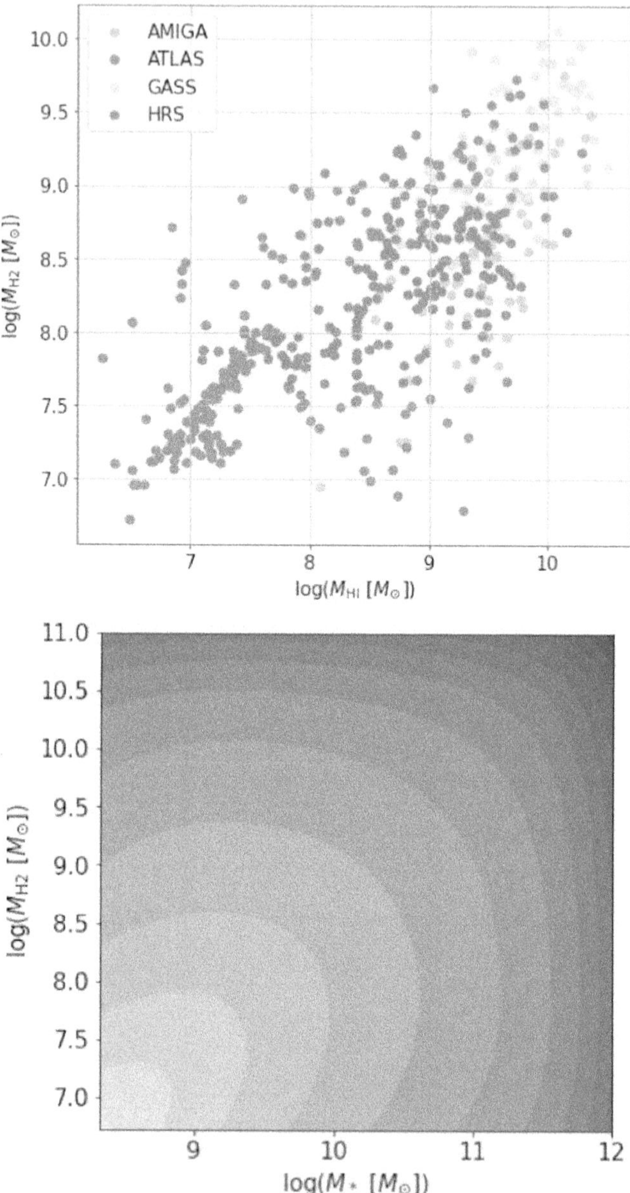

Figure 10.12 Relation between M_{HI} and $M_{\mathrm{H_2}}$ (Takeuchi & Kono, 2020). Left: the data distribution on the M_{HI}–$M_{\mathrm{H_2}}$ plane. Right: the corresponding bivariate PDF. The selected copula is the Frank copula with a parameter of $\delta = 8.80 \pm 0.43$. See ebook for the color version of this figure. Credit: Takeuchi, T. T. & Kono, K. T., 2020, Monthly Notices of the Royal Astronomical Society, 498, 4365, Fig. 8.

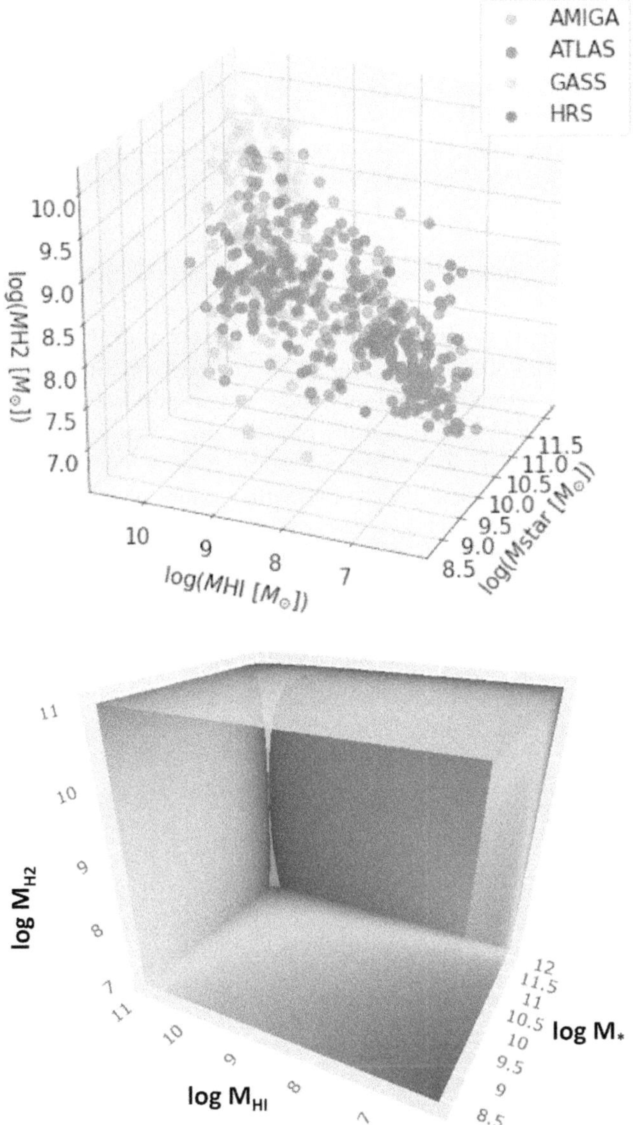

Figure 10.13 The M_*-M_{HI}-M_{H_2} 3-dim mass function which was obtained from the vine copula likelihood estimation (Takeuchi & Kono, 2020). Left: the data distribution in the mass space. Right: The estimated 3-dim mass function (PDF). See ebook for the color version of this figure. Credit: Takeuchi, T. T. & Kono, K. T., 2020, Monthly Notices of the Royal Astronomical Society, 498, 4365, Fig. 9.

two variables, reflected in $\tau = 0.63$.

For the conditional copula between M_* and M_{HI} for given $M_{\mathrm{H_2}}$, the Student-t copula

$$C(u_1,u_2;\rho,v) = \int_{-\infty}^{u_1}\int_{-\infty}^{u_2} \frac{1}{2\pi\sqrt{1-\rho^2}}\left[1+\frac{s^2+t^2-2\rho st}{v(1-\rho^2)}\right]^{-\frac{v+2}{2}} \mathrm{d}s\mathrm{d}t,$$
$$(\rho \in (-1,1),\ v > 2), \tag{10.88}$$

was selected as the appropriate copula with $\rho = 0.29 \pm 0.03$ and $v = 30.0$, respectively.

Though it is not easy to visualize the 3-dim structure of the PDF, its heavily asymmetric structure is well described in Fig. 10.13. Of course, it requires further analysis for a physical interpretation, the copula method can provide us with a fundamentally an important tool to understand the physical processes behind the multivariate LF/MF. The result here is just a demonstration of how the vine copula performs well for the multidimensional LF/MF estimation. More physical discussion should be done as the next step.

If we go to much higher dimensions, the human-intuitive approach may not work anymore even for a simple visualization. Very plausibly, we will confront the need for such a tremendously large data analysis. A machine-aided method will be a promising way to address such issues. We will discuss such methods in Part III.

11 High-Dimensional Statistical Analysis

11.1 INTRODUCTION

In the past decade, the term "Big Data" has been widely recognized and observed almost all the fields of scientific researche, of course also in astronomy. Current astronomical instruments provide us with overwhelmingly large data, providing quantitative information of their physical condition. The most typical big data in astronomy is the product of large surveys, such as the Sloan Digital Sky Survey (SDSS)[1], the Dark Energy Spectroscopic Instrument (DESI) Survey[2], Gaia[3], the Panoramic Survey Telescope And Rapid Response System (Pan-STARRS) Surveys[4], among many others.

We should note, however, that there is another type of big data in astronomy. Detailed observations are often very time-consuming, and it is not easy to map objects to obtain many independently sampled measurements. Such a situation is frequently found in the integral field spectroscopy (e.g., UVES at the Very Large Telescope (VLT)[5], SAURON on the William Herschel Telescope[6], SDSS MaNGA data[7], among others. Another typical example of this kind is spectroscopic mapping surveys by radio interferometers (e.g., Atacama Large Millimeter/Submillimeter Array) or by satellite instruments at X-ray, ultraviolet or mid/far-infrared wavelengths. In such a case, if we denote the dimension in the wavelength (or frequency) with d and the number of samples with n, we often find that $n \ll d$.

Classical statistical analysis implicitly assumed a sample size n is much larger than data dimension d ($n \gg d$). As we mentioned, this is not true for many cases in scientific researche. Traditionally in astrophysics, such a situation is regarded as an ill-posed problem, and there was no choice but to throw away most of the information in wavelength direction to let $d < n$. It was more or less the same for other research fields before 2010. Many researchers have simply given up further analysis or even did not find a need to deal with such problems before the advent of the Big Data era. For example, Hand et al. (1994) published a well-known compilation of statistical data, containing more than 500, but most of them satisfied the classical condition of the assumption $d < n$. However, in the late 1990s, data with $d \gg n$ suddenly started to appear along with the development of information science. One of the

---

[1] https://www.sdss.org/surveys/.

[2] https://www.desi.lbl.gov/.

[3] https://www.cosmos.esa.int/web/gaia.

[4] https://outerspace.stsci.edu/display/PANSTARRS/.

[5] https://www.eso.org/sci/facilities/paranal/instruments/uves.html.

[6] https://www.ing.iac.es/PR/SH/SH2007/sauron.html.

[7] https://www.sdss.org/surveys/manga/.

DOI: 10.1201/9781003104315-11

epoch-making example of such kind of researche was Golub et al. (1999). They have used gene expression monitoring by DNA microarrays applied to human acute leukemias as a test case. The dimension of the DNA microarray was $d = 7129$ and the sample size $n = 72$. Statistics in 1990s could not guarantee the accuracy for the estimation in such a case, since it was on the basis of $d < n$.

Various fundamental problems were known for data with the condition of $n \ll d$:

1. Sample covariance matrix is usually numerically unstable or does not exist, and classical multivariate analysis cannot work.
2. We meet various types of "curse of dimity".
3. Calculation cost is often unreachably tremendous.

A method to handle such a problem have been desired since then. In order to study such type of data with a very high dimension and small sample data, first of all we had to push back the boundaries of the traditional technique of multivariate statistical analysis. This is because traditional methods are not able to capture the characteristics of the high-dimensional data space and hence, not able to make use of rich information which the data intrinsically contain.

The situation drastically changed in 2000s. A turning point was brought from the field of probability theory and theoretical physics based on the random matrix theory. In these studies, the asymptotic behavior of eigenvalues of the sample covariance matrix in the limit as $d \longrightarrow \infty$ was explored under Gaussian or Gaussian-type assumptions on the population distribution when d and n increase at the same rate, i.e., $n/d \longrightarrow c > 0$ (c: const.) (e.g., Baik et al., 2005; Baik & Silverstein, 2006; Johnstone, 2001; Paul, 2007). In our current interest, however, the assumptions as $n/d \longrightarrow c > 0$ and Gaussian(-type) are actually unrealistically restrictive. A seminal paper by Hall et al. (2005) introduced a new concept, high dimension low sample size (HDLSS) asymptotics[8]. The limits they have studied were taken as the dimension of the data $d \longrightarrow \infty$, while the sample size n was fixed. At that time, this was a revolutionary departure from the traditional asymptotics based on $n \longrightarrow \infty$ with d fixed.

Hall et al. (2005) and Ahn et al. (2007) discussed an important representation of the HDLSS data, but still under the assumption of Gaussian(-type). In contrast, Yata & Aoshima (2012) developed a new theory by clearing away this assumption and discovered a completely different type of representation when the Gaussian assumption is not satisfied (see Sec.11.2 for the details). These works were the foundation of research in mathematical statistics on very high dim data analysis, and evolved into an important part of the currently fashionable research area, the Big Data Science. The new insight shed light on various areas of studies as genomics, medical research, neuroscience, and image and shape analysis. Then, in their sequence of papers, M. Aoshima, K. Yata, A. Ishii and collaborators further developed a new framework of statistical methodology to tackle this type of problem (e.g., Aoshima & Yata, 2011, 2014, 2015, 2019; Ishii et al., 2016; Yata & Aoshima, 2012, 2013,

---

[8]The first consideration on the HDLSS asymptotics, however, dates back to 1980s by the work of Casella & Hwang (1982)

and references therein). Readers who are interested in the mathematical background of the high-dimensional statistical analysis itself are guided to some reviews,(e.g., Aoshima et al., 2018; Aoshima & Yata, 2017).

In spite of the great developments in other research fields, this new methodology of statistics seemed to be completely unknown in astronomy. High dim data as the ALMA data are regarded as a sum of informative signals and noise, but we cannot separate them quantitatively without prior information. Supposing that we know the frequencies of emission lines is a kind of such prior information which has been traditionally used in traditional astronomy, but this is not what we prefer. While some frequency regions are considered to contain no information, we proceeded extracting information not by traditional procedure, but by an objective statistical method without a prior information. The high-dimensional statistical analysis is one of such an objective methods.

Thus, in this work, we introduce the method of high-dimensional statistical analysis for the HDLSS data in astronomy. In order to examine the performance of the high-dimensional methods to the astronomical data, we apply it to a spectroscopic map of the central region of NGC 253. NGC 253 is a nearby prototypical starburst galaxy (Rieke et al., 1980), and has been observed in various wavelengths. We apply the high-dimensional principal component analysis (PCA) to the ALMA spectral map (Ando et al., 2017). Ando et al. (2017) have presented an 8 pc $\times$ 5 pc resolution view of the central ~ 200 pc region of NGC 253, based on the ALMA Band 7 observations. As we introduce in detail, the ALMA maps can be regarded as being typically HDLSS. The spectrum of NGC 253 is very rich in molecular lines (e.g., Ando et al., 2017; Martín et al., 2021). Another focus of this work is to classify the spectra full of spectral line features by the PCA. The evolution of galaxies is mainly driven by star formation, a transition from interstellar medium (ISM) to stars. Various phases of the ISM are related, and the evolution of the ISM is a key to complete the understanding of the galaxy evolution. Spectroscopic observations are of vital importance to extract and interpret the information of matter in galaxies(e.g., Martín et al., 2021). Then, this classification is potentially a powerful tool for the studies of galaxy evolution if it turns out to be useful.

11.2 METHOD: HIGH-DIMENSIONAL STATISTICAL ANALYSIS

11.2.1 HIGH-DIMENSIONAL STATISTICAL ANALYSIS

The HDLSS data have a sample size of n and data dimension d with $n \ll d$. We mean the newly developed statistical methodology to deal with the HDLSS data by "high-dimensional statistical analysis". When we have HDLSS data, classical statistical analysis such as introduced in Chapter 8 is not guaranteed to be valid. Particularly, traditional limit theorems do not hold for the HDLSS data.

11.2.1.1 Unusual Behavior of the High-Dimensional Statistics

As a pedagogical example, we demonstrate a typical "unusual" properties only appear in the high-dimensional statistical analysis, known as the strong inconsistency.

We consider a sample mean $\bar{\bar{x}}$,

$$\bar{\bar{x}}_n \equiv \frac{1}{n} \sum_{i=1}^{n} \vec{x}_i \, . \tag{11.1}$$

Suppose that $\vec{x}_i$s are drawn from a d-dim distribution with a mean $\vec{\mu}$. Under the usual condition often supposed in the traditional statistics, $d/n \longrightarrow 0$, we have

$$\left\| \bar{\bar{x}} - \vec{\mu} \right\| \xrightarrow{\text{P}} 0 \tag{11.2}$$

where the superscript P stands for the convergence in probability[9]. Equation (11.2) is referred to as the consistency of the sample average. However, in the high-dimensional statistics, $d/n \longrightarrow \infty$, and shockingly we have

$$\left\| \bar{\bar{x}} - \vec{\mu} \right\| \xrightarrow{\text{P}} \infty \, . \tag{11.3}$$

This is the so-called *strong inconsistency* in the HDLSS framework. What eq. (11.3) mean is that the noise in the observed data drastically increases with the increase of dimension d. Traditional statistical techniques suppose that the noise is negligible compared to the data information. Hence, traditional statistics can never handle the HDLSS data.

In order to have a more concrete image of such "counter-intuitive" phenomena, we examine the behavior of this huge noise in high dimensions. For this, however, some mathematical background is necessary. In the following, we introduce the background mathematical concepts. We do not try to introduce them in a strict manner, either not try to give rigorous mathematical proofs, even if the following sections seem to be rather mathematically inclined.

11.2.2 GEOMETRIC REPRESENTATIONS OF $\tilde{S}_D$

In this section, we give geometric representations of $\tilde{S}_D$ when $d \longrightarrow \infty$ while n is fixed (Yata & Aoshima, 2010a, 2012). The geometric representations are keys to deal with the HDLSS data by "high-dimensional statistical analysis.

We consider the following sphericity condition

$$\frac{\text{tr}\left(\tilde{\Sigma}^2\right)}{\left(\text{tr}\,\tilde{\Sigma}\right)^2} = \frac{\sum_{i=1}^{d} \lambda_i^2}{\left(\sum_{i=1}^{d} \lambda_i\right)^2} \longrightarrow 0 \quad \text{as} \quad d \longrightarrow \infty \, . \tag{11.4}$$

---

[9]More precisely, $\lim\limits_{d/n \to 0} \mathbb{P}\left(\|\bar{\bar{x}}_n - \vec{\mu}\| > \varepsilon\right) = 0$ for any $\varepsilon > 0$. Here $\|\cdot\|$ stands for the Euclid norm, and $\mathbb{P}(A)$ is the probability of an event A. Similarly, $\lim\limits_{d/n \to \infty} \mathbb{P}\left(\|\bar{\bar{x}}_n - \vec{\mu}\| < \varepsilon\right) = 0$ for any $\varepsilon > 0$ for eq. (11.3).

When $\tilde{X}$ is Gaussian, Ahn et al. (2007), Jung & Marron (2009) and YA12 showed a geometric representation as

$$n \left(\sum_{i=1}^{d} \lambda_i \right)^{-1} \tilde{S}_{\mathrm{D}} \xrightarrow{\mathrm{P}} \tilde{I}_n \quad \text{as} \quad d \longrightarrow \infty . \tag{11.5}$$

Here the λ_is are eigenvalues of the covariance matrix $\tilde{\Sigma}$, i.e., not the sample covariance $\tilde{S}$. Let

$$\vec{w}_i = n \left(\sum_{i=1}^{d} \lambda_i \right)^{-1} \tilde{S}_{\mathrm{D}} \vec{u}_i = n \left(\sum_{i=1}^{d} \lambda_i \right)^{-1} \hat{\lambda}_i \vec{u}_i \tag{11.6}$$

and $\mathscr{S}^{n-1} \equiv \{ \vec{x} \in \mathbb{R}^n \mid \|\vec{x}\| = 1 \}$: a unit $(n-1)$-sphere[10]. YA12 showed that

$$\vec{w}_i \in \mathscr{S}^{n-1} \quad \text{in probability as} \quad d \longrightarrow \infty , \tag{11.7}$$

that is, the surface concentration of the unit n-sphere. From the geometric representation by eq. (11.7), we observe that *the sample eigenvalues become deterministic* and converge to the same value. This means that it becomes impossible to estimate the eigenvalues and the eigenvectors by using $\tilde{S}_{\mathrm{D}}$ (or $\tilde{S}$) as in the conventional method.

When $\tilde{X}$ is non-Gaussian, Yata & Aoshima (2010a) and YA12 discovered a different type of geometric representation,

$$n \left(\sum_{i=1}^{d} \lambda_i \right)^{-1} \tilde{S}_{\mathrm{D}} - \tilde{D}_n \xrightarrow{\mathrm{P}} \tilde{O}_n \quad \text{as} \quad d \longrightarrow \infty . \tag{11.8}$$

where $\tilde{D}_n$ is a diagonal matrix whose diagonal elements are of $O_{\mathrm{P}}(1)$, and $\tilde{O}_n$ is a zero matrix of dimension $n \times n$. Let

$$\mathscr{A}^n \equiv \left\{ \begin{pmatrix} 1 \\ 0 \\ \vdots \\ 0 \end{pmatrix} \cup \begin{pmatrix} 0 \\ 1 \\ \vdots \\ 0 \end{pmatrix} \cup \cdots \cup \begin{pmatrix} 0 \\ 0 \\ \vdots \\ 1 \end{pmatrix} \right\} , \tag{11.9}$$

($\mathscr{A}^n \subset \mathbb{R}^n$). When $\tilde{X}$ is non-Gaussian, YA12 showed that

$$\vec{u}_i \in \mathscr{A}^n \quad \text{in probability as} \quad d \longrightarrow \infty , \tag{11.10}$$

that is, the axis concentration of the n-dim dual space. From the geometric representation by eq. (11.10), we can see that the eigenvectors of the data matrix converge to the unit vector of one of the axes of $\mathbb{R}^n$ and the eigenvalues would be indefinite.

We consider $(z_{1k}^2 - 1, \ldots, z_{dk}^2 - 1)$ ($k = 1, \ldots, n$) and its covariance matrix $\Phi = (\phi_{ij})$. YA12 considered a boundary that separates the two cases described by

---

[10]YA12 denoted $\mathscr{S}^{n-1}$ as $\mathbf{R}^n$. We used a more familiar notation in geometry in this work.

eq. (11.7) and eq. (11.10) as follows.

$$\frac{\displaystyle\sum_{i,j=1}^{d} \lambda_i \lambda_j \phi_{ij}}{\left(\displaystyle\sum_{i=1}^{d} \lambda_i\right)^2} \longrightarrow 0 \quad \text{as} \quad d \longrightarrow \infty. \tag{11.11}$$

Note that eq. (11.11) is equivalent to eq. (11.4) under Condition (11.24). Based on this YA12 proposed the following important theorem.

Theorem 5. [Yata & Aoshima (2012) I]
Assume the sphericity condition [eq. (11.4)]. If the elements of $\tilde{Z}$ satisfy eq. (11.11), we have eqs. (11.5) and (11.7) as $d \longrightarrow \infty$. Otherwise, eq. (11.8) holds as $d \longrightarrow \infty$, and under some regularity conditions, we have eq. (11.10) as $d \longrightarrow \infty$.

Now we are ready to examine the geometric behavior of the noise in the dual space. Consider a sample with a d-dim Gaussian distribution with a mean $\vec{\mu} = \vec{0}$ and covariance matrix $\tilde{\Sigma}_d = \tilde{I}_d$. If we generate a sample $n = 3$ from this distribution, we obtain a sample distribution displayed in Fig. 11.2. For a low-d, we have a sample distribution not very far from our intuition, i.e., the sample data concentrate on the origin and have a certain dispersion around it. However, for a higher-d, the sample data concentrate on a sphere with a radius of $(d/n)^{\frac{1}{2}}$ (Left panel of Fig. 11.2). This clearly shows the situation of high-dimensional data, namely that the intrinsic feature of the data is completely overwhelmed by a tremendous disturbance from the huge noise sphere.

The behavior of the sample is even more counter-intuitive for a non-Gaussian case. If we generate a sample from a d-dim t-distribution $t_d(0, I_d, v)$, with a mean $\vec{\mu} = \vec{0}$, a covariance matrix I_d and a degrees of freedom $v = 5$, we observe a very particular behavior (Right panel of Fig. 11.2). All the data concentrate on one of the axes for high-d cases. This is referred to as the axis concentration, typically seen for a non-Gaussian density distribution in high dimensions.

Thus, we understood the mathematical reason why we have the counter-intuitive behaviors of the HDLSS data in Fig. 11.2. This also gives the important basis of the central method we apply in this work.

11.2.3 GEOMETRIC REPRESENTATION OF THE DATA

Consider a parent distribution of dimension d, with a mean vector $\vec{\mu} = \vec{0}$ and a covariance matrix $\tilde{\Sigma}_d > \tilde{O}$ (i.e., all the eigenvalues are positive). We can adopt this constraint for the sake of simplicity. Let λ_i $(i = 1, \ldots, d)$ be the eigenvalues of $\tilde{\Sigma}$, and sort them out as

$$\lambda_1 \geq \lambda_2 \geq \cdots \geq \lambda_d \ (> 0). \tag{11.12}$$

Then, we can make an eigenvalue decomposition of

$$\tilde{\Sigma}_d \;=\; \tilde{H}\tilde{\Lambda}\tilde{H}^\top \tag{11.13}$$

$$\tilde{\Lambda} \;\equiv\; \mathrm{diag}\,(\lambda_1,\lambda_2,\ldots\lambda_d)\;. \tag{11.14}$$

Here, $\tilde{H} = \left[\vec{h}_1,\ldots,\vec{h}_d\right]$ is an orthogonal matrix of the corresponding eigenvectors. Note that these are all outrageously huge matrices.

Now we draw a set of n samples from this parent population $(d > n)$, $\vec{x}_1, \vec{x}_2, \ldots, \vec{x}_n$. Suppose that we describe this set of data by a $(d \times n)$ data matrix

$$\tilde{X} = (\vec{x}_1, \vec{x}_2, \ldots, \vec{x}_n)\;, \tag{11.15}$$

and $\vec{x}_i$ $(i = 1,\ldots,n)$ are independently and identically distributed (i.i.d.). We should note that the i.i.d. condition is not always satisfied in practice. However, this condition is often required for the rigorous construction of the mathematical methodology. The sample covariance matrix $(d \times d)$ is expressed as

$$\tilde{S} = \frac{1}{n}\tilde{X}\tilde{X}^\top\;. \tag{11.16}$$

Here, we define the dual sample covariance $(n \times n)$,

$$\tilde{S}_{\mathrm{D}} \equiv \frac{1}{n}\tilde{X}^\top\tilde{X}\;. \tag{11.17}$$

A schematic description of a sample covariance matrix $\tilde{S}$ and its dual matrix $\tilde{S}_{\mathrm{D}}$ is shown in Fig. 11.1. If we consider the sample eigenvalues of $\tilde{S}_{\mathrm{D}}$, its n eigenvalues

$$\hat{\lambda}_1 \geq \hat{\lambda}_2 \geq \cdots \geq \hat{\lambda}_n > 0 \tag{11.18}$$

are common with the first n eigenvalues of $\tilde{S}$.

An important piece of the initial exploration of HDLSS asymptotics in Hall et al. (2005) was the revelation of a surprisingly rigid geometric structure that naturally emerges from the asymptotics in an increasingly random setting, referred to as the geometric representation by them. Recall the eigenvalue decomposition of the covariance matrix $\tilde{\Sigma} = \tilde{H}\tilde{\Lambda}\tilde{H}^\top$. Let

$$\tilde{X} \equiv \tilde{H}\tilde{\Lambda}^{\frac{1}{2}}\tilde{Z}. \tag{11.19}$$

Then, $\tilde{Z}$ is a $(d \times n)$ sphered data matrix from a distribution with the identity covariance matrix. Here, we write $\tilde{Z} = [\vec{z}_1,\ldots,\vec{z}_d]^\top$ and $\vec{z}_i = (z_{i1},\ldots,z_{in})^\top$, $(i = 1,\ldots,d)$. Note that $\mathbb{E}(z_{ij}z_{i'j}) = 0$ $(i \neq i')$ and $\mathbb{V}(\vec{z}_i) = \tilde{I}_n$, where $\mathbb{E}$ and $\mathbb{V}$ denote the expectation and variance of a random variable, and $\tilde{I}_n$ is the n-dim identity matrix. We assume that the fourth moments of each variable in $\tilde{Z}$ are uniformly bounded. Note that if $\tilde{X}$ is Gaussian, the z_{ij}s are i.i.d. standard normal random variables. We also define the eigenvalue decomposition of $\tilde{S}_{\mathrm{D}}$ by

$$\tilde{S}_{\mathrm{D}} = \sum_{i=1}^{n} \hat{\lambda}_i\vec{u}_i\vec{u}_i^\top\;. \tag{11.20}$$

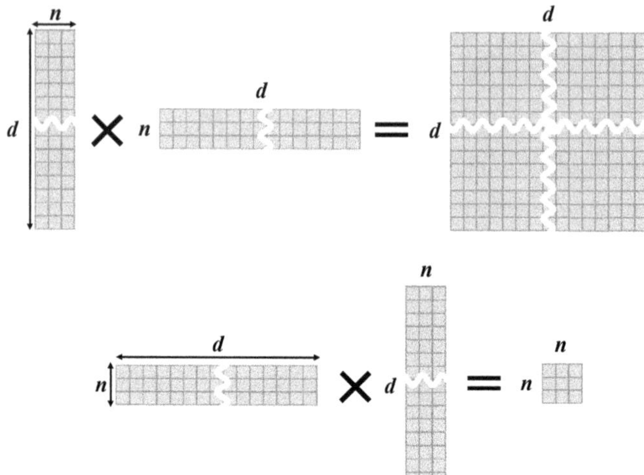

Figure 11.1 Schematic description of the relation between a sample covariance matrix $\tilde{S}$ and its dual matrix $\tilde{S}_{\mathrm{D}}$ (Takeuchi et al., 2024). Top panel describes the sample covariance matrix, and the Bottom panel is the dual matrix, respectively. Credit: Takeuchi, T. T., et al. 2024, Astrophysical Journal Supplement Series, 271, id.44, Fig. 1, reproduced by permission of AAS.

Here, $\vec{u}_i$ denotes a unit eigenvector corresponding to $\hat{\lambda}_i$.

Let $\hat{\vec{h}}_i$ be the i-th eigenvector of $\tilde{S}$. Note that $\hat{\vec{h}}_i$ can be calculated by

$$\hat{\vec{h}}_i = (n\hat{\lambda}_i)^{-1/2}\tilde{X}\vec{u}_i. \tag{11.21}$$

We note that the sample covariance matrix $\tilde{S}$ and the dual sample covariance $\tilde{S}_{\mathrm{D}}$ share their non-zero eigenvalues. We also note that the eigenvectors of $\tilde{S}$ can be calculated by the eigenvectors of $\tilde{S}_{\mathrm{D}}$. See eq. (11.21). Thus the merit of the dual sample covariance ($n \times n$) is rather obvious:

1. We can reduce the computational cost significantly.
2. We can visualize the behavior of data more easier than the original data matrix ($d \times d$).
3. Asymptotic theory can be developed on the dual space.

11.3   HIGH-DIMENSIONAL PRINCIPAL COMPONENT ANALYSIS

We adopt the same setting as the previous sections for the data distribution: we suppose a random variable with a mean $\vec{\mu} = \vec{0}$ and a covariance matrix $\tilde{\Sigma}_d > 0$. We assume the following model, referred to as "the generalized spiked model", for the eigenvalue distribution of $\tilde{\Sigma}_d$,

$$\lambda_i = \begin{cases} a_i d^{\alpha_i} & \text{for} \quad i = 1,\dots,m \\ c_i & \text{for} \quad i = m+1,\dots,d \end{cases} \tag{11.22}$$

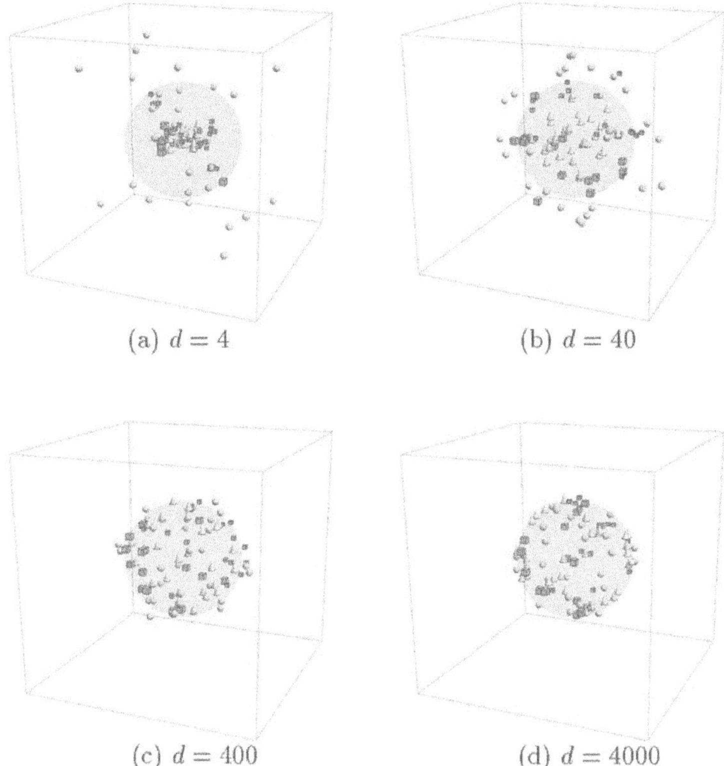

Figure 11.2 A sample distribution of eigenvectors generated from a d-dim Gaussian distribution with a mean $\vec{\mu} = \vec{0}$ and a covariance matrix $\tilde{\Sigma}_d = \tilde{I}_d$ ($\tilde{I}_d$: a d-dim identity matrix) (Takeuchi et al., 2024). Credit: Takeuchi, T. T., et al. 2024, Astrophysical Journal Supplement Series, 271, id.44, Fig. 20, reproduced by permission of AAS.

where $a_i(> 0), c_i(> 0)$ and α_i ($\alpha_1 \geq \cdots \geq \alpha_m > 0$) are unknown constants that preserves the order $\lambda_1 \geq \cdots \geq \lambda_d$, and m is an unknown positive fixed integer[11]. The m stands for the number of spiked eigenvalues in the sense of quadratic contribution ratio. The details of the spiked model including the generalized spiked model are considered and given in Yata & Aoshima (2013). The eigenvalues λ_i ($i \leq m$) describe the intrinsic information of the data, and λ_i ($i \geq m + 1$) represents the noise.

Note that

$$\frac{\displaystyle\sum_{i=m+1}^{d} \lambda_i^2}{\left(\displaystyle\sum_{i=m+1}^{d} \lambda_i\right)^2} \xrightarrow{\text{P}} 0 \quad \text{as} \quad d \longrightarrow \infty \tag{11.23}$$

---

[11]One may feel suspicious if such a restrictive model can really describe the general properties of eigenvalue distribution of real data. However, as we see in Sec.11.5 and 11.6, the generalized spiked model really well represents the actual eigenvalues.

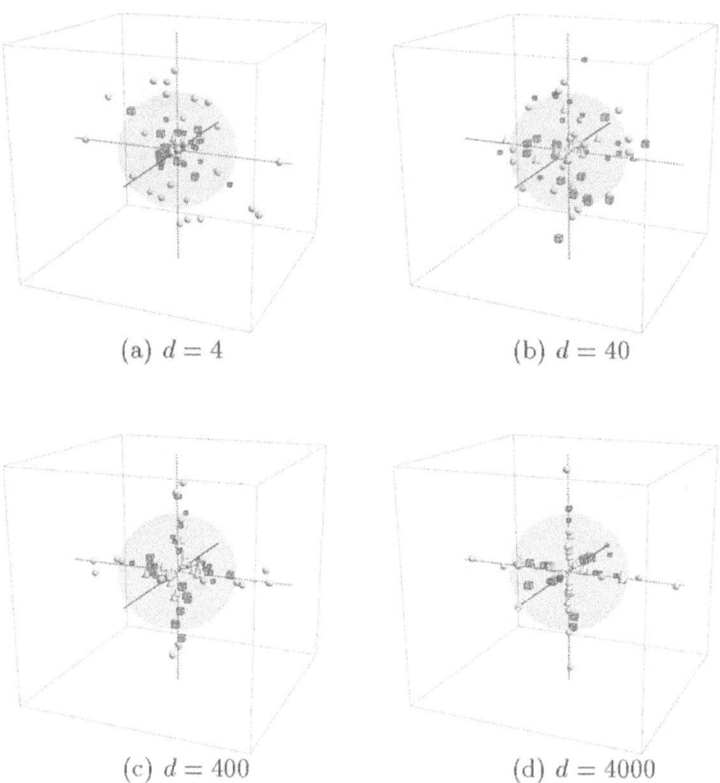

(a) $d = 4$ (b) $d = 40$

(c) $d = 400$ (d) $d = 4000$

Figure 11.3 A sample distribution of eigenvectors generated from a d-dim t-distribution with $t_d(0, I_d, v)$, with a mean $\vec{\mu} = \vec{0}$, a covariance matrix I_d and a degrees of freedom $v = 5$ (Takeuchi et al., 2024). Credit: Takeuchi, T. T., et al. 2024, Astrophysical Journal Supplement Series, 271, id.44, Fig. 20, reproduced by permission of AAS.

i.e., the noise satisfies the sphericity condition [eq. (11.4)]. We set the sample co-variance matrix $\tilde{S} = n^{-1}\tilde{X}\tilde{X}^{\top}$ as above. Here, we define $n(d)$ as the size of a sample depending on d.

11.3.1 LIMITATION OF THE TRADITIONAL PRINCIPAL COMPONENT ANALYSIS (PCA)

Yata & Aoshima (2009) investigated the properties of the traditional PCA when it is applied to HDLSS data, and demonstrated the condition so that the estimation by the traditional PCA would have the consistency in the form of $n(d) = d^{\gamma}$ [12] for a positive

---

[12] In this context, consistency stands for the situation as follows: Consider an estimator $t(n)$ (n: sample size) is a consistent estimator for a parameter θ if and only if, for all $\varepsilon > 0$, we have

$$\lim_{n \to \infty} \mathbb{P}\left\{|t(n) - \theta| < \varepsilon\right\} = 1.$$

constant γ and the limitation of the traditional PCA. They have given the following two theorems on the eigenvalues, depending on the condition would satisfy

$$z_{ij} \quad (i=1,\ldots,d,\ j=1,\ldots,n) \text{ are independent.} \tag{11.24}$$

Theorem 6. [Yata & Aoshima (2009) I]
If Z satisfies Condition (11.24), for $i \le m$,

$$\frac{\hat{\lambda}_i}{\lambda_i} = 1 + o_{\mathrm{P}}(1)\,, \tag{11.25}$$

under the conditions

1. *$d \longrightarrow \infty$ and $n \longrightarrow \infty$ for i that satisfies $\alpha_i > 1$.*
2. *$d \longrightarrow \infty$ and $d^{1-\alpha_i}/n(d) \longrightarrow 0$ for i that satisfies $\alpha_i \in (0,1]$.*

Next we give the consistency eq. (11.25) when Z does not satisfy Condition. (11.24).

Theorem 7. [Yata & Aoshima (2009) II]
If Z does not satisfy Condition (11.24), eq. (11.25) holds for $i \le m$ under the conditions

1. *$d \longrightarrow \infty$ and $n \longrightarrow \infty$ for i that satisfies $\alpha_i > 1$.*
2. *$d \longrightarrow \infty$ and $d^{2-2\alpha_i}/n(d) \longrightarrow 0$ for i that satisfies $\alpha_i \in (0,1]$.*

Theorems 6 and 7 give the condition that the estimation by the traditional PCA would have consistency. In (2), n is heavily depending on d when the components of $\tilde{Z}$ do not satisfy Condition (11.24). For the non-Gaussian case, Yata & Aoshima (2010b) proposed a new robust PCA called the cross-data-matrix methodology.

Theorem 6 gives the condition that the estimation by the traditional PCA would have the consistency. The condition described by $d \longrightarrow \infty$ and $n \longrightarrow \infty$ in (1) of these theorems is a mild condition in the sense that one can choose n free from d or that n may be much smaller than d such as $n = \log d$. However, in (2), n is heavily depending on d. Hence, we should be very careful to apply the conventional PCA to the HDLSS data. In order to overcome this fundamental problem of the conventional PCA, Yata & Aoshima (2010b) and YA12 proposed two new PCAs for HDLSS data. In this work, we focus on one of them, the noise-reduction methodology (YA12).

11.3.2 NOISE-REDUCTION PCA (NRPCA): A NEW METHOD FOR HDLSS DATA

Here we introduce the high-dimensional PCA. First, we informally present the concept of the high-dimensional PCA, which is schematically described in Fig. 11.4. Because of the enormous noise that concentrates on the sphere in the data space, the intrinsically important characteristics of the data are completely buried under the noise sphere and obscured. However, it is not hopeless to estimate the characteristics even under such a situation, thanks to the fact that the noise behavior around the

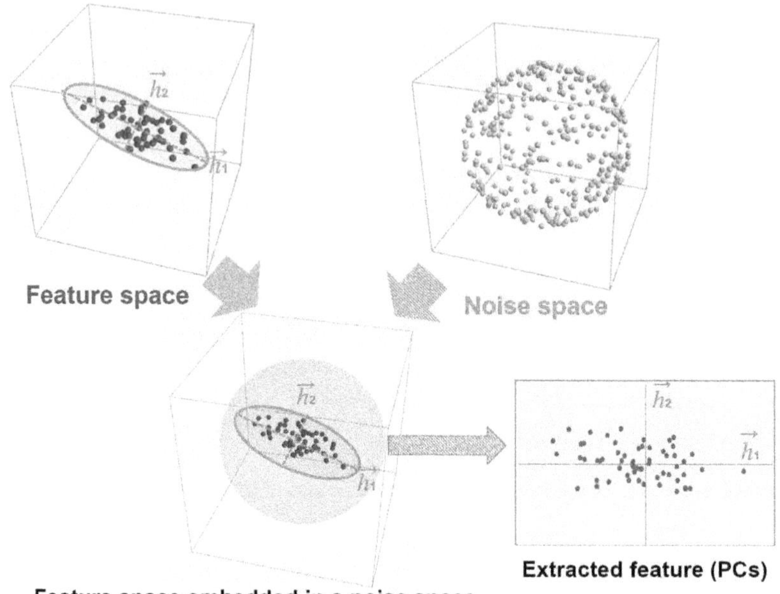

Figure 11.4 A schematic description of the high-dimensional principal component analysis (PCA) (Takeuchi et al., 2024). Even if we have significant important features, because of the existence of a huge noise sphere, they are completely embedded in the observed data. The high-dimensional PCA can subtract the noise sphere and enable to extraction the desired features. Credit: Takeuchi, T. T., et al. 2024, Astrophysical Journal Supplement Series, 271, id.44, Fig. 2, reproduced by permission of AAS.

sphere can be well analyzed and essentially *subtracted*. The method to concretely realize this idea is the noise-reduction (NR) methodology proposed by YA12.

We try to explain it in a more rigorous manner. YA12 proposed the NR methodology derived from the geometric representation [eq. (11.5)]. First we decompose $\tilde{S}_{\mathrm{D}}$ as

$$n\tilde{S}_{\mathrm{D}} = \sum_{i=1}^{m} \lambda_i \vec{z}_i \vec{z}_i^{\top} + \sum_{i=m+1}^{d} \lambda_i \vec{z}_i \vec{z}_i^{\top} \tag{11.26}$$

The second term represents the noise. Then, from Theorem 5, under Condition (11.24), we obtain the following geometric representation for the noise

$$\frac{\displaystyle\sum_{i=m+1}^{d} \lambda_j \vec{z}_i \vec{z}_i^{\top}}{\displaystyle\sum_{i=m+1}^{d} \lambda_j} \xrightarrow{\mathrm{P}} \tilde{I}_n \quad \text{as} \quad d \longrightarrow \infty, \tag{11.27}$$

i.e., the noise part becomes deterministic under the HDLSS condition. This result enables us to remove the noise part and obtain the formula

$$\check{\lambda}_i = \hat{\lambda}_i - \frac{1}{n-i}\left(\operatorname{tr}\tilde{S}_{\mathrm{D}} - \sum_{j=1}^{i}\hat{\lambda}_i\right) \qquad (i = 1,\ldots,n-1)\,. \tag{11.28}$$

Estimation of m is given by Aoshima & Yata (2018). We have the following theorem.

Theorem 8. [Yata & Aoshima (2012) II].
When the components of $\tilde{Z}$ satisfy Condition (11.24), it holds that for $i \le m$,

$$\frac{\check{\lambda}_i}{\lambda_i} = 1 + o_{\mathrm{P}}(1)\,, \tag{11.29}$$

under the conditions that

1. *$d \longrightarrow \infty$ and $n \longrightarrow \infty$ for i such that $\alpha_i > 1/2$,*
2. *$d \longrightarrow \infty$ and $d^{1-2\alpha_i}/n(d) \longrightarrow 0$ for i such that $\alpha_i \in (0,1/2]$.*

If we compare Theorem 8 with Theorem 6, we see that the consistency holds for $\check{\lambda}_i$s under milder conditions than $\hat{\lambda}$ [under Condition (11.24)] [13].

Next, we apply the NR methodology to the PC direction vectors. Define $\check{\vec{h}}_i \equiv (n\check{\lambda}_i)^{-1/2}\tilde{X}\vec{u}_i$, and examine its performance as the estimator of the PC direction vector $\vec{h}_i$. Note that $\check{\vec{h}}_i = (\hat{\lambda}_i/\check{\lambda}_i)^{-1/2}\hat{\vec{h}}_i$. YA12 showed the following theorem.

Theorem 9. [Yata & Aoshima (2012) III]
Assume λ_i ($i \le m$) has multiplicity one. Under conditions in Theorem 8, and the components of $\tilde{Z}$ satisfy Condition (11.24), it holds that

$$\check{\vec{h}}_i^{\top}\vec{h}_i = 1 + o_{\mathrm{P}}(1). \tag{11.30}$$

We also consider the PC scores. The i-th PC score of x_i is given by $\vec{h}_i^{\top}\vec{x}_j = z_{ij}(\lambda_i)^{\frac{1}{2}} \equiv s_{ij}$. Let $\vec{u}_i \equiv (\hat{u}_{i1},\ldots,\hat{u}_{in})^{\top}$ ($i = 1,\ldots,d$). Consider we estimate the PC score by $\hat{u}_{ij}(n\check{\lambda}_i)^{\frac{1}{2}} \equiv \check{s}_{ij}$. We define the mean squared error (MSE) of the i-th PC score by

$$\mathrm{MSE}(\check{\vec{s}}_i) = \frac{1}{n}\sum_{j=1}^{n}(\check{s}_{ij} - s_{ij})^2\,. \tag{11.31}$$

Then we have yet another important theorem.

Theorem 10. [Yata & Aoshima (2012) V]
Assume that λ_i ($i \le m$) has multiplicity one. Then, under the same conditions as Theorem 9,

$$\frac{\mathrm{MSE}(\check{s}_i)}{\lambda_i} = o_{\mathrm{P}}(1) \tag{11.32}$$

---

[13]If $\alpha_i > 1/2$, the NR methodology has consistency even when n is much smaller than d such as $n = \log d$.

holds.

All these theorems hold for the cases with nonzero mean, by replacing the setting with zero-centered variables. Thus, now we have demonstrated that the NR methodology yields high performances for the estimation of the PC eigenvalues (contributions), PC vectors and PC scores for HDLSS data. In the following discussions, the presented eigenvalues are obtained by the NRPCA.

11.3.3 AUTOMATIC SPARSE PCA (A-SPCA)

In this section, we introduce (thresholded) sparse PCA methods for high-dimensional data. The sparse PCA (SPCA) can be summarized as follows: Let $\hat{\vec{h}}_i = (\hat{h}_{i(1)}, ..., \hat{h}_{i(d)})^\top$ for all i. Given a sequence of threshold values $\zeta > 0$, define the thresholded entries as

$$\hat{h}_{i*(s)} = \begin{cases} \hat{h}_{i(s)} & \text{if } |\hat{h}_{i(s)}| \geq \zeta, \\ 0 & \text{if } |\hat{h}_{i(s)}| < \zeta \end{cases} \quad \text{for } s = 1, ..., d. \tag{11.33}$$

Let $\hat{\vec{h}}_{i*} = (\hat{h}_{i*(1)}, ..., \hat{h}_{i*(d)})^\top$. The thresholded estimator of $\vec{h}_i$ is defined by

$$\hat{\vec{h}}_{i**} = \hat{h}_{i*}/\|\hat{h}_{i*}\|. \tag{11.34}$$

However, the estimator heavily depends on a choice of ζ and does not hold consistency properties for HDLSS data. Recently, Yata & Aoshima (2022) proposed a SPCA method by using the NR method as follows: For the PC direction by the NR method, $\check{\vec{h}}_i = (\check{h}_{i(1)}, ..., \check{h}_{i(d)})^\top$, we assume $|\check{h}_{i(1)}| \geq \cdots \geq |\check{h}_{i(d)}|$ for the sake of simplicity. For a given constant $\omega \in (0, 1]$, we consider PC directions whose cumulative contribution ratio is greater than or equal to ω. There exists a unique integer $\check{k}_{i\omega} \in [1, d]$ such that

$$\sum_{s=1}^{\check{k}_{i\omega}-1} \check{h}_{i(s)}^2 < \omega \quad \text{and} \quad \sum_{s=1}^{\check{k}_{i\omega}} \check{h}_{i(s)}^2 \geq \omega. \tag{11.35}$$

We define that

$$\check{h}_{i\omega(s)} = \begin{cases} \check{h}_{i(s)} & \text{if } |\check{h}_{i(s)}| \geq |\check{h}_{i(k_{i\omega})}|, \\ 0 & \text{otherwise} \end{cases} \quad \text{for } s = 1, ..., d.$$

Let $\check{\vec{h}}_{i\omega} = (\check{h}_{i\omega(1)}, ..., \check{h}_{i\omega(d)})^\top$. Yata & Aoshima (2022) called the new PCA method that uses $\check{h}_{i\omega(s)}$s as the "automatic SPCA (A-SPCA)". They showed some consistency properties of $\check{\vec{h}}_{i\omega}$ in the HDLSS data and investigated the performance in actual data analyses. Here the A-SPCA uses the same eigenvalues estimated by the NRPCA. Here we set $\omega = 1/2$ [14]. Intuitively, the A-SPCA can extract the most important components of the eigenvectors obtained by the NRPCA. Namely, these components practically control the PCs.

---

[14] The value of $\omega = 1/2$ was obtained by simulations.

11.4 EXAMPLE OF DATA ANALYSIS

we show an example of the high-dimensional PCA. NGC 253, is a nearby barred spiral galaxy at a distance of 3.5 Mpc, i.e., $1''$ corresponds to 17 pc (Rekola et al., 2005). Morphological type Sc is assigned. It has a systemic heliocentric velocity of $243 \pm 2 \, \mathrm{km \, s^{-1}}$ ($z = 0.000811$: Kamphuis et al. 2015), mainly due to the cosmological expansion. This galaxy is one of the brightest sources of far infrared emission except the Magellanic Clouds and has been extensively studied at many wavelengths. It is known as a pure starburst, whose star formation rate (SFR) in the central molecular zone is estimated to be SFR $\sim 2 \, M_\odot \, [\mathrm{yr^{-1}}]$ (Bendo et al., 2015; Keto et al., 1999; Rieke et al., 1980). Massive, dense, and warm ISM (e.g. Sakamoto et al., 2011) were identified in the central region. Superstar clusters are identified at optical and near-infrared (NIR) bands (Fernández-Ontiveros et al., 2009). An intense outflow is also observed (Bolatto et al., 2013; Matsubayashi et al., 2009). The superstar clusters and outflow are also associated with the starburst activity.

11.4.1 ALMA MAP OF NGC 253

The data used in this work is a spectroscopic map of a nearby archetype starburst galaxy NGC 253, obtained by Ando et al. (2017), at Band 7. Various emission lines are detected in the spectra. Details of the original data are found in Takeuchi et al. (2024). The number of spectroscopic resolution units (sampling number) is 2248.

The original map consists of 864×864 points in the spatial domain (R.A and dec.), and 2248 components along the frequency domain (in units of [Hz]) in total. However, since the pixel scale was smaller than the synthesized beam size of the map $0.''45 \times 0.''3$ (oversampled), each spatial grid in the map is not independent of each other. To overcome this issue, we gridded the map in the spatial dimension. The equivalent circular beam size corresponding to the elliptical beam of $0.''45 \times 0.''3$ is $0.''37$. We used this effective beam size to regrard the image map. The resulting map with independent grids consists of 230×230 pixels. We then should cut out low-signal regions. For this, instead of estimating S/N for some lines, we integrated the intensity over the whole range of frequency and calculated the total intensity $I \, [\mathrm{Jy \, beam^{-1}}]$ and chose the regions with $I > 1.0$, based on the rms level. The number of remaining pixels n is 231.

11.4.2 DOPPLER SHIFTS DUE TO THE SYSTEMIC HUBBLE VELOCITY AND ROTATION

The central region of NGC 253 is strongly inclined ($i \sim 78°$). This causes a fairly coherent rotation pattern on the spectroscopic map through the Doppler shift. We made a Gaussian fit to three prominent emission lines HCN(4–3), HNC(4–3), and CS(7–6) in the spectra to estimate the line center wavelengths. The map of the velocity field is presented in Fig. 11.5. We determined the amount of the shift by averaging the shifts obtained from three lines. Then we shifted the frequencies with respect to the systemic velocity of NGC 253 caused by the Hubble velocity. After this correction, we

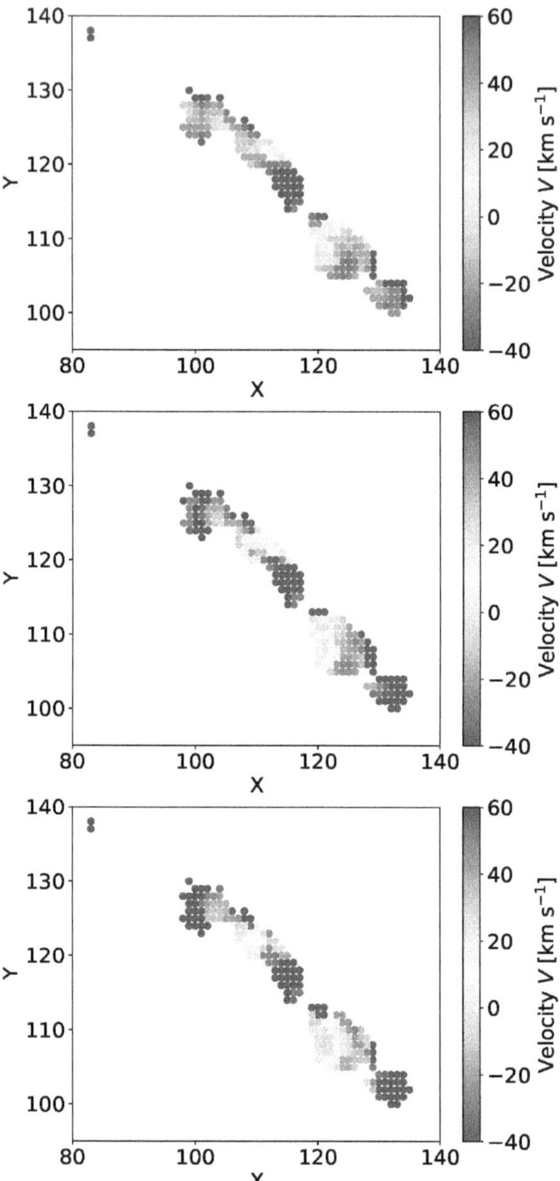

Figure 11.5 The velocity map of NGC 253 estimated from HCN(4–3) (Top), HNC(4–3) (Middle), and CS(7–6) (Bottom) (Takeuchi et al., 2024). The systemic velocity of 243 km s^{-1} is subtracted. Credit: Takeuchi, T. T., et al. 2024, Astrophysical Journal Supplement Series, 271, id.44, Fig. 4, reproduced by permission of AAS.

have a map without the systematic effect of the coherent rotation and we can analyze more subtle properties of the submillimeter emission lines.

We have frequency gaps in our ALMA data cube. If we try to shift the spectrum according to the velocity, the boundaries of the spectra are also shifted. We simply adopted frequency ranges common to all the pixels. The final size of the data is 231 along the spatial position, and 1971 in spectral dispersion after all corrections.

11.4.3 SPECTRAL MAP AS HDLSS DATA

The resulting map of NGC 253 has spatial dimension $231 \times$ spectral dimension 1971, i.e., $n = 231$ and $d = 1971$, and clearly $n \ll d$. Though it is not as extreme as DNA microarray data (typically $n \sim 10$–100 and $d \sim 10^4$–10^5), our ALMA map is typical HDLSS data [15]. Problems from the astrophysical side is that a very large amount of information is contained in the spectra, and a large variety of spectral lines are observed compared to n. This situation is not special in the high-dimensional analysis.

However, the application of high-dimensional methods to spectroscopic data is not completely straightforward. Spectral lines are broadened and/or shifted by a turbulent motion and other physical mechanisms in the ISM. Then, the information carried by each spectral resolution unit is not independent[16]. Such a complication has never been examined in other field of science about HDLSS data. However, since it is commonly seen in spectroscopic data in any field of astronomy, before going to the detailed analysis, we first examine if the high-dimensional statistical analysis really works for the astronomical spectroscopic map, with line shifts and broadening.

11.5 PREPARATORY ANALYSIS

As we have already mentioned above, there are some potential problems specific to astronomical spectroscopic data, which are different from typical examples of HDLSS data in genomics and other fields. The most fundamental issue is that, when we regard a spectroscopic data as a vector data $\vec{x} = (x_1, \ldots, x_d)^\top$, the information of neighboring frequency (or wavelength) index i and $i+1$ ($i = 1, \ldots, d-1$) is not independent. Namely, we often have internal motion in the ISM of a galaxy that leads to a Doppler broadening along the frequency axis. We also find a systemic rotation of the system over the whole observed region of an object (Sec.11.4). In order to examine if the high-dimensional statistical analysis would work for such type of astronomical spectroscopy data, we made a preparatory analysis of the original spectral mapping data of NGC 253 with the high-dimensional PCA. By the term "original", we mean that we started this analysis without correcting the Doppler shift on the map due to the internal systemic rotation. Namely, we blindly applied the high-dimensional PCA methods to the data and examined what feature would be detected. For this, we sort

---

[15] Note that $n > \log d = 7.59$ when $d = 1971$ and $n = 231$, so that the NR methodology works well for the HDLSS data. See Footnote 12.

[16]The existence of continuum radiation makes the situation even more complicated, but we do not try to examine this problem in the current work.

the 2-dim coordinates along the R.A.- direction and then dec. direction, and renumbered so that we can regard the data as a 1-dim vector. Though we can apply the PCA directly to a 2-dim image, we adopted this procedure throughout this chapter, as a first simplistic analysis.

11.5.1 EIGENSPECTRA

In the following part of this work, we mainly focus on the A-SPCA results. However, since the NRPCA deals with all the spectral features without extracting only the most important ones, we also present the eigenspectra (eigenvectors constructed from the observed spectra by PCA) from the NRPCA. Upper panel of Fig. 11.6 shows the eigenspectra constructed by the NRPCA from the ALMA spectral map of NGC 253[17]. We observe that specific features are commonly found in similar frequency ranges of the spectra for all the PCs. This suggests that some spectral features contain richer information on the whole spectra than other spectral regions.

The A-SPCA specifies and localizes the most important components that correspond to these particular features. The lower panel of Fig. 11.6 shows the eigenspectra obtained by the A-SPCA. As seen in Fig. 11.6, the A-SPCA only leaves the values if the features are important to determine the spectra, and forces the other components to be zero. Hence, the A-SPCA eigenspectra are not used for the reconstruction of the spectra with all features, and the NRPCA is used for this purpose. We make use of this outstanding advantage of the A-SPCA to specify the controlling spectral features of the PCA in the following analysis.

11.5.2 DISTRIBUTION OF THE CONTRIBUTION OF PCS

First we show the eigenvalue distribution of the PCs sorted with their contribution obtained by the NRPCA in Fig. 11.7. First several eigenvalues are prominently larger than the rest. This behavior is referred to as "spiked" in the context of the high-dimensional statistical analysis. Recall that the eigenvalues of the PCA represent the contribution of each PC, namely, the PCA decomposes the whole covariance matrix into contributions from each PC. Since the obtained eigenvalue distribution shows this spiked structure, we see that the high-dimensional PCA detected some characteristics of the ALMA spectral map of NGC 253. This guarantees that the current data can be safely regarded as typically HDLSS, and the assumptions of the high-dimensional PCA apply.

Astrophysically, Fig. 11.7 demonstrates a very promising possibility of the high-dimensional statistical analysis to the spectral mapping data. Ando et al. (2017) identified more than 30 molecular/radical lines on the map (see their Fig. 1), as well as some broad or continuum features. Even if we restrict our interest to the lines, it is implausible that we classify them only by physical intuition. Traditionally, for example, a line ratio has been used to extract physical information from spectral

---

[17]We evaluated the advantage of the NRPCA compared to the traditional PCA through the eigenvalue estimation.

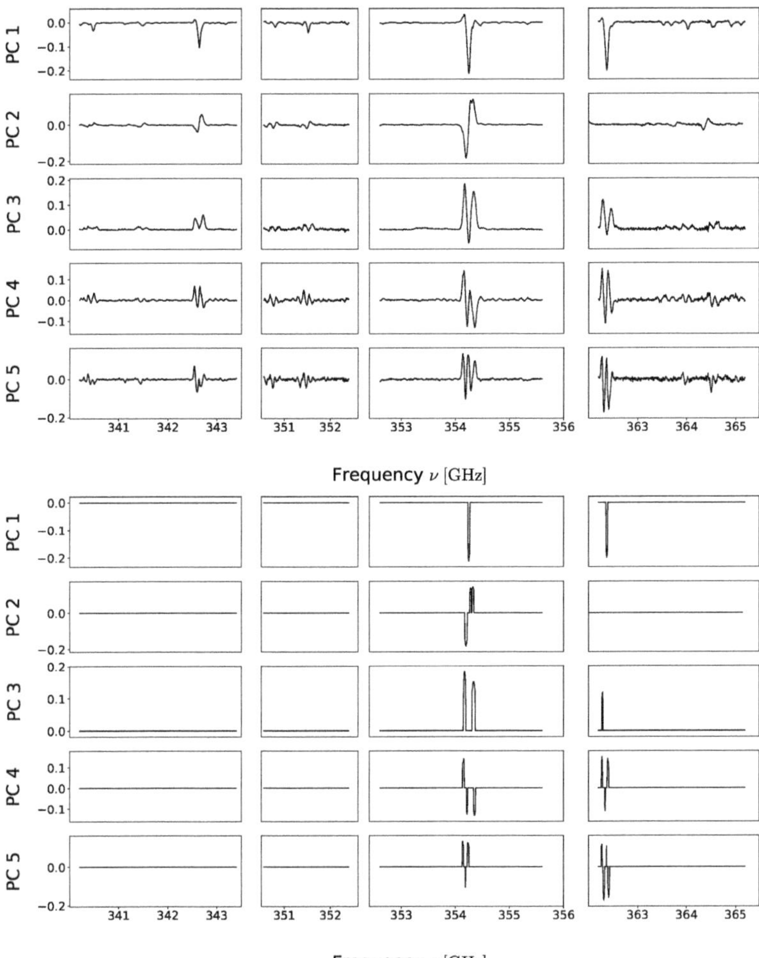

Figure 11.6 The eigenspectra corresponding to the 1st–5th principal components (PCs) constructed from the original ALMA spectral map of NGC 253 (Takeuchi et al., 2024). Upper five panels: eigenspectra obtained by the noise-reduction principal component analysis (NR-PCA). Lower five panels: same as Upper panels, but obtained by the automatic sparse principal component analysis (A-SPCA). Only the controlling features have nonzero values. Credit: Takeuchi, T. T., et al. 2024, Astrophysical Journal Supplement Series, 271, id.44, Fig. 5, reproduced by permission of AAS.

data. However, brute-force trials may not work if we try to diagnose many lines simultaneously, because the number of line ratios would increase prohibitively with a number of features, known as a combinatorial explosion. In a clear contrast, Fig. 11.7 implies that the whole information of the complicated spectral map can be represented by first several PCs. This means that the complicated spectral features including more than 30 lines are controlled by several bases represented by the PCs. This

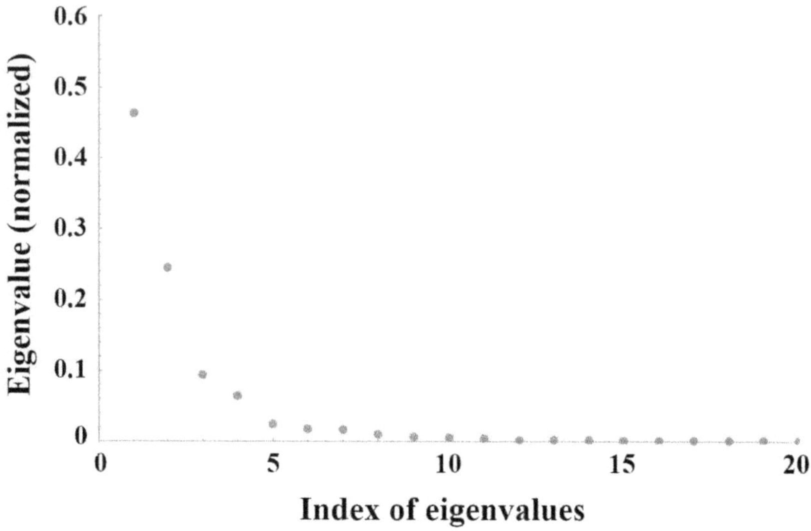

Figure 11.7 Normalized eigenvalues (contributions) obtained from the analysis of the ALMA spectroscopic map of NGC 253 by the NRPCA (Takeuchi et al., 2024). Credit: Takeuchi, T. T., et al. 2024, Astrophysical Journal Supplement Series, 271, id.44, Fig. 6, reproduced by permission of AAS.

advantage itself is a well-known benefit to using the PCA in general as a dimity reduction method (e.g., Galaz & de Lapparent, 1998; Pace et al., 2019; Portillo et al., 2020; Ronen et al., 1999; Wang et al., 2011, among many others), but all these previous applications were on integrated 1-dim spectra with a sufficiently large number of n, namely, not HDLSS data. We stress that the analysis in this work is substantially different from these works: the current method is able to be applied to a spectral map which is a typical HDLSS dataset and traditional PCA would not have worked due to the huge noise sphere, as mentioned in Section 11.2.

Since we confirmed that the methodology of high-dimensional statistical analysis worked well for the astrophysical spectral amp data, we safely proceeded to further analysis by the high-dimensional PCA.

11.5.3 PHYSICAL MEANING OF FIRST TWO PCS

Next step is to examine what PC1 and 2 represent. Figure 11.8 shows the scatter of PC1 and PC2. The most striking feature is the butterfly-like pattern symmetric with respect to the x-axis. It reminds us of the coherent rotation of the central region of NGC 253. We reconstructed the 2-dim structure to map PC1 and 2 in Fig. 11.9.

It is clear that PC1 represents the intensity of lines, and PC2 represents the symmetric and coherent pattern of the center of NGC 253. We observe a nice agreement between Lower panel of Fig. 11.9 and with directly estimated velocity field of the same region in Fig. 11.5.

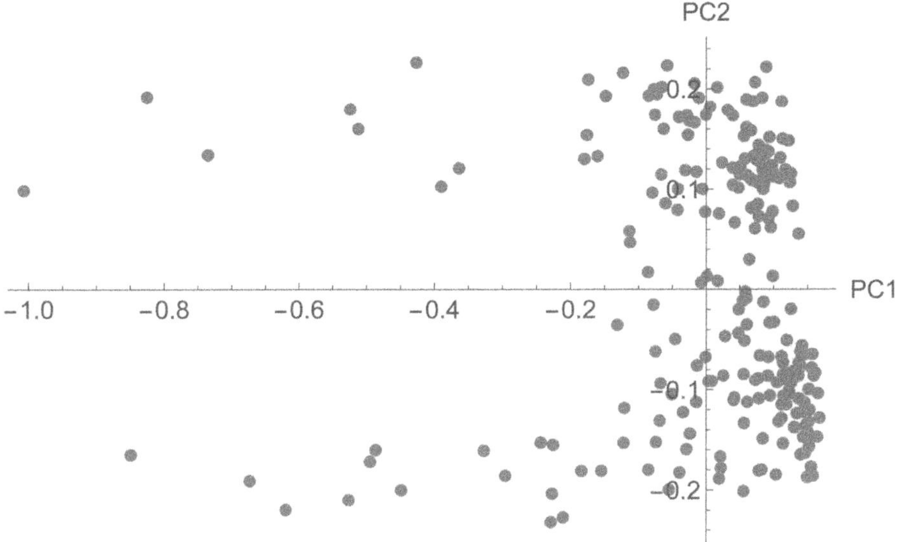

Figure 11.8 The distribution of PC1 and PC2 of the ALMA map of NGC 253 by the A-SPCA (Takeuchi et al., 2024). Credit: Takeuchi, T. T., et al. 2024, Astrophysical Journal Supplement Series, 271, id.44, Fig. 7, reproduced by permission of AAS.

To have a deeper look at the relation, we present the correlation between PCs and observed properties in Fig. 11.10. Upper panel of Fig. 11.10 is the scatter plot between PC1 and the total intensity integrated over the whole frequency range for each pixel. We indeed observe a good correlation between them (Fig. 11.10). The correlation is not linear, because the PCA extracts only the linear features of the data, while the relation to the physical quantities would not be necessarily linear. We also examined the relation between the peculiar velocity and PC2 in Right panel of Fig. 11.10. As we already presented in 2-dim map, PC2 approximately delineate the velocity field in Fig. 11.10. However, we see some outliers where the PC2 and the peculiar velocity are not coherent. These regions may be affected by some local phenomenon, for example, a strong outflow. We will revisit this issue later with a detailed analysis.

11.5.4 SPECTRAL FEATURES CORRESPONDING TO PCS: ORIGINAL DATA

It is interesting to examine what features are concretely responsible for PCs. Actually the A-SPCA can specify the responsible features to characterize the PCs. We show the responsible spectral features in Figs. 11.11. The spectra are integrated over the whole analyzed regions for both figures. In Fig. 11.11, stars represent the features responsible for PC1, and triangles are for PC2. We do not show higher-order contributions for clarity. PC1-related features are assigned to the HCN(4–3) and HNC(4–3) lines, particularly close to the central part of the lines. PC2-related features are

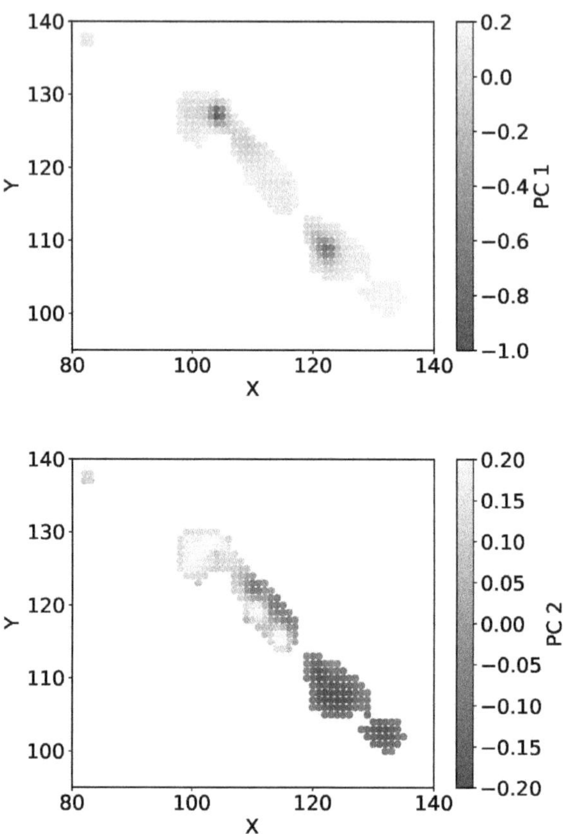

Figure 11.9 The 2-dim structure to map PC1 and PC2 obtained by the A-SPCA (Takeuchi et al., 2024). Top panel shows the map of PC1, and the Bottom panel describes the map of PC2, respectively. Credit: Takeuchi, T. T., et al. 2024, Astrophysical Journal Supplement Series, 271, id.44, 21 Fig. 8, reproduced by permission of AAS.

related to HCN(4–3), but in contrast to PC1, they are connected to the wing part of the line. Since the spectra are integrated over the analyzed regions, the wing is made actually by the Doppler shifts. This is clearly consistent with the fact that the PC2 map (Fig. 11.9) agrees well with the Doppler velocity field of NGC 253 map (Fig. 11.5).

The performance of the A-SPCA to the ALMA spectroscopic map as HDLSS data is certainly guaranteed by this preparatory analysis. We can safely proceed to the detailed analysis of the molecular lines, as in the next Section.

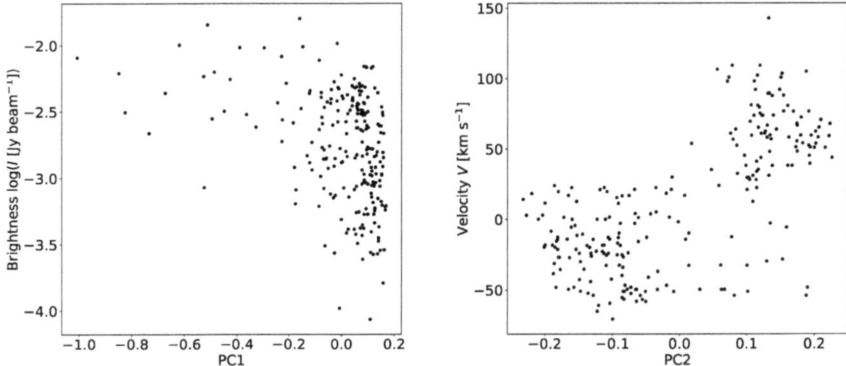

Figure 11.10 The scatter plot between PCs and observed physical properties (Takeuchi et al., 2024). Left: the correlation between PC1 and the integrated intensity over the whole frequency range I [Jy beam^{-1}] on the ALMA map of NGC 253. Right, the correlation between PC2 and the peculiar velocity on the ALMA map. Credit: Takeuchi, T. T., et al. 2024, Astrophysical Journal Supplement Series, 271, id.44, Fig. 9, reproduced by permission of AAS.

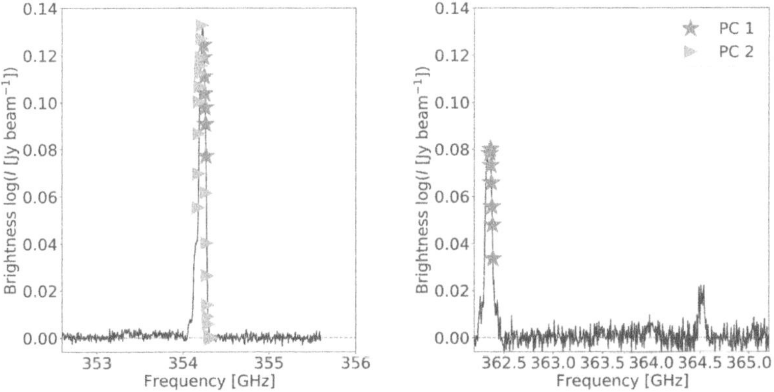

Figure 11.11 Responsible features to characterize PCs from the A-SPCA for the ALMA map of NGC 253 (Takeuchi et al., 2024). Stars represent the features responsible for PC1, and triangles are for PC2. Credit: Takeuchi, T. T., et al. 2024, Astrophysical Journal Supplement Series, 271, id.44, Fig. 10, reproduced by permission of AAS.

11.6 RESULTS AND DISCUSSION

After examining the performance of the high-dimensional statistical analysis in Sec. 11.5, now we can proceed to a further physical analysis.

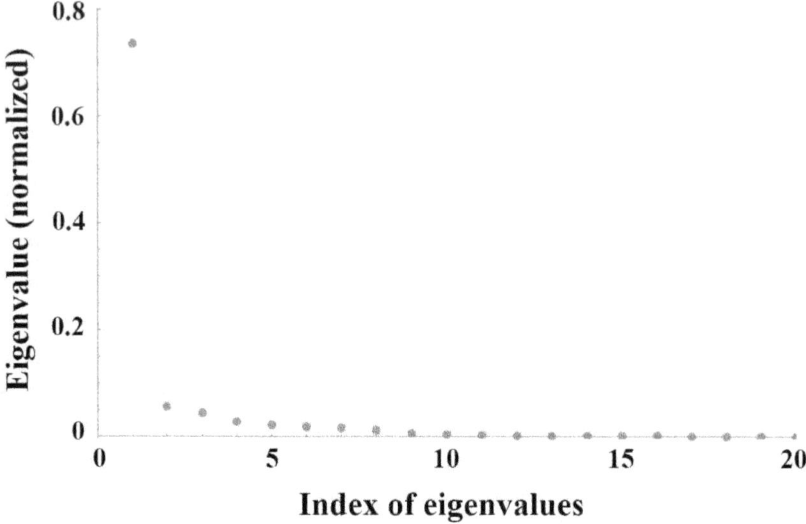

Figure 11.12 Normalized eigenvalues (contributions) obtained from the analysis of the ALMA spectroscopic map of NGC 253 by the NRPCA, after correcting the effect of the Doppler shifts of the systemic rotation (Takeuchi et al., 2024). Credit: Takeuchi, T. T., et al. 2024, Astrophysical Journal Supplement Series, 271, id.44, Fig. 13, reproduced by permission of AAS.

11.6.1 APPLICATION OF THE HIGH-DIMENSIONAL PCA TO THE DOPPLER SHIFT-CORRECTED MAP OF NGC 253

We apply the high-dimensional PCA to the Doppler-corrected map of NGC 253, exactly in the same manner as we did in Sec.11.5. Note that, though we use the same notation for PC2, now PC2 is the value obtained from the Doppler-corrected map and different from those in Sec.11.5 from the NRPCA. We should also note that since we applied the NRPCA and A-SPCA to the Doppler-corrected map, all the contributions of PCs are affected by the correction.

Same as the preparatory analysis, we first show the eigenspectra of the Doppler-corrected spectral map in Fig. 11.13. Comparing with Fig. 11.6, the meaning of the shape of each eigenspectrum becomes clearer. We stress that since we have already subtracted the global shift of the whole spectra, *it does not represent the systemic velocity of the position anymore*. Thus, the velocity structure is always the first-order deviation from the systemic rotation pattern, and appears as a broadening or asymmetric distortion of the line profiles. Another important difference is that an important spectral line is identified in PC5 (see PC5 of the Bottom panel in Fig. 11.13) which was not specified from the original data. This is because the dominant effect of the systemic rotation was successfully removed from the map, and more subtle features appear more prominent in the Doppler-corrected map. Now very clearly PC2 represents the wing of the HCN and some other lines, indicating that it reflects a very

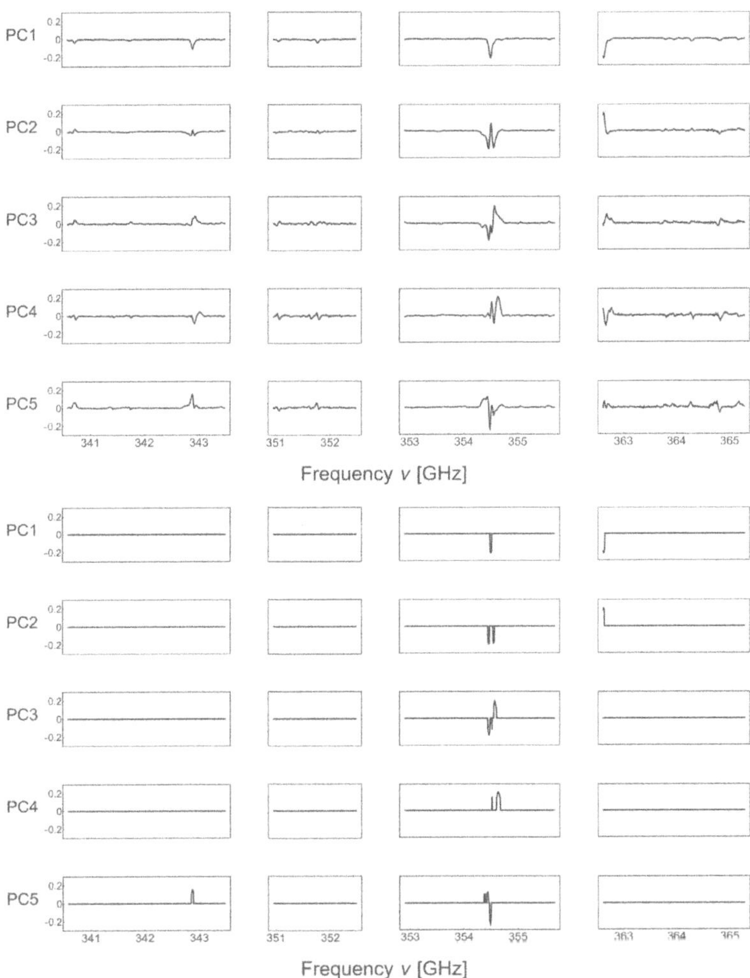

Figure 11.13 The eigenspectra corresponding to the 1st–5th principal components (PCs) constructed from the Doppler-corrected ALMA spectral map of NGC 253 (Takeuchi et al., 2024). Upper five panels present the eigenspectra obtained by the NRPCA, and the Lower five panels show that obtained by the A-SPCA. Credit: Takeuchi, T. T., et al. 2024, Astrophysical Journal Supplement Series, 271, id.44, Fig. 11, reproduced by permission of AAS.

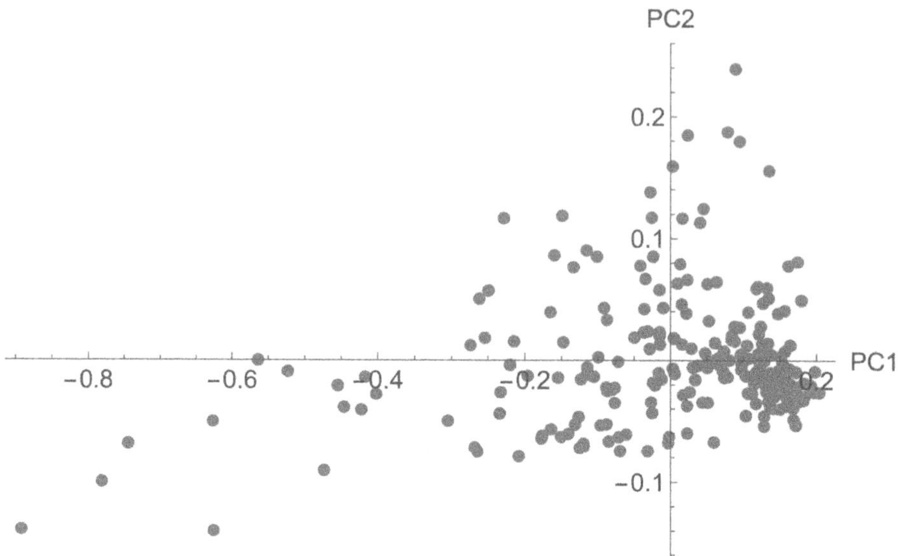

Figure 11.14 The 2-dim structures of the distribution of PC1 and PC2 of the NGC 253 obtained by the A-SPCA, after correcting the effect of the Doppler shifts of the systemic velocity (Takeuchi et al., 2024). Bottom: same as Top panel, but with PC1 and PC3. Note that the PC2 in this figure is different from that in Fig. 11.9. Credit: Takeuchi, T. T., et al. 2024, Astrophysical Journal Supplement Series, 271, id.44, Fig. 14, reproduced by permission of AAS.

local expansion of the gas. PC3 and 4 describe the blueward distortion of the lines, caused by a gas motion toward us faster than the expansion velocity. Though PC5 is less prominent, but similarly to PC3 and 4, it represents the redward distortion of the lines.

We present the distribution of the contribution of PCs in Fig. 11.12. Now the contribution of the new PC2 is much smaller than old PC2 in Fig. 11.7. This is also because of the successful removal of the the systemic rotation. Instead, we have a smaller but non-negligible contribution of new PC2. Now PC2 would describe much more subtle and local features of the spectral line map of NGC 253, which is consistent with the implications from the eigenspectra.

This is more drastically represented on the scatter plot of PCs. We demonstrate the relations between PCs in Figs. 11.14–11.16. Figure 11.14 shows the 2-dim distribution of PC1 and PC2. The butterfly-like pattern seen in Fig. 11.8 completely disappeared. Figures 11.15 and 11.16 are the relations between PC1 and PC3, and PC2 and PC3, respectively. Again we do not find any symmetric pattern in Figs. 11.15 and 11.16. From these figures, we do not have an obvious symmetric pattern in the intrinsic spectral features in the map. Both in 2- and 3-dim, we observe a continuous distributions among PCs.

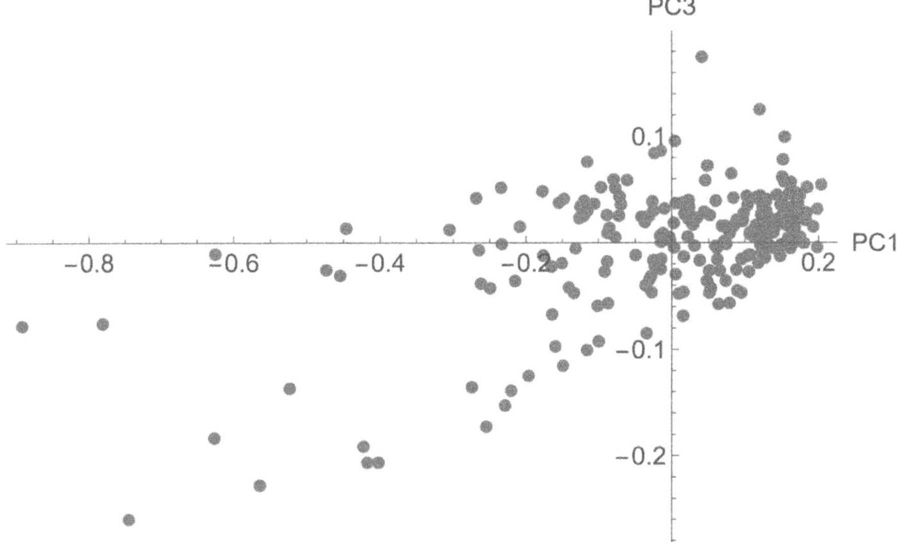

Figure 11.15 Same as Fig. 11.14, but for PC1 and PC3 (Takeuchi et al., 2024). Credit: Takeuchi, T. T., et al. 2024, Astrophysical Journal Supplement Series, 271, id.44, Fig. 15, reproduced by permission of AAS.

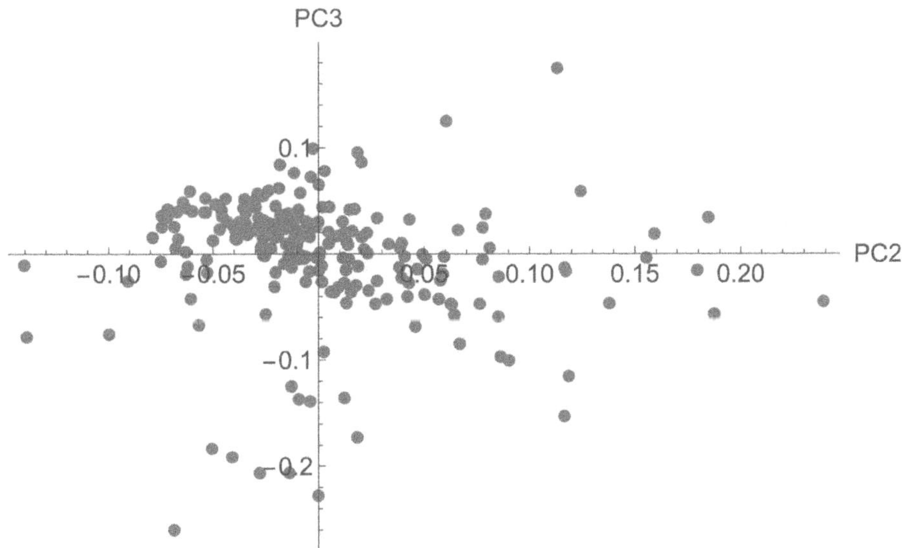

Figure 11.16 Same as Fig. 11.14, but for PC2 and PC3 (Takeuchi et al., 2024). Credit: Takeuchi, T. T., et al. 2024, Astrophysical Journal Supplement Series, 271, id.44, Fig. 16, reproduced by permission of AAS.

11.6.2   MAP OF PCS OF THE NGC 253 SPECTRA

We reconstructed the 2-dim structure to map PC1, 2, and 3 in Fig. 11.17. Again, PC1 represents the intensity of lines. In contrast to Fig. 11.9, new PC2 shows a more complicated, localized pattern in NGC 253. The map of PC3 also shows smaller-scale structures. Recalling that PC2 suggest the existence of small-scale outflow. According to Ando et al. (2017), the spatial scale (diameter) of the most prominent region with large PC2 is ~ 10 pc, namely it corresponds to a small-scale phenomenon, indeed.

Particularly interesting is the region with negative PC3 in the Bottom panel of Fig. 11.17. The eigenspectra show that PC3 and partially PC4 describe a local but larger-scale flow toward us, which appears in the blueshift of the lines. Bolatto et al. (2013) reported a global molecular outflow from the center of the radio continuum. More recently, Walter et al. (2017) and Krieger et al. (2019) performed detailed ALMA observations of this region. They made a detailed analysis and named this outflow as "SW streamer". This outflow is going out toward us, and the position of the root of the SW streamer precisely agrees with the region with negative PC3 in Fig. 11.17 (Krieger et al., 2019, see Fig. 1). The anomalous region specified by the peculiarity in velocity in Fig. 11.18 also agrees well with the root of the SW streamer. In addition to the molecular observations, an outflow is also observed in Hα and other optical emission lines by Fabry-Perot spectroscopy (Matsubayashi et al., 2009). Though this delineates the flow of ionized gas, the position, shape, and direction of the outflow agree fairly well with the molecular outflow. All these clues indicate that the high-dimensional PCA efficiently extracts an outflow phenomenon from an ALMA spectroscopic map, in a purely objective way.

11.6.3   SPECTRAL FEATURES OF THE DOPPLER-CORRECTED DATA CORRESPONDING TO THE PCS

In Fig. 11.19, stars and triangles represent PC1- and PC2-related responsible spectral features similar to Fig. 11.11, and filled circles are PC3-related. First we observe a global shift of lines compared to Fig. 11.11. This is, of course, due to the recession velocity. Since the boundary between ALMA Bands are fixed at the observer's frame, some parts of the spectra go out of the boundary after the Doppler shift correction. As a result, we lose significant parts of the spectra in Fig. 11.19. After the Doppler correction, PC1 describes the intensity of HCN(4–3) and HNC(4–3) lines more specifically, which is reflected to the fact that responsible features concentrate on the peak of these lines. PC2 is assigned to the part of the profile slightly far from the line center for HCN(4–3). However, it is related to the lower-frequency side of HNC(4–3) line, though it is difficult to see clearly because this side is cut out by the correction. PC3 delineate further side of the higher-frequency side of HCN(4–3) line, while it has some overlap with PC2-related features on the lower-frequency side. These properties were already suggested by eigenspectra (Fig. 11.13), and we confirm this by the detailed information from the A-SPCA. Thus, even if the spectra look formidably complicated, the high-dimensional PCA methods like NRPCA and

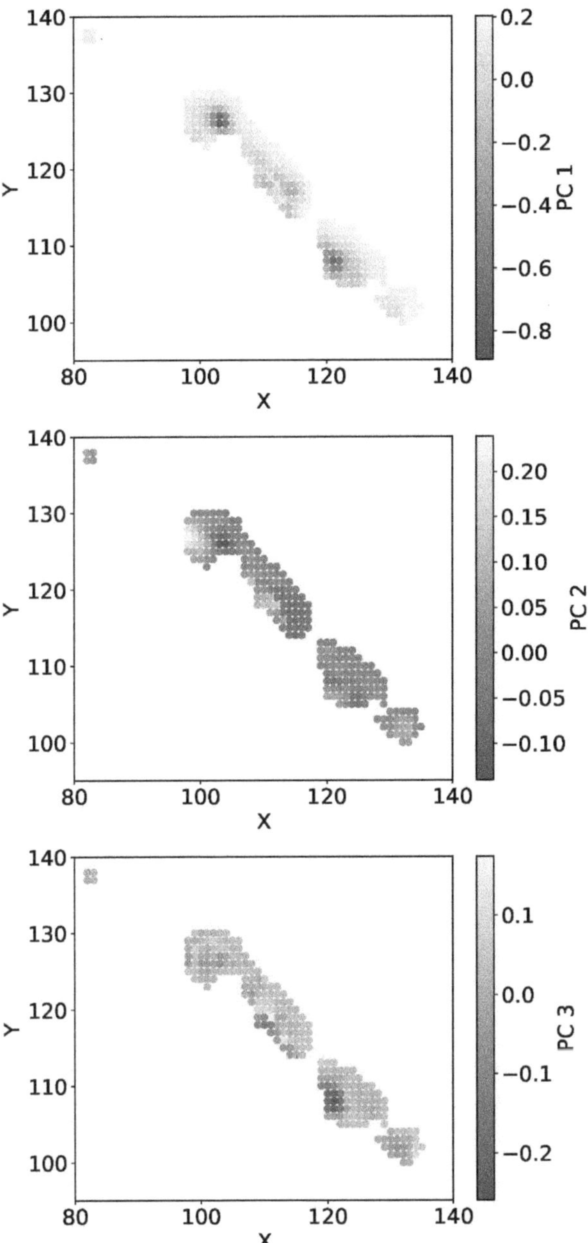

Figure 11.17 The 2-dim structure to map PC1, PC2, and PC3 obtained by the NRPCA (Takeuchi et al., 2024). Credit: Takeuchi, T. T., et al. 2024, Astrophysical Journal Supplement Series, 271, id.44, Fig. 17, reproduced by permission of AAS.

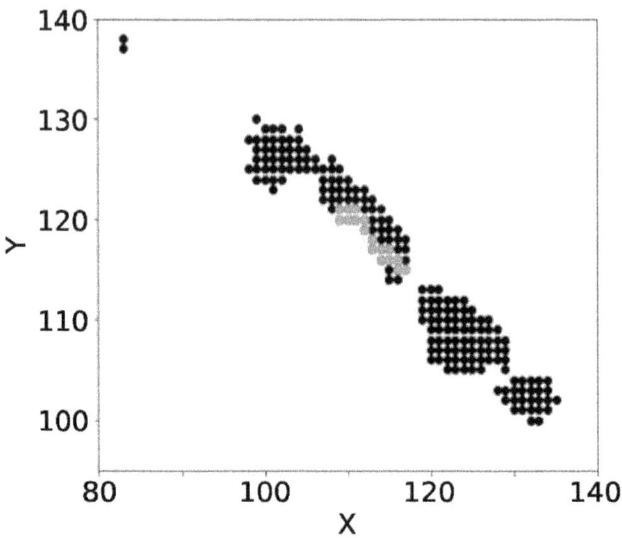

Figure 11.18 The region with velocity peculiarity in Fig. 11.10 (Takeuchi et al., 2024). Credit: Takeuchi, T. T., et al. 2024, Astrophysical Journal Supplement Series, 271, id.44, Fig. 18, reproduced by permission of AAS.

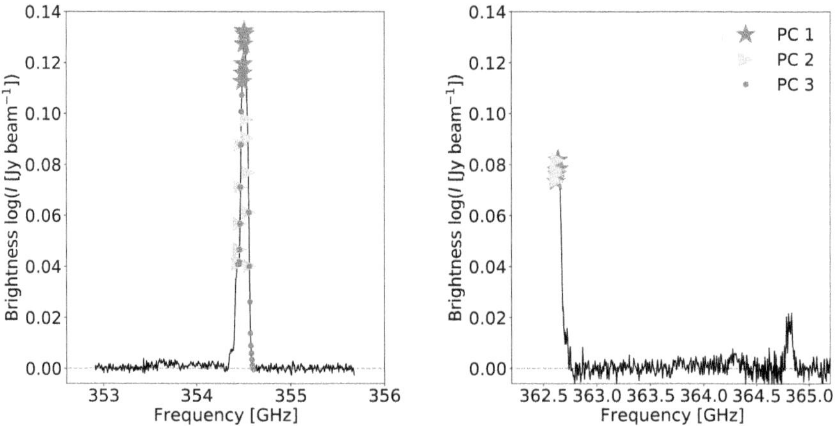

Figure 11.19 Responsible features to characterize PCs from the A-SPCA for the ALMA map of NGC 253, after the Doppler shift correction due to the systemic rotation (Takeuchi et al., 2024). Stars and triangles represent PC1- and PC2-related responsible spectral features similarly to Fig. 11.11, while filled circles are PC3-related. Credit: Takeuchi, T. T., et al. 2024, Astrophysical Journal Supplement Series, 271, id.44, Fig. 19, reproduced by permission of AAS.

A-SPCA can disentangle the information and pick up the controlling spectral features from the spectroscopic map. Rather surprisingly, the detailed profile of HCN and HNC lines govern almost all spectral characteristics of the ALMA map of NGC 253, including the outflow phenomenon. Further investigation on detailed astrophysical processes from the ALMA map of NGC 253 is left to our future work.

11.7 SUMMARY AND PROSPECT

Galaxy evolution is primarily driven by star formation. Since the interstellar medium evolves through star formation and the subsequent nucleosynthesis, understanding this process is key to comprehending galactic evolution. Spectroscopic observations are an important astronomical method for elucidating this process, but detailed spectroscopic mapping requires considerable time, making it difficult to obtain independent data from many observational points. As a result, spectroscopic map data become high-dimensional low sample size (HDLSS) data. HDLSS data involve difficulties unique to high dimensions, and new statistical methods, which replace classical techniques, are required for their analysis.

To solve this problem, high-dimensional statistical analysis has been developed. The analysis of HDLSS data faces many issues stemming from the "curse of dimity," and classical statistical methods often lead to incorrect results. high-dimensional statistical analysis overcomes this problem by using dual matrices and their geometric representations, providing a framework that leverages the unique behavior of HDLSS data for analysis. We applied high-dimensional PCA to ALMA spectroscopic map data of NGC 253 and identified spectral features that characterize the spectral map. Without assuming a model, high-dimensional PCA successfully extracted the global rotation of the central region of NGC 253. Furthermore, by applying high-dimensional PCA to data corrected for the Doppler effect of the extracted rotation, we obtained eigen-spectra indicating gas outflow phenomena and velocity field anomalies caused by explosive star formation.

These results were obtained not by manually selecting features based on physical intuition but by utilizing the full information from high-dimensional data. In this way, high-dimensional statistical analysis offers a new methodology for the spectroscopic map analysis of future space telescopes. It is also useful, for example, in the classification of extremely rare astronomical objects' spectroscopic data. Another example is the observation of the HI forest (HI forest) by SKA, where the radio continuum from active galactic nuclei is absorbed by hydrogen atoms in the intergalactic medium. We aim to apply high-dimensional statistical analysis to the spectral data of rare celestial objects and absorption line systems in the intergalactic medium to elucidate the process of galaxy formation. In this way, high-dimensional statistics are expected to open entirely new doors to a wide range of astronomical challenges.

Section III

Machine Learning Approach

12 Basics of Machine Learning

12.1 MACHINE LEARNING

The concept of machine learning is broad. The term "artificial intelligence (AI)" first appeared in 1956. AI refers to systems or software on computers designed to mimic intelligent tasks typically performed by humans, such as understanding spoken language, making logical inferences from data, or learning from experience.

AI is classified into two categories:

- Artificial General Intelligence (AGI): AI that can think like a human.
- Narrow AI: AI that can perform a specific task as well as or better than a human.

Machine learning is a technology that enhances the performance of narrow AI.

$$\text{AI} \supset \text{machine learning} \tag{12.1}$$

More specifically, machine learning enables computers to learn from experience, much like humans. It can iteratively learn from data, identify patterns and characteristics, and make predictions for new, unseen data.

For example, machine learning excels at image recognition and classification. It can learn the distinguishing features of images of specific objects, such as cars, and identify that an image or a hand-drawn illustration is of a car. Furthermore, machine learning can classify objects in images into predefined categories. In general, machine learning can recognize and classify images, predict future outcomes based on current data, and group data into clusters (see Fig. 12.1). A specialized type of unsupervised machine learning called topological data analysis (TDA) is discussed in Chapter 15.

Machine learning can be classified as follows:

Unsupervised machine learning This approach identifies patterns in a dataset and uncovers the characteristics of the data without requiring labeled training data. Examples:

- Clustering: k-means, Gaussian mixture model, etc.
- Dimensionality reduction: principal component analysis, autoencoder, manifold learning, etc.

Supervised machine learning This approach predicts unknown outcomes by using labeled data. It requires training data with correct answers. Examples:

- Regression: ridge regression, etc.
- Classification: support vector machines, etc.

DOI: 10.1201/9781003104315-12

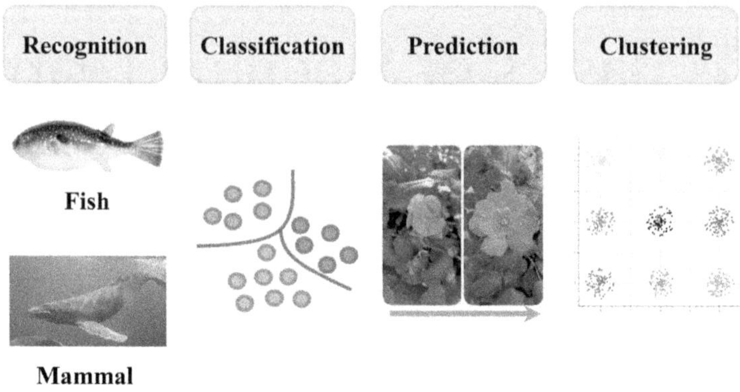

Figure 12.1   Functions of machine learning.

Other methods These include various other methods developed for specific data characteristics or purposes. Examples include recommendation systems, reinforcement learning, deep learning, etc.

12.1.1   UNSUPERVISED MACHINE LEARNING

In unsupervised learning, only input data is provided without the correct answers, and patterns or rules in the data are identified. The goal is to discover the underlying regularities within the data. Examples include:

Model Estimation Finding a general pattern that governs the entire dataset.

Pattern Mining Extracting deeply embedded regularities from numerous features within the dataset or subsets.

We have already discussed the Gaussian mixture model in Chapter 9 and the principal component analysis in Chapter 11. Variational autoencoder (VAE) is applied to galaxy morphology in Chapter 13. Applications related to k-means and manifold learning will be discussed in Chapter 14.

12.1.2   SUPERVISED MACHINE LEARNING

In contrast to unsupervised learning, supervised machine learning creates a model using data with known correct answers. The training data consists of input-output pairs, where the output is either a label or a numerical value. For example, labels could be categories like {female/male}, while numerical values could represent quantities such as length or temperature. If the output consists of labels, the problem is a classification task, whereas if the output is a number, it is a regression problem. Yet another type of algorithm based on supervised machine learning is the generative model. An application of supervised machine learning is explored in Chapter 16.

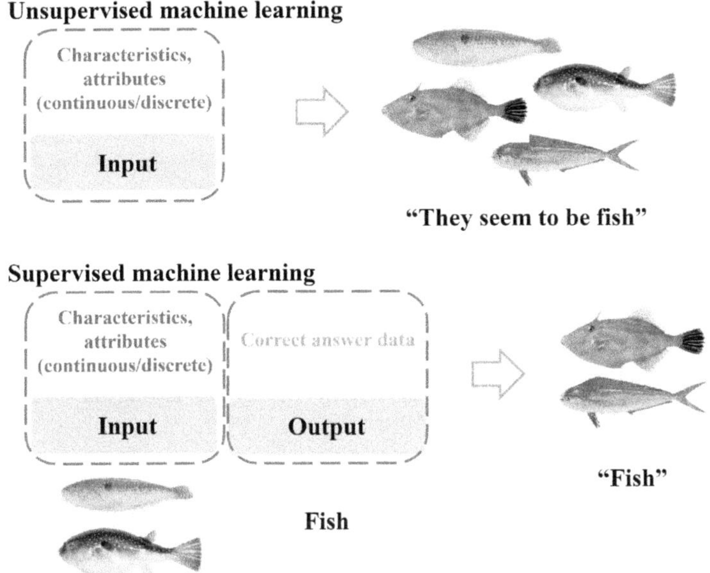

Figure 12.2   Concept of unsupervised and supervised machine learning.

12.2   NEURAL NETWORKS

12.2.1   THE CONCEPT OF NEURAL NETWORKS

Neural networks are a key tool in supervised machine learning. These algorithms are inspired by how the human brain processes patterns. They transmit information between layers by adjusting parameters, mimicking how neurons in the brain communicate with one another (see Fig. 12.3). By feeding large amounts of data into a neural network, it can learn common features for classification tasks through parameter optimization. While originally believed to closely resemble human learning, it is now understood that computer neural networks are quite different from the human brain, though the concept remains central to machine learning.

12.2.2   CONVOLUTIONAL NEURAL NETWORKS (CNN)

Convolutional Neural Networks (CNNs) are particularly effective for processing image data. A CNN is a type of neural network that leverages local features in large datasets. Unlike fully connected neural networks (FCNs), where all units are connected between layers, CNNs limit connections to spatially local regions, making them well-suited for learning features in images that are spatially close.

12.2.3   LEARNING METHOD

In machine learning, data is typically divided into three categories: training data, validation data, and test data. The learning process proceeds as follows:

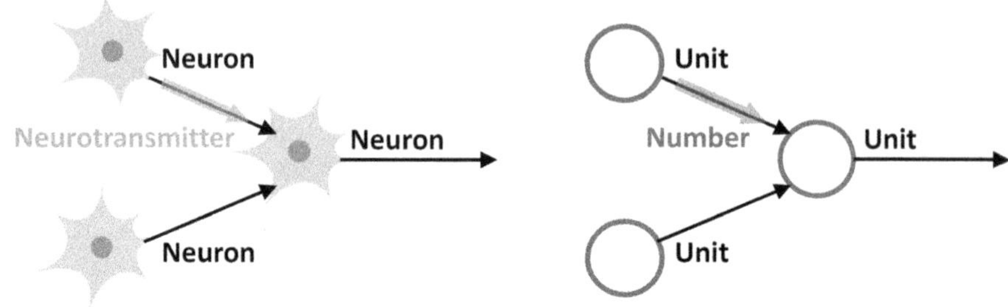

Figure 12.3 Schematic of neural networks. Left: a biological neural network where signals are transmitted between neurons through neurotransmitters along the axon. Right: computer neural network where signals are transmitted between units by transferring numerical values.

1. **Training Phase**

   The model is built using the training and validation data. Training data is used to optimize the parameters between layers, while validation data is used to tune hyperparameters, which determine the model's performance. Hyperparameters are typically adjusted manually, with common ones listed below:

   - **Number of Epochs**

     The number of times the training data is passed through the model. More epochs improve model accuracy but increase computational cost.

   - **Batch Size**

     The size of subsamples is input into the model. Powers of 2 are commonly used, with batch sizes like 512 or 1024 frequently employed for large datasets.

   - **Number of Filters in the Convolutional Layer**

     Filters detect local features during convolution operations. In CNNs, filters are learned from data and determine the number of output feature maps.

   - **Number of Units in the Fully Connected Layer**

     The number of units in the final, fully connected layer.

   - **Dropout Rate**

     The percentage of connections randomly deactivated between layers to prevent overfitting.

   - **Activation Function**

     The function used to determine the output of each neuron in the hidden layers, influencing the transmission of information to subsequent layers.

   - **Optimization Function**

     The function used to minimize the loss between predicted and actual values by adjusting parameters.

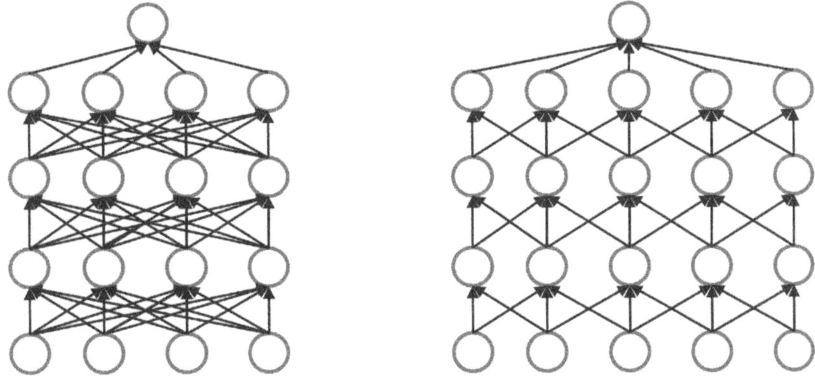

Figure 12.4 Comparison of fully connected neural networks (FCN) and convolutional neural networks (CNN). Left: schematic diagram of FCN. Right: schematic diagram of CNN, where connections are spatially localized.

The training data is used to learn the model parameters, while the validation data is used to optimize hyperparameters and ensure accuracy and low loss. The model that achieves near-perfect accuracy is considered optimal.

2. **Testing Phase**

   In the testing phase, data that was not used during training is fed into the model to evaluate its generalization ability. If the model performs well on test data, it is considered capable of making accurate predictions on new data. However, if test accuracy is low and loss is high, it indicates overfitting, where the model has adapted too closely to the training data and struggles to generalize to new data. Measures such as dropout can help prevent overfitting.

An illustration of how training, validation, and test data are divided is shown in Fig. 12.5.

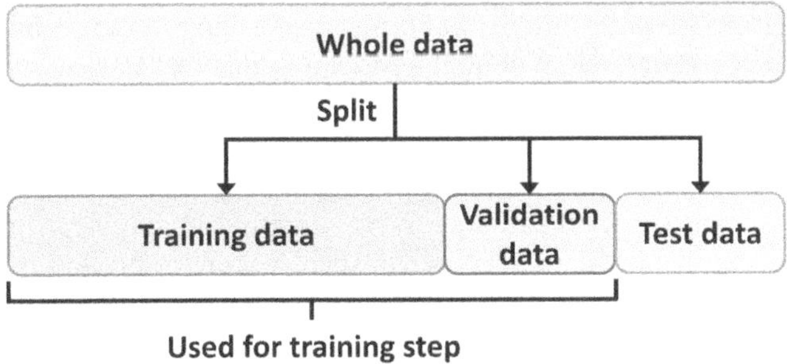

Figure 12.5 Illustration of the division of data into training, validation, and test sets. Training data is used to train the model, while validation data is used to tune hyperparameters.

13 Galaxy Face

13.1 INTRODUCTION

The distribution of stars in galaxies, and in turn its light distribution, is called its 'morphology' (see Chapter 1). This morphological structure is what we observe visually when we look at galaxies and comes in various shapes and sizes. However, while galactic morphology is largely unique for each galaxy, there are certain shared features commonly observed: bulges, spiral arms, and disks. Morphological classification systems such as the tuning fork diagram are based on such shared features. Galaxies which contain uncommon features are generally classified as "peculiar" galaxies.

Structure of a galaxy has been shown to be strongly related with various physical properties, such as stellar mass (Bundy et al., 2005) and star formation rate (SFR). For example, spiral galaxies tend to have higher associated SFRs when compared with elliptical galaxies. This is due to their spiral arms, which help trigger shocks in molecular gas clouds and cause them to undergo gravitational collapse when they pass through, hence accelerating star formation activity (Chapter 3). In addition, the morphology of a galaxy is also a reflection of its evolution, under the 'bottom-up' theories of galaxy formation, the earliest galaxies formed initially with spiral or disk structures, before eventually undergoing merger to become the larger elliptical galaxies (Chapter 7). Morphological properties of galaxies could hence be viewed as a proxy for estimating its underlying or hidden properties.

To do so, we require an understanding of the relationship between morphology of a galaxy and its physical properties. One way of doing so is by understanding how the structure of a galaxy can be decomposed into more fundamental 'latent features' which exhibit correlations with different properties. Previous works by Uzeirbegovic et al. (2020) have shown that the principal component analysis (PCA) is one way of doing this, and has shown that most galaxies can be represented simply as unique sums of a small set of latent features. Some of these are immediately recognisable as well-understood features like spiral arms, center bulges, etc., while other latent features are more novel. We will attempt to apply this to a dataset of near-infrared spatially resolved galactic images, and examine the relationship between the principal components of a galaxy and its physical properties.

However, such methods are limited as they can only find linear manifolds of the higher dimensional space. If the underlying manifold is nonlinear in nature, PCA will fail to find suitable latent vectors. As such, we are motivated to explore nonlinear methods like neural networks too. In particular, we will make use of various autoencoder architectures (a deep unsupervised learning method) to study the latent features of the same dataset.

DOI: 10.1201/9781003104315-13

13.2 DATA

13.2.1 GALAXY CATALOGUE

We utilize the Reference Catalog of galaxy SEDs (RCSED) dataset, which is a value-added join between the Sloan Digital Sky Survey (SDSS), Galaxy Evolution Explorer (GALEX), and UKIRT Infrared Deep Sky Survey (UKIDSS) datasets, as our main galaxy catalogue. In particular, this catalogue consists of low to intermediate redshift galaxies only ($0.007 < z < 0.6$).

We further used the Galaxy Zoo 2 dataset taken from the Galaxy Zoo project to filter and keep only face-on galaxies. The Galaxy Zoo 2 dataset was collected by showing human collaborators images of galaxies, and having them vote for whether a feature is applicable to the galaxy. If more than 80 % of the votes for a classification is true, a 'clean' flag is raised. We used this clean flag to determine if a galaxy was face-on. This was done as the more interesting features of an edge-on galaxy are not visible to us due to our position, causing most of them to have similar sharp elliptical profiles.

Galaxies with 'oddities', such as those undergoing merging, containing dust lanes, or whose position overlaps with other galaxies behind it, were also removed in the same way, as they were considered to be outliers.

13.2.2 GALAXY IMAGES

Example Images

Figure 13.1 Grayscale examples of images found in our dataset.

SDSS Data Release 17 i-band images were retrieved through NASA SkyView API. i-band images were chosen as these wavelengths, which are near-infrared, are sensitive to the emissions of long-living stars. As such, as the long-term structure of a galaxy is defined by the distribution of such long-living stars, this band would give us the best representation of our morphological analysis of galaxies. The downloaded single-channel gray scale images were set to be 150×150 pixels in dimension (Fig. 13.1).

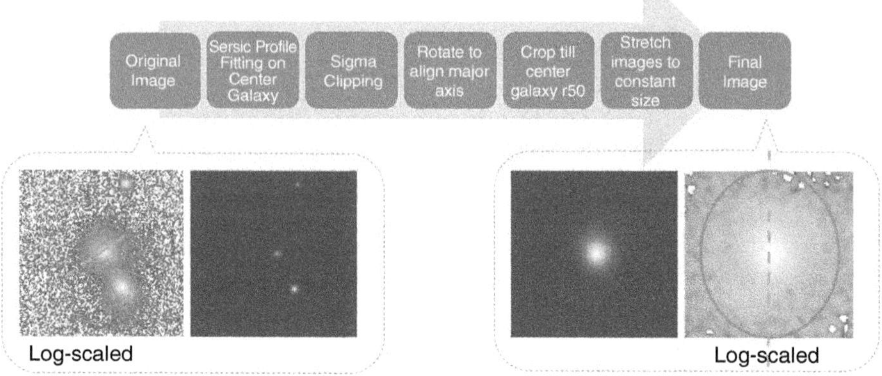

Figure 13.2   Image preprocessing of this analysis.

13.2.3   IMAGE PREPROCESSING

Since the relative size of the galaxies in each image was different, we had to preprocess them to ensure that the focal galaxy took up the same proportion of dimensions of each image. To do so, we first conducted Sérsic fitting via the statmorph python package on all the objects present in each image, before isolating the Sérsic function of the center-most object. Each fit generates a 2D Sérsic profile (i.e., 2 Sérsic fits along perpendicular directions), which allows us to create ellipticals with the longest and shortest edges being along each perpendicular Sérsic fit.

We then rotated the images according to the angle between the y-axis and the major axis of this fitted elliptical, such that the major axis of all galaxies are vertically aligned. We then cropped down each image to 1.5 times the Petrosian radius of a galaxy, before finally scaling all the images to a size of 100×100 pixels. A visualization of the steps can be seen in Fig. 13.2. We then flattened each image to 10000-dimensional vectors, which is required for passing into our neural network. Finally, each image vector is then stacked on each other to form a 12539 rows $\times$ 10000 columns image data matrix 13.3.

13.3   METHOD

13.3.1   PRINCIPAL COMPONENT ANALYSIS

Consider a data set matrix, where each row is a data point and each column is one of the dimensions of a data point. The principal components (PCs) of such a dataset is the set of orthogonal unit vectors where the average squared distance from the data points vector line is minimized. Importantly, when data is expressed in coordinates defined by these principal components, each dimension of the data is linearly uncorrelated with one another. The information held by each of these dimensions does not overlap with each other.

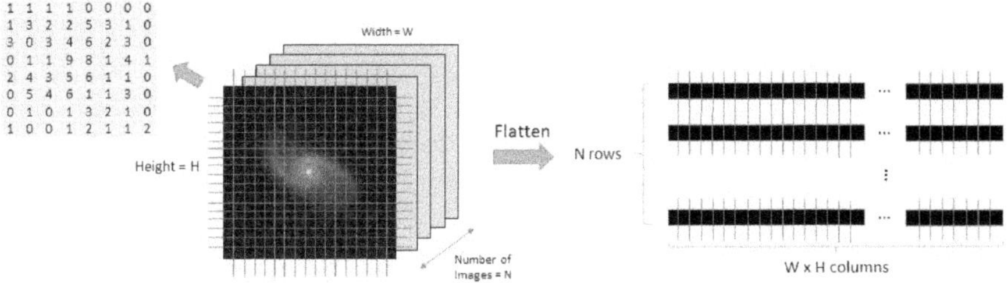

Figure 13.3 Stacking of image data in a matrix form.

Furthermore, the principal component vectors are defined such that when the data is scalar projected, the greatest possible variance is achieved precisely when the data is projected onto the first principal component. The next highest variance results from projecting onto the second principal component, and so on. Indeed, by looking at the example in Fig. 13.4 we see that PC1 is precisely the direction along which the variance of the data is maximum, followed by PC2.

We also note that if we were to change our coordinate frame to that defined by PC1 and PC2, the data points are scattered with minimal variance along PC2. We interpret this as PC2 containing very little information that uniquely defines each data point, especially in comparison with PC1. As such, we can choose to ignore this dimension altogether, effectively reducing the number of dimensions of our dataset and making it simpler to work with.

In the case of our galactic images, we also want to find a lower dimensional representation that has minimal loss in this 'information'. We hence look for the principal components of our image data matrix, and choose a suitable number of principal components that still make up a reasonable explained variance ratio.

To calculate the principal components, one method is to finding the eigenvectors of the covariance matrix of the dataset. For a $n \times p$ dataset matrix B, its $p \times p$ covariance matrix Σ is given by

$$\Sigma = \frac{1}{n-1} B^* B, \tag{13.1}$$

where a factor $1/(n-1)$ is used for Bessel's correction.

One can then calculate the eigenvectors $\vec{v}_i$ of the covariance matrix to get the principal components. The eigenvalues λ_i of each eigenvector will be precisely the explained variance of that particular principal component. The explained variance ratio of ith PC is given by $\lambda_i / \sum_{j=1}^{n} \lambda_j$.

13.3.2 AUTOENCODERS

Autoencoders are a type of unsupervised neural network that:

- Learns the optimal way of encoding an image into a compressed, low-dimensional representation.

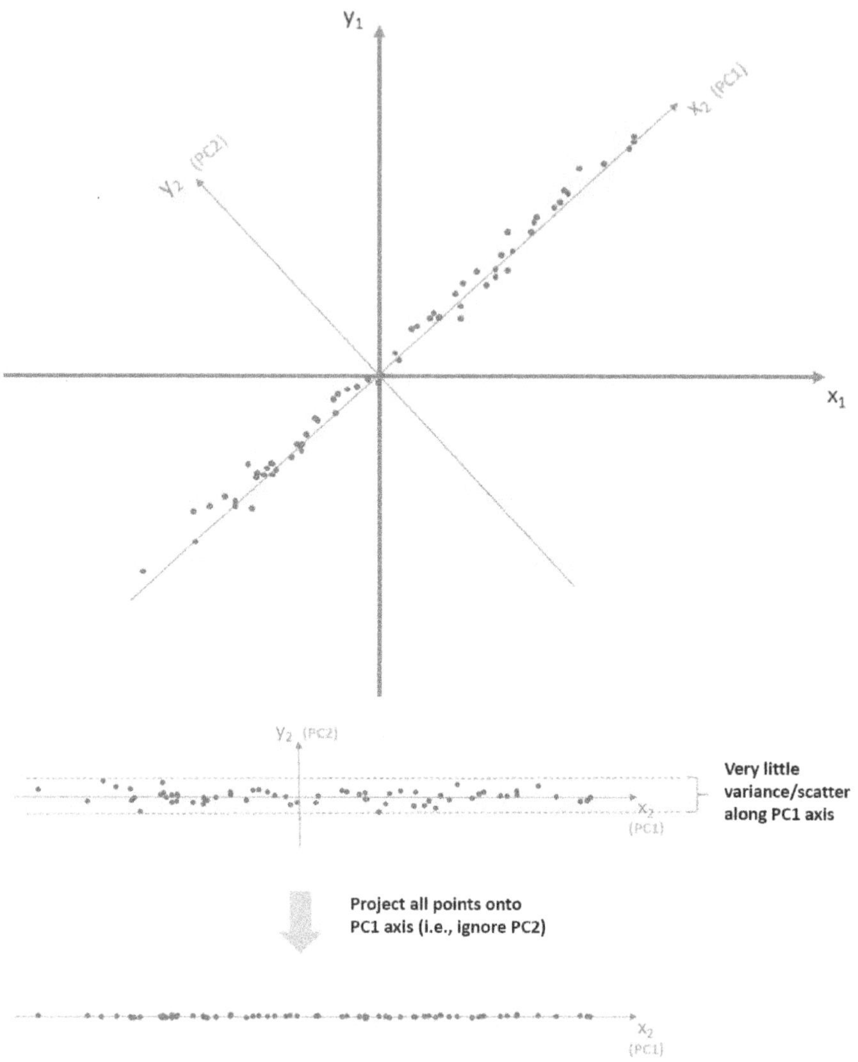

Figure 13.4   PCA in the simple case of reducing a dataset from 2 to 1 dimensional.

- Learns to decode the encoded image such that the final output is similar to the input.

Mathematically, this means that we are searching for two functions, ϕ and ψ, where

$$\phi : X \longrightarrow F \qquad\qquad \psi : F \longrightarrow \hat{X} \qquad\qquad (13.2)$$

such that the loss function $L(X, \hat{X})$ between input data set X and output data set $\hat{X}$ is minimised.

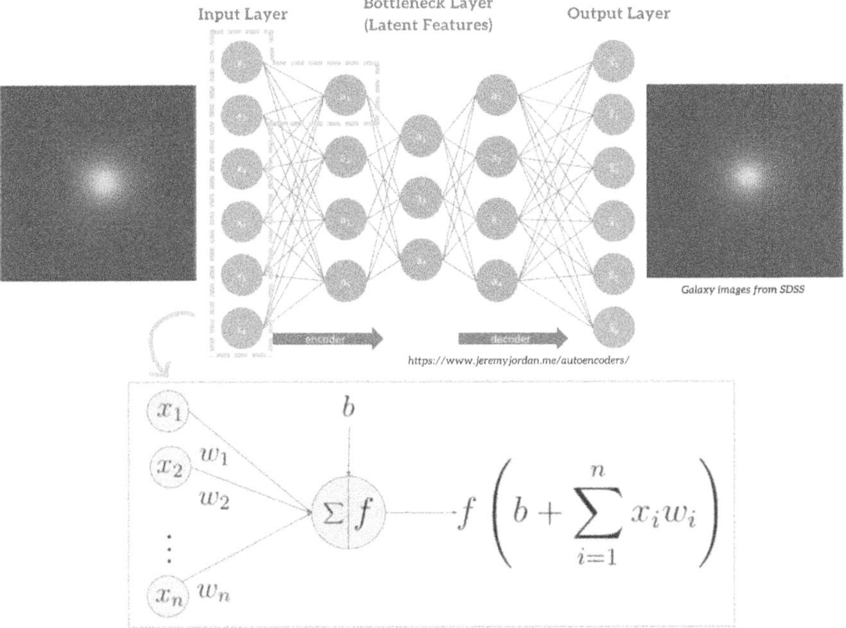

$$f\left(b + \sum_{i=1}^{n} x_i w_i\right)$$

Figure 13.5 Forward propagation in a simple fully connected neural network

Each node in the input layer corresponds to the value of one dimension from our input data. In the case of our $n \times 1$-dimensional image data, each node would contain the value of a row in the flattened image vector (which in turn corresponds with one pixel from the original image array). The concept of autoencoders is described in Fig. 13.5.

For each neuron j in the following second layer, we repeat the following steps. Firstly, the value from each node i of the input data x, x_i, is multiplied by a weight $w_i j$, before the weighted values from all the nodes are then summed up. This output is then passed into a nonlinear activation function σ.

$$\text{Output}_j = \sigma\left(\sum_i (w_{ij} x_i)\right) \tag{13.3}$$

The output from this activation function is then finally passed as the value for one of the destination nodes in the next hidden layer. This is then repeated for every layer in the network. Note that the output from the final layer will have the same dimensions as the input layer. In particular, the dimensions of the hidden layers (number of nodes in the layer) are intentionally constrained in the middle. This forces the neural network to learn how to compress the N number of dimensions initially present at the input layer into a smaller number of dimensions with minimal loss in information.

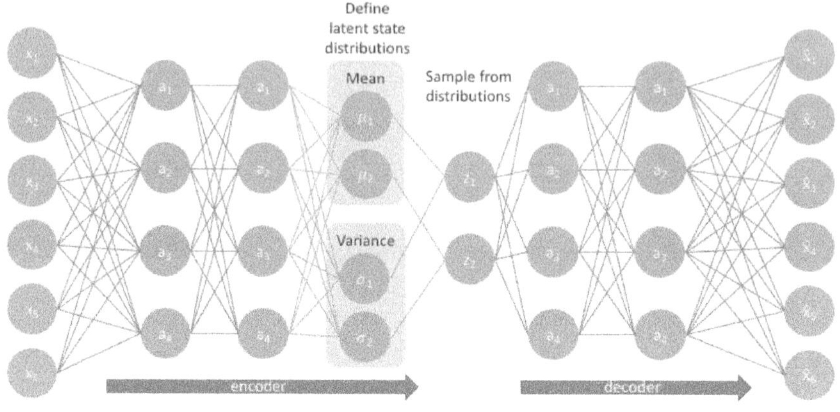

Figure 13.6   Schematic description on the variational autoencoder (VAE).

The neural network trains itself by first checking the error between the output and input values. Through a process called 'backpropagation', it then decides how the weights w need to be adjusted to reduce this error. In particular, we will make use of a loss function called the mean squared error (MSE), which for a single input vector y and output vector $\hat{y}$ is defined as

$$\text{MSE} = (y - \hat{x})^2 \tag{13.4}$$

For a given dataset Y of n vectors $y_1, y_2, ..., y_n$, the overall loss function is defined as

$$L(Y, \hat{Y}) = \frac{1}{n} \sum_i^n (y_i - \hat{y}_i)^2 \tag{13.5}$$

We then check how this loss function changes when we make changes to the weights w (and biases) by finding the gradients with respect to them, and search for its local or global minimum.

13.3.3   VARIATIONAL AUTOENCODERS

Unlike a basic autoencoder, the main feature of a variational autoencoder (VAE) is that it finds a probabilistic distribution for the latent features of each input vector, rather than a fixed value. In practice, we assume that our latent distributions are Gaussian, such that the encoded input data is a set of means and variances, μ_i and σ_i, for each latent dimension i. Our decoder then takes a random sample from each distribution and attempts to reconstruct the original image.

However, it is important to note that sampling directly from the $N_i(\mu_i, \sigma_i)$ will prevent us from finding the partial derivative of the loss function with respect to the weights attached to the latent layer. This means that we will be unable to train our neural network as we are unable to perform backpropagation. To get around this, we implement the 'reparameterization trick', where we instead randomly sample from a unit Gaussian $N(0,1)$, and shift it by the mean μ_i before scaling it by the variance σ_i. Since we are now taking fixed values from the latent features μ_i and σ_i, we are able to take the partial derivatives and perform backpropagation normally.

The main advantage of a VAE is that we are able to find smoother latent distributions when compared to a basic autoencoder. The loss function of a VAE contains an additional term on top of the MSE, called the Kullback–Leibler (KL) divergence term (cf. 8).

$$L(Y,\hat{Y}) = \frac{1}{n}\sum_{i=1}^{n}(y_i - \hat{y}_i)^2 + D_{\mathrm{KL}}\left[N(\mu_Y,\sigma_Y)|\mathcal{N}(\vec{0},\mathbb{I})\right] \tag{13.6}$$

(see Chapter 8). The KL divergence $D_{\mathrm{KL}}(P|Q)$ measures how different the probability distribution P is from Q. By imposing that the distribution of our data points in latent space should be close to a unit Gaussian $\mathcal{N}(0,1)$ via the KL divergence term, we are able to 'smooth' our distribution. This allows for increased interpretability of our results since there are no empty gaps in latent space between clusters of points. An extension of this idea, the β-VAE, adds a hyperparameter coefficient in front of the KL divergence term, which allows one to adjust the balance between reconstruction accuracy and distribution smoothness.

13.4 RESULTS AND DISKUSSION

13.4.1 PRINCIPAL COMPONENT ANALYSIS

With reference to Fig. 13.7 and 13.8, the first 5 principal components were sufficient for explaining 85 % of the explained variance ratio in our data set, while 12 would explain 90 % of the explained variance ratio. This represents a significant compression from 10000-dimensional image data (one dimension for each pixel) to just 12 dimensions. Among these principal components, PC1 is visually similar to the nucleus (sometimes denoted as morphology 'N') that we would find at the core of many galaxies, which explains its high explained variance ratio (40 %). We note that the nature of our principal components are very different from those of Uzeirbegovic et al. (2020).

In particular, while the principal components in Uzeirbegovic et al. (2020) were radially symmetrical and decomposed galactic images into superposed 'eigengalaxies', this application yielded individual structural features instead.

Reconstruction of images from varying numbers n of principal components gave the results in Figs. 13.9 and 13.10. Reconstructions from the first 2, 4, and 8 PCs yielded largely elliptical-like galaxies. However, at $n > 16$, as more spiral-like PCs (like PC10, 11 and 16), the reconstructed images began showing clearer spiral-like features. The mean-squared error loss of the output and input images fell from 0.007 with 2 PCs, to < 0.0015 at 64 PC.

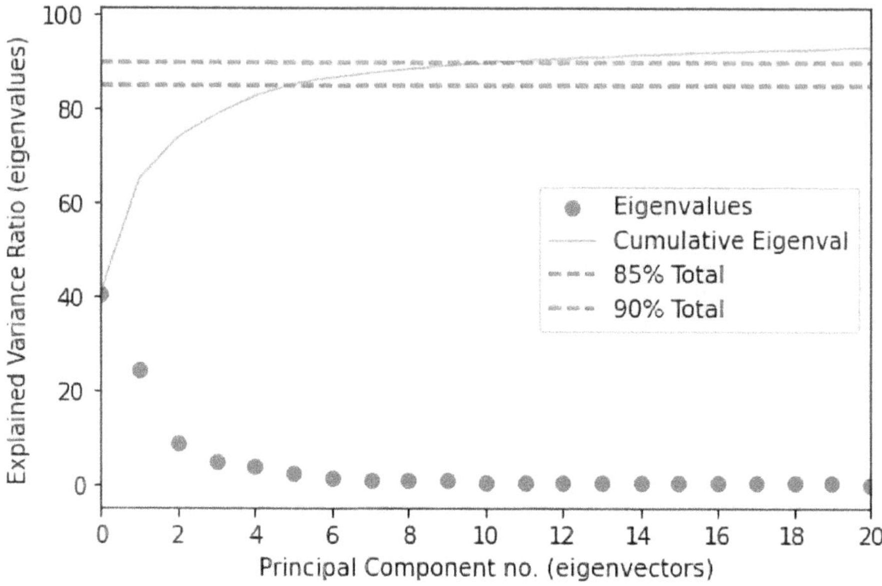

Figure 13.7 Cumulative variance across principal components.

13.4.1.1 Relation of Principal Components to Morphological Features

From Fig. 13.11, we note that if we only consider the first 3 PCs (explained variance ratio $\sim 77\,\%$), the distribution of galaxies in PC-space is such that different morphological classes are distinctly clustered. Spiral galaxies tend to be found in regions of higher PC3, while the opposite is true for elliptical galaxies. However, there is little clustering along PC1 and PC2, indicating that these features are not particularly unique to either of these two broad classes.

The principal components are also linearly correlated with the Galaxy Zoo 2 vote data. For example, as shown in Table 13.1, the Pearson correlation matrix indicates that higher amounts of PC1 is moderately correlated with a greater probability of a galaxy having a smooth profile, and a lower probability of having spiral or disk features. This is indicative of a class of elliptical galaxies. Similarly, PC3 is moderately positively correlated with the probability of the galaxy having spiral/disk features, which is characteristic of spiral galaxy classes.

13.4.1.2 Relation of principal components to physical features

The principal components also show moderate-strong correlations with many of the physical properties of galaxies, as seen in Figure 13.12. In particular, PC1 exhibits a significant negative correlation with Petrosian's half-light radius r_{50} of a galaxy ($\rho = -0.645$). As discussed before, PC1 represents the presence of a bright nucleus. As such, a large nucleus would lead to most of its emissions being concentrated in the center, which explains the negative correlation with r_{50}. PC3 also exhibits moderate correlations with many of the physical properties. In particular, PC3 is positively

First 16 Principal Components

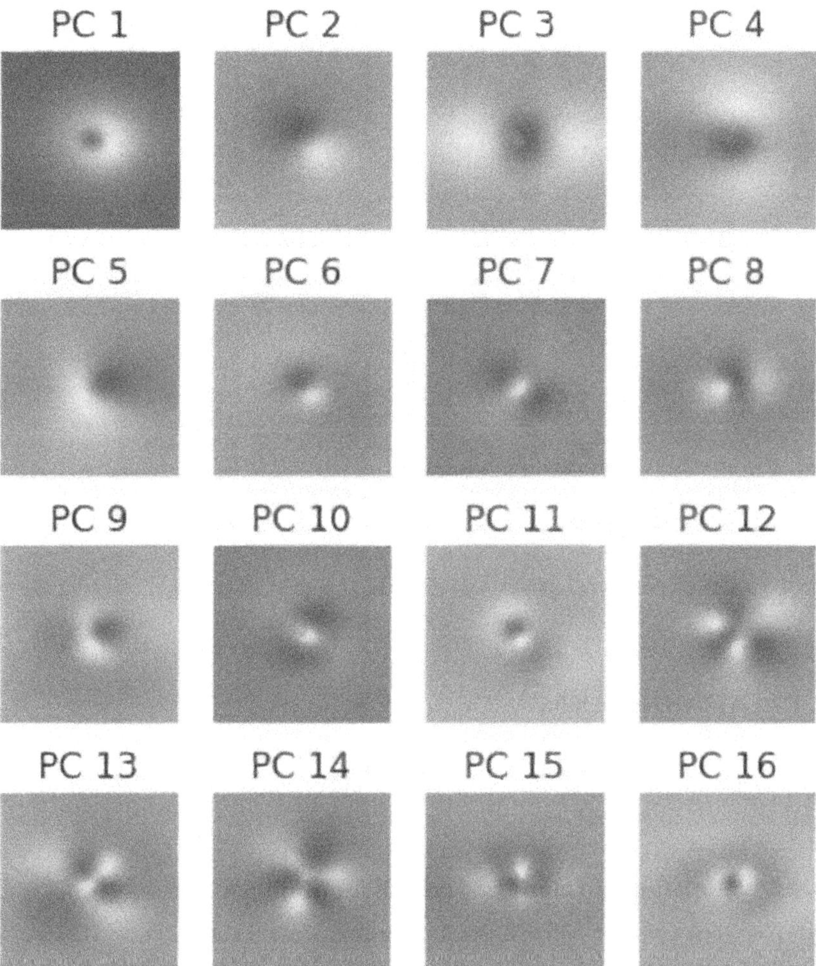

Figure 13.8 First 16 principal components of dataset.

correlated with the logarithm of the specific star formation rate (log SSFR). Since this structure drives the star formation activity and surrounds the core, it could be posited that this is an underlying substructure found within the spiral arm structures. Despite this, the 3rd principal component does not exhibit linear correlations to either log SFR nor stellar mass $\mathcal{M}_*$.

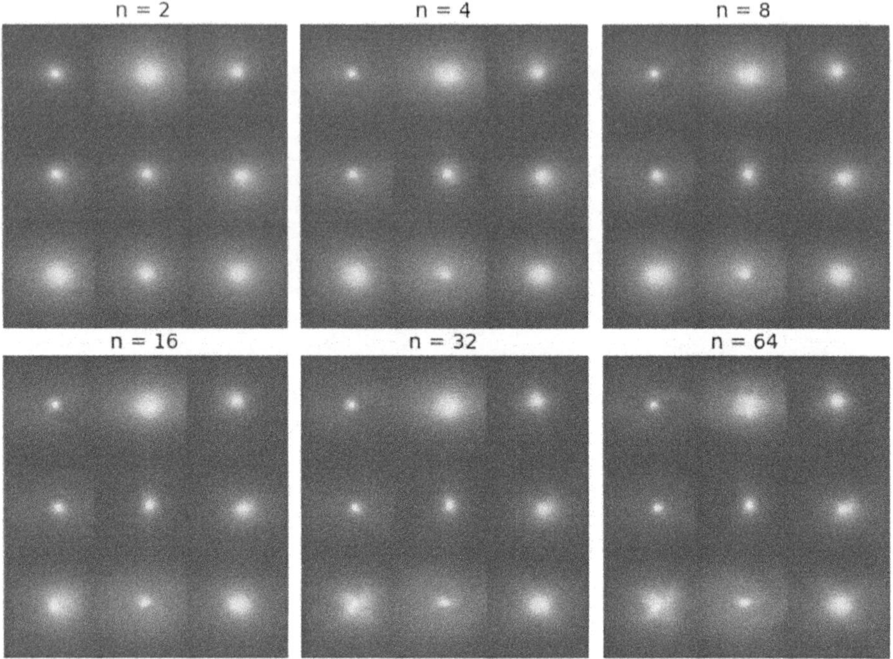

Figure 13.9 Reconstructed galaxies with varying numbers of principal components.

13.4.2 AUTOENCODERS

The image data was scaled logarithmically prior to input into the neural network, in order to dynamically scale the pixel values and ensure better representation of less bright features like spiral arms.

13.4.2.1 Relation of Latent Features to Morphological Properties

For a basic autoencoder with a 3-dim bottleneck, we were able to train it to learn a latent space where the galaxies are again distributed in distinct morphological clusters. From Fig. 13.15, however, we note that there is likely a degeneracy between latent Feature 1 and Feature 2, as they exhibit close to extremely high linear correlations. This implies that the actual underlying latent space is likely to be 2 dimensional, i.e., we can compress our image data into just 2 dimensions while retaining most of the important defining information.

The 2-dim latent space, along with galactic images generated by linear interpolation of the features, can be seen in Fig. 13.16. We note that, again, there is a distinct clustering of classes, but no more degeneracy. The lower amount of Feature 1 tends to lead to the galaxies having more diffuse disk-like structures, while higher amounts of Feature 2 are related to brighter and larger center bulges. However, the boundary

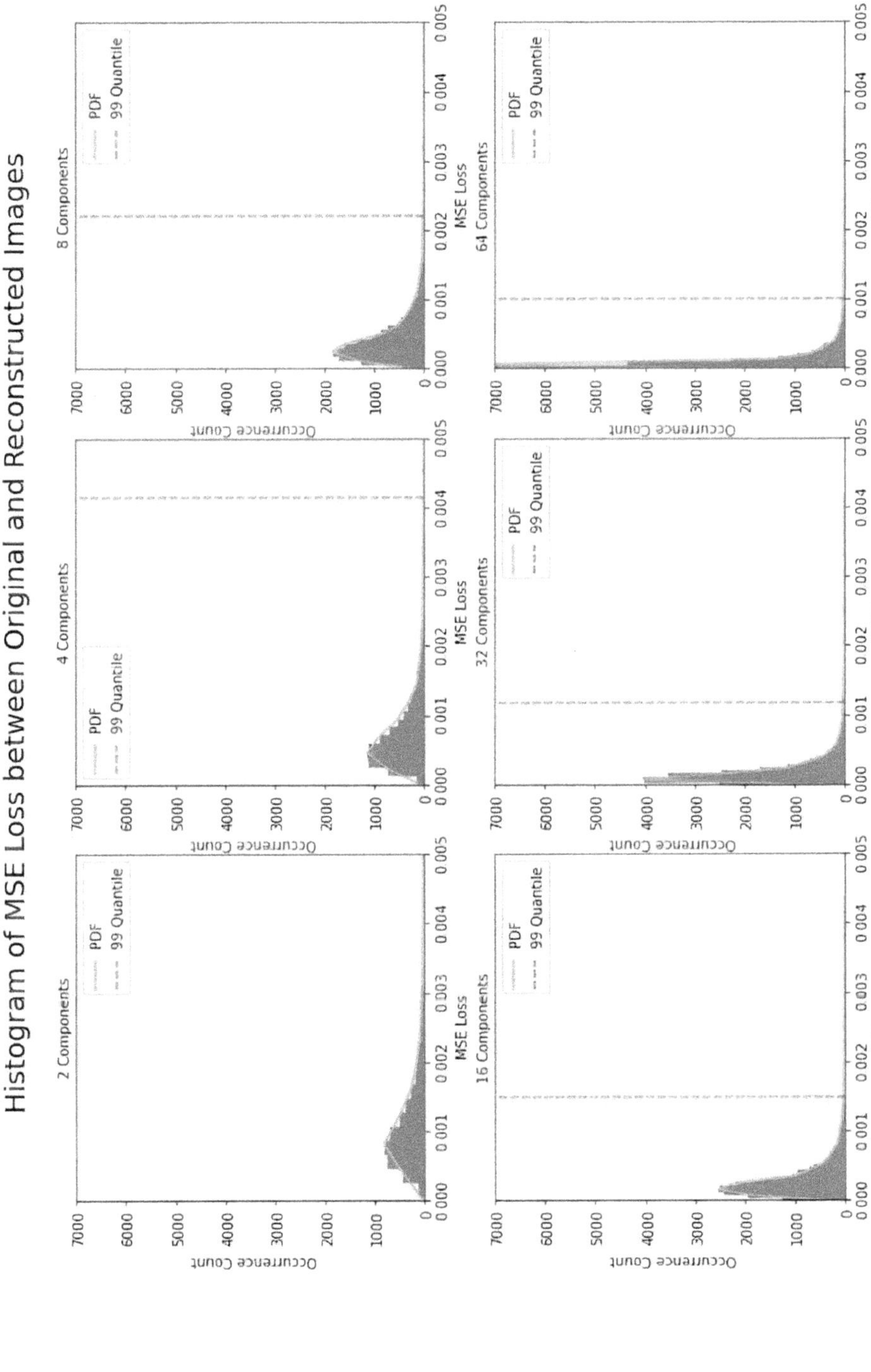

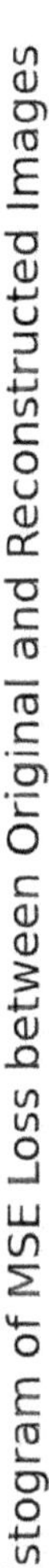

Figure 13.10 Reconstruction loss with different numbers of principal components.

Morphological Class Distribution in Latent Space

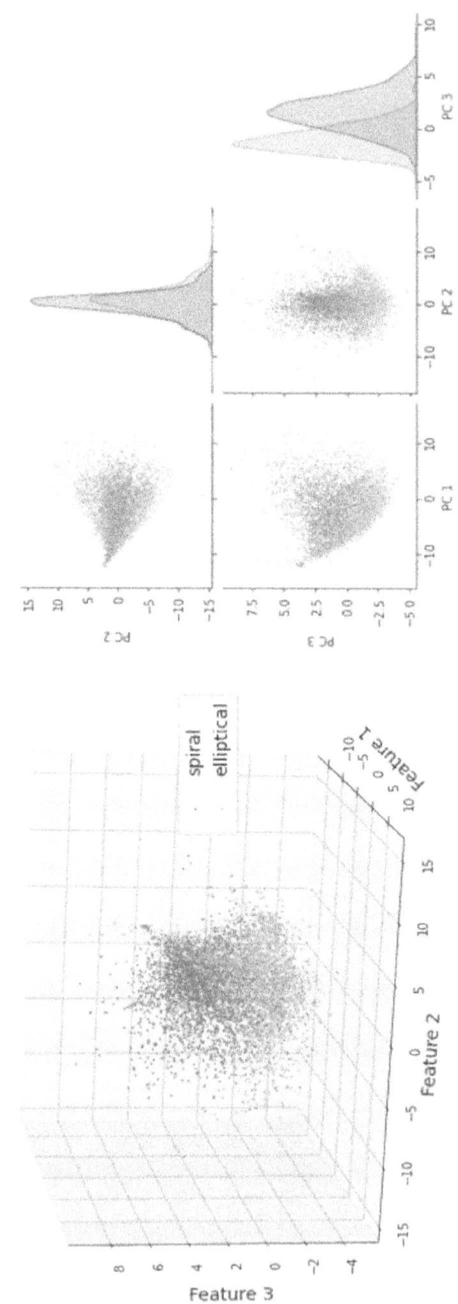

Figure 13.11 Morphological class distribution in PC space. See ebook for the color version of this figure.

Table 13.1

Pearson's linear correlations between the principal components of galaxies and flagged Galaxy Zoo features.

Pearson correlations between PCs and Galaxy Zoo votes			
Galaxy Feature	PC 1	PC 2	PC 3
Smooth profile	0.392331	−0.010117	−0.422918
Features or disk	−0.394702	0.008477	0.421957
Spiral arms	−0.211708	−0.009526	0.454300
Obvious bulge	−0.152348	0.042601	−0.447374
Completely round	−0.092058	0.049670	−0.411865

between the galaxy classes along each axis is not straightforward. Rather, we observe that the boundary value for each feature is linearly dependent on the other. We also note that Feature 2 is moderately positively correlated with the likeliness of a galaxy to have a smooth profile, and negatively correlated with the likeliness of having spiral/disk features. Further, the likelihood of a galaxy being spiral is negatively correlated with Feature 2 (Table 13.2).

13.4.2.2 Relation of Latent Features to Physical Properties

We note that the various physical properties gradate smoothly across the latent space. We note that there are strong negative linear correlations between Feature 2 and r_{50}, which agrees with the idea discussed above that Feature 2 represents the brightness and size of the bulge. It is observed that there is a general trend of negative and positive gradation when moving toward higher Feature 1 and Feature 2. This area of latent space tends to contain older galaxies with lower SSFR, which is characteristic of elliptical galaxies. Unlike with PCA, Fig. 13.17 and Table 13.2 shows that the latent features here actually exhibit bivariate linear correlations with SFR and $\mathcal{M}_*$.

13.4.3 VARIATIONAL AUTOENCODERS

We set the latent space of our variational autoencoder to be 2-dimensional. We immediately note that the distribution of our galaxies is smoother than when we used a basic autoencoder, and is close to a 2-dim Gaussian.

13.4.3.1 Relation of Latent Features to Morphological Properties

We note that higher amounts of latent Feature 1 is are representative of a smaller bulge, while Feature 2 is proportional to the relative size of a diffused disk structure. However, unlike the latent features of autoencoder, there is less overlap between the physical appearance of Feature 1 and Feature 2, which indicates that the variational

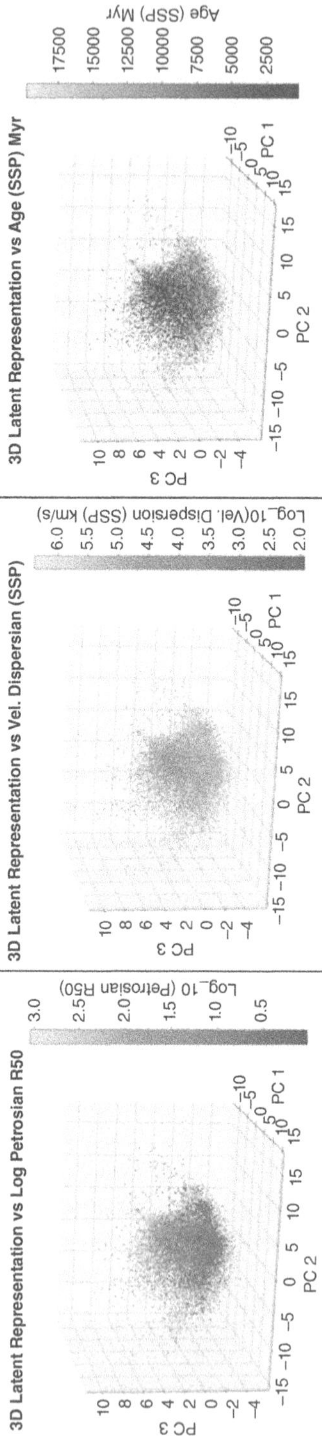

Figure 13.12 Distributions of various galaxy physical properties in 3-dim PC Space.

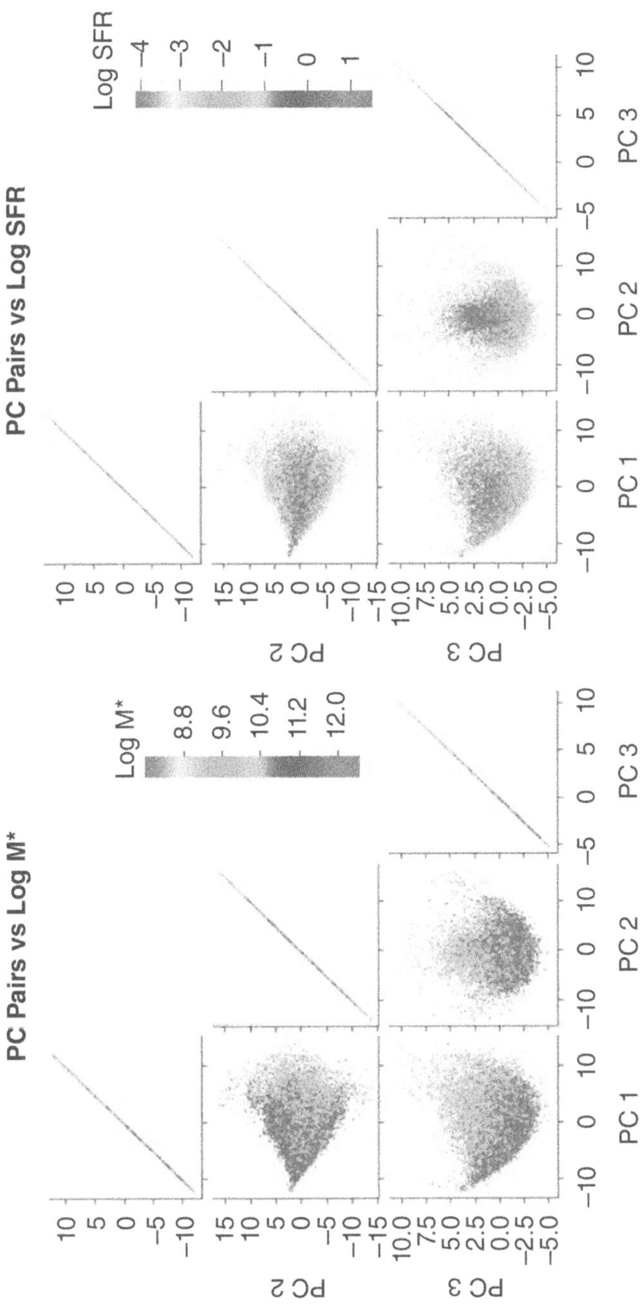

Figure 13.13 Distribution of SFR$/\mathcal{M}_*$ in PC space.

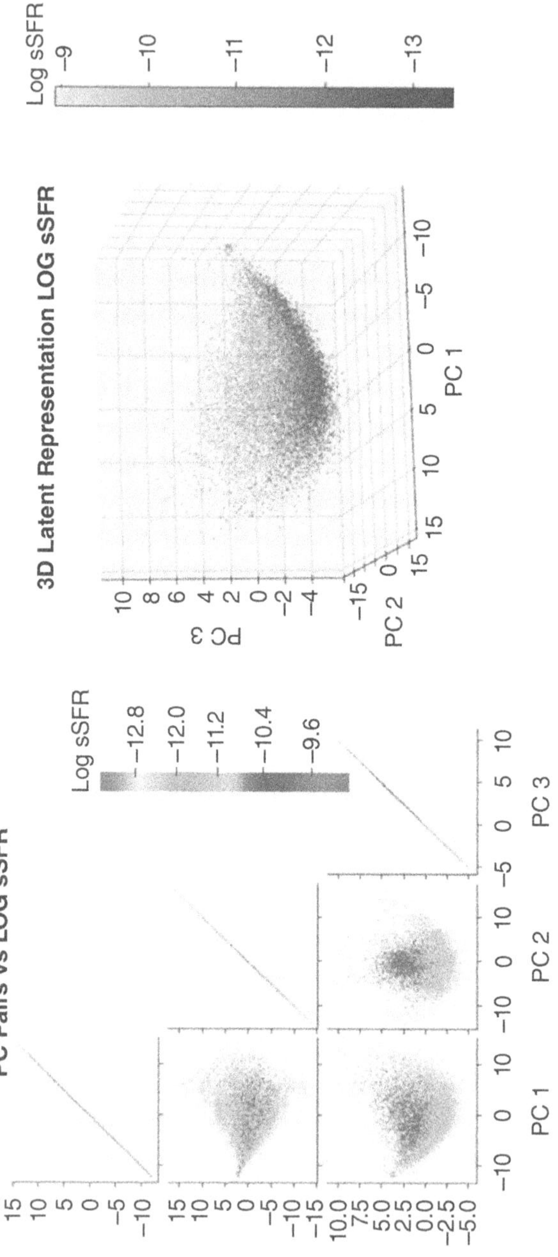

Figure 13.14   Distribution of galaxy SSFR in the autoencoder 3D latent space.

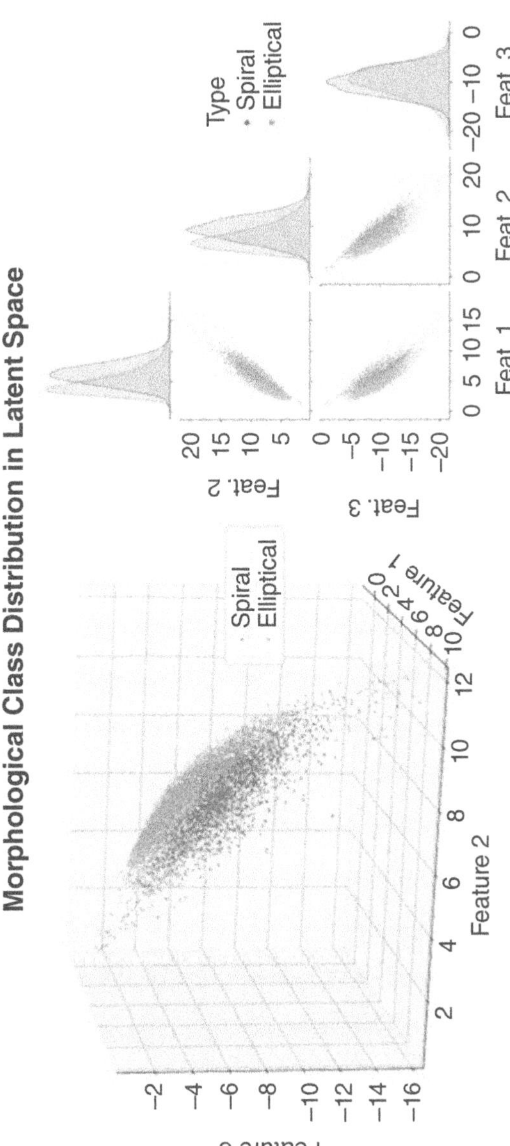

Figure 13.15 Distribution of galaxy morphologies in AE 3D latent space. See ebook for the color version of this figure.

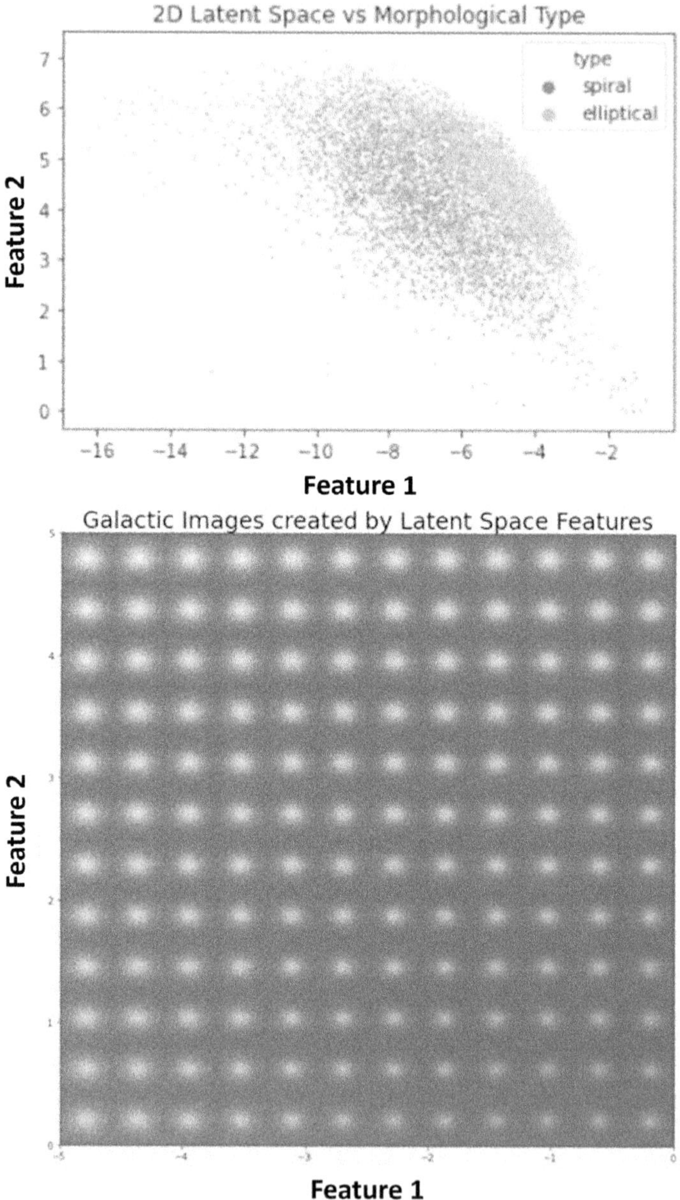

Figure 13.16 Top: distribution of galaxy morphologies in latent space of autoencoder with 2-dim bottleneck layer. Bottom: galaxy image reconstructions using different ratios of the latent features. See ebook for the color version of this figure.

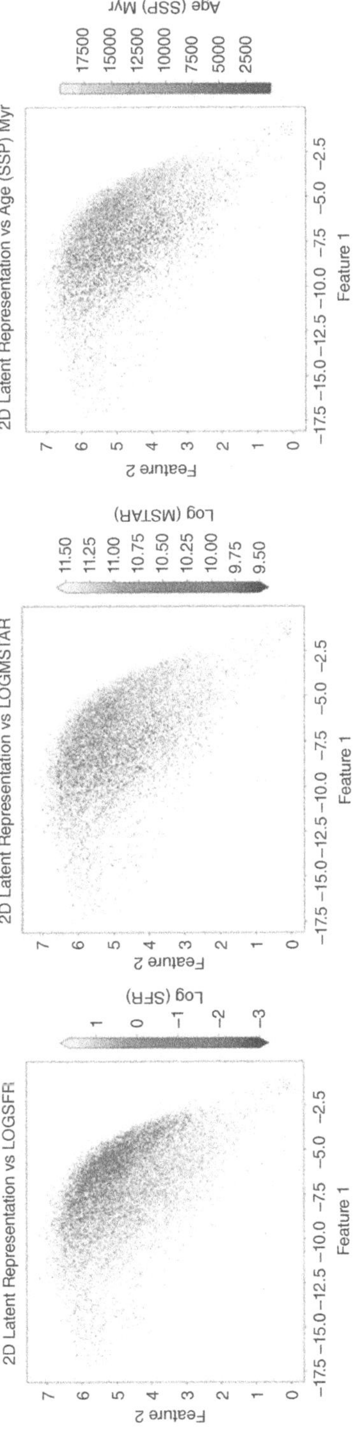

Figure 13.17 Distribution of various galaxy physical properties in 2D AE latent space. See ebook for color version of this figure.

Table 13.2

Pearson's linear correlation coefficients between basic AE latent space representation and morphological/physical properties

Correlation between features and Galaxy Zoo votes		
Galaxy feature	Feature 1	Feature 2
Smooth profile	−0.123796	0.563262
Features or disk	0.126552	−0.561337
Spiral arms	−0.022620	−0.425911
Obvious bulge	0.292862	0.088417
Completely round	0.172631	0.014948

Correlations between feats and Galaxy Zoo classification		
Galaxy Feature	Feature 1	Feature 2
Spiral	0.018326	−0.390460
Elliptical	0.282897	−0.045579
Uncertain	−0.249939	0.376698

Correlations between latent feats and physical properties		
Galaxy feature	Feature 1	Feature 2
Velocity dispersion (SSP) [km s^{-1}]	0.321578	0.135615
Age (SSP) [Myr]	0.326011	0.025508
Metallicity (SSP)	0.321414	0.049316
Velocity dispersion (exp SFH) [km s^{-1}]	0.322730	0.142525
Age (exp SFH) [Myr]	−0.369470	0.013757
Metallicity (exp SFH)	0.339767	0.001796
Petrosian 50 radius [arcsec]	0.361540	−0.746581
$\log \mathrm{SFR}(\mathrm{SED})$	−0.058724	0.055595
$\log \mathcal{M}_*$	−0.024831	0.075587
$\log \mathrm{SSFR}$	−0.190473	−0.172685

autoencoder latent features are more linearly independent. We note that while the morphological classes of our galaxies exhibit distinct clustering, there is some overlap around the origin. This is likely because, in this region of low Feature 1 and Feature 2, neither the bulge nor disk features are the pronounced enough to conclusively define the morphological class of a galaxy.

13.4.3.2　Relation of Latent Features to Physical Properties

We note that the physical properties of the galaxies gradates in varying ways across latent space. The r_{50} of a galaxy increases smoothly and proportionately across Feature 1, but exhibits less variance along Feature 2. This can be explained as the r_{50} of a galaxy being greatly dependent on the size of its center bulge, given that it is

Table 13.3

Pearson's correlations between VAE latent features and galaxy physical properties.

Correlations between PCs and physical properties		
Galaxy Feature	Feature 1	Feature 2
Velocity dispersion (SSP) [km s^{-1}]	-0.051947	-0.465508
Age (SSP) [Myr]	0.034974	-0.417873
Metallicity (SSP)	0.019536	-0.455701
Velocity dispersion (exp SFH) [km s^{-1}]	-0.058639	-0.472199
Age (exp SFH) [Myr]	-0.049727	0.466863
Metallicity (exp SFH)	0.057441	-0.445452
Petrosian 50 radius [arcsec]	0.717488	-0.033900
$\log$ SFR(SED)	-0.058900	0.045080
$\log \mathcal{M}_*$	-0.069458	-0.008703
$\log$ SSFR	0.110948	0.334421

typically much brighter than its diffused disk. On the other hand, the metallicity of a galaxy proportionately increases along Feature 2, but exhibits less variance along Feature 1.

The SSFR and age, however, exhibit a clearer bivariate relationship with the latent features. In particular, galaxies of high age and low SSFR are found in regions of higher Feature 1 and Feature 2. This can be understood intuitively, as these regions are representative of galaxies with big diffused disks (characteristic of older galaxies) and relatively small bulges (which correlates with low star formation activity).

We also note that the SSFR and r_{50} of galaxies are inversely related with each other in the spiral clusters, which agrees with the theory that the star formation rate is linearly related to the surface density for spiral galaxies.

13.4.3.3 Merging and Irregular Galaxies in VAE Latent Space

We also explored how merging and irregular galaxies, which were excluded in the training dataset, would be distributed in this latent space. Merging galaxies were largely found along the 'border' between the elliptical and spiral galaxies. When coupled with the distribution of galaxy ages, we get a picture of how a galaxy morphology evolves with time. This agrees with the existing theory, where the younger spiral galaxies merge eventually to form ellipticals. Comparing this distribution with the SSFR along the border also indicates that SSFR drops rapidly upon merging, which supports the idea that galaxy mergers can initiate quenching (Davies et al., 2022).

Irregular galaxies, on the other hand, are found largely in the regions with a high Feature 2 and positive Feature 1, where spirals are located. More interestingly, the irregular galaxies are largely found in the regions with a very high SSFR, which agrees with the existing theory that irregular galaxies can have comparable SSFR with spirals, though the exact mechanism requires further investigation.

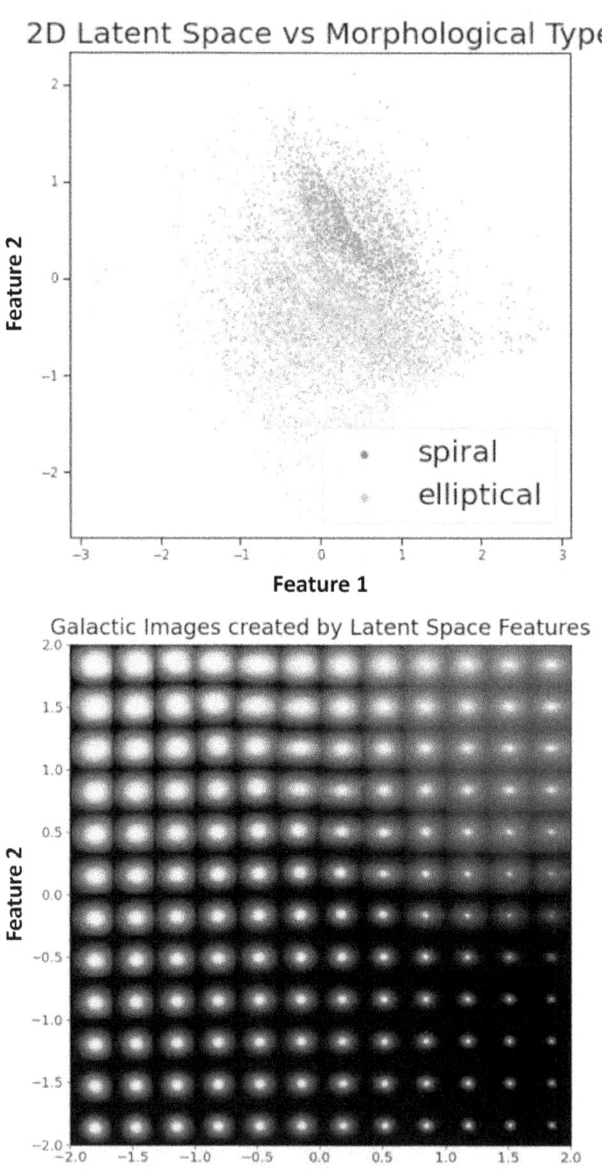

Figure 13.18   Top: distribution of galaxy morphologies in latent space of VAE with 2-dim bottleneck layer. Bottom: galaxy image reconstructions using different ratios of the latent features. See ebook for the color the version of this figure.

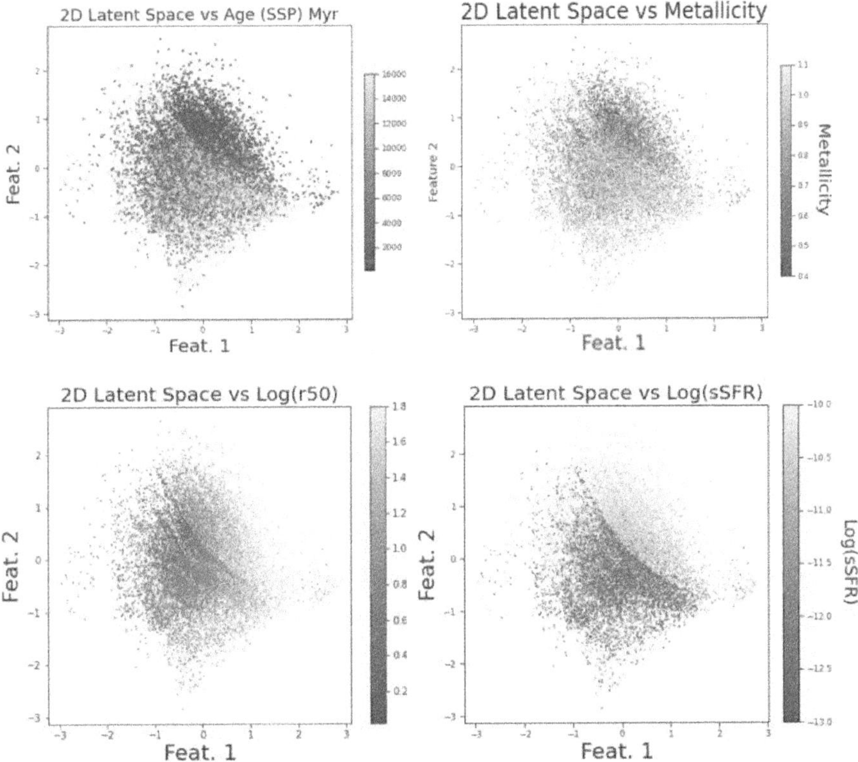

Figure 13.19 Map of various physical properties to different areas in VAE latent space.

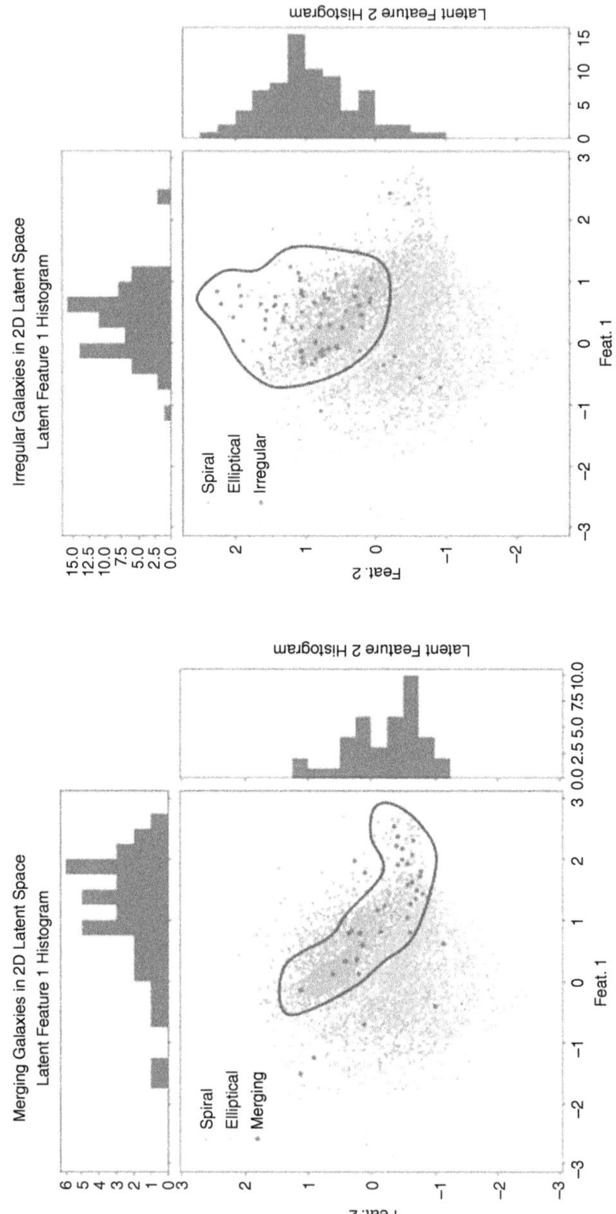

Figure 13.20   Distribution of merging and irregular galaxies in VAE latent space. See ebook for the color version of this figure.

14 New Quantification of Galaxy Evolution by Manifold Learning

14.1 INTRODUCTION

14.1.1 GALAXY EVOLUTION IN THE ERA OF LARGE-SCALE GALAXY SURVEYS

The attempt to quantitatively explain galaxy evolution based on physical laws began in the 1970s. Under the assumption that galaxies formed from a single massive gas cloud, theories were developed to address the history of star formation and related heavy element synthesis. Although this line of research was theoretically completed by Tinsley (1980) in the early 1980s, it did not mark the end of galaxy evolution research. Cosmological studies conducted concurrently revealed that galaxies grow through mergers. This indicates that galaxy evolution is a highly complex process that heavily depends on the density of surrounding galaxies and gas. Thus, it has become clear that galaxy evolution is a very complex process, greatly influenced by the density of surrounding galaxies and gas. The equation describing the new galaxy evolution can symbolically be written as follows:

$$
\begin{aligned}
\mathrm{SFR}(t) &= f_1(\mathrm{SFR}, \mathcal{M}_*, \mathcal{M}_{\mathrm{mol}}, \mathcal{M}_{\mathrm{HI}}, \mathcal{M}_{\mathrm{dust}}, \mathcal{M}_{\mathrm{halo}}, \delta_{\mathrm{gal}}, \cdots), \\
\mathcal{M}_*(t) &= f_2(\mathrm{SFR}, \mathcal{M}_*, \mathcal{M}_{\mathrm{mol}}, \mathcal{M}_{\mathrm{HI}}, \mathcal{M}_{\mathrm{dust}}, \mathcal{M}_{\mathrm{halo}}, \delta_{\mathrm{gal}}, \cdots), \\
\mathcal{M}_{\mathrm{mol}}(t) &= f_3(\mathrm{SFR}, \mathcal{M}_*, \mathcal{M}_{\mathrm{mol}}, \mathcal{M}_{\mathrm{HI}}, \mathcal{M}_{\mathrm{dust}}, \mathcal{M}_{\mathrm{halo}}, \delta_{\mathrm{gal}}, \cdots), \\
\mathcal{M}_{\mathrm{HI}}(t) &= f_4(\mathrm{SFR}, \mathcal{M}_*, \mathcal{M}_{\mathrm{mol}}, \mathcal{M}_{\mathrm{HI}}, \mathcal{M}_{\mathrm{dust}}, \mathcal{M}_{\mathrm{halo}}, \delta_{\mathrm{gal}}, \cdots), \\
\mathcal{M}_{\mathrm{dust}}(t) &= f_5(\mathrm{SFR}, \mathcal{M}_*, \mathcal{M}_{\mathrm{mol}}, \mathcal{M}_{\mathrm{HI}}, \mathcal{M}_{\mathrm{dust}}, \mathcal{M}_{\mathrm{halo}}, \delta_{\mathrm{gal}}, \cdots), \\
\mathcal{M}_{\mathrm{halo}}(t) &= f_6(\mathrm{SFR}, \mathcal{M}_*, \mathcal{M}_{\mathrm{mol}}, \mathcal{M}_{\mathrm{HI}}, \mathcal{M}_{\mathrm{dust}}, \mathcal{M}_{\mathrm{halo}}, \delta_{\mathrm{gal}}, \cdots), \\
\delta_{\mathrm{gal}}(t) &= f_7(\mathrm{SFR}, \mathcal{M}_*, \mathcal{M}_{\mathrm{mol}}, \mathcal{M}_{\mathrm{HI}}, \mathcal{M}_{\mathrm{dust}}, \mathcal{M}_{\mathrm{halo}}, \delta_{\mathrm{gal}}, \cdots), \\
&\vdots
\end{aligned}
$$

Here, $\mathrm{SFR}(t)$, $\mathcal{M}_*(T)$, $\mathcal{M}_{\mathrm{mol}}(t)$, $\mathcal{M}_{\mathrm{HI}(t)}$, $\mathcal{M}_{\mathrm{dust}}(t)$, $\mathcal{M}_{\mathrm{halo}}(t)$, $\delta(t)$ represent the star formation rate, stellar mass, molecular gas mass, atomic hydrogen gas mass, dust mass, dark matter halo mass, and the excess density of surrounding galaxies at time t, respectively. The variables on the right-hand side are written symbolically to indicate that they depend on the entire past history of each variable.

To formulate galaxy evolution, it is necessary to determine such a vast system of equations. Astrophysicists have constructed governing equations based on first principles of physical laws, but when the dimensional space exceeds ten dimensions, such methods become impractical. From the 1970s to the mid-1980s, classical

DOI: 10.1201/9781003104315-14

multivariate analysis methods such as principal component analysis (PCA) were used to combine physical quantities of galaxies in high-dimensional space. This led to the discovery of various (logarithmic) linear relationships, known as galaxy scaling relations. The concept of the galaxy manifold emerged from the efforts to unify scaling relations and find fundamental relationships (Brosche, 1973; Djorgovski, 1992). However, classical PCA could only handle linear relationships, and although it is still useful for exploratively verifying the (logarithmic) linear relationships of galaxies, the concept of the galaxy manifold remained extremely limited and was almost forgotten for a while (Ginolfi et al., 2020; Hunt et al., 2012; Zhang & Zaritsky, 2016). Time passed, and in the 21st-century galaxy surveys, hundreds of physical quantities for hundreds of millions of galaxies were obtained, resulting in typical big data both in quality and quantity. The feature space of galaxies to be analyzed exceeds 100 dimensions. Therefore, characterizing galaxy evolution with traditional methods relying on physical intuition is impossible, necessitating fundamentally different new approaches.

Therefore, we have started discussing galaxy evolution using modern methods (Siudek et al., 2018). Specifically, we constructed a 13-dimensional feature space including 12 wavelengths from ultraviolet to near-infrared (wavelength $\lambda = $ 150 nm–2.2 μm) and luminosities for each cosmic age, and applied the Fisher EM algorithm (FEM), an unsupervised machine learning method (FEM: Bouveyron & Brunet, 2013). Consequently, FEM successfully discovered the star-forming main sequence, a relationship observed between the total stellar mass $\mathcal{M}_*$ and the star formation rate SFR of galaxies, without arbitrary sample selection (Fig. 14.1). FEM also discovered that the star-forming main sequence continuously connects to a series of galaxies that have ceased star formation beyond a certain total stellar mass. This structure, contrary to the hypothesis that galaxy star formation halts abruptly and transitions discontinuously to galaxies that have stopped star formation, was only discovered by fully utilizing the information in the multi-wavelength luminosity space. This galaxy continuum represents one projection of the galaxy manifold that embodies the fundamentals of galaxy evolution. The galaxy manifold in the multi-wavelength luminosity space could not have been discovered in previous studies based on classical PCA due to its nonlinear spatial structure.

However, astrophysical research does not settle for merely a quantitative description of the galaxy manifold. It is tasked with fully understanding its structure and elucidating its dependence on the physical parameters governing galaxy evolution (probably several). For this further goal, more sophisticated methods are needed.

14.1.2 GALAXY MANIFOLD IN THE MULTI-WAVELENGTH LUMINOSITY SPACE

The temporal evolution of the star formation rate is called the star formation history, which is one of the crucial factors determining galaxy evolution. In the wavelengths from ultraviolet to near-infrared, the radiation spectrum of a galaxy is dominated by contributions from stars and gas. The temperature and lifespan of stars strongly

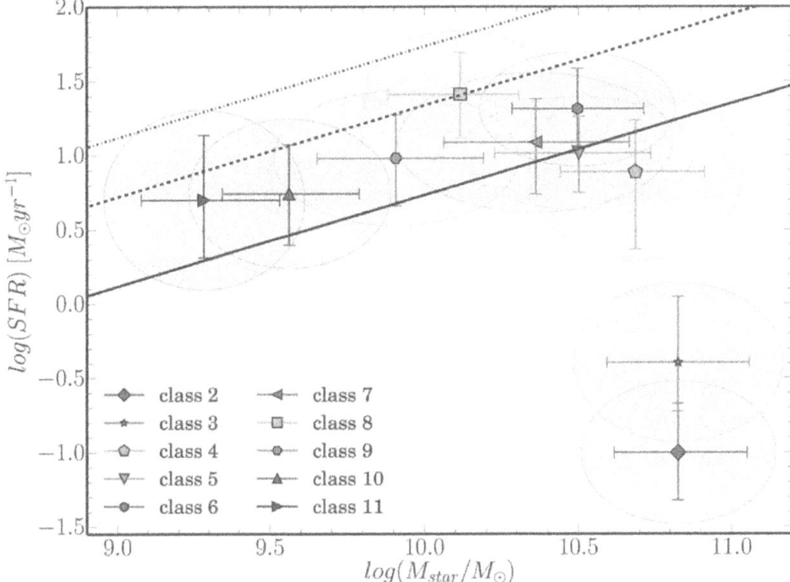

Figure 14.1   The relationship between stellar mass ($\mathcal{M}_*$) and SFR obtained by the Fisher EM algorithm (FEM). The median and variance of log SFR and $\log \mathcal{M}_*$ for galaxy classes classified by FEM are shown. The area of the ellipse corresponds to the median absolute deviation. The black solid line represents the star-forming main sequence at $z = 0.7$ (Whitaker et al., 2012). The dashed and dotted lines correspond to the star-forming main sequence of active star-forming galaxies and starburst galaxies, respectively (Rodighiero et al., 2011). Credit: Siudek, M., et al. 2018, Astronomy and Astrophysics, 617, A70, Fig. 7.

depend on their mass, with more massive stars being brighter, hotter, and having shorter lifespans. Hot stars emit a large amount of ultraviolet radiation, while cool stars are faint in the ultraviolet and shine in the near-infrared. Quantitatively, the time a star can remain as a main-sequence star, where it stably undergoes nuclear fusion, is denoted as τ_{SF}, the surface temperature of the star as T, and the luminosity of the star as L, which can be approximated as:

$$\tau_{\mathrm{MS}} \propto \mathcal{M}^{-2.5}, \tag{14.1}$$

$$L \propto \mathcal{M}^{3.5}, \tag{14.2}$$

$$L \propto T^4 \tag{14.3}$$

(recall eq. (4.23)). As a consequence, high-temperature stars exhaust their lifespans first, thus the star formation history is directly reflected in the spectrum of the galaxy. In other words, the star formation history is expected to be appropriately characterized in the space spanned by the multi-wavelength (band) luminosities of galaxies.

In traditional astronomy, the evolution in the multi-wavelength luminosity space has been characterized by the ratios of luminosities at various wavelengths. These ratios are referred to as colors in astronomy. Consider the pair of monochromatic luminosities L_{λ_1} and L_{λ_2} at two different wavelengths λ_1 and λ_2 ($\lambda_1 < \lambda_2$). If $L_{\lambda_1} < L_{\lambda_2}$, the object is described as "red", and if $L_{\lambda_1} > L_{\lambda_2}$, it is described as "blue". When plotting the relationship between galaxy luminosity (absolute magnitude and color (color-magnitude diagram), two distinct sequences become apparent. This is known as the bimodality of galaxy colors. Specifically, there exists a tight sequence of red galaxies (red sequence) and a broader sequence of blue galaxies (blue cloud) universally. Galaxies in the region between the red sequence and the blue cloud are relatively few and are sometimes referred to as the green valley (e.g., Blanton, 2006). Galaxies in the blue cloud are actively forming stars and contain massive, short-lived, high-temperature stars, whereas the red sequence is dominated by low-mass, low-temperature stars with ceased star formation. It is believed that galaxy evolution involves a transition from the blue cloud to the red sequence, but how this transition occurs has long remained an unresolved issue. Recent studies have suggested that there exists a continuous structure in the three-dimensional color–color–magnitude space connecting the blue cloud and the red sequence (e.g., Chilingarian & Zolotukhin, 2012).

However, traditional methods of evaluating galaxy evolution based on colors have several potential issues. One common aspect of all astronomical survey data is that they only include objects brighter than the detection limit of the observational instruments. Considering the magnitude m_ν of an object at frequency ν, the observational data only include objects with $m_\nu < m_\nu^{\text{lim}}$. This is known as the magnitude selection effect. As mentioned earlier, color is the ratio of luminosities at two wavelengths, so the selection effects in observations appear in a complex manner, making a simple completeness check almost impossible. Studies of galaxy evolution on the color-magnitude diagram have been plagued by this complex selection effect, which could not be separated from the physical characteristics, leading to confused discussions. However, since color is the ratio of luminosities, it is also possible to discuss it in the multi-dimensional space spanned by the multi-wavelength luminosities (absolute magnitudes). This method has the advantage of directly evaluating selection effects. For example, the bimodality observed in the color-magnitude diagram should correspond to peak structures in the original multi-dimensional luminosity space. Therefore, we focus on the structures formed by galaxies in the high-dimensional space of multi-wavelength luminosities.

The galaxy manifold we discovered by Siudek et al. (2018) has a nonlinear structure. Surprisingly, the spectra of the sample galaxies constituting this galaxy manifold could be distinguished using only information from several broadband luminosities[1], without requiring more complex combinations of physical quantities. This fact suggests that the multi-wavelength luminosities of galaxies from ultraviolet to visible and near-infrared can be explained with at most a few physical quantities. This represents a new characterization of galaxy evolution that could never have been found with traditional methods. Triggered by this discovery, we have begun research to further explore the galaxy manifold, elucidate the dependence of the physical parameters governing galaxy evolution (probably only a few), and derive the governing equations of galaxy evolution. To this end, we are focusing on manifold

---

[1]Luminosities measured over a wide wavelength range are called broadband luminosities.

learning, one of the latest methods in data science (e.g., Ma & Fu, 2012), which is fundamentally different from traditional astronomical methodologies, and are advancing further analyses. We introduce the current status and future prospects of this series of studies.

14.2 MANIFOLD LEARNING DATA

The data used in this study is from the Reference Catalog of Galaxy Spectral Energy Distribution (RCSED: Chilingarian et al., 2017). RCSED integrates the all-sky survey catalog from the ultraviolet astronomy satellite *GALEX*, the catalog from the large-scale optical spectrophotometric survey project SDSS, and the near-infrared wide-area survey UKIDSS catalog, constructed using the latest astronomical spectral analysis methods. RCSED covers approximately 25% of the entire sky and includes k-corrected photometric data in 11 bands (FUV, NUV, u,g,r,i,z,Y,J,H,K) for millions of galaxies, along with information on related physical quantities. Additionally, reprocessed related information from several public datasets is added to the photometric catalog. The base object list is the spectroscopic sample of non-active galaxies within the redshift range $0.007 < z < 0.6$ from the SDSS Data Release 7 (DR7) (Abazajian et al., 2009)[2]. This dataset includes 800,299 galaxies.

From the entire sample, extracting galaxies with photometric values in all 11 bands yield 90,565 galaxies. Removing galaxies with a redshift reliability of ≤ 0.5 results in 90,460 galaxies. The significant reduction in the number of galaxies compared to the parent sample is mainly due to the limited sky area that can be cross-matched with the UKIDSS sample. The main objective of this study is to discover and quantify universal relationships in the galaxy luminosity space. To avoid magnitude selection effects, we constructed a complete sample based on the SDSS g-band magnitude[3]. Using the limiting magnitude $m_{\mathrm{AB},g} = 18.3$[4], we calculated the limiting absolute magnitude curve to maximize the number of galaxies in the final sample. As a result, a sample consisting of 27,056 galaxies within the range $z_{\mathrm{lim}} < 0.097$, $M_{\mathrm{lim},g} \leq -20.016$ was constructed. All subsequent analyses are based on this volume-limited sample.

In manifold learning, preprocessing to optimize the range of data values is crucial. In this study, we performed analyses by centering the galaxy luminosities in each band using the simple mean of absolute magnitudes and normalizing the variance to 1, as well as using the raw values of absolute magnitudes without rescaling. The results of these two analyses were almost identical quantitatively (when rescaled to absolute magnitudes again for each axis). Therefore, we only show the results of the analysis using the raw absolute magnitudes as the feature space of the data. This

---

[2]Specifically, galaxies marked as GAL'EM and GALAXY in the SDSS spectroscopic classification indicator (specclass). In other words, galaxies dominated by black hole radiation, such as quasars and Type 1 Seyfert galaxies, are excluded from the sample.

[3]In astronomy, such data are referred to as volume-limited. Although there is no fixed translation, it means a sample that includes all objects brighter than a certain luminosity L_v^{lim} within the considered volume.

[4]The subscript AB indicates that the magnitude is in the AB magnitude system, defined based on physical principles.

is because, in this sample, the contributions from stars within the galaxy dominate the radiation in each band, resulting in the absolute magnitudes falling within similar ranges across bands. However, it should be noted that appropriate normalization will be necessary for future analyses that include other physical quantities such as redshift.

14.3 METHOD

14.3.1 GALAXY MANIFOLD IN THE MULTI-WAVELENGTH LUMINOSITY SPACE

To validate the sample constructed in Section 14.2, we first applied FEM to the data, as in Siudek et al. (2018), confirming that a low-dimensional structure formed by galaxies can also be obtained in the 11-dimensional multi-wavelength luminosity space. This galaxy manifold forms a two-dimensional surface in the feature space of multi-wavelength luminosities. The RCSED galaxy manifold is shown in Fig. 14.2.

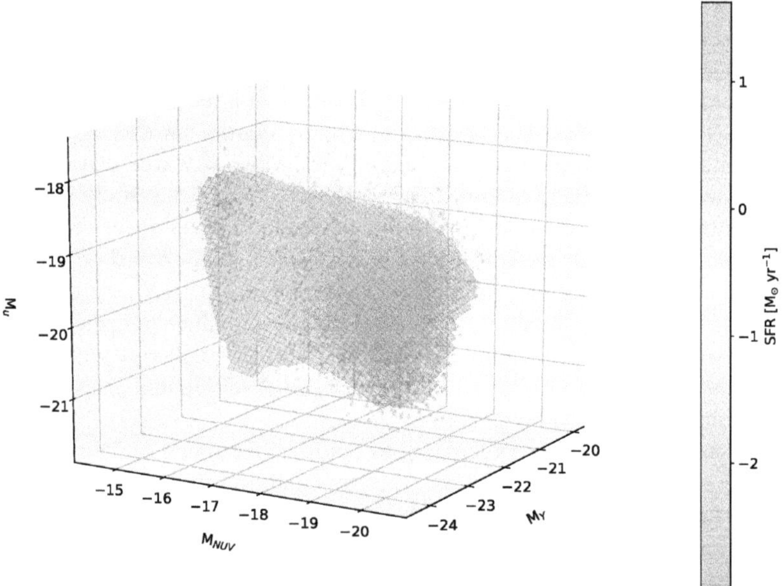

Figure 14.2 Galaxy manifold which was discovered in the multi-dimensional feature space. The original feature space is 11-dimensional, but the manifold has only a 2-dimensional structure, essentially embedded in the 3-dimensional space of ultraviolet, visible, and near-infrared luminosities. The manifold has a curved shape, which could never be detected using methods analyzing linear relationships like classical PCA. The color coding represents the star formation rate of the sample galaxies.

Our galaxy manifold, represented as a low-dimensional subspace within the 11-dimensional luminosity space has a curved structure that is difficult to visualize and even harder to quantify. The multimodality and dispersion in classical scaling laws often result from suboptimal projections that do not reflect the nonlinear structure of the galaxy manifold. As mentioned in 14.1, observational data related to galaxy evolution continue to grow. Therefore, concise methods to utilize the information to its fullest extent are needed for future research.

To quantify the galaxy manifold more effectively, we focused on a group of methods called dimensionality reduction. Specifically, we applied manifold learning to elucidate the dependence on the physical quantities (probably a few at most) that govern the physics of galaxy evolution from the galaxy manifold. Using this method, we attempt a new quantification of galaxy evolution that is fundamentally different from the first-principles theoretical constructions in classical astronomy. Another significant advantage of quantitatively representing the galaxy manifold is that it allows for the direct estimation of observational quantities, such as missing luminosities, and physical quantities, such as SFR and stellar mass $\mathcal{M}_*$, from the positions on the manifold. This can be achieved by mapping the galaxy manifold back into the luminosity space. For this purpose, it is convenient to describe the distances on the manifold using the metric of the original multi-wavelength luminosity space.In astronomical observations, which often involve difficult data analysis close to detection limits, the quantitative representation of the galaxy manifold that can be used for prediction, estimation, and interpolation of observational quantities will become a powerful tool in future astronomical research.

14.3.2 MANIFOLD LEARNING

In manifold learning, data is considered as a finite set of points $\{y_i\}$ $(i = 1,\ldots,N)$ randomly sampled from a smooth d-dimensional manifold $\mathcal{M}$ with a metric determined by the geodesic distance $d^{\mathcal{M}}$ (details of the notation and definitions related to manifolds are discussed in Appendix B.2). These data points are embedded into a feature space (or input space) $\mathcal{X} = \mathbb{R}^n$ $(d \ll n)$ with Euclidean metric $\| \cdot \|_{\mathcal{X}}$ via a smooth mapping ψ. Considering the data points in the embedded feature space as $\{x_i\}$ $(i = 1,\ldots,N)$, the embedding map is $\psi : \mathcal{M} \longrightarrow \mathcal{X}$, and the points $y_i \in \mathcal{M}$ on the manifold is represented as

$$y_i = \psi^{-1}(x_i), \; x_i \in \mathcal{X}. \tag{14.4}$$

The goal of manifold learning is to find the specific form of the manifold $\mathcal{M}$ and the mapping ψ given the set of input data points $\{x_i\} \in \mathcal{X}$, and to reconstruct the original set of data points $\{y_i\} \in \mathcal{M}$. By applying a manifold learning algorithm to the input data points, the data set in $\mathbb{R}^n$ is mapped to a lower-dimensional space $\mathbb{R}^d (d \ll n)$ while preserving the relationships between neighboring points. This can be expressed as

$$x_i \mapsto \hat{y}_i = \left(\psi_1^{-1}(x_i),\ldots,\psi_d^{-1}(x_i)\right)^{\top} \in \mathbb{R}^{\hat{d}} \tag{14.5}$$

where $\{\hat{y}_i\} \subset \mathbb{R}^{\hat{d}}$ are the estimates of the original $\{y_i\} \subset \mathbb{R}^d$. Here, $\top$ denotes the transpose of the vector. If $\hat{d} = d$, the estimation is considered successful. This method of reducing the dimensionality of the data, based on the assumption that the data set is distributed on a low-dimensional submanifold within a high-dimensional feature space (the "manifold hypothesis" (e.g., Goodfellow et al., 2016)), is known as non-linear dimensionality reduction.

The origins of research on manifold learning can be traced back to sporadic publications in the 1990s, but it gained popularity and active research interest following the publication of two seminal papers (Roweis & Saul, 2000; Tenenbaum et al., 2000). Manifold learning algorithms can "unfold" manifolds with complex shapes in the feature space and provide local coordinate systems on them (Roweis & Saul, 2000; Tenenbaum et al., 2000). The key is that the relationships between data points in the reduced-dimensional space should faithfully represent the relationships between data points in the original high-dimensional space. The algorithm must learn the shape of the data, which is the essence of the term "manifold learning."

While classical linear methods like PCA are effective in representing the global structure of data, nonlinear methods excel in representing local structures. However, many nonlinear methods focus on low-dimensional representations of the positional relationships of neighboring data points, which can sometimes lead to the loss of global structure. Therefore, it is essential to choose the appropriate algorithm depending on the objective when using manifold learning. Additionally, it should be noted that the coordinate system provided by the low-dimensional representation from manifold learning does not necessarily guarantee an intuitive or physical meaning (e.g., Liu et al., 2017). This point will be discussed in Sec. 14.4.

14.3.3 `Isomap` AND `UMAP` ALGORITHMS

In this study, we adopt the `Isomap` (isometric feature mapping: Tenenbaum et al., 2000) and `UMAP` (uniform manifold approximation and projection: McInnes et al., 2020, 2018) algorithms for manifold learning. The aim of this study is to quantify the dependence of galaxy evolution on physical quantities, requiring that the connected structure in the original high-dimensional feature space is also mapped to a connected structure in the manifold. Both algorithms possess the property of preserving the connectivity of the original data point distribution, making them ideal for our purposes. These computations were performed using `scikit-learn` (Pedregosa et al., 2011).

14.3.3.1 `Isomap`

The `Isomap` algorithm assumes that a smooth manifold $\mathcal{M}$ is a geodesically convex region in $\mathbb{R}^d$ ($d \ll n$), and the embedding map $\psi : \mathcal{M} \longrightarrow \mathcal{X}$ is an isometry. First, geodesic convexity is defined as follows (Tenenbaum et al., 2000).

Definition 10. Geodesically convex
Let $(\mathcal{M}, g)$ be a Riemannian manifold. A subset $\mathcal{U}$ of $\mathcal{M}$ is said to be geodesically convex if, for any two points in $\mathcal{U}$, there exists a unique shortest geodesic within $\mathcal{U}$ connecting them.

A geodesically convex Riemannian manifold is also a convex metric space concerning the geodesic distance.

Thus, the assumptions of Isomap can be expressed as follows:

Convexity $\mathcal{M}$ is a geodesically convex subset of $\mathbb{R}^d$.

Isometry Geodesic distances are preserved under the mapping ψ. For any two points $y, y' \in \mathcal{M}$ on the manifold, the geodesic distance between them equals the Euclidean distance between the corresponding embedded points $x = \psi(y), x' = \psi(y') \in \mathcal{X}$ in $\mathbb{R}^n$. That is,

$$d^{\mathcal{M}}(y, y') = \|x - x'\|_{\mathcal{X}} . \tag{14.6}$$

Isomap generalizes the multidimensional scaling (MDS) algorithm using the assumptions that $\mathcal{M}$ is a geodesically convex region and ψ is an isometric transformation. MDS is a method that searches for a lower-dimensional subspace in which data points are distributed while preserving the Euclidean distances between pairs of data points (see Appendix B.3.1). MDS is a linear dimensionality reduction method and does not work well in curved regions. Isomap follows the MDS philosophy and approximates the geodesic distances on $\mathcal{M}$ for all pairs of data, thereby maximizing the preservation of the global geometric structure of the nonlinear manifold. In this sense, Isomap is both a local and a global method for manifold learning. Concrete algorithm is presented in Appendix B.3.

Due to its structure, Isomap preserves the "surface density" of the data points on the manifold by maintaining the metric between pairs of points. That is, regions where data points are densely populated in the feature space will also be dense on the manifold, and sparse regions will be sparse on the manifold. Because Isomap assumes that the manifold $\mathcal{M}$ is a geodesically convex submanifold of Euclidean space, it does not work well in cases of excessive curvature, holes in the manifold, or non-convexity. Practically, if there is noise—meaning data points are not necessarily distributed on the manifold—the performance of Isomap depends on how the neighborhoods are defined. Provided the noise is not too large, Isomap is generally robust to noise. In this study, the neighborhood size for Isomap was set to $K = 5$.

14.3.3.2 UMAP

UMAP (uniform manifold approximation and projection) is a relatively new method proposed in 2018, based on differential geometry and algebraic topology. In UMAP, data points that are close in the original feature space remain close on the manifold. Its fast execution time reduces computational costs, and it is capable of dimensionality reduction to manifolds with four or more dimensions. UMAP is based on topological data analysis and Riemannian geometry. This algorithm relies on the following three assumptions: 1) data is uniformly distributed on a Riemannian manifold, 2) the Riemannian metric is locally constant (or can be approximated as such), 3) the manifold is locally connected. From these assumptions, it is possible to model a manifold with a fuzzy topological structure. It is important to note that, unlike Isomap, UMAP does not preserve the surface density of data points because the manifold is defined to make the data points as uniformly distributed as possible. The UMAP algorithm consists of the following three stages:

1. Estimation of the Riemannian manifold,
2. Representation by fuzzy topology in the distance space,
3. Dimensionality reduction.

The core concept of UMAP is the fuzzy topological representation, but due to the use of category theory, a concise description is difficult, so details are provided also in Appendix B.3. Because of its construction, UMAP is more robust to noise compared to Isomap. In this study, the neighborhood size for UMAP was set to $K = 50$.

14.4   RESULTS AND DISCUSSION

14.4.1   RESULTS: Isomap AND UMAP GALAXY MANIFOLDS

The galaxy manifolds obtained using Isomap and UMAP are shown in Fig. 14.3. It is noteworthy that the different algorithms Isomap and UMAP yield qualitatively very

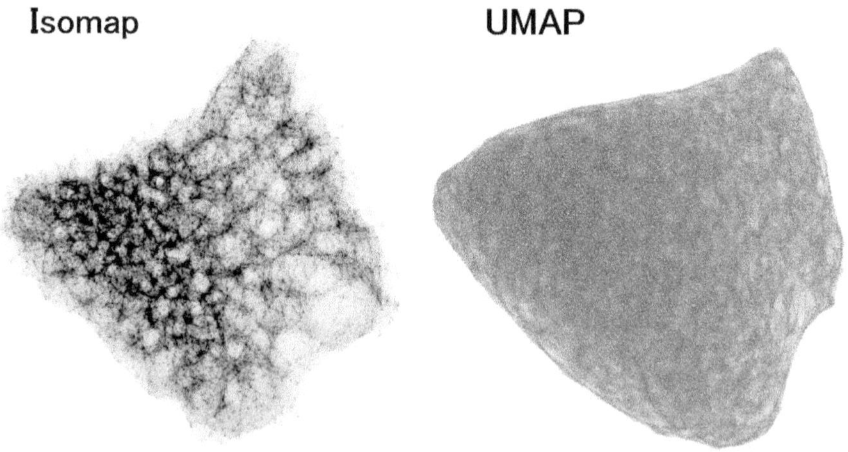

Figure 14.3   The "unfolded" galaxy manifold using manifold learning algorithms Isomap and UMAP. The left and right panels show the manifolds obtained from Isomap and UMAP, respectively. The structure of the manifold in this space is much easier to recognize than in Fig. 14.2. While the overall shapes differ slightly, the features on the manifold are consistent.

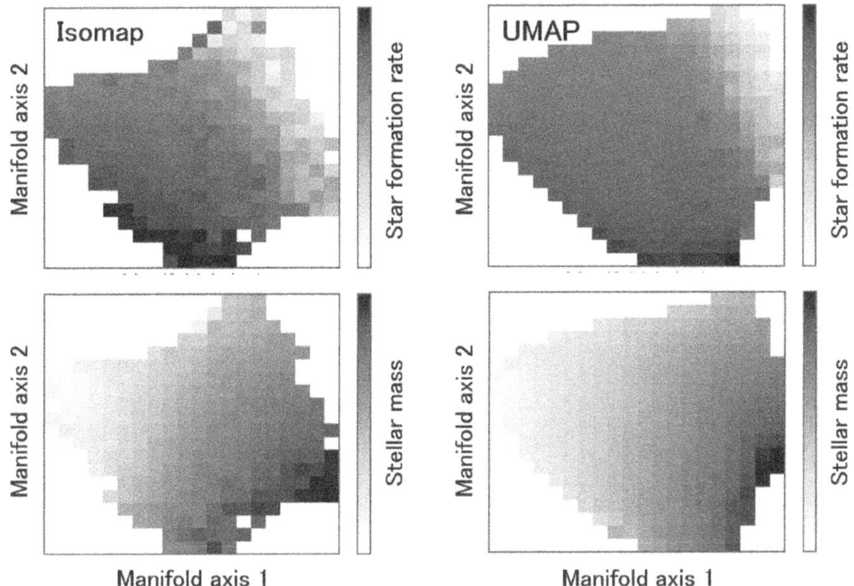

Figure 14.4 Galaxy manifolds obtained by two different manifold learning algorithms, Isomap and UMAP. The star formation rate (SFR) and stellar mass M_* are represented as functions on the manifold. The left panel shows the star formation rate and stellar mass on the manifold obtained by Isomap. The right panel shows the manifold obtained by UMAP, with the same color coding as the left panel.

similar 2-dim manifolds (Fig. 14.3). The differences in the galaxy manifolds estimated by the two methods are clearly visible in Fig. 14.3. Since Isomap preserves the density of the data point cloud, the manifold shows density structures, meaning there are dense and sparse regions on the manifold. In contrast, UMAP tries to make the density as uniform as possible when constructing the manifold, resulting in a much more uniform density distribution in the UMAP manifold. Thus, regions with high density in the Isomap manifold appear enlarged in the UMAP manifold.

To examine how information about galaxy evolution appears on the galaxy manifold, we compare the SFR and stellar mass as functions on the manifold in Fig. 14.4. Figures 14.5 and 14.6 show the correlations between the 2-dim galaxy manifold coordinates 1 and 2 and the star formation rate and stellar mass for Isomap and UMAP, respectively. The behavior of the star formation rate (SFR) and stellar mass (M_*) is qualitatively very similar in both figures, suggesting that the estimated manifold structures are robust. This indicates that manifold learning indeed "learns" the essential features of galaxy evolution in the multi-wavelength luminosity space. Figures 14.5 and 14.6 show that manifold coordinate 1 is strongly correlated with stellar mass, and coordinate 2 is strongly correlated with the star formation rate. We have

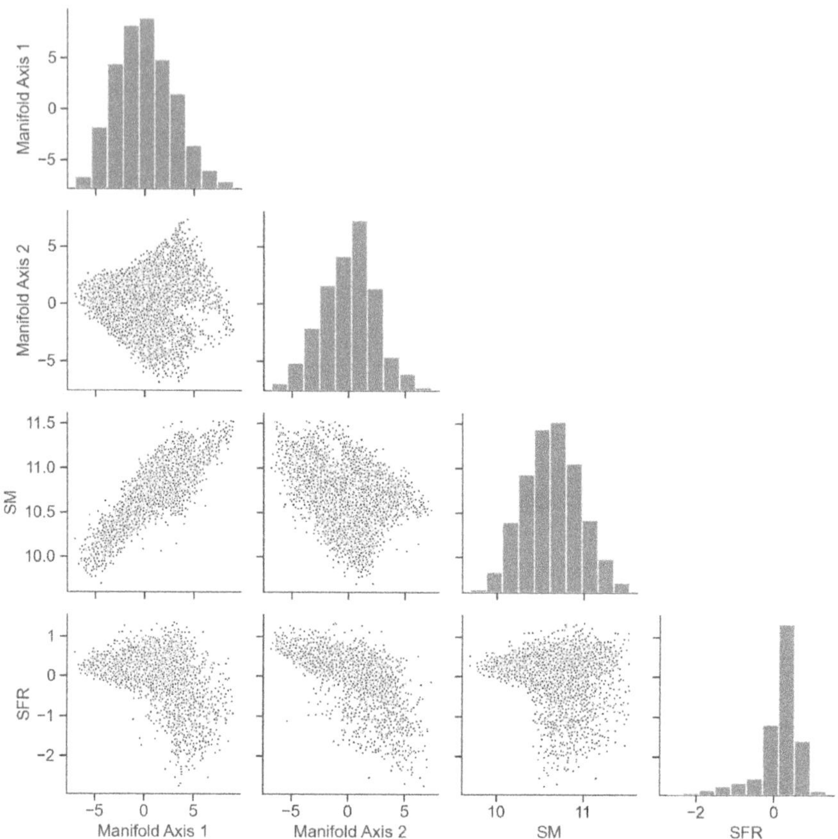

Figure 14.5   Scatter plot of the 2-dim galaxy manifold coordinates and the star formation rate (SFR) and stellar mass (SM) obtained by `Isomap`. Manifold coordinate 1 is strongly correlated with stellar mass, and coordinate 2 is strongly correlated with the star formation rate.

already seen that the galaxy manifold in the optical luminosity space is essentially 2-dim. This means that galaxy evolution in the wavelengths of ultraviolet, visible light, and near-infrared can be adequately described by only two physical quantities, star formation rate and stellar mass, imposing significant constraints on theories of galaxy evolution (Cooray et al., 2023).

Thus, manifold learning can connect the galaxy manifold with physical quantities such as star formation rate and stellar mass. By further advancing this, it is theoretically possible to parameterize galaxy evolution on the manifold. When stars form, the stellar mass, i.e., the total mass of accumulated stars, increases. This is one of the fundamental aspects of galaxy evolution, and it can be visualized as a vector field on the manifold. The vector fields of star formation rate evolution on the galaxy manifold are shown in Figs. 14.7 and 14.8. The "velocity field" of galaxy evolution

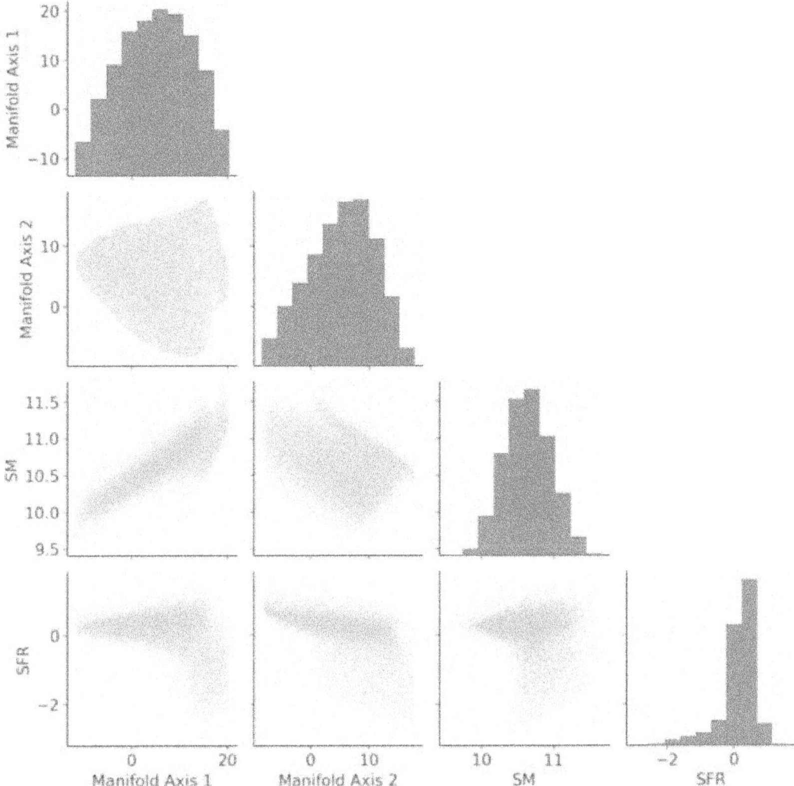

Figure 14.6   Scatter plot of the 2-dim galaxy manifold coordinates and the star formation rate (SFR) and stellar mass (SM) obtained by UMAP. Similar to Figure 14.5, manifold coordinate 1 is strongly correlated with stellar mass, and coordinate 2 is strongly correlated with the star formation rate.

is clearly visible in these two figures. Low-mass galaxies evolve rapidly, with a decreasing star formation rate and increasing stellar mass (direction from the top left to the bottom right in the figure). High-mass galaxies evolve slowly, staying in the same position on the manifold (top right) for a longer time. The next step is to construct a systematic way to describe the vector field on the manifold in an interpretable form, like a set of equations. This remains as a future work.

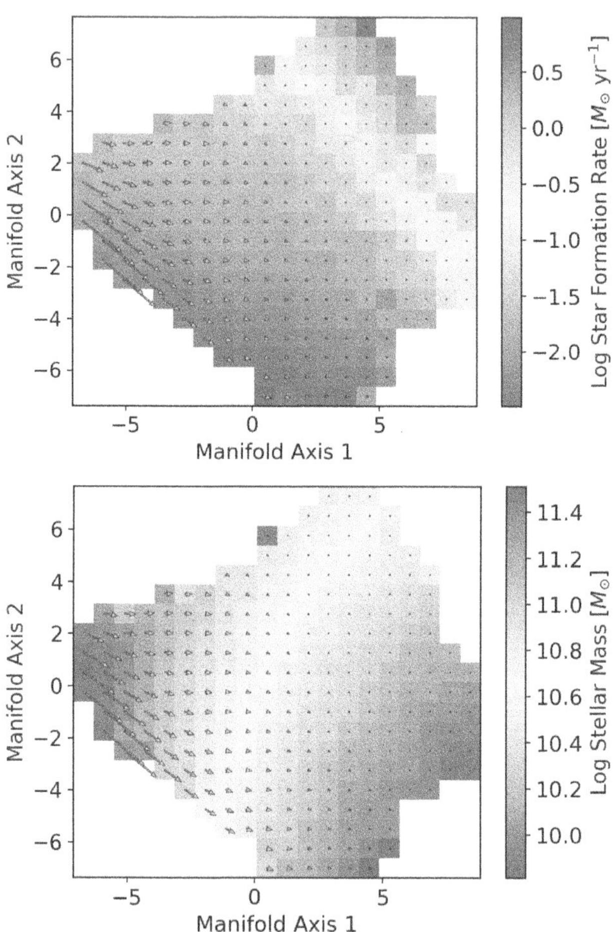

Figure 14.7 Vector fields of star formation rate and stellar mass evolution on the `Isomap` galaxy manifold. The color bar in the top panel represents the current star formation rate of galaxies, and the bottom panel represents stellar mass.

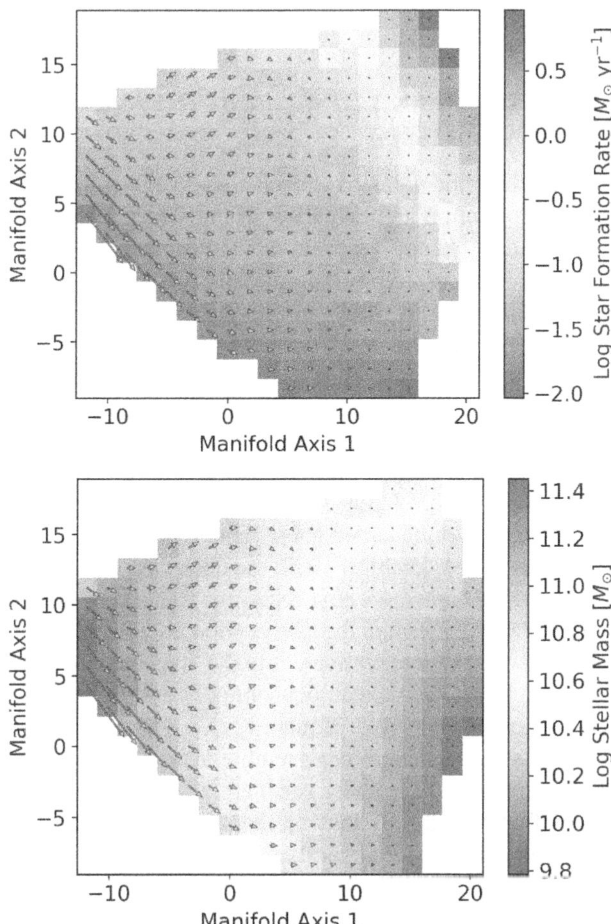

Figure 14.8 Vector fields of star formation rate and stellar mass evolution on the UMAP galaxy manifold.

15 Topological Data Analysis of the Large-Scale Structure

15.1 INTRODUCTION

15.1.1 CHARACTERIZING AND DESCRIBING THE MATTER DISTRIBUTION IN THE UNIVERSE

The distribution of matter and galaxies at each cosmic age contains fundamental information about the formation and evolution of the structure of the Universe (e.g., Bernardeau et al., 2002; Efstathiou & Silk, 1983; Peebles, 1980). One key fact that must be emphasized is that cosmology theoretically predicts the statistical properties of matter or galaxy distributions, but it does not have the predictive power to determine where exactly in space-time any specific galaxy will form. Therefore, comparisons between cosmological theory and observations can only be made through statistical descriptions. For this reason, numerous sophisticated methods have been proposed to characterize the statistical properties of fluctuations in the matter distribution (e.g., Bernardeau et al., 2002; Martinez & Saar, 2001; Peebles, 1980). Among these methods, the n-point correlation function is the most commonly studied and widely used in the analysis of observed galaxy distributions (e.g., Bernardeau et al., 2002; Martinez & Saar, 2001; Peebles, 1980, see Chapter 6.). To understand the new method we adopt in this chapter, we first recall the method of correlation function, introduced in Chapter 6. We measure the number density of galaxies at a position $\vec{x}$ in the Universe, denoted as $n(\vec{x})$, and compare it to the statistical average $\langle n \rangle$. While we can only observe one Universe, we can conceptually assume an ensemble of many Universes and consider the average of a quantity Q across many Universes. This average is referred to as the statistical or ensemble average, denoted by $\langle Q \rangle$. On the other hand, the average taken over space is called the volume average, denoted by $\bar{Q}$. In cosmology, we often base discussions on the assumption that these two averages are equal (the cosmological ergodic hypothesis).

With this preparation, consider two small volumes d^3x_1 and d^3x_2. The joint probability that galaxies are found in both of these small volumes is given by $\mathscr{P}(\vec{x}_1,\vec{x}_2)\mathrm{d}^3x_1\mathrm{d}^3x_2$. If the galaxy distribution is uniform, i.e., there is no clustering (no correlation), the probability of finding a galaxy within a small volume is

$$\mathscr{P}(\vec{x})\mathrm{d}^3x = \langle n \rangle \mathrm{d}^3x \,, \tag{15.1}$$

so the joint probability of finding galaxies in volumes d^3x_1 and d^3x_2 is

$$\mathscr{P}(\vec{x}_1,\vec{x}_2)\mathrm{d}^3x_1\mathrm{d}^3x_2 = \langle n \rangle^2 \mathrm{d}^3x_1\mathrm{d}^3x_2. \tag{15.2}$$

DOI: 10.1201/9781003104315-15

However, if galaxies are clustered, the joint probability becomes

$$\mathscr{P}(\vec{x}_1,\vec{x}_2)\mathrm{d}^3 x_1\mathrm{d}^3 x_2 = \langle n(\vec{x}_1)n(\vec{x}_2)\rangle\mathrm{d}^3 x_1\mathrm{d}^3 x_2 \equiv \langle n\rangle^2\left[1+\xi(\vec{x}_1,\vec{x}_2)\right]\mathrm{d}^3 x_1\mathrm{d}^3 x_2, \quad (15.3)$$

which is higher by $\langle n\rangle^2\xi(\vec{x}_1,\vec{x}_2)\rangle\mathrm{d}^3 x_1\mathrm{d}^3 x_2$ compared to the uniform case. This $\xi(\vec{x}_1,\vec{x}_2)$ is called the two-point correlation function of the galaxy distribution. Similarly, in general, the n-point correlation function is defined using the joint probability of finding galaxies within n small volumes.

So far, we have formulated the galaxy distribution as a point set, but when discussing galaxy distributions, it is often useful to smooth out the distribution of number density and treat it as a density field (recall the discussion in Chapter 7). Although the mathematical treatment of point sets and density fields differs, in astrophysics, this distinction is often not emphasized, and we will follow this convention here. For cosmological formulations, it is convenient to consider the density fluctuation of the galaxy distribution field $\rho(\vec{x})$.

$$\delta(\vec{x}) \equiv \frac{\rho(\vec{x})-\bar{\rho}}{\bar{\rho}}, \quad (15.4)$$

where $\bar{\rho}(\vec{x})$ represents the average density. The density $\rho(\vec{x})$ can represent either the number density of galaxies as a point set or the mass density of a smoothed galaxy distribution. Using the density fluctuation $\delta(\vec{x})$, the correlation function $\xi(\vec{x}_1,\vec{x}_2)$ can be expressed as

$$\xi(\vec{x}_1,\vec{x}_2) = \frac{\langle[\rho(\vec{x}_1)-\bar{\rho}]\,[\rho(\vec{x}_2)-\bar{\rho}]\rangle}{\bar{\rho}^2} = \langle\delta(\vec{x}_1)\delta(\vec{x}_2)\rangle. \quad (15.5)$$

In cosmology, we assume that the Universe is homogeneous and isotropic, so it is generally treated as

$$\xi(\vec{x}_1,\vec{x}_2) = \xi(|\vec{x}_1-\vec{x}_2|) \equiv \xi(r). \quad (15.6)$$

In the form of eq. (15.5), we can see that the correlation function is the second-order cumulant of the density field. In general, a random variable $X(\vec{x})$ indexed by a vector $\vec{x} = (x_1,\dots,x_n)^\top$ in space is called a random field. Mathematically, any random field is uniquely determined by its infinite moments or cumulants, which can be represented by n-point correlation functions due to the cumulant expansion theorem of the characteristic function of a random field (e.g., Daley & Vere-Jones, 2003). Thus, in principle, measuring the correlation function allows one to determine the statistical properties of the cosmic density field.

However, in practice, estimating higher-order n-point correlations is extremely difficult and unrealistic, so alternative approaches to handling information from higher-order correlations have been explored. The discovery of the 3-dim structure of galaxy distributions (e.g., de Lapparent et al., 1986) has led to ideas for quantifying large-scale structures and identifying the underlying physical processes using the geometry (topology) of the smoothed density field (e.g., Boerner & Mo, 1989; Gott et al., 1986, 1987, 1989; Shandarin & Zeldovich, 1983). One of the earliest methods which was used to characterize topology through the genus (Gott et al., 1986). By using the threshold of ν times the standard deviation σ of the density fluctuation $\delta(\vec{x})$, we can define upper-level sets for each ν. These are called excursion sets, denoted here as E_ν, and the isodensity surfaces where $\rho(\vec{x}) = \nu\sigma$ form their boundaries. Since we are dealing with a density field, E_ν is a 3-dim set. The genus of the galaxy distribution is defined as $-1/2$ of the Euler characteristic of the smoothed galaxy density field.

$$G = -\frac{1}{2} \int K \mathrm{d}A \,, \tag{15.7}$$

where K is the Gaussian curvature of the isodensity surface, and $\mathrm{d}A$ is an area element on the isodensity surface. Note that this definition in cosmology differs from the usual mathematical definition by one. The genus curve, $G(\nu)$, expresses the genus per unit volume as a function of ν. Using the genus curve as a summary statistic of the galaxy distribution, cosmological research can be conducted through genus statistics. This approach has been developed in connection with perturbation theory for describing the growth of large-scale structures, which treats gravitational growth as small deviations (perturbations) from the initial density distribution (e.g., Gay et al., 2012; Matsubara, 1994, 1996, 2003; Matsubara et al., 2022; Matsubara & Suto, 1996; Matsubara & Yokoyama, 1996; Pogosyan et al., 2009). By using perturbation theory, the genus curve can be analytically written as a function of moments on large scales (e.g., Matsubara, 1994, 1996; Matsubara & Suto, 1996; Matsubara & Yokoyama, 1996), allowing the expression of topological information quantitatively. The genus curve is particularly useful for distinguishing Gaussian fields from other random fields, establishing itself as a general tool for analyzing large-scale structures.

The genus statistics approach is part of a broader framework known as Minkowski functionals (e.g., Hikage et al., 2003; Matsubara, 2003; Sullivan et al., 2019). In applications of Minkowski functionals to large-scale structures, we also consider the smoothed galaxy density field's fluctuation, as defined in eq. (15.4). For $k = 0$, the Minkowski functional M_k is given by

$$M_0 = \frac{1}{V} \int_{E_\nu} \mathrm{d}V \,, \tag{15.8}$$

and for $k = 1, 2, 3$, it is defined by integrating the curvature on the isodensity surface at threshold ν. Thus, in principle, the galaxy density field can be fully characterized by Minkowski functionals (Schmalzing & Buchert, 1997; Schmalzing & Gorski, 1998) (throughout this chapter, $k = 0, 1, 2, 3$ refers only to the dimensionality). This method has been widely applied to cosmological analyses (density field analysis: e.g., Beisbart et al. 2002; Hikage et al. 2003; Kerscher & Tikhonov 2010; Kerscher et al. 1997; Mecke et al. 1994; Wiegand et al. 2014; Wiegand & Eisenstein 2017; redshift-space distortions: e.g., Appleby et al. 2018; weak gravitational lensing cosmology: e.g., Parroni et al. 2020; Spurio Mancini & otehrs 2018; Vicinanza et al. 2019; cosmic reionization: e.g., Bag et al. 2019, 2018; Chen et al. 2019;

Yoshiura et al. 2017; testing cosmology/gravity theories: e.g., Fang et al. 2017; Junaid & Pogosyan 2015; Ling et al. 2015; Shiraishi et al. 2016; and cosmic microwave background (CMB): e.g., Chingangbam et al. 2017; Ganesan & Chingangbam 2017; Joby et al. 2019; Matsubara 2010; Planck Collaboration et al. 2016; Santos et al. 2016).

15.1.2 FROM MINKOWSKI FUNCTIONALS TO TOPOLOGICAL DATA ANALYSIS

As discussed, the galaxy density field can be quantitatively evaluated through the topology of the density field. The genus statistics described earlier characterize the properties of point cloud data. Unlike genus, which represents the entire point cloud data with a single characteristic related to its hole structure, persistent homology groups provide a representation that preserves features such as the scale of each individual hole structure contained in the point cloud data (TDA: e.g., Edelsbrunner et al., 2002; Wasserman, 2018; Zomorodian & Carlsson, 2004). In what follows, we briefly explain TDA, a data analysis method based on persistent homology groups.

Using TDA, which evaluates each individual hole structure within the data, it is possible to further advance data analysis in cosmology. The hole structures of the point cloud data studied in this research are formed by a set of n-dimensional spheres with radius r placed at each point of the N-body system consisting of galaxies or arbitrary celestial objects. Early studies in cosmology using this approach were essentially applications of random percolation (Gott et al., 1986). In other words, spheres were placed around each galaxy, and by increasing the radius r, the geometric structure of the connected components within the considered volume was examined as a function of the radius r. This is precisely the concept employed in TDA.

The emergence of the concept of TDA in a rudimentary sense dates back to the 1990s, but its substantial birth occurred in the 2000s (see e.g., Edelsbrunner & Harer, 2010; Edelsbrunner et al., 2002). In TDA, data sets are often represented as persistent homology groups. Similar to random percolation, persistent homology groups consider a (non-hollow) sphere of radius r around each point in the point cloud data and handle the generation and disappearance of hole structures of various dimensions. This process is schematically explained in Fig. 15.1. Consider a set of three points in a 2-dim plane. When the radius r is small, three disks are placed around the points (Fig. 15.1: left). As r increases beyond a certain value, the disks connect, forming a cycle structure (Fig. 15.1: center), which is regarded as the birth of a hole. As r continues to grow, the interior of the cycle structure is filled with disks, and the hole disappears (Fig. 15.1: right). A persistence diagram, which is a central representation method of persistent homology groups, is constructed by plotting the birth radius r on the x-axis and the death radius r on the y-axis for all hole structures present in the data. For point cloud data in n-dimensional space, the persistence diagram can be defined separately for 0-dim holes (connected components), 1-dim holes (ring structures), 2-dim holes (shell structures), and so on. More rigorous definitions and methodologies are provided in Sec. 15.2.

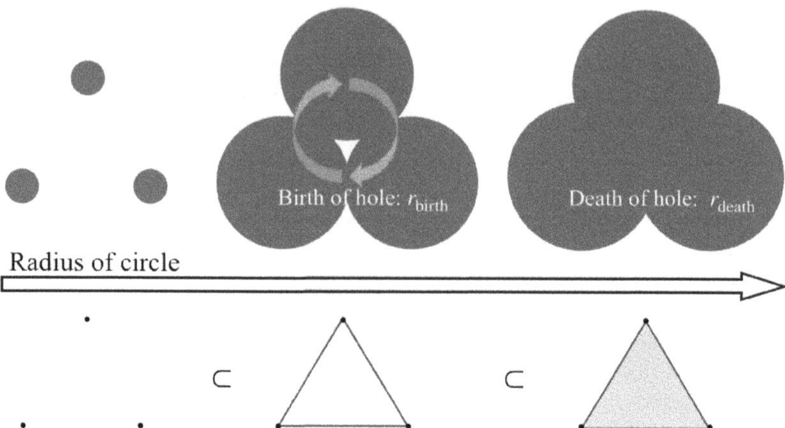

Figure 15.1   Schematic diagram of persistent homology groups (Takeuchi et al., 2023). Topological information is characterized by "holes" formed from n-dimensional spheres with radius r around discrete data points. Credit: Takeuchi, T. T., et al. 2023, Proceedings of the Institute of Statistical Mathematics, 71(2), 159–187.

Applications of modern TDA in astrophysics and cosmology have been gradually increasing in recent years. Pranav et al. (2017, 2019a), Feldbrugge et al. (2019), Wilding & otehrs (2021), and Biagetti et al. (2021) have formulated probabilistic homology for Gaussian and non-Gaussian random fields, with applications in mind for general random fields. Using this method, Pranav et al. (2019b) detected a unique signal in the cosmic microwave background (CMB). Cole & Shiu (2018) developed a method to evaluate the non-Gaussianity of the CMB by applying TDA. Xu et al. (2019) applied TDA to a large galaxy dataset generated from N-body simulations (Libeskind et al., 2018), detecting 23 voids in the simulation data. This demonstrated the usefulness of TDA in void detection and showed good agreement with results from conventional methods (e.g., CLASSIC: Kitaura & Angulo, 2012). Furthermore, they tested a method to constrain cosmological parameters and resolve degeneracies between parameters by comparing them with the MassiveNu simulations (Liu et al., 2018). Elbers & van de Weygaert (2019) and Giri & Mellema (2021) discussed the effectiveness of TDA in quantitatively estimating the structure and evolution of ionized bubbles during cosmic reionization, as well as their size distribution and topology.

15.2   TOPOLOGICAL DATA ANALYSIS (TDA)

In TDA, the topological and geometric properties of a finite point cloud P in an N-dimensional Euclidean space $\mathbb{R}^N$ are examined ($\mathbb{R}$, $\mathbb{N}$, and $\mathbb{N}_0$ represent the sets of real numbers, natural numbers, and non-negative natural numbers, respectively). Since P is a finite set, simply applying traditional topology would not allow for the extraction of meaningful information regarding the characteristics of P. TDA extracts significant topological and geometric features of the point cloud data by investigating properties that remain persistent across varying scales (e.g., Boissonnat et al., 2018; Edelsbrunner et al., 2002). For this reason, we introduce such geometric models that play a fundamental role in TDA. The primary tool in this analysis is the persistent homology group.

15.2.1 PERSISTENT HOMOLOGY GROUPS

15.2.1.1 The "Shape" of Point Data

In the analysis of large, complex datasets, it is essential to represent the data by reducing it to a few key features that characterize its properties. Dimensionality reduction plays a central role in this. For example, consider a point cloud data set in Euclidean space as follows:

$$\mathscr{P} \equiv \left\{ x_i \in \mathbb{R}^N \middle| i = 1, \ldots, m \right\}. \tag{15.9}$$

If the data $\mathscr{P}$ has a periodic structure, a Fourier transform works very well, and if the data points are clustered, analysis via the correlation function is effective (e.g., Peebles, 1980).

However, such approaches are not effective for complex data that is non-periodic or lacks a clear cluster structure. The schematic shown in Fig. 15.1 is an example of such data. Traditional analysis methods using Fourier transforms or correlation functions struggle to characterize the shape of such point clouds. Instead, a method that characterizes data by placing spheres centered on each data point and introducing connections based on the intersections of these spheres is more effective. For example, characterizing the connections by the number of holes corresponds to representing the point cloud data in a drastically reduced form using homology groups, which are invariant under homeomorphisms, such as identifying a donut and a coffee cup as the same. However, this approach loses information about the scale, such as the size of the holes. Thus, it becomes useful to extract the birth and death of hole structures as the radius of the spheres placed on each data point increases. This yields a reduced representation of the point cloud data that retains information about the shape and size of the holes. This extended homology group is called a "persistent homology group."

15.2.1.2 Algebraic Representation of "Holes" in Point Cloud Data

As mentioned earlier, in point cloud data analysis, it is useful to place spheres around each point, increase the radius, and investigate the hole structures of the connected components within the volume, as is done in random percolation. The set of spheres constructed from such a point cloud can be viewed as a set of convex closed sets. When a structure is covered by a finite collection of convex closed sets, it can be transformed into a simplicial complex, a representation that preserves topological features (homotopy equivalence) while allowing for algebraic operations. A simplicial complex is a representation of the structure of a shape and is constructed as a combination of simplices, which are generalizations of triangles to k dimensions.

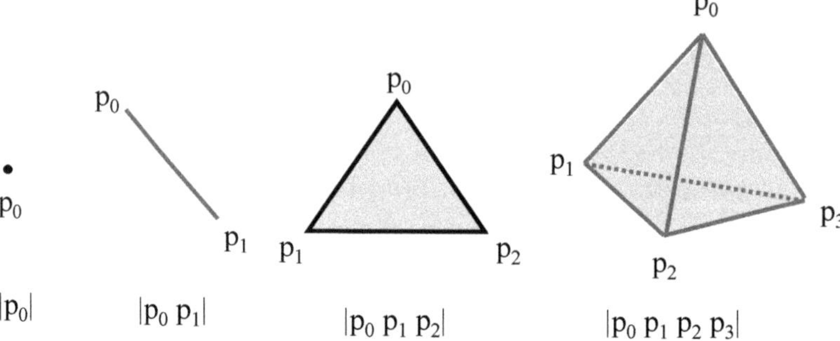

Figure 15.2   An example of a simplex (Takeuchi et al., 2023). The shaded areas in the figure indicate that the interior is filled. Credit: Takeuchi, T. T., et al. 2023, Proceedings of the Institute of Statistical Mathematics, 71(2), 159–187.

In general, a simplicial complex does not have the structure of a topological space, but we first explain the concept of a simplicial complex embedded in Euclidean space, which is easier to visualize. In Euclidean space, a k-simplex is defined as follows:

Definition 11 (Simplex in Euclidean space).
A k-simplex $|p_0, p_1, \ldots, p_k|$ is defined by $k+1$ points $p_0, p_1, \ldots, p_k$ in $\mathbb{R}^n$, where the k vectors $\overrightarrow{p_0 p_1}, \overrightarrow{p_0 p_2}, \ldots, \overrightarrow{p_0 p_k}$ are linearly independent:

$$|p_0, p_1, \ldots, p_k| := \left\{ \sum_{i=0}^{k} \lambda_i p_i \,\middle|\, \lambda_i \geq 0 (i = 0, 1, \ldots, k), \sum_{i=0}^{k} \lambda_i = 1 \right\}$$

Specifically, a 0-dimensional simplex is a point, a 1-dimensional simplex is a triangle, and a 2-dimensional simplex is a tetrahedron (see Fig. 15.2). A simplicial complex is formed by gluing these simplices together without overlap. Based on this gluing operation, a simplicial complex in Euclidean space is defined as follows:

Definition 12 (Simplicial complex in Euclidean space).
A finite collection K of simplices in $\mathbb{R}^n$ is called a simplicial complex if it satisfies the following conditions:

1. *If σ is a face of a simplex τ in K, then σ is also in K.*
2. *The intersection of two simplices $\tau, \sigma \in K$ is either empty or a face of both τ and σ.*

Here, the face of a simplex refers to a lower-dimensional simplex, such as the triangular faces of a tetrahedron (3-dim simplex) or the endpoints of a line segment (1-dim simplex). An example of a simplicial complex in Euclidean space is shown in Fig. 15.3.

As mentioned earlier, a simplicial complex itself does not depend on the topological space and is defined as a combinatorial structure as follows:

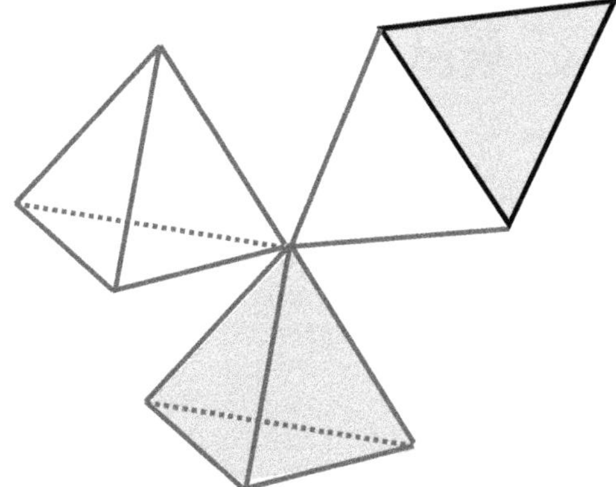

Figure 15.3   An example of a simplicial complex (Takeuchi et al., 2023). As in Fig. 15.2, the shaded areas indicate that the figures are solid, not hollow. Credit: Takeuchi, T. T., et al. 2023, Proceedings of the Institute of Statistical Mathematics, 71(2), 159–187.

Definition 13 (Simplicial Complex).
A simplicial complex X on a finite set V is a family of subsets of V that satisfies the following conditions:

   1. $\{a\} \in X$ for all $a \in V$,
   2. If $\sigma \in X$ and $\tau \subset \sigma$, then $\tau \in X$.

The maximum dimension of the simplices in X is called the dimension of the complex and is written as dim X. The set of k-simplices in X is denoted as X_k.

Next, we explain the procedure for constructing a simplicial complex from point cloud data. More specifically, the extraction of "holes" in this study is based on the construction of a Čech complex, a type of simplicial complex.

The Čech complex on $\mathscr{P}$ is defined as follows:

Definition 14 (Čech complex).
Consider a point cloud $\mathscr{P} \in \mathbb{R}^N$. Around each x_i, place a sphere of radius r

$$B_r(x_i) = \left\{ x \in \mathbb{R}^N; \|x - x_i\| \leq r \right\}, \tag{15.10}$$

where $\|x\|$ is the Euclidean norm. The Čech complex $\mathscr{C}(P,r)$ is a simplicial complex on $\mathscr{P}$, and the simplices that constitute it are defined as

$$\left\{ x_{i_0}, \ldots, x_{i_k} \right\} \in \mathscr{C}(\mathscr{P}, r) \Leftrightarrow \bigcap_{j=0}^{k} B_r(x_{i_j}) \neq \emptyset. \tag{15.11}$$

That is, simplices are formed based on the relationships (overlaps) between the spheres placed around the points in $\mathscr{P}$. The behavior of the Čech complex as the

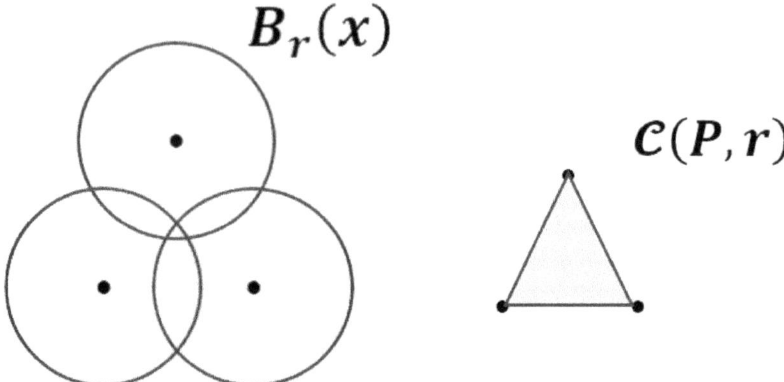

Figure 15.4   An example of a Čech complex (Takeuchi et al., 2023). Left panel: The union of spheres $\mathscr{B}$ around x_i, Right panel: The corresponding Čech complex $\mathscr{C}$. Credit: Takeuchi, T. T., et al. 2023, Proceedings of the Institute of Statistical Mathematics, 71(2), 159–187.

radius r changes is shown in Fig. 15.1. As can be seen from Fig. 15.1, the Čech complex corresponds to the union of the spheres, which can be continuously transformed into one another. This follows from one of the fundamental theorems in topology, the Nerve Theorem (e.g., Kozlov, 2008). In other words, the union of spheres $\bigcap_{j=0}^{k} B_r(x_{i_j})$ and the simplicial complex $\mathscr{C}(P, r)$ are guaranteed to be transformable into each other by continuous deformation (homotopy equivalence). Thus, information about holes is preserved in the simplicial complex. In this way, an algebraic representation of "holes" that allows for algebraic operations is realized from point cloud data.

The Čech complex is determined by the data $\mathscr{P}$ and the radius r. When the radius r is small, the Čech complex has the same discrete structure as the original point cloud $\mathscr{P}$. As r increases, the data points begin to connect through the spheres, and when r is sufficiently large, the entire structure eventually becomes connected. Thus, as mentioned earlier, r can be viewed as the "resolution" for handling the connectivity between data points. Corresponding to radii $r_1 < r_2 < \ldots, < r_n$, we obtain a growing sequence of Čech complices $\mathscr{C}(\mathscr{P}, r_1) \subset \mathscr{C}(\mathscr{P}, r_2) \subset \cdots \subset \mathscr{C}(\mathscr{P}, r_n)$. In general, such an increasing sequence of simplicial complices is called filtration.

In practical applications, alpha complices, which share properties with Čech complices but are more computationally feasible, are often used. Alpha complices are an improved version of the Čech complex, using Voronoi tessellation. For simplicity, we will not provide a rigorous introduction to alpha complices here but will only show the concept in Fig. 15.5. In the analysis that follows, we adopt alpha complices.

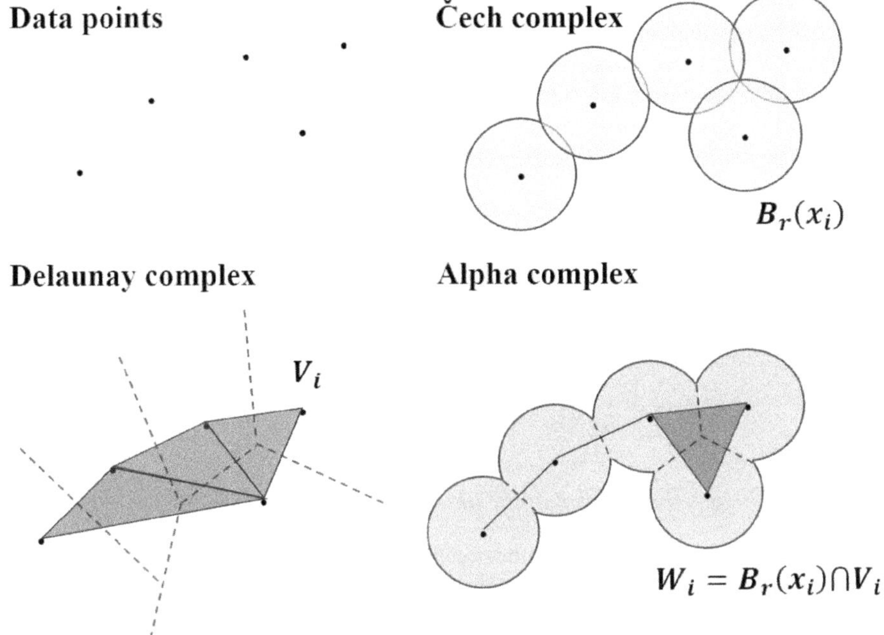

Figure 15.5 Example of a point data set (top left), its Čech complex (top right), Delaunay complex (bottom left), and alpha complex (bottom right) (Takeuchi et al., 2023). Credit: Takeuchi, T. T., et al. 2023, Proceedings of the Institute of Statistical Mathematics, 71(2), 159–187.

15.2.1.3 Homology

Next, we explain how to algebraically calculate the "hole" structures from the resulting simplicial complex. Hole structures are expressed as homology groups. A homology group is an algebraic object that corresponds to the geometric meaning of holes. There are several methods for the algebraic calculation of homology groups, and one such method is outlined below.

Let $V = \{1,\ldots,m\}$ represent the set of all vertices in a simplicial complex X. Let X_k represent the set of all k-simplices in X. We assign an order to the vertices of a given k-simplex $\sigma_k = |v_0,\ldots,v_k|$ in X_k. In other words, an order $v_0 < \cdots < v_k$ is assigned to the set of vertices. The oriented k-simplex is then denoted as $\langle \sigma_k \rangle \equiv \langle v_0,\ldots,v_k \rangle$. Using this oriented k-simplex as a basis, we introduce a vector space with integer coefficients $\mathbb{Z}$:

$$C_k(X) \equiv \left\{ \sum_{\sigma_k \in X_k} a_{\sigma_k} \langle \sigma_k \rangle \,\middle|\, a_{\sigma_k} \in \mathbb{Z} \right\}, \tag{15.12}$$

where $C_k(X) = 0$ if $k > \dim X$ or $k < 0$.

To define the holes in a simplicial complex, we introduce the concept of cycle structures. A cycle structure refers to a closed path formed by traversing connected simplices and returning to the starting point. A k-dimensional hole is a cycle structure formed by k-dimensional simplices in X_k that are not the boundary of a $(k+1)$-simplex in X_{k+1}. Thus, to extract the cycle structure, it is sufficient to find the set of k-simplices in X_k that do not form boundaries. In this way, calculating the boundary of simplices is necessary for extracting hole structures. The boundary of an oriented k-simplex $\langle \sigma_k \rangle = \langle v_0, \ldots, v_k \rangle$ is obtained by applying the following boundary operator ∂_k:

Definition 15 (Boundary operator of a simplex).
The boundary operator ∂_k is defined as

$$\partial_k \langle v_0, \ldots, v_k \rangle \equiv \sum_{i=0}^{k} (-1)^i \langle v_0, \ldots, \check{v}_i, \ldots, v_k \rangle, \tag{15.13}$$

$$\partial_k \sum_{\sigma_k \in X_k} a_{\sigma_k} \langle \sigma_k \rangle = \sum_{\sigma_k \in X_k} a_{\sigma_k} \partial_k \langle \sigma_k \rangle \tag{15.14}$$

(where $\check{v}_i$ indicates the removal of the ith component).

As mentioned earlier, a k-dimensional hole in a simplicial complex X is a set of k-simplices without boundaries, i.e., those that are not the boundaries of $(k+1)$-simplices. The set of boundaryless k-simplices, which correspond to cycles in X, is given by

$$\operatorname{Ker} \partial_k := \{ c \in C_k(X) \,|\, \partial_k(c) = 0 \}. \tag{15.15}$$

Meanwhile, the set of boundaries of $(k+1)$-simplices in X is given by

$$\operatorname{Im} \partial_{k+1} := \{ c \in C_k(X) \,|\, c = \partial_{k+1}(c'), c' \in C_{k+1}(X) \}. \tag{15.16}$$

Therefore, the k-dimensional cycles that are not boundaries of $(k+1)$-simplices are given by taking the quotient $\operatorname{Ker} \partial_k / \operatorname{Im} \partial_{k+1}$.

Thus, the kth homology group, which carries the mathematical meaning corresponding to the k-dimensional holes in a simplicial complex X, is defined as follows:

Definition 16 (Homology group).
The quotient vector space

$$H_k(X) = \operatorname{Ker} \partial_k / \operatorname{Im} \partial_{k+1} \tag{15.17}$$

is the kth homology group of the simplicial complex X. Note that H_0 here refers to the 0th homology group and should not be confused with the Hubble parameter H_0.

In this way, information about the holes inherent in the data $\mathscr{P}$ can be extracted by calculating the homology groups of the Čech complex $\mathscr{C}(\mathscr{P}, r)$. However, this method does not extract information about the scale of the holes, nor can it examine the persistence of holes, i.e., their birth at one radius r_2 and disappearance at another radius r_3. To extract such information, we must introduce the persistent homology group.

15.2.1.4   Persistent Homology Group

The persistent homology group, which encodes information about the birth and death of holes, can be algebraically computed in the same manner as homology groups. To do this, we first increase the radius r and record the radius at which new simplices are added. We also record the simplicial complex at the time a simplex is added, denoted as $X^t := \mathscr{C}(\mathscr{P}, r_t)$ for the radius r_t where the tth simplex $(0 \leq t \leq n)$ is added. Thus, we obtain a set of radii r_t and simplicial complices X^t, where n represents the number of new simplices added. Next, using r_t and X^t, we algebraically compute the hole structures that are generated and disappear as the radius r increases. Since $X^t \subset X^{t+1}$, the disappearance of a hole always occurs due to the birth of a higher-dimensional simplex that bounds the hole. Thus, recording the time when a simplex is born suffices to capture the changes in the geometric structure. At the time of birth, we define a basis space corresponding to the simplex. For example, if a k-simplex $\langle \sigma_k \rangle$ is born at r_t, it can be expressed as

$$
e_{\langle \sigma_k \rangle} = \begin{matrix} (0, & 0, & \dots, & \langle \sigma_k \rangle, & \dots, & 0, & 0) \\ 1 & 2 & & t & & n-1 & n \end{matrix} \tag{15.18}
$$

The simplicial complex X^t can be algebraically represented as the sum of simplices expressed in this basis space.

As with the calculation of homology groups, we compute the birth and death of cycles and boundaries using the simplices expressed in this basis space. The important point here is that the radius at which a boundary appears differs from the radius at which a cycle (a geometric structure) appears. To capture the existence of cycles in the basis space, which only holds the birth information of simplices, we introduce an operator x that acts on the basis to evolve time (increase t). For instance, applying x to the simplex $e_{\langle \hat{\sigma}_k \rangle}$ results in the following transformation,

$$
xe_{\langle \sigma_k \rangle} = \begin{matrix} x(0, & 0, & \dots, & \langle \sigma_k \rangle, & \dots, & 0, & 0) \\ 1 & 2 & & t & & n-1 & n \end{matrix} = \begin{matrix} (0, & 0, & \dots, & 0, & \langle \sigma_k \rangle, & \dots, & 0, & 0) \\ 1 & 2 & & t & t+1 & & n-1 & n \end{matrix}
$$

Using this operator, we can calculate the birth and death of boundaries, as follows:

$$
\partial_k(e_{\langle \sigma_k \rangle}) = \sum_{i=0}^{k} \left[x^{b(\sigma_k) - b(\sigma^i_{k-1})} \right] e_{\langle \sigma^i_{k-1} \rangle}, \quad \sigma_k \in X_\lambda, \tag{15.19}
$$

where $b(\cdot)$ is a function that returns the time t when the simplex was born, and $\langle \sigma^i_{k-1} \rangle$ represents the face $\langle v_0, \dots, \check{v}_i, \dots, v_k \rangle$ of the k-simplex $\sigma_k = \langle v_0, \dots, v_k \rangle$. As with homology groups, we can take the quotient of $\mathrm{Ker}\, \partial_k$ (cycles) and $\mathrm{Im}\, \partial_{k+1}$ (boundaries of $(k+1)$-simplices) to extract hole structures, preserving information about the birth time and persistence (lifespan) until death. In this way, the persistent homology group, which encodes the birth and death of holes, can be computed.

In general, the persistent homology group $PH_k(\mathscr{X})$ is defined as a sequence of mappings between the homology groups $H_k(X^i)$ of X^i:

Definition 17 (Persistent homology group).
The kth persistent homology group $PH_k(\mathscr{X})$ is defined as

$$PH_k(\mathscr{X}) : H_k(X^1) \longrightarrow \cdots \longrightarrow H_k(X^i) \longrightarrow \cdots \longrightarrow H_k(X^n). \qquad (15.20)$$

Here, the "time" $t_\sigma \equiv \min\{t \mid \sigma \in X^t\}$ refers to the birth time of a simplex $\sigma \in X^n$.

15.2.1.5 Persistence Diagram

For the persistent homology group $PH_k(\mathscr{X})$, there exist pairs $b_i, d_i \in \{1,\ldots,n\}$ ($b_i \leq d_i$) and non-negative integers ℓ_k, such that $PH_k(\mathscr{X})$ can be uniquely decomposed while preserving the structure of the homology groups as follows:

$$PH_k(\mathscr{X}) \simeq \bigoplus_{i=1}^{\ell_k} I[b_i, d_i], \qquad (15.21)$$

where the interval $I[b,d]$ indicates that a k-dimensional hole is born at X_b and dies at X_d, and $\simeq$ denotes homotopy equivalence.

From the direct sum decomposition in eq. (15.21), we see that the persistent homology group is uniquely determined by the birth-death pairs (b_i, d_i) (called persistent pairs). A persistence diagram, which plots these persistent pairs, can be defined as follows:

Definition 18 (Persistence diagram).

$$D_k(\mathscr{X}) = \{(b_i, d_i) \mid i = 1,\ldots,\ell_k\} \qquad (15.22)$$

is called the kth persistence diagram of $\mathscr{X}$.

In persistent pairs (birth-death pairs), $b_i < d_i$, so all points in the persistence diagram lie above the diagonal. By definition, points near the diagonal have short "lifespans" and disappear soon after being born. In contrast, points far from the diagonal represent long-lived holes (generators). This is illustrated in Fig. 15.6. In this study, to discuss the spatial structure of galaxy distributions, we refer to the radii corresponding to birth and death as r_{birth} and r_{death}, instead of b and d. Intuitively, r_{birth} represents the average distance between the points that form H_k, while r_{death} represents the size of the structure.

One important property of persistence diagrams is worth mentioning. Consider point cloud data with a ring structure. Figure 15.7 shows the persistence diagrams for two cases of such data. The top panel shows the persistent homology group and corresponding persistence diagram for low-density data, while the bottom panel shows the same for high-density data. The low-density data is essentially a subset of the high-density data in Fig. 15.7. In other words, both datasets share the same information regarding the ring structure. However, the interval between r_{birth} and r_{death} is smaller for the high-density data compared to the low-density data. This suggests

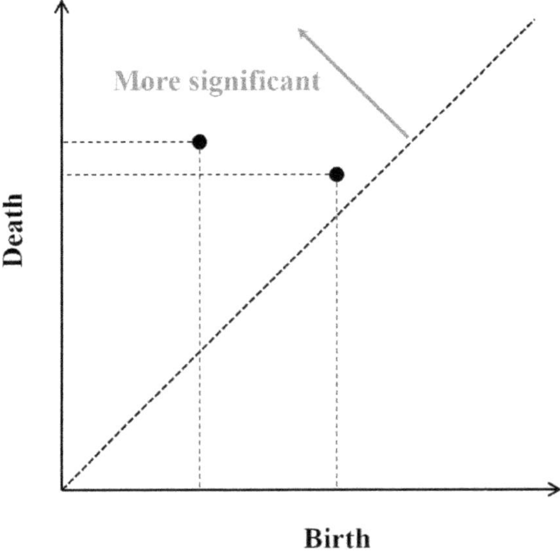

Figure 15.6 Schematic of a persistence diagram (Takeuchi et al., 2023). Points near the diagonal are considered topologically insignificant (noise), while points far from the diagonal represent significant, robust structures in the data. Credit: Takeuchi, T. T., et al. 2023, Proceedings of the Institute of Statistical Mathematics, 71(2), 159–187.

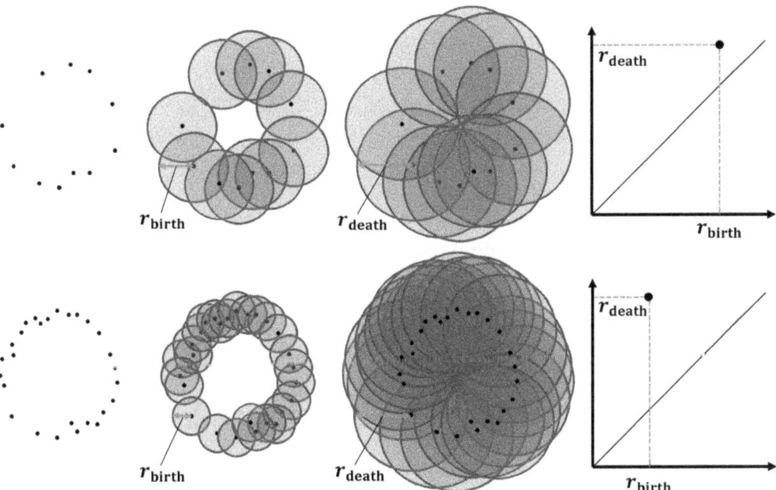

Figure 15.7 Conceptual diagram showing the influence of point data density on the birth and death of structures in a persistence diagram (Kono et al., 2020). From left to right: point data, diagram with balls of radius r_{birth}, diagram with balls of radius r_{death}, and corresponding H_1 persistence diagram. The top row shows low-density data points, and the bottom row shows high-density data points.

that for the same underlying structure, features detected from higher-density data (i.e., higher sampling rates) are more likely to have a significantly longer lifespan. in Fig. 15.7, it can be seen that the r_{birth} of the low-density data is large and close to r_{death}. This is a direct consequence of r_{birth} representing the average distance between data points. In contrast, r_{death}, which represents the size of the structure of interest, remains relatively unchanged. This fact is related to the discussion of the significance of detected features in the analysis of simulation and observational data presented in later chapters.

15.2.1.6 Bottleneck Distance

In various applications, it is necessary to compare the differences between two persistence diagrams D_1 and D_2. Since persistence diagrams are not vectors, care must be taken in this comparison. One method that allows this comparison is the bottleneck distance (e.g., Chazal et al., 2008; Cohen-Steiner et al., 2007; Edelsbrunner & Harer, 2010; Kerber et al., 2017).

Definition 19 (Bottleneck distance).
The bottleneck distance is defined as

$$d_{\text{b}}(D_1, D_2) \equiv \inf_{m} \sup_{x_1 \in |D_1|} \|x_1 - m(x_1)\|_{\infty}, \tag{15.23}$$

where $\|x\|_{\infty}$ is the Chebyshev distance $L^{\infty} \equiv \max\{|x_1|, |x_2|\}$, $|D|$ is the set of generators in the persistence diagram D, augmented with infinitely many generators on the diagonal (x, x), denoted Δ, and m is a bijection between $|D_1|$ and $|D_2|$. The diagonal multiset Δ consists of generators with zero lifespans, provided to match generators in one diagram that has no counterparts in the other. This allows for direct comparison between persistence diagrams with different numbers of generators.

The concept of bottleneck distance is illustrated in Fig. 15.8. When two persistence diagrams are identical, the bottleneck distance takes its minimum value. Next, we define the stability of the distance between persistence diagrams:

Definition 20 (Stability of the distance between persistence diagrams). *Let the Hausdorff distance D_{H} be defined as*

$$D_{\text{H}} = \max\{\sup_{x \in X}(x, Y), \ \sup_{y \in Y}(y, X)\}. \tag{15.24}$$

A distance d between persistence diagrams is said to be stable if there exists a constant L such that for any point sets $X = \{x_i\}$ and $Y = \{y_j\}$, the inequality

$$d(D_1, D_2) \leq L D_{\text{H}}(X, Y) \tag{15.25}$$

holds (Lipschitz continuity).

The bottleneck distance is guaranteed to be stable in this sense (Cohen-Steiner et al., 2007) and can be used to compare persistence diagrams.

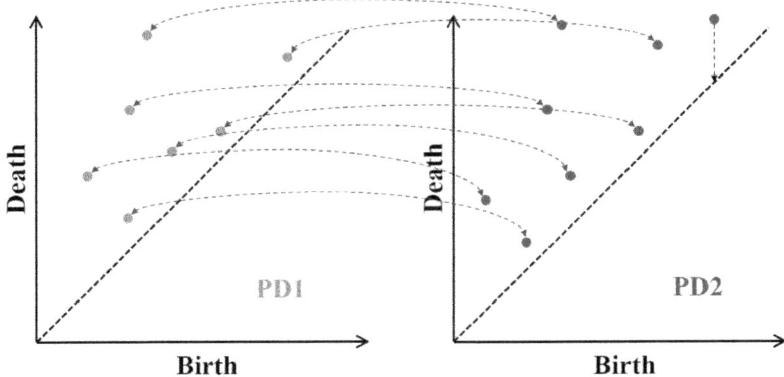

Figure 15.8 Concept of calculating the bottleneck distance between two persistence diagrams (Takeuchi et al., 2023). Points with no corresponding points in the other persistence diagram are considered to lie on the diagonal. Credit: Takeuchi, T. T., et al. 2023, Proceedings of the Institute of Statistical Mathematics, 71(2), 159–187.

15.2.2 INVERSE ANALYSIS OF PERSISTENCE DIAGRAMS

In traditional BAO analysis using the two-point correlation function, it is impossible to identify the positions or shapes of cosmological structures in the data. This is because such analyses represent only the statistical features of correlations, extracting only the average characteristics of the data. In TDA, it is possible to identify the geometric structures corresponding to each generator in the resulting persistence diagram. Such inverse analysis is useful for advancing the physical interpretation of the results. Through filtration with respect to radius r, the topological information of the corresponding simplicial complex $\sigma(r)$ is preserved. In this case, the goal of inverse analysis is to identify the simplicial complices $\sigma(r_{\mathrm{birth},i})$ and $\sigma(r_{\mathrm{death},i})$ that correspond to the specific long-lived holes (generators) $(r_{\mathrm{birth},i}, r_{\mathrm{death},i})$ that appear as significant points in the persistence diagram. For example, by examining the simplicial complices and radius $r_{\mathrm{death},i}$ when a hole disappears, we can infer information about the structure, position, and scale of the hole. In this study, inverse analysis was performed using the R TDA package. R TDA can perform not only inverse analysis but also the calculation of distances between different persistence diagrams and hypothesis testing on the existence of hole structures in persistence diagrams. R TDA also provides interfaces to topological data analysis algorithms written in C++, such as GUDHI (The GUDHI Project, 2015), Dionysus (Morozov, 2007), and PHAT (Bauer et al., 2012).

For the inverse analysis of three-dim hole (shell) structures detected from the three-dim point cloud data under study, the R package SCHU (URL: https://github.com/xinxuyale/SCHU; (Xu et al., 2019)) was used. SCHU is optimized for topological data analysis using Dionysus. This allows the identification of the location and shape of the geometric structures corresponding to the generators detected in the persistence diagram.

Due to the convenience of analysis, this combination of algorithms was adopted in this study. Since the field of inverse analysis is rapidly evolving, various other sophisticated methods have been proposed (see, for example, Obayashi et al. 2018). A comparison of different methods is beyond the scope of this study and will be addressed in future research.

15.3 PERFORMANCE EVALUATION OF THE ANALYSIS METHOD USING COSMOLOGICAL SIMULATIONS

15.3.1 DATA

To validate the performance of the method developed in this study, we first conducted a series of dark matter-only cosmological N-body simulations and generated data. These simulations were performed using the publicly available N-body simulation code $\texttt{Gadget-2}$ (Springel, 2005), and the initial conditions were generated using the method of Crocce et al. (2006). The initial conditions were set at a redshift of $z = 20$, and the simulation box was set to 2 Gpc per side, which is sufficiently larger than the typical BAO scale of ~ 150 Mpc. Since the mass resolution is not critical for BAO detection, the simulation box was populated with 256^3 particles, corresponding to a particle mass of $6.4 \times 10^{13} M_\odot$ ($M_\odot$ is the solar mass). Although this simulation only includes cold dark matter (CDM) particles, the era of interest in this study is long after the photon-baryon decoupling, and since we are primarily concerned with gravitational effects, it is generally sufficient to imprint BAO features in the initial conditions. The power spectrum of the initial fluctuations was calculated using the $\texttt{CLASS}$ code (Lesgourgues, 2011), which numerically solves the linear Boltzmann equations. A snapshot of the simulation at $z = 0$ was then taken, and by performing a power spectrum analysis, we confirmed the presence of BAO in the data.

As a control sample, we generated an identical set of simulations but replaced the initial power spectrum with one that excludes BAO. This was achieved by reducing the baryon fraction to an extreme minimum without changing the total mass of the matter (CDM + baryons). In the standard model, $\Omega_B = 0.049$, whereas the control sample was set to $\Omega_B = 0.002204$. For convenience, we will refer to these simulations as BAO-present/absent or baryon-present/absent, respectively. Figure 15.9 shows snapshots of the simulations with and without baryons. Since the same random seed was used in both simulation sets, it is very difficult to visually distinguish the differences. However, Figure 15.10 shows the power spectrum of both simulation sets, with the spectra normalized by the smoothed power spectrum without BAO features (Eisenstein & Hu, 1998). As expected, the control sample's power spectrum does not show a BAO signal. Here in this work, due to computational constraints, we used a subsample of 2000 particles randomly selected from the parent sample of 256^3 particles. In the future, we plan to parallelize the code and upgrade the computing system to handle much larger datasets.

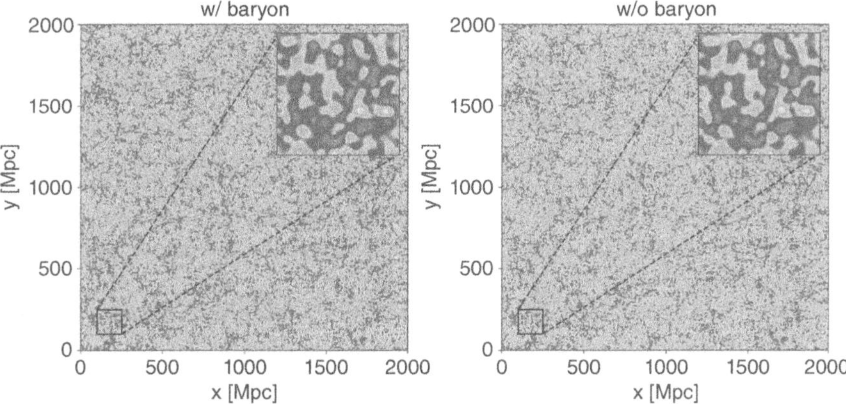

Figure 15.9   Simulation data with baryons (right) and without baryons (left) at redshift $z = 0$ (Takeuchi et al., 2023). The inset shows a zoomed-in view of a 100–250 Mpc box region. The density fluctuations are projected along the z-axis over a 100 Mpc thick region, and the values shown are transformed by `arcsinh` for clarity. Credit: Takeuchi, T. T., et al. 2023, Proceedings of the Institute of Statistical Mathematics, 71(2), 159–187.

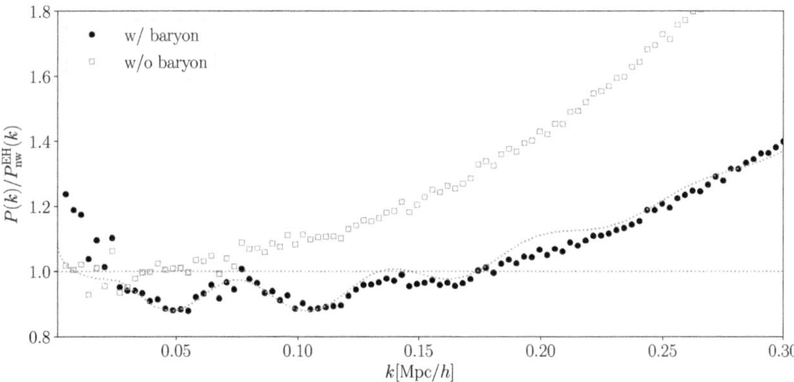

Figure 15.10   Power spectrum at $z = 0$ from the simulation (Takeuchi et al., 2023). The spectrum is normalized using the linear power spectrum without baryonic oscillations from Eisenstein & Hu (1998). The black dots represent simulation data with baryon oscillations, the red dashed line is the analytic solution calculated using the same cosmological parameter set and the halofit model (Takahashi et al., 2012), and the blue squares represent simulation data without baryon oscillations. As expected, the spectrum without baryons shows no oscillations for $k < 0.2 \mathrm{Mpc}/h$. Credit: Takeuchi, T. T., et al. 2023, Proceedings of the Institute of Statistical Mathematics, 71(2), 159–187.

15.3.2 ANALYSIS RESULTS OF THE SIMULATION DATA

Next, we examine the persistence diagrams obtained from the simulation data. To evaluate the statistical significance of the detected signals, it is necessary to compute confidence intervals. However, estimating confidence intervals for persistence diagrams is non-trivial, and several methods have been proposed (e.g., Fasy et al., 2014). In this study, we adopt a method that uses the bottleneck distance. First, we generate a set of samples using bootstrap resampling. From this, we extract pairs of persistence diagrams and estimate confidence intervals by measuring the bottleneck distance between the two diagrams.

Figure 15.11 shows the persistence diagrams obtained from the simulation. The left panel shows the result without baryons, and the right panel shows the result with baryons. Black dots, red triangles, and blue diamonds represent the 0th (H_0), 1st (H_1), and 2nd (H_2) persistent homology groups, respectively. The red and blue dashed lines represent the 90-% confidence intervals for H_1 and H_2, respectively. These confidence intervals were calculated using the bootstrap resampling method with $N_{\mathrm{boot}} = 30$. In this study, points above the corresponding confidence intervals are considered statistically significant.

As shown in the lower panel of Fig. 15.11, four significant generators corresponding to 3-dim holes (blue diamonds) were detected in H_2 from the baryon-present data. In contrast, no significant holes at these scales were detected in the baryon-absent data. The r_{birth} and r_{death} values for the significant holes detected in the baryon-present sample are listed in Table 15.1. The average radius r_{death} of the detected hollow holes was 150.16 ± 8.46 [Mpc]. Although only four holes were detected, the calculated $\bar{r}_{\mathrm{death}}$ matches the expected BAO radius. Figure 15.12 shows the position and shape of the structures in the point cloud data corresponding to the detected generators. This 3-dim structure forms a hollow cavity.

For the generators belonging to H_1 (red triangles), significant features were obtained in both datasets. These generators correspond to 2-dim loop structures. In the baryon-present data, 17 significant H_1 loops with $p < 0.2$ were detected. The average r_{death} of these significant generators was 99.00 ± 2.26 [Mpc]. In the baryon-absent data, 34 generators were detected, with $\bar{r}_{\mathrm{death}} = 100.49 \pm 3.24$ [Mpc]. The agreement of characteristic scales in the loop structures detected in H_1 suggests the presence of 2-dim loops that are unaffected by the presence of baryons. Large-scale filaments are possible candidates for such structures. In contrast, $\bar{r}_{\mathrm{birth}}$ differed depending on the presence of baryons, with 46.13 ± 2.24 [Mpc] for the baryon-present case and $\bar{r}_{\mathrm{birth}} = 62.34 \pm 2.84$ [Mpc] for the baryon-absent case. This analysis suggests that the distances between galaxies forming loops decrease in the presence of baryons.

Finally, the generators belonging to H_0 (black dots) showed a clear difference in distribution on the persistence diagram depending on the presence of baryons. H_0 represents connected components and corresponds to 1D continuous structures such as galaxy clusters or filaments. In the baryon-absent data, r_{birth} is widely distributed along the diagonal, while in the baryon-present data, r_{birth} is concentrated at low values. This localized distribution of r_{birth} in the baryon-present case indicates that connected structures were formed at a specific radius. This suggests that nearby galaxies forming galaxy clusters or filaments are evenly spaced, causing them to rapidly merge and form connected structures at a particular r_{birth}. Meanwhile, the

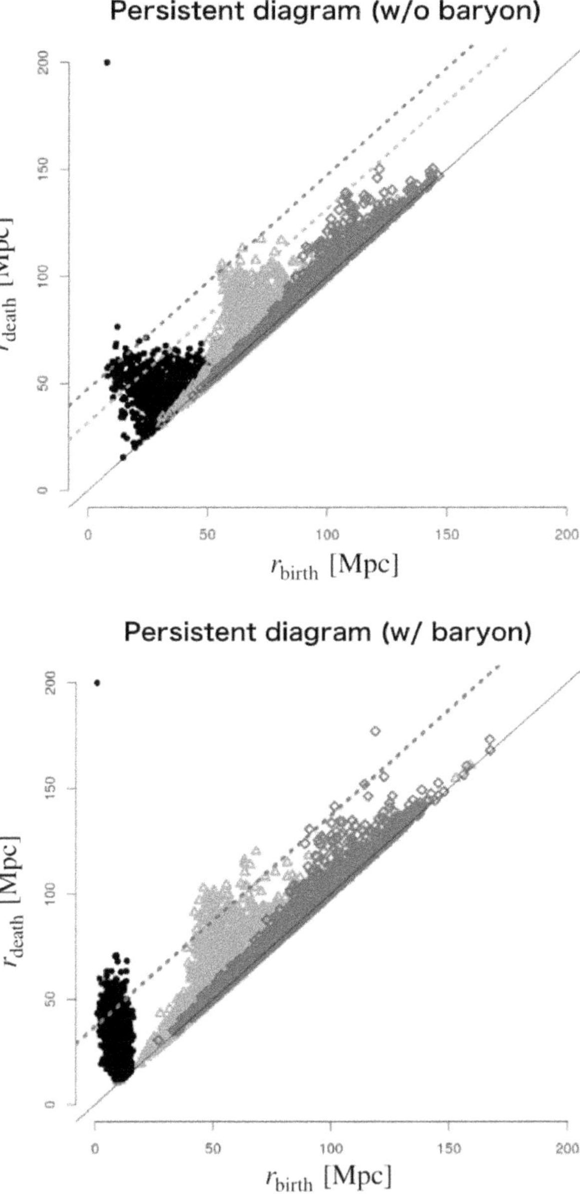

Figure 15.11   Persistence diagrams for the simulation data (Takeuchi et al., 2023). Filled dots, triangles, and diamonds represent H_0, H_1, and H_2, respectively. The black diagonal line represents $r_{\text{birth}} = r_{\text{death}}$. The left panel shows the results for the baryon-absent simulation, and the right panel shows the baryon-present simulation results. Credit: Takeuchi, T. T., et al. 2023, Proceedings of the Institute of Statistical Mathematics, 71(2), 159–187.

Table 15.1

r_{birth} **and** r_{death} **of the significant holes** $(p < 0.2)$ **detected in the simulation data.**

No.	r_{birth} [Mpc]	r_{death} [Mpc]
1	116.34	130.65
2	127.42	141.80
3	140.17	151.70
4	144.60	176.47

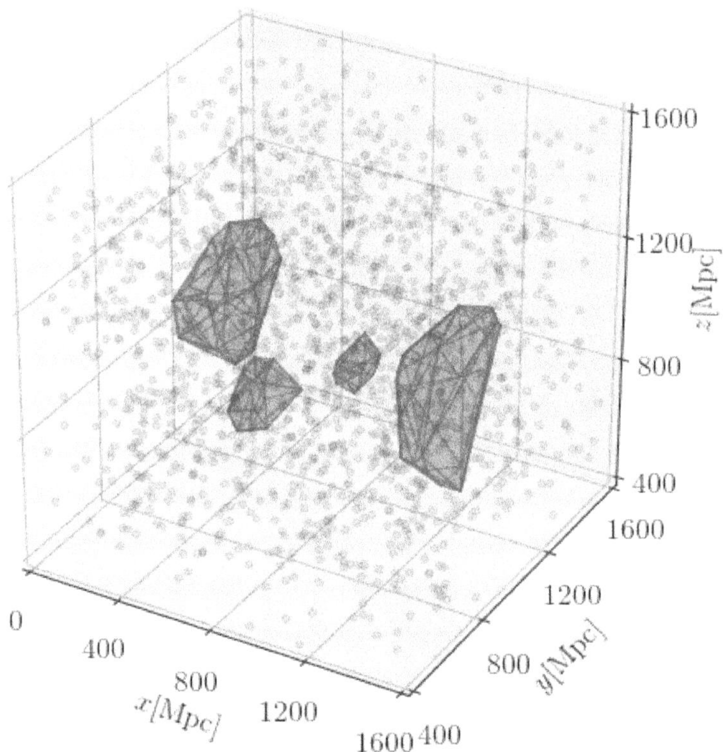

Figure 15.12 Example of inverse analysis (Takeuchi et al., 2023). The structure corresponding to the four most significant H_2 generators in the simulation data is shown in real space. The shaded region represents the convex hull of all particles forming the H_2 structure, revealing a hollow shape. Credit: Takeuchi, T. T., et al. 2023, Proceedings of the Institute of Statistical Mathematics, 71(2), 159–187.

distribution of r_{death} was broad regardless of the presence of baryons. This implies that the distances between galaxy clusters or filaments are not influenced by baryons. Thus, the features of the H_0 persistence diagram suggest that the presence of baryons imposes some constraint on the distances between nearby galaxies within galaxy

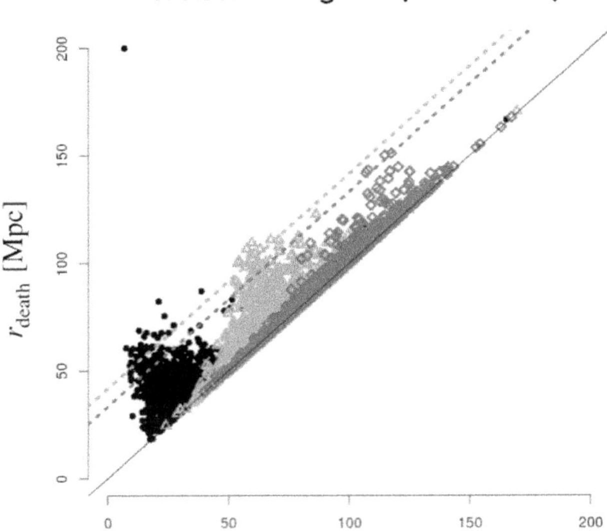

Figure 15.13 Persistence diagram of SDSS data (Takeuchi et al., 2023). The symbols are the same as in Fig. 15.11. Credit: Takeuchi, T. T., et al. 2023, Proceedings of the Institute of Statistical Mathematics, 71(2), 159–187.

clusters or filaments. Further analysis is expected to yield a physical interpretation of the characteristics observed in the H_0 and H_1 persistence diagrams.

15.4 ANALYSIS USING SLOAN DIGITAL SKY SURVEY DATA RELEASE 14

15.4.1 SDSS GALAXY SURVEY DATA

In this study, we used quasar data from the extended Baryon Oscillation Spectroscopic Survey (eBOSS) (Dawson et al., 2016) large-scale structure catalog (Ata et al., 2018) from Sloan Digital Sky Survey (SDSS) Data Release 14 (DR14). This is part of SDSS-IV Blanton et al. (2017). The observed bands (wavelength ranges) are u, g, r, i, and z. The survey area on the celestial sphere covers 2044 deg^2, with redshifts ranging from $0.8 < z < 2.2$. The effective area is 1288 deg^2 in the Northern Galactic Cap (NGC) and 995 deg^2 in the Southern Galactic Cap (SGC). The quasar sample was selected based on data from Myers et al. (2015). The presence of BAO in this sample has already been confirmed by Ata et al. (2018).

The entire sample includes 147,000 quasars, but for this study, we randomly selected 2000 galaxies, similar to the simulation data.

15.4.2 SDSS DATA ANALYSIS RESULTS

Figure 15.13 shows the persistence diagram of the SDSS DR14 data. As shown in Fig. 15.13, four significant generators, i.e., "shells," belonging to H_2 were detected in the SDSS DR14 data. The average r_{death} was 146.6 ± 2.0 [Mpc] (Table 15.2).

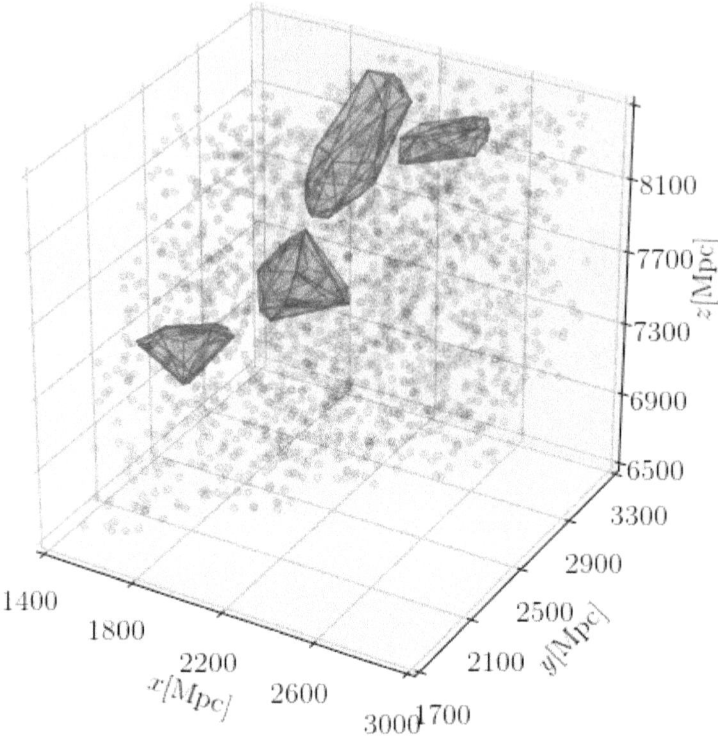

Figure 15.14   Real-space structures corresponding to the four significant H_2 generators detected in the actual galaxy survey (SDSS) data (Takeuchi et al., 2023). The shaded area represents the same meaning as in Fig. 15.12. Credit: Takeuchi, T. T., et al. 2023, Proceedings of the Institute of Statistical Mathematics, 71(2), 159–187.

Although only four significant generators were detected, $\bar{r}_{\mathrm{birth}}$ matches the expected radius from BAO predictions. The results of the inverse analysis for H_2 are shown in Fig. 15.14.

In this analysis, 19 generators belonging to H_1 were detected as significant holes (loops). The average r_{death} for these generators was 101.82 ± 3.54 [Mpc],

Table 15.2

The r_{birth} and r_{death} values of the significant holes ($p < 0.2$) detected in the SDSS DR14 observational data.

No.	r_{birth} [Mpc]	r_{death} [Mpc]
1	107.00	141.98
2	107.83	143.30
3	114.86	150.20
4	117.56	150.98

consistent with the $\bar{r}_{\text{birth}}$ obtained from the simulation data. The observed r_{birth} was 57.92 ± 3.05 [Mpc]. Interestingly, the distribution of the H_0 homology matches that of the baryon-absent simulation. Further detailed studies are needed to interpret these lower-dimensional homologies.

15.5 SUMMARY AND DISCUSSION

15.5.1 DIFFERENCE OF THE LARGE-SCALE STRUCTURE WITH AND WITHOUT BARYONS

In this study, we examined the usefulness of topological data analysis (TDA) for quantifying the galaxy distribution. The persistent homology groups targeted by the topological data analysis in this study form filtrations defined by a sequence of simplicial complices corresponding to different scales. This property is particularly useful for analyzing the scale and shape of holes inherent in the data.

First, using $N = 256^3$ N-body simulations, we examined the effect of baryons on persistent homology groups. It was shown that the persistence diagrams differ significantly between the simulations with and without baryons. From the baryon-present data, four significant generators (hole structures) were detected from the second persistence diagram, but no such structures were detected from the baryon-absent data. Clear differences were also observed in the distribution of the generators belonging to H_1 and H_0 in the persistence diagrams. The $\bar{r}_{\text{death}}$ of the detected one-dim holes took similar values, but the $\bar{r}_{\text{birth}}$ differed greatly. Similarly, the $\bar{r}_{\text{birth}}$ of the 0-dim holes showed a significant tendency to differ depending on the presence of baryons.

Next, we successfully detected the BAO signal from a subsample of up to 2000 galaxies extracted from the quasar survey in the SDSS observational data. The obtained $\bar{r}_{\text{death}}$ for H_2 was 146.6 ± 2.0 [Mpc].

15.5.2 INTERPRETATION AS THE BARYONIC ACOUSTIC OSCILLATION

Clearly the existence of baryons change the persistent homology of the resulting large-scale structure in the Universe. Recall that the visible Universe, the one we can observe through electromagnetic radiation, is shaped by the structures of ordinary matter that interacts with electromagnetic radiation. In cosmology, ordinary matter is commonly referred to as baryons, so we will also use the term baryons to refer to ordinary matter. Baryons undergo complex evolution through electromagnetic interactions (e.g., radiative heating/cooling, gas pressure, and hydrodynamic processes). A typical example of such non-trivial phenomena is the baryon acoustic oscillations (BAO) generated in the baryon–photon fluid in the early Universe (Peebles & Yu, 1970; Sunyaev & Zeldovich, 1970).

The sound speed at this time is given by eq. (7.91) as

$$c_{\text{s}} = \frac{c}{\sqrt{3\left(1 + \dfrac{3\rho_{\text{B}}}{4\rho_{\text{R}}}\right)}} \tag{15.26}$$

which leads to the final radius of the sound horizon, r_s,

$$
\begin{aligned}
r_s &= \int_0^{t_{dec}} (1+z)c_s \, dt \\
&= \int_{z_{dec}}^{\infty} \frac{c_s}{H_0 \sqrt{\Omega_{R,0}(1+z)^4 + \Omega_{M,0}(1+z)^3 + \Omega_{\Lambda,0}}} \, dz .
\end{aligned} \tag{15.27}
$$

What we observe is the superposition of many sound waves generated from primordial perturbations imprinted on large-scale structures. Although BAO is a phenomenon that occurs in baryons, it is known that the final states of baryon and dark matter density fluctuations match due to gravitational interaction (the baryon catch-up: e.g., Eisenstein et al., 2007; Ma & Bertschinger, 1995).

We note that the BAO scale remains constant in comoving coordinates (coordinates that account for the expansion of the Universe). As a result, the BAO signal can be detected primarily in the two-point correlation function of galaxies. However, since the BAO scale is extremely large compared to typical galactic scales, the BAO signal that appears in two-point correlations is relatively weak. Therefore, detecting the BAO signal requires large-scale galaxy surveys with densely sampled data over a vast observational volume. The Sloan Digital Sky Survey (SDSS) is the most extensive photometric and spectroscopic survey, covering one-third of the entire sky. Through SDSS, a BAO signal of approximately 150 Mpc was detected in the two-point correlation function (Eisenstein et al., 2005).

Nevertheless, BAO analysis using traditional methods has only become feasible with large-scale, wide-area survey data like SDSS. Thus, developing methods for detecting and quantifying the BAO signal that place fewer demands on the data is a critical issue for constraining cosmological models. We apply the TDA to the spatial distribution of galaxies, and discuss methods for significance testing, demonstrating that TDA is useful for evaluating the BAO scale.

BAO detection using the method in this study contrasts sharply with traditional two-point correlation function analysis. While estimating the two-point correlation function requires densely sampled galaxy data, topological data analysis can detect weak signals like BAO from sparsely sampled data of ~ 2000 galaxies. This suggests that even sparsely sampled galaxy surveys could be used for cosmological studies by leveraging topological data analysis. This could provide new directions for designing and strategizing next-generation cosmological galaxy surveys.

Furthermore, a notable advantage of topological data analysis compared to conventional methods is its ability to visualize BAO structures through inverse analysis. The function of inverse analysis in persistence diagrams is a clear feature that distinguishes it from traditional methods and is worth emphasizing. For example, as a direct extension of this study, the real-space galaxy distribution structure contributing to the BAO signal can be easily identified, allowing for much easier cosmological discussions compared to traditional methods.

However, a simple topological data analysis has the drawback of not being robust to noise, which needs to be addressed in future detailed applications. Robust methods for topological data analysis are currently being researched (e.g., Vishwanath et al., 2022). Such a method is definitely necessary to give a solid conclusion if our detected BAO-like signal is indeed the BAO.

16 Radio Morphology of Galaxies with Machine Learning

16.1 INTRODUCTION: RADIO GALAXIES AND THEIR CLASSIFICATION

16.1.1 RADIO GALAXIES

Galaxies composed of ordinary stars do not emit strong radiation at radio wavelengths. The primary sources of radio emissions in the Universe are ionized gas clouds, supernova remnants, and the radio galaxies studied here. Radio galaxies are a special class of galaxies that have a strong radio emission with luminosities ranging from 10^{41}–10^{46} erg s^{-1}. For comparison, the radio luminosity of the Milky Way is 10^{37} erg s^{-1}, highlighting the extremely high radio luminosity of radio galaxies. Most radio galaxies are hosted by elliptical galaxies, with an active galactic nucleus (AGN) at their center, from which jets are ejected. An AGN refers to an active phenomenon occurring in the central region of galaxies, not driven by stellar processes (1). The central engine of an AGN is thought to be a black hole, which shines as matter is accreted onto it. The radio emissions of radio galaxies primarily originate from synchrotron radiation produced by the AGN, the jets ejected from it, and the radio lobes formed at the extremities of the jets. Synchrotron radiation is electromagnetic radiation emitted when electrons, accelerated close to the speed of light, are captured by magnetic field lines and undergo rotational motion (e.g., Rybicki & Lightman, 1985).

The jets emitted by the AGN collide with the surrounding intergalactic gas and creates shockwaves. These shockwaves cause the gas, electrons, and magnetic fields to mix violently, generating synchrotron radiation. The morphology of observed radio galaxies is shaped by this synchrotron radiation, reflecting how far the energy released from the galactic center has traveled and been dissipated.

16.1.2 FANAROFF–RILEY (FR) CLASSIFICATION

The Fanaroff–Riley (FR) classification categorizes AGNs with extended radio sources into two types (Fanaroff & Riley, 1974). FR I galaxies represent weaker radio sources ($\ll 10^{25}$ [W Hz^{-1}], normalized at frequencies other than $\nu \simeq 1.4$ [GHz]), with the central region being the brightest and the radio luminosity gradually decreasing toward the edges (edge darkening). Additionally, these galaxies exhibit radio lobes with steep spectral slopes. Jets are emitted continuously in two directions

DOI: 10.1201/9781003104315-16

Figure 16.1   Schematic illustration of the FR I and FR II classification.

and are brighter than the lobes. The host galaxies are the brightest class of elliptical galaxies or cD galaxies (cf. Chapter 1).

FR II galaxies, on the other hand, are brighter radio sources (above 10^{35} [W s^{-1}] in the GHz range), with the outer edges of the radio structure being the brightest (edge brightening). FR II galaxies often feature kpc-scale hotspots, and their spectral slope steepens toward the inner regions. Quasars, which are a type of AGN, are typically classified as FR II. The jets are asymmetric and clumpy. Furthermore, the contrast between the jets and lobes is lower. The host galaxies are usually ordinary elliptical galaxies.

16.2   MORPHOLOGY OF RADIO GALAXIES WITH CNN

Here, we present an example of using CNNs to classify the morphology of radio galaxies. This study aims to classify the morphology of active galactic nuclei observed through radio waves from images using machine learning. Previous research has applied machine learning to morphological classification, and by using the same data, we seek to verify prior studies while achieving classification with simpler algorithms. Since the morphology of radio galaxies is closely tied to energy transport from the galactic center, it is believed that large-scale systematic classification using machine learning can help reveal the physics behind these powerful central energy sources in galaxies.

16.2.1   DATA ACQUISITION

The galaxy catalog used in this study comes from the earlier research by Ma et al. (2019). That study conducted morphological classification using machine learning and classified 15,687 objects based on their images. In this study, we used the positional coordinates and classification labels from the catalog. Below, we explain the galaxy image data and classification labels used in this chapter.

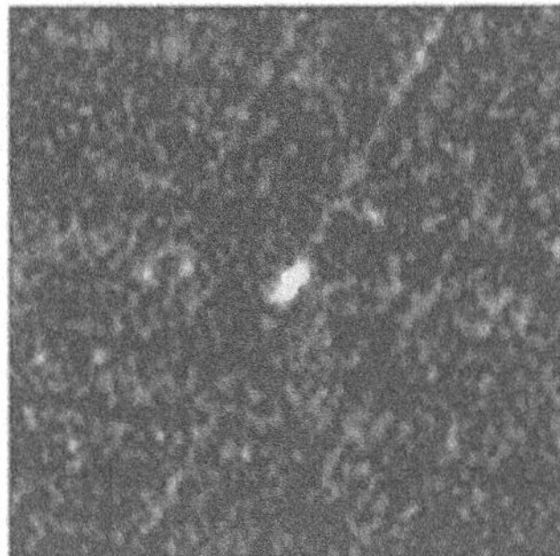

Figure 16.2 Examples of omitted and accepted images. Left: omitted due to boundary effects, Right: accepted.

16.2.1.1 Image data

This study used data from VLAFIRST (Faint Images of the Radio Sky at Twenty-cm). The VLA (Very Large Array) is a radio interferometer operated by the National Radio Astronomy Observatory (NRAO) in the United States, consisting of 27 antennas, each with a diameter of 25 meters. Using the positional coordinates of galaxies with SkyView, a Python module in Astroquery, we obtained the FITS files of regions containing the objects (approximately $4.5' \times 4.5'$ based on previous research) and converted them to JPG format (150×150 pixels), which served as our image dataset. Next, we excluded images that were clipped due to boundary effects in the survey region during the cross-matching of galaxy coordinates (Fig. 16.2).

16.2.1.2 Classification Label Data

The morphological classification of radio galaxies used in this study includes six types. We adopted the classifications from Ma et al. (2019) and applied two methods for assigning labels. First, machine learning labels were assigned based on the results of the prior study. The dataset contains 14,245 objects, which we used in the learning phase (training data and validation data). These were selected from Best & Heckman (2012), with a redshift range of $0.01 < z < 0.70$ and a flux density range of $5.0\,\mathrm{mJy} < S < 2.3 \times 10^4$ mJy at 1.4 GHz. Second, labels were assigned from six catalogs from other studies, consisting of 1,442 objects, which were used for model evaluation (test data). The redshift range is $0.06 < z < 3.90$, and the flux density range is $6.0\,\mathrm{mJy} < S < 1.5 \times 10^4$ mJy at 1.4 GHz.

The catalogs used for labeling from Ma et al. (2019) are shown in Table 16.1[1]. The classification labels consist of six categories, from Class 1 to Class 6 (Fig. 16.3).

Class 1 consists of compact galaxies, which are bright only near the center and

---

[1] References for Table 16.1: FR0CAT: Baldi et al. (2018); FRICAT: Capetti et al. (2017a); FRIICAT: Capetti et al. (2017b); Cheung (2007): Cheung (2007); Proctor (2011): Proctor (2011); CoNFIG 1–4: Gendre et al. (2010).

Table 16.1

Catalogs used to construct the sample (Ma et al., 2019).

Catalog	Number and Type of Sources
FR0CAT	108 (compact: 104)
FRICAT	233 (FR I: 192; compact: 1; BT: 5)
FRIICAT	123 (FR II: 80; BT: 8; XRG: 3)
Cheung (2007)	100 (XRG: 81)
Proctor (2011)	475 (BT: 285; RRG: 32)
CoNFIG 1–4	856 (FR I: 19; FR II: 351; BT: 9; compact: 272)

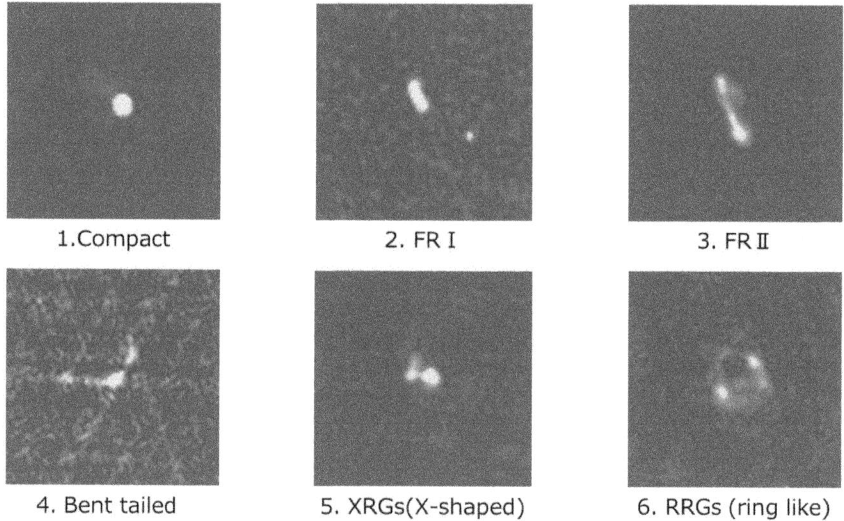

Figure 16.3   Examples of galaxies in each class.

lack extended radio lobes. Class 2 represents typical FR I galaxies, which have extended lobes that are brighter near the center and fade sharply toward the ends. Class 3 represents FR II galaxies, where the lobes are brighter at the edges than at the center. Class 4 consists of galaxies with bent tails, where the extended lobes are bent and irregular in shape. Class 5 comprises X-shaped radio galaxies (XRG), where the extended lobes form an X shape. Class 6 consists of ring-like radio galaxies (RRG), where the extended lobes have a ring-like structure.

From Table 16.2, we see that Classes 1 and 2 have more samples than the other classes. To avoid bias due to sample size, we randomly selected 2,000 images each from Class 1 and Class 2.

Table 16.2
Summary of the training data.

Class	Label	Number	Sample	Augmentation
Class 1	Compact sources (FR0)	5541	2000	4000
Class 2	Typical FR I sources	4745	2000	4000
Class 3	Typical FR II sources	2292	2292	4584
Class 4	Bent-tailed (BT) sources	1212	1212	4848
Class 5	X-shaped radio galaxies	289	289	5780
Class 6	Ring-like radio galaxies	85	85	3400
	Total	14164	7878	26612
	Discarded	81		

Table 16.3
Summary of the test data.

Class	Label	Number	Test Sample
Class 1	Compact sources (FR0)	405	30
Class 2	Typical FR I sources	187	30
Class 3	Typical FR II sources	429	30
Class 4	Bent-tailed (BT) sources	303	30
Class 5	X-shaped radio galaxies	81	30
Class 6	Ring-like radio galaxies	32	30
	Total	1437	180
	Discarded	5	

16.2.1.3   Summary of Data

The number of data samples used is summarized in Tables 16.2 and 16.3. Of the 26,612 images collected, 70 % were used as training data, and 30 % were used as validation data. Due to the significant imbalance in sample size across different classifications, we selected 30 objects from each class for the test samples, resulting in a total of 180 images used.

16.3   DATA PROCESSING

16.3.1   SIGMA CLIPPING

In radio observations, various types of noise can affect the data, such as electrical currents in the observation equipment, interference from mobile phones, or distant

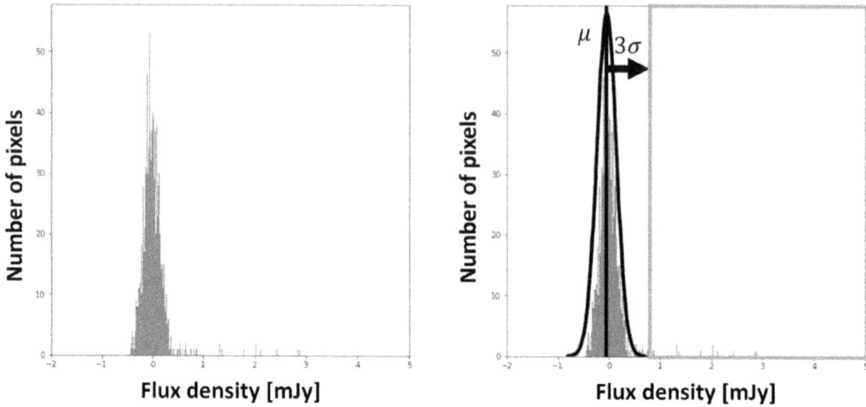

Figure 16.4   Schematic description of the sigma clipping.

thunderstorms. In this study, since significant background noise was superposed on the images, sigma clipping was applied to remove it. A histogram was generated using the data from the obtained FITS files, with the horizontal axis representing radio intensity [mJy] and the vertical axis representing the number of pixels (Fig. 16.4, left). Elements with low radio intensity and a large number of pixels were considered background noise. If the background light is random noise, it is expected to follow a Gaussian distribution. Therefore, a Gaussian fitting was performed, with eq. (16.1).

$$f(x) = \frac{1}{\sqrt{2\pi\sigma^2}} \exp\left[-\frac{(x-\mu)^2}{2\sigma^2} \right] \tag{16.1}$$

Here, σ is the standard deviation, and μ is the mean. Based on the μ obtained from the fitting, elements beyond 3σ were excluded, and the fitting process was repeated. This process was continued until the values converged.

The average value after convergence was considered to be the background noise, which was then subtracted from the image. Any negative values generated during this process were set to zero. The images before and after sigma clipping are shown in Fig. 16.5. From Fig. 16.5, it can be seen that the background light was successfully removed, resulting in a clearer image with reduced noise.

16.3.2   DATA AUGMENTATION

Data augmentation refers to the process of increasing the number of samples by adding rotated or enlarged images to the dataset. This study applied data augmentation for two main purposes.

First, to enhance the recognition ability of the algorithm. By adding rotated and enlarged images to the dataset, the algorithm can be trained to classify these images just like the original ones, thus improving its recognition ability. Second, to reduce

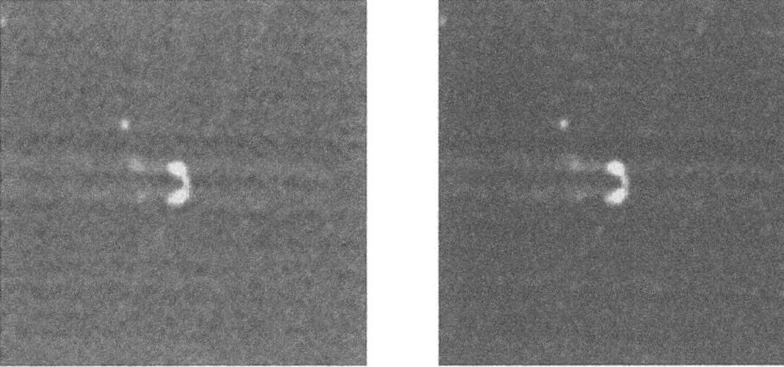

Figure 16.5   Comparison of sigma clipping (Left: before sigma clipping, Right: after sigma clipping).

Figure 16.6   Example of data augmentation. (1): Original image, (2)–(4): Flipping operations (horizontal, vertical, both), (5): Enlarged by 1.5 times and randomly rotated.

the imbalance in the number of data samples. By increasing the number of images in the smaller classes using augmented images, the imbalance in data volume can be mitigated. In this study, we used flipping operations (vertical, horizontal, and both vertical and horizontal) and rotation/enlargement operations (enlarging by 1.5 times and then randomly rotating). Examples of these operations are shown in Fig. 16.6.

Data augmentation was applied to ensure that the number of images in each class approached approximately 4,000. The operations applied to each class are shown in Table 16.4. This approach helped to reduce biases caused by differences in the amount of data available in each class. The final number of images for each class is shown in the rightmost column of Table 16.2.

16.4   BUILDING THE CLASSIFICATION MODEL

When constructing the model, it is necessary to determine the hyperparameters. Optuna, a Python hyperparameter optimization module, was used to find these values. By providing the range of hyperparameters and the number of trials, Optuna searches within the given range and outputs the best hyperparameters based on the trial results. The model provided to Optuna, as well as the hyperparameters

Table 16.4

Data augmentation operations for each class (○: applied, ×: not applied).

Class	1	2	3	4	5	6
Amount of Augmentation	×2	×2	×2	×4	×20	×40
Horizontal Flip	○	○	○	○	○	○
Vertical Flip	×	×	○	○	○	○
Horizontal/Vertical Flip	×	×	×	○	○	○
Enlargement/Rotation	×	×	×	×	○	○

Table 16.5

Model provided to Optuna. **The number of convolutional layers in Table 16.6 refers to the repetition of Step A.**

	Layer
	Convolutional layer: number of filters
	Batch normalization
A	Convolutional layer: number of filters
	Batch normalization
	Maxpooling
	Repeat steps A
	Flatten layer
	Fully connected layer: number of units, activation function
	Dropout rate
	Batch normalization
	Fully connected layer: number of units 6, activation function: softmax

determined using Optuna and their respective ranges, are shown in Tables 16.5 and 16.6. Using the optimal hyperparameters determined from this range, the model was created.

16.5 RESULTS AND DISCUSSION

The optimal model and its hyperparameters, obtained after 15 epochs and 100 trials, are shown in Tables 16.7 and 16.8. The results of training with 15 epochs and a batch size of 256 are shown in Fig. 16.7. Using both the training and validation data, the model achieved 99% accuracy (i.e., an accuracy of 0.99). Generalization was then tested using the test data, which yielded an accuracy of 0.56. Since the accuracy was less than 60 %, the model cannot be considered well-generalized, and overfitting is

Table 16.6

Range for hyperparameter tuning.

Hyperparameters	Range
Number of convolutional layers	Integer 2–5
Filters in convolutional layers	5,10,15,20
Number of units in fully connected layers	Integer 70–90
Dropout rate	0–1
Activation function	relu, sigmoid
Optimization function	RMSprop, Adam

Table 16.7

Model used in this work.

Conv2d: filter $= 15$
Batch normalization
Conv2d: filter $= 15$
Batch normalization
Maxpooling
Conv2d: filter $= 15$
Batch normalization
Maxpooling
Flatten
Dense: units $= 83$, activation $=$ 'relu'
Dropout: dropout rate $= 0.0199185793114493$
Batch normalization
Dense: units $= 6$, activation $=$ 'softmax'

suspected. Therefore, we proceed to examine how CNN classified the data in each class.

16.5.1 CLASS 1: COMPACT GALAXIES

For Class 1, 16 out of 30 objects were correctly classified, while the remaining 14 were classified into other classes (Fig. 16.8). Most misclassifications occurred between Class 1 and Class 2 (FR I galaxies), with 10 objects misclassified. Many misclassified galaxies appeared compact, without the extended radio lobes typical of FR I galaxies. Therefore, it is concluded that the CNN did not classify Class 1 galaxies very accurately.

Table 16.8

Optimal hyperparameters.

Hyperparameters	Optimal
Number of convolutional layers	3
Filters in convolutional layers	15, 15, 15
Number of units in fully connected layers	83
Dropout rate	0.019919
Activation function	relu
Optimization function	Adam

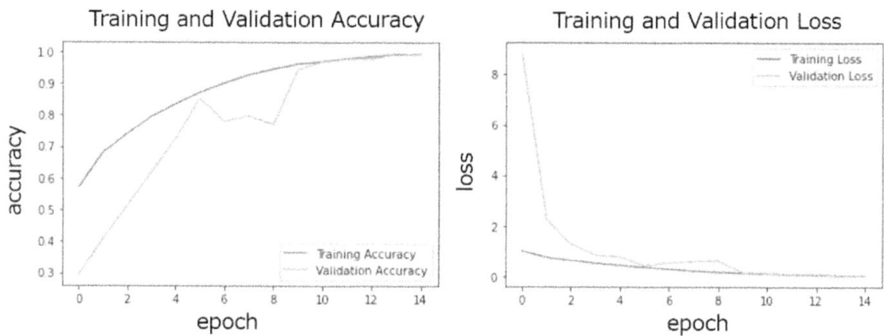

Figure 16.7   Changes in accuracy and loss per epoch.

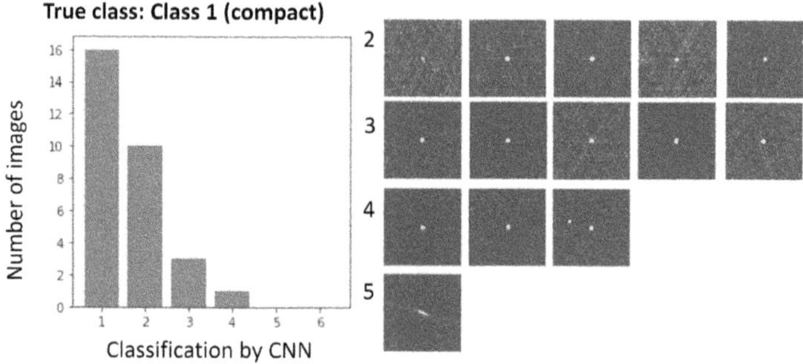

Figure 16.8   Classification results of Class 1 by CNN. Left: Number of correctly and incorrectly classified sources. Right: Examples of misclassified images.

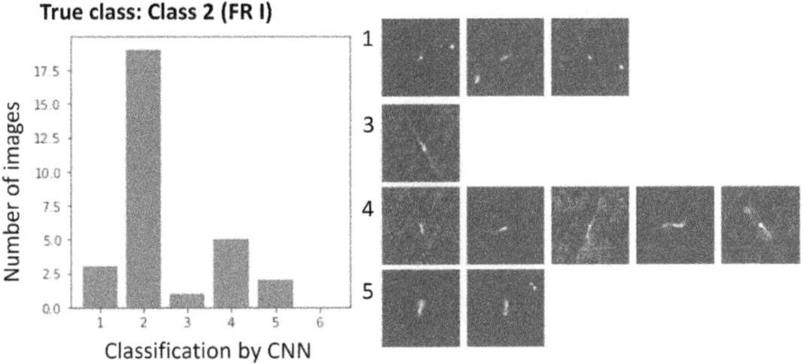

Figure 16.9 Same as Fig. 16.8, but for Class 2 (FR I galaxies).

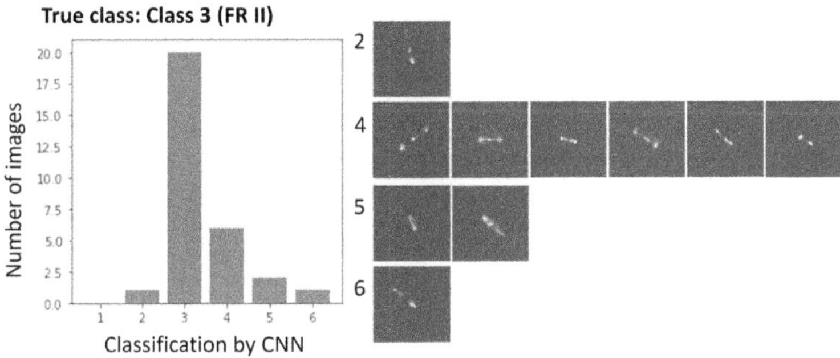

Figure 16.10 Same as Fig. 16.8, but for Class 3 (FR II galaxies).

16.5.2 CLASS 2: FR I GALAXIES

For Class 2, 18 out of 30 objects were correctly classified as Class 2, while the remaining 12 were classified into other classes (Fig. 16.9). Class 2 galaxies were most frequently misclassified as Class 4 (galaxies with bent tails), with 5 misclassifications. Upon reviewing the misclassified galaxies, many displayed FR I-like features, but some also exhibited bent-tail features. Thus, the CNN generally classified these images correctly.

16.5.3 CLASS 3: FR II GALAXIES

For Class 3, 20 out of 30 objects were correctly classified as Class 3, while the remaining 10 were classified into other classes (Fig. 16.10). Class 3 galaxies were most frequently misclassified as Class 4 (galaxies with bent tails), with 6 misclassifications. Upon reviewing the misclassified images, many of the FR II-type galaxies exhibited radio emission not only at the edges but also near the center.

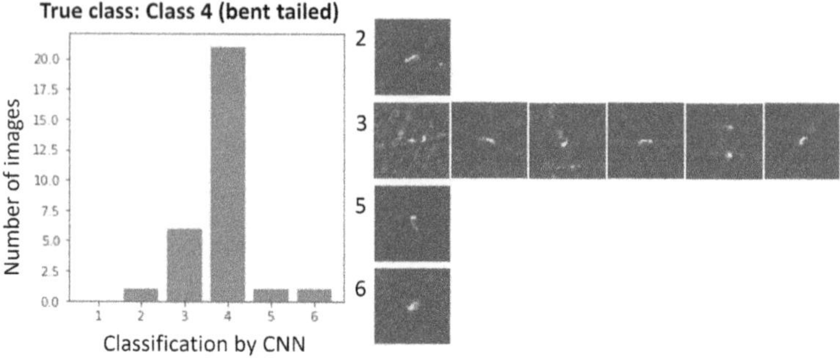

Figure 16.11   Same as Fig. 16.8, but for Class 4 (BT galaxies).

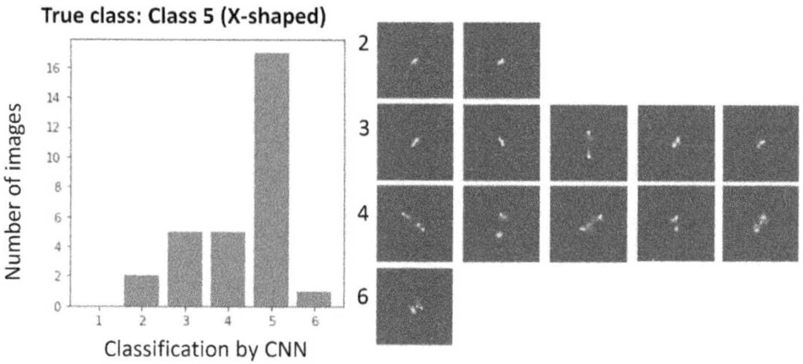

Figure 16.12   Same as Fig. 16.8, but for Class 5 (XRGs).

16.5.4   CLASS 4: BENT-TAILED (BT) GALAXIES

For Class 4, 21 out of 30 objects were correctly classified, while 9 were classified into other classes (Fig. 16.11). Class 4 galaxies were most frequently misclassified as Class 3 (FR II galaxies), with 6 misclassifications. Of the misclassified images, only one exhibited features consistent with Class 3, suggesting the CNN did not classify Class 4 galaxies very well.

16.5.5   CLASS 5: X-SHAPED RADIO GALAXIES (XRGS)

For Class 5, 17 out of 30 objects were correctly classified, while 13 were classified into other classes (Fig. 16.12). Class 5 galaxies were most frequently misclassified as Class 3 (FR II) or Class 4 (bent-tailed galaxies), with 5 misclassifications into each of these classes. Of the 13 misclassified images, only one clearly exhibited an X-shaped feature, while the rest appeared to resemble other classifications, such as bent tails or two radio sources. Thus, the classification by CNN was reasonably accurate.

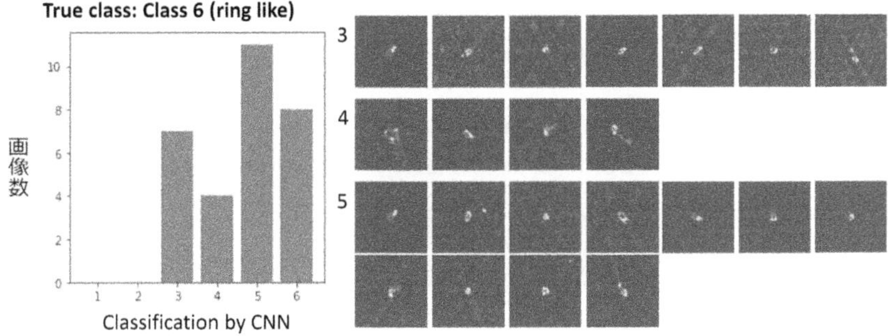

Figure 16.13 Same as Fig. 16.8, but for Class 6 (RRGs).

16.5.6 CLASS 6: RING-LIKE RADIO GALAXIES (RRGS)

For Class 6, only 8 out of 30 objects were correctly classified, while the remaining 22 were misclassified into other classes. The most frequent misclassification was into Class 5 (X-shaped), with 11 objects, surpassing the 8 correctly classified in Class 6 (Fig. 16.13).

Upon reviewing the misclassified images, it was found that the ring structures were collapsed, or the radio intensity in parts of the ring was low, making them appear less like rings, which led to their classification as Class 5. Additionally, the second most frequent misclassification was into Class 3 (FR II galaxies), with 7 objects, but these objects did not exhibit the bright edge characteristic of Class 3, and their appearance resembled the images misclassified into Class 5. Thus, the CNN did not work very well for these 7 objects.

16.5.7 CROSS VALIDATION

For further analysis, we applied cross-validation (CV) to evaluate the performance of the model. CV is a technique used to assess generalization performance while minimizing overfitting. In this chapter, we applied the commonly used K-fold CV. In K-fold CV, the data are divided into K parts, where one part is used for validation, and the remaining $K-1$ parts are used for training. This process is repeated K times, with each part serving once as validation data. The average accuracy across all runs is then calculated (Fig. 16.14). Since all data are used once as validation data, this method helps improve the generalization performance. In this study, we adopted $K = 5$.

The optimal model and its hyperparameters, obtained after 15 epochs and 50 trials using CV, are presented in Tables 16.9 and 16.10. The results of training with 15 epochs and a batch size of 256 are shown in Fig. 16.15. The CV resulted in a model with an accuracy of 0.82. When generalization was tested with the test data, an accuracy of 0.62 was achieved, an improvement over the previous accuracy of 0.56 before CV. We will now examine how CNN classified each class after applying CV.

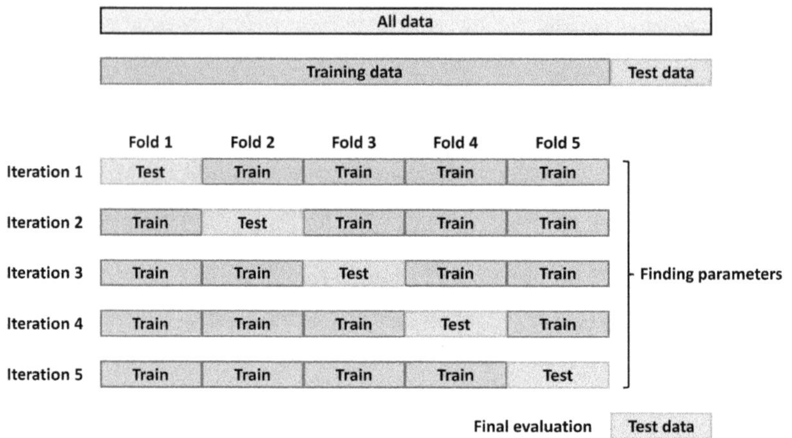

Figure 16.14 Concept of 5-fold cross-validation.

Table 16.9
Model structure after cross-validation.

Conv2d: filter $= 5$
Batch normalization
Conv2d: filter $= 20$
Batch normalization
Maxpooling
Conv2d: filter $= 10$
Batch normalization
Maxpooling
Conv2d: filter $= 20$
Batch normalization
Maxpooling
Conv2d: filter $= 15$
Batch normalization
Maxpooling
Flatten
Dense: units $= 80$, activation $=$ 'relu'
Dropout: dropout rate $= 0.301287740058092$
Batch normalization
Dense: units $= 6$, activation $=$ 'softmax'

Table 16.10

Optimal hyperparameters after cross-validation.

Hyperparameters	Optimal
Number of convolutional layers	5
Filters in convolutional layers	5, 20, 10, 20, 15
Number of units in fully connected layers	80
Dropout rate	0.30128774
Activation function	relu
Optimization function	Adam

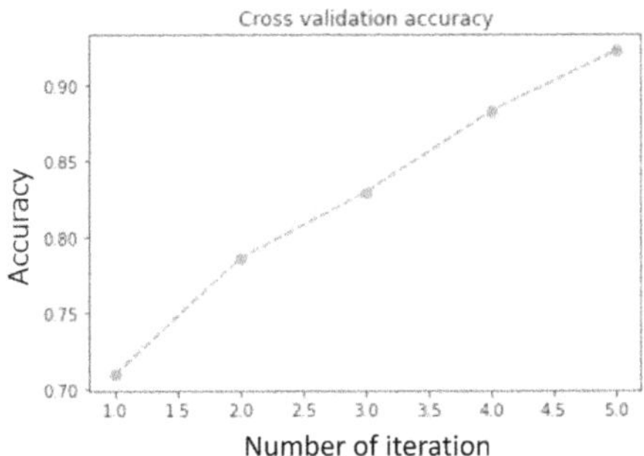

Figure 16.15   Accuracy improvement with the number of iterations during cross-validation.

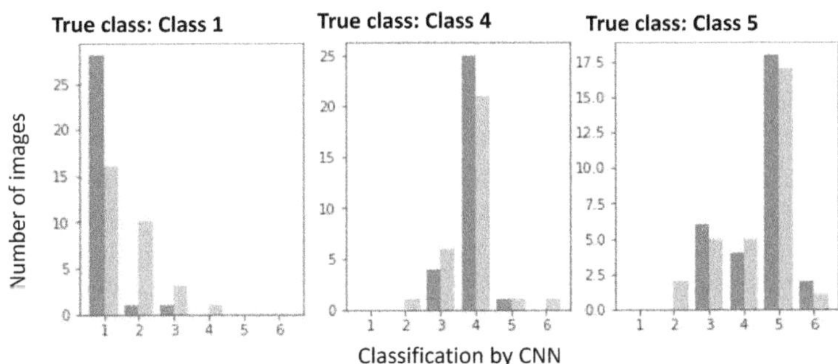

Figure 16.16   Classification results for classes 1, 4, and 5 after CV. Blue histograms represent the results after applying CV, and orange histograms show the results before CV.

For Class 1, 28 out of 30 objects were correctly classified; for Class 4, 25 out of 30; and for Class 5, 18 out of 30 were correctly classified. The accuracy improved for these three classes after CV (Fig. 16.16).

For Classes 1 and 4, images that should have been correctly classified were misclassified before CV. However, after CV, the results matched the expected

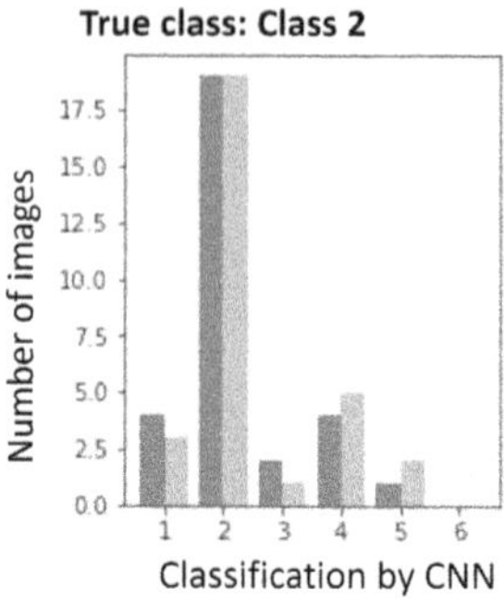

Figure 16.17   Same as Fig. 16.16 but for Class 2 (FR I galaxies).

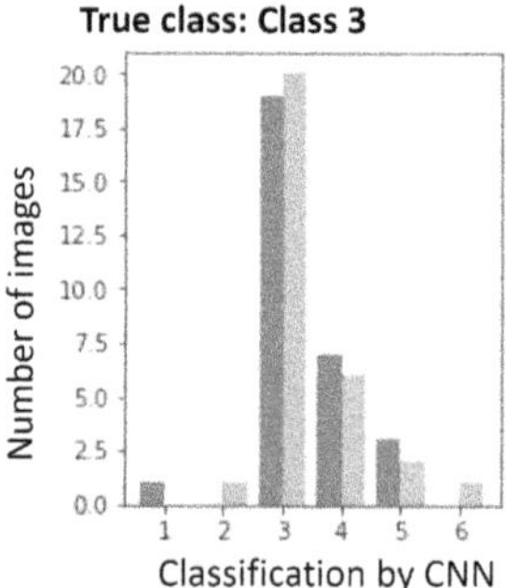

Figure 16.18   Same as Fig. 16.16 but for Class 3 (FR II galaxies).

classifications, showing improved accuracy. Both Class 1 and Class 4 achieved over 80% accuracy. For Class 5, one additional image was correctly classified, and misclassification into Class 2 was eliminated, though the accuracy remained at 60 %.

For Class 2, 18 out of 30 objects were correctly classified, while the remaining 12 were misclassified into other classes (Fig. 16.17). For Class 2, the number of correctly classified images did not change before and after CV. However, CV reduced misclassifications into classes 4 (bent tails) and 5 (XRG), while increasing misclassifications into classes 1 (compact) and 3 (FR II). Misclassifications into classes 1 and 4 were reasonable, as some images did not clearly show extended lobes or appeared bent, indicating that classification was based on image features. However, images classified into Class 3 did not exhibit the characteristic bright edges of Class 3.

For Class 3, 19 out of 30 objects were correctly classified, while the remaining 11 were misclassified into other classes (Fig. 16.18). After CV, one additional misclassification occurred. The most frequent misclassification was still into Class 4 (bent-tail galaxies). Upon reviewing the misclassified images, it was found that some galaxies were correctly classified after CV (Fig. 16.19).

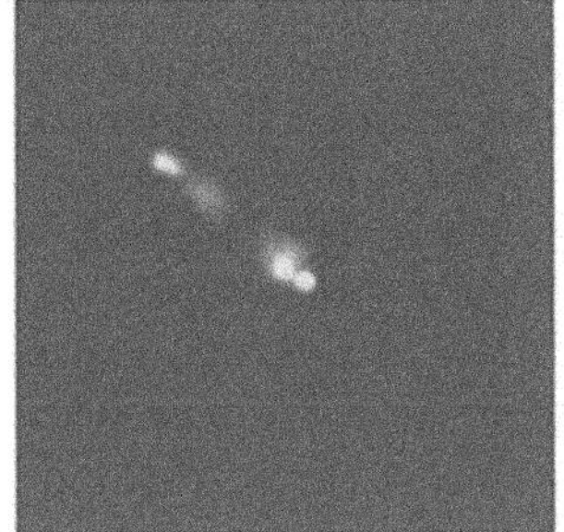

Figure 16.19 Examples of misclassified images that were correctly classified after CV.

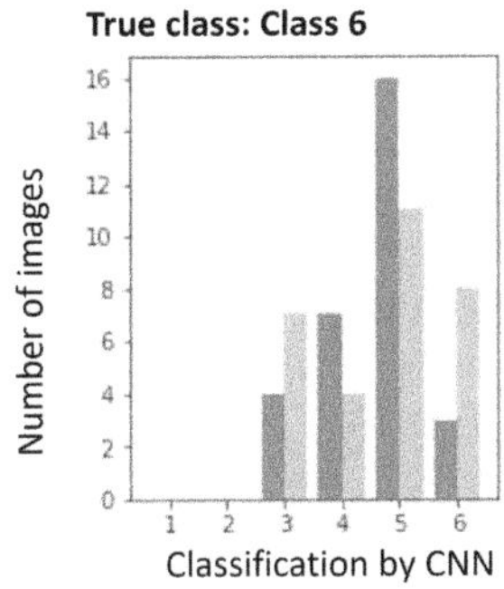

Figure 16.20 Same as Fig. 16.16 but for Class 6 (RRGs).

For Class 6, only 3 out of 30 objects were correctly classified, while the remaining 27 were misclassified into other classes. The number of misclassifications increased by 5 compared to before CV. The most frequent misclassification was into Class 5 (XRGs), with 16 objects (Fig. 16.20).

Upon reviewing the misclassified images, it was found that, as before, the ring structures were collapsed, or the radio intensity in part of the ring was too low to appear as a ring, leading to misclassification as Class 5. Before CV, Class 3 (FR II) was the second most common misclassification with 7 objects, but after CV, only 4 were misclassified into Class 3, showing a decrease in misclassification.

16.5.8 COMPARISON WITH PREVIOUS STUDIES

In light of the above results, we confirmed that classification accuracy improved with the application of CV. However, the results for Class 6 did not significantly improve. First, we note the small data size for Class 6. Among the 14,164 images,

Class 6 contains only 85 samples. Due to the limited data, it is possible that the CNN did not sufficiently learn the typical characteristics of RRGs. Second, the low radio emission in the ring regions may have hindered the appearance of ring-like structures, contributing to poor classification accuracy.

In a previous study by Ma et al. (2019), who used the same sample, 14,245 objects without labels were input into an autoencoder for pre-training, followed by learning on 1,442 objects, adjusting the parameters and assigning labels to the 14,245 objects. Unlike this work, their study reversed the data usage for learning and validation, and did not perform pre-training with an autoencoder, making this study a simpler model. After data augmentation, the ratio of Class 1 to Class 6 was 8:2:2:2:1, showing a greater imbalance compared to this study. Moreover, the previous study also used augmented images for learning and validation during pre-training and parameter adjustment.

The classification accuracy of Ma et al. (2019) for classes 5 and 6 (XRG and RRG) was below 50 % when compared with human-labeled data (46.15 % and 35.52 %, respectively). The previous study pointed out the small sample size and ambiguous morphology as reasons. Although human labeling may not always be certain, using these labels for training in this study may have resulted in insufficient accuracy for the test data. Despite issues with the training data, this study achieved high classification accuracy for Class 1 (compact galaxies) and Class 4 (bent-tail galaxies) with approximately 93.3 % and 83.3 %, respectively. These results are more accurate than the classification results of the previous study (88.86 % for Class 1 and 60.38 % for Class 4). Additionally, this work achieved 60 % accuracy for Class 5 (XRG), higher than the previous study.

16.6 SUMMARY AND PROSPECTS

We performed morphological classification using 26,792 radio galaxy images. The learning phase achieved an accuracy of 82 %, and the testing phase achieved an accuracy of 62 %. The model struggled with the classification of galaxies with ring structures (RRGs), likely due to unreliable labels in the training data. However, the model achieved over 80 % accuracy for compact galaxies and bent-tail galaxies, outperforming previous studies with a simpler model.

A review of the training data and classification labels is necessary. Comparing this work with other labeled data, such as those used in the Radio Galaxy Zoo (https://blog.galaxyzoo.org/category/radio-galaxy-zoo/) by Wu et al. (2019), will allow for more accurate labeling. Additionally, utilizing visualization tools like Grad-CAM can help identify which morphological features the CNN focuses on, enabling improvements in classification. Expanding the classification to include multi-wavelength data could help categorize galaxies in relation to their host galaxies and surrounding environments, which is an avenue for future research.

Section IV

Appendix

A Cosmological Basics

A.1 HOMOGENEOUS UNIVERSE

Galaxies and all the other structures in the Universe have formed from the homogeneous energy distribution in space, and co-evolved with the cosmic expansion. Namely, the evolution of the spacetime and structure in the Universe are intimately bound up with each other. The main tool to handle such a tight relation between spacetime and matter is the general relativity and cosmology based on it. We see that the cosmic expansion is regulated by the state of energy (matter and radiation) in the Universe.

A.1.1 FRIEDMANN-LEMAÎTRE-ROBERTSON-WALKER METRIC

In the differential geometry, a line element (or metric) is the basic tool to describe local properties of a manifold. General Relativity showed that the spacetime is well characterized by its geometrical structure, and the differential geometry is the fundamental framework to use. Friedmann-Lemaître-Robertson-Walker metric was invented to represent the (local) geometry of homogeneous and isotropic spacetime:

$$
\mathrm{d}s^2 = g_{\mu\nu}\mathrm{d}x^\mu\mathrm{d}x^\nu = -c^2\mathrm{d}t^2 + a^2(t)\left[\frac{\mathrm{d}r^2}{1-Kr^2} + r^2\left(\mathrm{d}\theta^2 + \sin^2\theta\mathrm{d}\phi^2\right)\right] \quad \text{(A.1)}
$$

where r is the comoving radial distance, θ and ϕ are angular components of the spherical coordinates, K is the Gaussian curvature, and $a(t)$ is the scale factor. We set $a_0 \equiv a(t_0) = 1$. Here, there are two standard ways for the definition of K.

1. One can normalize $K = \{+1, 0, -1\}$. In this case, $a(t)$ has a dimension of length, and r is dimensionless. When $K = \pm 1$, $a(t)$ represents the radius of curvature of the space. In this convention, the curvature is often denoted by a lowercase k rather than K, and the scale factor is denoted as $R(t)$ (see e.g., Weinberg, 1972).
2. Alternatively, we can define so that K has a dimension length^{-2}. In this case, r has a dimension of length, and $a(t)$ becomes dimensionless.

Of course, the two give the equivalent results. We adopt the latter definition in the following.

DOI: 10.1201/9781003104315-A

A.1.2   FRIEDMANN EQUATIONS

With the FLRW metric and Einstein's equation, we obtain the Friedmann equations as

$$\left(\frac{\dot{a}}{a}\right)^2 = \frac{8\pi G\rho}{3} - \frac{c^2 K}{a^2} + \frac{c^2\Lambda}{3} \tag{A.2}$$

$$\frac{\ddot{a}}{a} = -\frac{4\pi G}{3}\left(\rho + \frac{3p}{c^2}\right) + \frac{c^2\Lambda}{3}, \tag{A.3}$$

where ρ is the mass density of matter, p is the pressure, and Λ is the cosmological constant. Equation (A.2) comes from the 00 component of the Einstein equations and eq. (A.3) is obtained from the trace. By using eq. (A.2), eq. (A.3) is more simplified as

$$\dot{\rho}c^2 = -3\frac{\dot{a}}{a}\left(\rho c^2 + p\right) \tag{A.4}$$

which describes the conservation of energy in the Universe. The equation of state of ordinary matter is

$$p = nk_{\mathrm{B}}T \simeq \frac{\rho k_{\mathrm{B}}T}{\mu}, \tag{A.5}$$

where μ is the mean mass of particles. If the matter is nonrelativistic, $p \ll \rho c^2$, and we can approximate as

$$p = 0. \tag{A.6}$$

This leads to a solution for the evolution of matter density, as

$$\rho_{\mathrm{M}} = \rho_{\mathrm{M0}}\, a^{-3}, \tag{A.7}$$

when the matter dominates the cosmic density (K and $\Lambda = 0$).

Though the form of eqs. (A.2) and (A.3) is sufficiently general in most of the part of this monograph, for completeness, we also consider other contributors to the density, especially the energy density of radiation. In this case, ρ should be replaced with a general energy density ρ_{E} divided by c^2, and eqs. (A.2) and (A.3) become

$$\left(\frac{\dot{a}}{a}\right)^2 = \frac{8\pi G\rho_{\mathrm{E}}}{3c^2} - \frac{c^2 K}{a^2} + \frac{c^2\Lambda}{3}, \tag{A.8}$$

$$\frac{\ddot{a}}{a} = -\frac{4\pi G}{3c^2}\left(\rho_{\mathrm{E}} + 3p\right) + \frac{c^2\Lambda}{3}. \tag{A.9}$$

If we consider a case where $\rho_{\mathrm{E}} = \varepsilon_{\mathrm{R}}$ (energy density of radiation: subscript R stands for radiation), since the equation of state of radiation is

$$p = \frac{\varepsilon_{\mathrm{R}}}{3}, \tag{A.10}$$

we have a solution

$$\varepsilon_R = \varepsilon_{R0} a^{-4} . \tag{A.11}$$

Considering both matter and radiation, we have

$$\left(\frac{\dot{a}}{a}\right)^2 = \frac{8\pi G(\varepsilon_R + \rho_M c^2)}{3c^2} - \frac{c^2 K}{a^2} + \frac{c^2 \Lambda}{3} , \tag{A.12}$$

$$\frac{\ddot{a}}{a} = -\frac{4\pi G}{3c^2}\left(\varepsilon_R + \rho_M c^2 + 3p\right) + \frac{c^2 \Lambda}{3} . \tag{A.13}$$

A.1.3 COSMOLOGICAL PARAMETERS AND EVOLUTION OF THE UNIVERSE

We define the following quantities, referred to as the cosmological parameters.

1. Hubble parameter:

$$H(t) \equiv \frac{\dot{a}(t)}{a(t)} \tag{A.14}$$

2. Matter density parameter:

$$\Omega_M(t) \equiv \frac{\rho(t)}{\rho_c(t)} \equiv \frac{8\pi G \rho(t)}{3H^2(t)} , \tag{A.15}$$

   where $\rho_c = 1.88 \times 10^{-29} h^2 \, [\mathrm{g\,cm^{-3}}]$ in the present-day Universe ($h \equiv H_0/100$).
3. Radiation density parameter:

$$\Omega_R(t) \equiv \frac{\varepsilon_R(t)}{c^2 \rho_c(t)} \equiv \frac{8\pi G \varepsilon_R(t)}{3c^2 H^2(t)} \tag{A.16}$$

4. Dimensionless cosmological parameter:

$$\Omega_\Lambda(t) \equiv \frac{c^2 \Lambda}{3H^2(t)} \tag{A.17}$$

5. Curvature parameter:

$$\Omega_K(t) \equiv -\frac{c^2 K}{a(t)^2 H^2(t)} \tag{A.18}$$

By using eqs. (A.14)–(A.18), we can rewrite eq. (A.12) as

$$\Omega_R(t) + \Omega_M(t) + \Omega_\Lambda(t) + \Omega_K(t) = 1 \tag{A.19}$$

However, since the radiation density in the present-day Universe is small, this reduces to

$$\Omega_M(t) + \Omega_\Lambda(t) + \Omega_K(t) = 1 . \tag{A.20}$$

Evaluating eq. (A.19) at the present epoch t_0 gives a relation

$$\Omega_{M0} + \Omega_{\Lambda 0} + \Omega_{K0} = 1 . \tag{A.21}$$

The current estimates of the cosmological parameters from the *Planck* collaboration are as follows: the present-day Hubble parameter $H_0 = 67.4 \pm 0.5$ [km s^{-1}Mpc^{-1}], and $\Omega_{M0} = 0.315 \pm 0.007$ ($\Omega_{\mathrm{DM0}} = 0.264 \pm 0.002$ and $\Omega_{\mathrm{B}} = 0.0496 \pm 0.0002$), $\Omega_{\Lambda 0} = 0.685 \pm 0.007$, and Ω_K 0.001 ± 0.002 (Collaboration et al., 2020). These values strongly supports the flat, accelerating, cosmological constant (or dark energy) dominated Universe.

A.1.4 DOMINANT EPOCH OF EACH COMPONENT

As seen in eq. (A.3), the expansion of the Universe is governed by the energy density ρ_E. In general, the energy density consists of some components. If one of them dominates the total energy density, the expansion is controlled only by the dominant component, referred to as the dominant epoch of the component. Also, if the curvature term dominates, it is called the curvature dominant epoch. In terms of the density parameters, let

$$\Omega_{\mathrm{tot}} = \sum_{i=1}^{n} \Omega_i = 1 . \tag{A.22}$$

Then, during the dominant epoch of the jth component,

$$\Omega_j \simeq 1, \quad \Omega_{i \neq j} \simeq 0, \quad \Omega_K \simeq 0 . \tag{A.23}$$

In the present-day Universe, the fraction of the radiation component is quite small. However, from eq. (A.11), it will dominate the total cosmic energy density if we go back in time. Also, the cosmological constant does not change in time while the matter density goes down as eq. (A.7). This means that the Universe in the future will be dominated by the cosmological constant (or the dark energy in general).

Unless we consider the very early Universe, usually we can treat each density component independently, without taking into account the interaction. If we consider the density of radiation, matter, and the cosmological constant (as well as the curvature), eq. (A.12) leads

$$\left[\frac{H(t)}{H_0} \right]^2 = \frac{\Omega_{R0}}{a(t)^4} + \frac{\Omega_{M0}}{a(t)^3} + \frac{\Omega_{K0}}{a(t)^2} + \Omega_{\Lambda 0} . \tag{A.24}$$

A.1.4.1 Radiation-dominated (RD) epoch

The time when the energy densities of radiation and matter become equal is referred to as the time of matter–radiation equality, denoted as t_{eq}. At $a \leq a(t_{\mathrm{eq}})$, eq. (A.24) becomes

$$\left[\frac{H(t)}{H_0} \right]^2 = \frac{\Omega_{R0}}{a(t)^4} \tag{A.25}$$

which has a solution

$$a = \left(2H_0\sqrt{\Omega_{\mathrm{R}0}}\,t\right)^{\frac{1}{2}}, \tag{A.26}$$

Here we dropped the integration constant since it can be neglected after some time lapses. We adopt this approximation all in the following. We also have

$$H(t) \simeq \frac{1}{2t}, \tag{A.27}$$

$$\rho(t) \simeq \frac{\varepsilon_{\mathrm{R}}(t)}{c^2} \simeq \frac{3}{32\pi G t^2}. \tag{A.28}$$

A.1.4.2 Matter-Dominated (MD) Epoch

At $a > a(t_{\mathrm{eq}})$, similar to the RD epoch, we have

$$\left[\frac{H(t)}{H_0}\right]^2 = \frac{\Omega_{\mathrm{M}0}}{a(t)^3} \tag{A.29}$$

which leads

$$a = \left(\frac{3}{2}H_0\sqrt{\Omega_{\mathrm{M}0}}\,t\right)^{\frac{2}{3}}. \tag{A.30}$$

As in the RD epoch, we obtain

$$H(t) \simeq \frac{2}{3t}, \tag{A.31}$$

$$\rho(t) \simeq \rho_{\mathrm{M}}(t) \simeq \frac{1}{6\pi G t^2}. \tag{A.32}$$

A.1.4.3 Curvature-Dominated Epoch

If and only if $K < 0$, i.e., $\Omega_{K0} > 0$, at some time the curvature dominates, as

$$\left[\frac{H(t)}{H_0}\right]^2 = \frac{\Omega_{K0}}{a(t)^2} \tag{A.33}$$

and we have

$$a = H_0\sqrt{\Omega_{K0}}\,t, \tag{A.34}$$

$$H(t) \simeq \frac{1}{t}, \tag{A.35}$$

$$\rho(t) \simeq \frac{3}{8\pi G t^2}. \tag{A.36}$$

This type of the cosmological model is referred to as the Milne Universe, even though his original model did not include the idea of the cosmic expansion (Milne, 1935).

A.1.4.4 Cosmological Constant-Dominated Epoch

At large t, the cosmological constant dominates, but for this $\Omega_{\Lambda 0}$ is required. In parallel with other cases,

$$\left[\frac{H(t)}{H_0}\right]^2 = \Omega_{\Lambda 0} \tag{A.37}$$

which immediately gives

$$a \propto \exp\left(H_0\sqrt{\Omega_{\Lambda 0}}t\right), \tag{A.38}$$

$$H(t) \simeq \sqrt{\Omega_{\Lambda 0}}H_0, \tag{A.39}$$

$$\rho(t) \simeq \rho_{\Lambda 0} = \frac{3H_0^2\Omega_{\Lambda 0}}{8\pi G}. \tag{A.40}$$

The most striking characteristic of this epoch is the exponential expansion of the scale factor. This type of the Universe is called 'de Sitter Universe' (and we should note that this is *not* the same as the Einstein–de Sitter Universe). Here we restricted our discussion to the cosmological constant, but the accelerating Universe solution also appears for a general dark energy.

From these results, we can see that $\ddot{a} < 0$ for the RD and RD epochs, $\ddot{a} =$ const. for the curvature-dominated epoch, and $\ddot{a} > 0$ for the cosmological constant-dominated epoch. Substituting the observed cosmological parameters, we immediately find that the present-day Universe is cosmological constant (more generally dark energy)-dominated. Since the observation strongly supports that $K = 0$, we obtain

$$a_\Lambda \equiv \left(\frac{\Omega_{M0}}{\Omega_{\Lambda 0}}\right)^{\frac{1}{3}} \simeq 0.7 \tag{A.41}$$

by equating the second and third terms in eq. (A.24). At this epoch, corresponding to $t \simeq 9$ Gyr, the MD epoch ended and the Λ-dominated epoch began. Similarly,

$$a_{\rm eq} \equiv \left(\frac{\Omega_{R0}}{\Omega_{M0}}\right) \simeq 3 \times 10^{-4} \tag{A.42}$$

is the equality time at which the Universe switched from RD to MD.

A.2 COSMOLOGICAL REDSHIFT

The light emitted at a point $(r, \phi, \theta) = (r_{\rm e}, 0, 0)$ in comoving coordinate, in the past $t = t_{\rm em}$ comes to the observer located at the origin $(r, \phi, \theta) = (0, 0, 0)$ at $t = t_0$. Since the light travels along the null geodesic $ds^2 = 0$, one can relate the distance and the time from eq. (A.1) by

$$\frac{cdt}{a(t)} = \frac{dr}{\sqrt{1 - Kr^2}}. \tag{A.43}$$

Changing variables r to the coordinate separation, χ by

$$\chi \equiv \int_0^{r_{\text{em}}} \frac{dr}{\sqrt{1 - Kr^2}}, \tag{A.44}$$

and t to the conformal time, η by

$$\eta \equiv \int^t \frac{dt}{a(t)}, \tag{A.45}$$

the equation for the light is

$$c(\eta_0 - \eta) = \chi, \tag{A.46}$$

where η_0 is the time when the light reaches the observer. Suppose the light emitted at $\eta = \eta_0$ and $\eta = \eta_0 + \delta\eta_0$, reached the observer at η_1 and $\eta_1 + \delta\eta_1$ respectively. Since the right-hand side of eq. (A.46) is independent of η, one can obtain

$$\delta\eta_0 = \delta\eta_1 \tag{A.47}$$

namely

$$\frac{\delta t_0}{a(t_0)} = \frac{\delta t_1}{a(t_1)} \, . \tag{A.48}$$

The conservation of the phases of the light waves within the time interval δt_0 and δt_1 leads to the relation

$$a(t_0)\nu_0 = a(t_1)\nu_1 \iff \frac{\lambda_0}{a(t_0)} = \frac{\lambda_1}{a(t_1)}. \tag{A.49}$$

Since $a(t_0) > a(t_1)$ in the expanding universe, the observed wavelength is longer than the emitted one, $\lambda_0 > \lambda_1$ (redshited). The redshift, z, is defined as

$$z \equiv \frac{\lambda_0 - \lambda_1}{\lambda_1} = \frac{a(t_0)}{a(t_1)} - 1. \tag{A.50}$$

Since we define the normalization of the scale factor, $a(t)$, as $a(t_0) = 1$, the relation between $a(t)$ and z is as

$$z = \frac{1}{a} - 1 \, , \tag{A.51}$$

or

$$a(t) = \frac{1}{1 + z} \, . \tag{A.52}$$

A.2.1 COSMOLOGICAL QUANTITIES AS A FUNCTION OF REDSHIFT

We derive the redshift dependence of some cosmologically important quantities. By using eqs. (A.24) and (A.52), the cosmic time–redshift relation is obtained as

$$\frac{\mathrm{d}z}{\mathrm{d}t} = -(1+z)H$$

$$= -H_0\,(1+z)\,\sqrt{\Omega_{\mathrm{R}0}(1+z)^4 + \Omega_{\mathrm{M}0}(1+z)^3 + \Omega_{K0}(1+z)^2 + \Omega_{\Lambda 0}}\,. \qquad \text{(A.53)}$$

Thus, the cosmic time at redshift z is

$$t = \int_0^t \mathrm{d}t' = \int_\infty^z \frac{\mathrm{d}t}{\mathrm{d}z'}\mathrm{d}z'$$

$$= \int_z^\infty \frac{\mathrm{d}z'}{(1+z')H(z')}\,. \qquad \text{(A.54)}$$

This directly leads to the cosmic age

$$t_0 = \int_0^\infty \frac{\mathrm{d}z'}{(1+z')H(z')}\,. \qquad \text{(A.55)}$$

As for the distance, we should note that there are many definitions of distance in cosmology. Here we introduce the formula for the comoving distance which is not observable. We derive observable distances in Sec. A.2.2. The Hubble–Lemaître law is written as

$$d = \frac{cz}{H_0} \qquad \text{(A.56)}$$

at low-redshift Universe. In general, the comoving distance $\chi(z)$ corresponding to a redshift z is expressed as

$$\chi(z) = \int_t^{t_0} \frac{c\,\mathrm{d}t'}{a(t')} = \int_0^z \frac{c\,\mathrm{d}z'}{H(z')}\,. \qquad \text{(A.57)}$$

To relate χ to a radial coordinate distance r, we use the FLRW metric eq. (A.1) with $\mathrm{d}\theta = \mathrm{d}\phi = 0$, and consider a photon traveling from an emitter to an observer. Since in this case $\mathrm{d}s^2 = 0$,

$$c\,\mathrm{d}t = a(t)\frac{\mathrm{d}r}{\sqrt{1-Kr^2}}\,. \qquad \text{(A.58)}$$

Combining eqs. (A.57) and (A.58), we have

$$\mathrm{d}\chi = \frac{\mathrm{d}r}{\sqrt{1-Kr^2}}\,. \qquad \text{(A.59)}$$

Then, by integrating it, we have

$$r = S_K(\chi) \equiv \begin{cases} \dfrac{\sinh(\sqrt{-K}\chi)}{\sqrt{-K}} & (K < 0) \\[2ex] \chi & (K = 0) \\[2ex] \dfrac{\sin(\sqrt{K}\chi)}{\sqrt{K}} & (K > 0) \end{cases}\,. \qquad \text{(A.60)}$$

A.2.2 OBSERVABLE PROPERTIES

A.2.2.1 Luminosity Distance

The definition of distance turns out to be nontrivial in cosmology. Since we cannot observe the very distant Universe at the same time with an observer, several definitions of observable distance are possible. More rigorously, the coordinate separation of two spacetime positions is not an observable value because they do not lie on the synchronous section of the spacetime but on the light-cone hypersurface.

If a source has an intrinsic luminosity L, it will emit energy $L\mathrm{d}t$ in a time interval $\mathrm{d}t$. This quantity is called the bolometric luminosity, integrated over the whole range of frequency or wavelength. This energy which will be received by the observer at position 0 in a time interval

$$\mathrm{d}t_0 = \frac{\mathrm{d}t}{a(t)} \tag{A.61}$$

(eq. (A.49)) would have undergone a redshift by a factor of $1/a(t)$ and will be distributed over a sphere of radius $4\pi r^2$, where r is the coordinate distance of the emitting object. The observed flux S will be

$$S = \left(\frac{\mathrm{d}t}{\mathrm{d}t_0}\right)\frac{a(t)L}{4\pi r^2} = \frac{a(t)^2 L}{4\pi r^2}. \tag{A.62}$$

Considering this relation, we naturally define the luminosity distance $d_{\mathrm{L}}(z)$ as

$$d_{\mathrm{L}}(z) \equiv \sqrt{\frac{L}{4\pi S}} = (1+z)r = (1+z)S_K(\chi)$$

$$= (1+z)S_K\left(\int_0^z \frac{c\mathrm{d}z'}{H(z')}\right). \tag{A.63}$$

In practice, taking into account the measured cosmological parameters (Section A.1.3), a flat Universe ($K = 0$) is safely assumed.

A.2.2.2 Angular Diameter Distance

In the static Euclidean space, the angular diameter of an object is inversely proportional to the distance. Then, if we know the physical size of the object l, we can determine the distance to the object by measuring its angular size θ as

$$l = r\theta \tag{A.64}$$

where $\Delta\theta \ll 1$ is assumed. In cosmology, it can be generalized as

$$l = \int_0^\theta ar\,\mathrm{d}\theta = ar\theta = \frac{r\theta}{1+z} \tag{A.65}$$

where the coordinates of the both sides of the object is taken as $(r,\theta,0)$ and $(r,0,0)$. The angular diameter distance d_A is defined as

$$d_A(z) \equiv \frac{l}{\theta} = \frac{r}{1+z}$$
$$= \frac{1}{1+z} S_K \left(\int_0^z \frac{c\,dz'}{H(z')} \right). \tag{A.66}$$

Actually d_A and d_L are simply related as

$$d_A(z) = \frac{d_L(z)}{(1+z)^2}. \tag{A.67}$$

The angular diameter distance between two positions at different redshifts is required in the discussion on gravitational lensing. Consider z_1 and z_2 ($z_2 > z_1$), then

$$d_A(z_2,z_1) = a(t_2)r_{21}$$
$$= \frac{S_K\left[\chi(z_2) - \chi(z_1)\right]}{1+z_2}. \tag{A.68}$$

We note that $d_A(z_2,z_1) \neq d_A(z_2) - d_A(z_1)$.

A.2.2.3 Comoving Volume Element

We often need to calculate an integral of a physical quantity over a cosmological comoving volume. We derive the comoving volume element per unit solid angle and unit redshift, as

$$\frac{d^2V}{d\Omega dz} = (1+z)^3 d_A(z)^2 \left|\frac{c\,dt}{dz}\right| = \frac{c}{H(z)}(1+z)^2 d_A(z)^2$$
$$= \frac{c}{H(z)} \frac{d_L(z)^2}{(1+z)^2}. \tag{A.69}$$

This is used in, e.g., galaxy number counts and cosmic radiation background.

A.2.2.4 Hubble Time and Radius

An important observable cosmological scale is the Hubble time and Hubble radius. Hubble time is defined as

$$t_{\text{Hubble}} \equiv \frac{a(t)}{\dot{a}(t)} = H(t)^{-1}. \tag{A.70}$$

If we neglect the curvature of the Universe,

$$t_{\text{Hubble}} \simeq \sqrt{\frac{3}{8\pi G\rho(t)}}, \tag{A.71}$$

namely, it can be regarded as the "dynamical timescale" of the Universe.

It also defines a rough measure of the size of the observable Universe,

$$r_{\text{Hubble}} \equiv ct_{\text{Hubble}} = \frac{c}{H(t)} . \tag{A.72}$$

A.2.2.5 Particle Horizon

Now we give a more rigorous definition of the horizon size. In cosmological discussion, the range within which particles can be causally connected is important. Let L_H be the comoving distance up to which a photon departed from the origin at the birth of the Universe ($t = 0$) and can travel before a time t. This is the particle horizon. The physical size of the particle horizon l_H is given by

$$l_H = a(t)L_H(t) = a(t) \int_0^t \frac{c\,dt'}{a(t')} \tag{A.73}$$

$$= \frac{1}{1+z} \int_z^\infty \frac{c\,dz'}{H(z')} . \tag{A.74}$$

The present-day horizon size l_{H0} is then

$$l_{H0} = \int_0^\infty \frac{c\,dz'}{H(z')} . \tag{A.75}$$

We evaluate l_H for the dominant epochs of the density components.

1. Radiation-dominated (RD) epoch
   From eq. (A.25),

$$a^2 H \simeq H_0 \sqrt{\Omega_{R0}} . \tag{A.76}$$

   Thus,

$$l_H \simeq \frac{ca^2}{H_0 \sqrt{\Omega_{R0}}} = 2ct . \tag{A.77}$$

2. Matter-dominated (MD) epoch
   Equation (A.29) yields

$$a^2 H \simeq H_0 \sqrt{\Omega_{M0}}\, a^{\frac{1}{2}} . \tag{A.78}$$

   We then obtain

$$l_H \simeq \frac{ca^{\frac{3}{2}}}{H_0 \sqrt{\Omega_{M0}}} = 3ct . \tag{A.79}$$

3. Cosmological constant (Λ)-dominated epoch
   Equation (A.37) leads

$$a^2 H \simeq H_0 \sqrt{\Omega_{\Lambda 0}}\, a^2 . \tag{A.80}$$

In this case,

$$l_{\rm H} \simeq \frac{c}{H_0\sqrt{\Omega_{\Lambda 0}}} = \text{const.} .$$ (A.81)

For the radiation and matter-dominated epochs, the horizon size increases, and matter that newly entered from outside the horizon starts to interact with that already existed within the horizon. In contrast, for the cosmological constant-dominated epoch, the *physical* horizon size is nearly constant. Since the cosmic expansion is exponential in this epoch, this means that the matter on the horizon goes out of it very rapidly. Thus, the horizon size plays an important role in the structure formation in the Universe (Chapter 7).

A.2.2.6 Monochromatic Flux Densities

In spite of the advent of a new era of multimessenger astronomy, still astronomical observations largely depend on electromagnetic waves at a certain range of wavelength (or frequency). Since a simultaneous observation at the whole range of wavelengths is unrealistic, we should redefine the formulae for cosmological observables. The flux observed at a single frequency ν is referred to as the flux density, which we denote as S_ν. The flux density at a single wavelength S_λ is similarly defined. Unlike eq. (A.62), we have to take into account the redshift of frequency or wavelength, as

$$S_{\nu_{\rm obs}} {\rm d}\nu_{\rm obs} = \frac{\mathscr{L}_{\nu_{\rm em}} {\rm d}\nu_{\rm em}}{4\pi d_{\rm L}(z)^2} = \frac{\mathscr{L}_{(1+z)\nu_{\rm obs}} {\rm d}(1+z)\nu_{\rm obs}}{4\pi d_{\rm L}(z)^2}$$
$$= \frac{(1+z)\mathscr{L}_{(1+z)\nu_{\rm obs}} {\rm d}\nu_{\rm obs}}{4\pi d_{\rm L}(z)^2}$$ (A.82)

therefore

$$S_{\nu_{\rm obs}} = \frac{(1+z)\mathscr{L}_{(1+z)\nu_{\rm obs}}}{4\pi d_{\rm L}(z)^2} .$$ (A.83)

For S_λ, we obtain

$$S_{\lambda_{\rm obs}} = \frac{\mathscr{L}_{\frac{\lambda_{\rm obs}}{(1+z)}}}{4\pi d_{\rm L}(z)^2(1+z)} .$$ (A.84)

A.2.2.7 *K*-Correction

We consider an astronomical object located at a redshift z, observed at the present-day Universe $z = 0$. We define $\mathscr{L}_\lambda$ as the monochromatic luminosity measured at wavelength λ in its rest frame, in units of $[{\rm erg\,s^{-1}\mathring{A}^{-1}}]$ (or any equivalent). We also define the observed band flux in a filter band with effective wavelength λ_0, $S_{[\lambda_0]}$, in units of $[{\rm erg\,s^{-1}cm^{-2}}]$,

$$S^{[\lambda_0]} \equiv \int S_\lambda R_\lambda^{[\lambda_0]} {\rm d}\lambda$$ (A.85)

where S_λ is the monochromatic flux density of the object $[\mathrm{erg\,s^{-1}cm^{-2}\,\mathring{A}^{-1}}]$ and $R_\lambda^{[\lambda_0]}$ is the system response function. The superscript $[\lambda_0]$ indicates its effective wavelength. Since the monochromatic flux density is related to the monochromatic luminosity by eq. (A.84), we have

$$
\begin{aligned}
S^{[\lambda_0]} &= \frac{1}{4\pi d_{\mathrm{L}}(z)^2(1+z)}\int \mathscr{L}_{\frac{\lambda}{(1+z)}}R_\lambda^{[\lambda_0]}\,\mathrm{d}\lambda \\
&= \frac{\int \mathscr{L}_\lambda R_\lambda^{[\lambda_0]}\,\mathrm{d}\lambda}{4\pi d_{\mathrm{L}}(z)^2(1+z)}\frac{\int \mathscr{L}_{\frac{\lambda}{(1+z)}}R_\lambda^{[\lambda_0]}\,\mathrm{d}\lambda}{\int \mathscr{L}_\lambda R_\lambda^{[\lambda_0]}\,\mathrm{d}\lambda}\,.
\end{aligned}
\tag{A.86}
$$

Here, the term

$$
\frac{\int \mathscr{L}_\lambda R_\lambda^{[\lambda_0]}\,\mathrm{d}\lambda}{4\pi d_{\mathrm{L}}(z)^2(1+z)} \equiv \tilde{S}^{[\lambda_0]}
\tag{A.87}
$$

is a hypothetical band flux which should have been obtained without cosmological redshift. Then, the rest of eq. (A.86) is the change of flux by observing a different wavelength regime of the spectrum of the object due to redshift.

In ultraviolet (UV), optical, and near-infrared (NIR) astronomy, magnitude has been traditionally used as a measure of the flux of an object. If we denote the apparent and absolute magnitudes of the object as m_{λ_0} and M_{λ_0}, and measure $d_{\mathrm{L}}(z)$ in [Mpc], we get

$$
\begin{aligned}
m_{\lambda_0} - M_{\lambda_0} &= -2.5\log S^{[\lambda_0]} + 2.5\log\left[\frac{4\pi d_{\mathrm{L}}(z)^2(1+z)}{4\pi(10\,[\mathrm{pc}])^2}\tilde{S}^{[\lambda_0]}\right] \\
&= 2.5\log\left[d_{\mathrm{L}}(z)^2(1+z)\right] - 2.5\log\left[\frac{\int \mathscr{L}_{\frac{\lambda}{(1+z)}}R_\lambda^{[\lambda_0]}\,\mathrm{d}\lambda}{\int \mathscr{L}_\lambda R_\lambda^{[\lambda_0]}\,\mathrm{d}\lambda}\right] + 25 \\
&= 5\log d_{\mathrm{L}}(z) + 2.5\log(1+z) - 2.5\log\left[\frac{\int \mathscr{L}_{\frac{\lambda}{(1+z)}}R_\lambda^{[\lambda_0]}\,\mathrm{d}\lambda}{\int \mathscr{L}_\lambda R_\lambda^{[\lambda_0]}\,\mathrm{d}\lambda}\right] + 25\,.
\end{aligned}
\tag{A.88}
$$

The second and third terms of eq. (A.88) are the K-correction (Oke & Sandage, 1968), and numerical factor 25 comes from the unit conversion between Mpc and pc (Peebles, 1993). The second corresponds to the effect of the stretch of bandwidth, and third term is the change of band-limited flux.

If the object is a galaxy, we should take into account the effect of the spectral evolution with time. This can be also included in a similar way, and is called the evolutionary (E-)correction. Examples of the actual K and E-correction can be found in e.g., Poggianti (1997). The evolution of galaxy spectra will be discussed in Chapter 4.

B Supplementary Information on Mathematics and Machine Learning

B.1 ALGEBRAIC PRELIMINARIES

We introduce basic terminologies to understand homotopy in algebra.

Definition A1 (Equivalence relation).
A given binary relation $\sim$ on a set X is an equivalence relation if and only if it is reflexive, symmetric and transitive. That is, for all a, b and $c \in X$,

1. *Reflexivity: $a \sim a$,*
2. *Symmetry: $a \sim b$ if and only if $b \sim a$,*
3. *Transitivity: if $a \sim b$ and $b \sim c$ then $a \sim c$.*

Definition A2 (Contractible).
When a set X is homotopy equivalent to a one-point space, X is contractible.

Definition A3 (Direct sum). *Let G_1, $G_2, \cdots, G_m$ be groups and let $g_1 \in G_1$, $g_2 \in G_2, \cdots, g_m \in G_m$ be elements from each group. The direct product G for these groups is defined as*

$$G \equiv G_1 \times G_2 \times \cdots \times G_m = \{(g_1, g_2, \cdots, g_m) \mid g_i \in G_i, i = 1, 2, \cdots, m\}. \quad \text{(B.1)}$$

Especially when G_1, $G_2, \cdots, G_m$ are additive, the direct product G would be an additive group as well.

$$(g_1, g_2, \cdots, g_m) + (g'_1, g'_2, \cdots, g'_m) \equiv (g_1 + g'_1, g_2 + g'_2, \cdots, g_m + g'_m) \quad \text{(B.2)}$$

In addition to this, the additive identity is $(0, 0, \cdots, 0)$ and $(-g_1, -g_2, \cdots, -g_m)$ is the inverse element for G. In this case, G is expressed as $G = G_1 \oplus G_2 \oplus \cdots \oplus G_m = \bigoplus_{i=1}^m G_i$ and is the definition of direct sum.

Definition A4 (Homotopy).
For continuous maps $f, g : X \to Y$, a continuous map F from a product space of a closed interval $I = [0, 1]$ and X to Y satisfying

$$F_{X \times \{0\}} = f, \quad F_{X \times \{1\}} = g \quad \text{(B.3)}$$

exists, f and g are said to be homotopic, denoted as $f \simeq g$. Here

$$F_{X \times \{t\}}(x) \equiv F(x, t)$$

F is referred to as the homotopy from f to g.

DOI: 10.1201/9781003104315-B

It follows that two homotopic maps f and g can be transformed continuously to each other by changing t via F. Then the binary relation $\simeq$ is an equivalence relation (see Definition A1).

Definition A5 (Homotopy equivalence). *A continuous map $f : X \to Y$ is a homotopy equivalence if there exist a continuous map $g : Y \to X$ such that*

$$g \circ f \simeq 1_X \text{ and } f \circ g \simeq 1_Y, \tag{B.4}$$

where 1_X and 1_Y are identity maps on X and Y, respectively. If a homotopy equivalence exits, X and Y are homotopy equivalent, denoted as $X \simeq Y$.

Intuitively, it means that X and Y are homotopy equivalent if they can be transformed into each other by bending, shrinking and expanding. The homotopy equivalence is fundamentally important because many concepts in algebraic topology are homotopy invariant, that is, they respect the relation of homotopy equivalence. Particularly in the TDA, the homology group is homotopy invariant. The homology group of a simplicial complex is compatible with computers and easy to calculate. Then, even if a homology group of a certain shape is difficult to calculate directly, we can calculate it by constructing the homotopy equivalent simplicial complex. This guarantees all the analysis presented in Chapter 15.

B.2 FUNDAMENTALS OF MANIFOLDS

Here we briefly introduce concepts related to manifolds that are used in the main text.

B.2.1 TOPOLOGICAL SPACE

Definition B6 (Topological space).
Given a set $\mathcal{X}$, let $\mathcal{O}$ be a subset of the power set $\mathfrak{P}(\mathcal{X})$ of $\mathcal{X}$. The family of subsets $\mathcal{O}$ of $\mathcal{X}$ is said to be a topology on $\mathcal{X}$ if it satisfies the following conditions 1–3:

1. *$\mathcal{X} \in \mathcal{O}$ and $\emptyset \in \mathcal{O}$,*

2. *For $m \in \mathbb{N}$, if $O_1, \ldots, O_m \in \mathcal{O} \Rightarrow \bigcap_{i=1}^{m} O_i \in \mathcal{O}$,*

3. *For any set Ξ, if for each element $\xi \in \Xi$, $O_\xi \in \mathcal{O} \Rightarrow \bigcup_{\xi \in \Xi} O_\xi \in \mathcal{O}$.*

When a topology $\mathcal{O}$ is given on a set $\mathcal{X}$, we say that $\mathcal{O}$ defines a topological structure on $\mathcal{X}$, and the elements of $\mathcal{O}$ are called open sets. A set $\mathcal{X}$ endowed with such a topological structure is called a topological space.

Here, we also define the important concept of embedding used in manifold learning.

Definition B7 (Embedding).
Consider topological spaces $\mathcal{X}$ and $\mathcal{Y}$. When $\mathcal{X}$ is homeomorphic (topologically equivalent) to a subspace of $\mathcal{Y}$, we say that $\mathcal{X}$ is embedded in $\mathcal{Y}$.

A topological space $\mathscr{X}$ is said to be locally Euclidean if there exists an integer $d \geq 0$ such that every point in $\mathscr{X}$ has a neighborhood homeomorphic to an open set in Euclidean space $\mathbb{R}^d$. A topological space $\mathscr{X}$ is called a Hausdorff space if any two distinct points have disjoint neighborhoods. A topological space $\mathscr{X}$ is said to be second-countable if there exists a countable collection of open sets $\mathscr{U} = \{U_i\}_{i=1}^{\infty}$ such that any open set in $\mathscr{O}$ can be expressed as a union of some subfamily of $\mathscr{U}$.

B.2.2 RIEMANNIAN MANIFOLD

Definition B8 (Topological manifold).
A topological space $\mathscr{M}$ is called a topological manifold of dimension d if $\mathscr{M}$ is a second-countable Hausdorff space and is locally Euclidean of dimension d. It is denoted as $\mathscr{M}^d$.

This definition suggests that any d-dimensional topological manifold $\mathscr{M}^d$ can be embedded in $\mathbb{R}^{2d+1}$ (Whitney's Embedding Theorem: Whitney, 1936). The next step is to define calculus on manifolds.

Definition B9 (Differentiable manifold).
A topological manifold $\mathscr{M}$ is called a differentiable manifold if it is continuously differentiable.

Consider two open sets $U \in \mathbb{R}^r$ and $V \in \mathbb{R}^s$, and let $g : U \longrightarrow V$ be a map such that $g(x) = y$ for $x \in U$ and $y \in V$. If the map g has first-order partial derivatives $\partial y_j / \partial x_i$ ($i = 1, \ldots, r$, $j = 1, \ldots, s$), then g is called a differentiable map on U. If all first-order partial derivatives are continuous, then g is called a $\mathscr{C}^1$-class function on U. If the map g is a homeomorphism from an open set in U to an open set in V, and if both g and g^{-1} are $\mathscr{C}^r$-class functions, then g is called a $\mathscr{C}^r$-diffeomorphism. If it is a $\mathscr{C}^{\infty}$-class diffeomorphism, it is simply called a diffeomorphism.

Let us consider a point $p \in \mathscr{M}$ on a manifold $\mathscr{M}$. The set of all vectors tangent to the manifold at the point p is called the tangent space at p, denoted by $T_p(\mathscr{M})$. The tangent space has the same dimension as $\mathscr{M}$. The tangent space $T_p(\mathscr{M})$ at the point p has an inner product $g_p = \langle \, \cdot \, \rangle : T_p(\mathscr{M}) \times T_p(\mathscr{M}) \longrightarrow \mathbb{R}$. The inner product satisfies the following properties for $x, y, z \in T_p(\mathscr{M})$:

Bilinearity For $a, b \in \mathbb{R}$, $g_p(ax + by, z) = a g_p(x, z) + b g_p(y, z)$,

Symmetry $g_p(x, y) = g_p(y, x)$,

Positive Definiteness $g_p(x, y) \geq 0$. If $x = 0$, and only then, $g_p(x, x) = 0$.

With these preparations, a Riemannian manifold is defined as follows.

Definition B10 (Riemannian manifold).
The set of inner products $g = \{g_p : p \in \mathscr{M}\}$ on a topological manifold $\mathscr{M}$ is called a Riemannian metric on $\mathscr{M}$, and the pair $(\mathscr{M}, g)$ is defined as a Riemannian manifold.

A curve on a Riemannian manifold $\mathcal{M}$ is defined as a smooth map from an open interval Ξ of $\mathbb{R}$ to $\mathcal{M}$. The point $\xi \in \Xi$ serves as the parameter of the curve. Let $c(\xi) \equiv (c_1(xi),\ldots,c_d(\xi))^\top$ be a curve in $\mathbb{R}^d$ parameterized by $\xi \in \Xi \subseteq \mathbb{R}$. If the coordinate functions $\{c_k(\xi)\}$ $(k = 1,\ldots,d)$ are differentiable to any order, then the curve c is called smooth. The velocity vector (or tangent vector) at the point ξ is defined as

$$c'(\xi) = \left(c_1'(\xi),\ldots,c_d'(\xi)\right)^\top \equiv \left(\frac{\mathrm{d}c_1(\xi)}{\mathrm{d}\xi},\ldots,\frac{\mathrm{d}c_d(\xi)}{\mathrm{d}\xi}\right)^\top. \tag{B.5}$$

The "speed" along the curve is defined as

$$\|c'(\xi)\| \equiv \left(\sum_{k=1}^{d} \left|c_k'(\xi)\right|^2\right)^{\frac{1}{2}}. \tag{B.6}$$

In general, the arc length $\ell(c)$ of a curve $c(\xi)$ from a point ξ_0 to a point ξ_1 is defined as

$$\ell(c) \equiv \int_{\xi_0}^{\xi_1} \|c'(\xi)\| \, \mathrm{d}\xi. \tag{B.7}$$

Now, let $C(p,q)$ be the set of all curves on $\mathcal{M}$ that connect p and q.

Definition B11 (Geodesic distance).
The distance $d^{\mathcal{M}}$ between points p and q on a Riemannian manifold $\mathcal{M}$ is defined as follows:

$$d^{\mathcal{M}}(p,q) \equiv \inf_{c\in C(p,q)} \ell(c). \tag{B.8}$$

$d^{\mathcal{M}}(p,q)$ *represents the shortest distance between two points (p,q) on $\mathcal{M}$.*

B.3 METHODS FOR DIMENSIONALITY REDUCTION AND MANIFOLD LEARNING

B.3.1 MULTIDIMENSIONAL SCALING (MDS)

Multidimensional scaling (MDS) is a method used to represent (visualize) individual similarities when there is a dataset, and distances between the data can be defined in some way. MDS arranges the data in Euclidean space using information about the distances between pairs of data when there are N data points. Specifically, given a distance matrix that includes the distances between each pair of data points in the dataset and a dimension d, MDS arranges the N points in a d-dimensional space in such a way that the distances between pairs of points are preserved as much as possible. In general, the dimension d is chosen to be lower than the original data dimension, so MDS is considered a dimensionality reduction method.

There are several algorithms for MDS, but the one used in the discussion in this paper is the classical MDS. Classical MDS is also known as Principal Co-ordinate Analysis (PCoA), Torgerson Scaling, or Torgerson–Gower scaling. When the number of data points is N, the distance matrix D has elements d_{ij} defined as the distance between the ith and jth data points. Due to the symmetry and non-degeneracy of the distances, D is a symmetric matrix of size $(N \times N)$ with diagonal elements equal to 0. Next, consider the matrix $S = ((d_{ij})^2)$, whose elements are the squared distances between the data points. MDS outputs the d-dimensional coordinates $(u_1, \ldots, u_N)$, $u_i \in \mathbb{R}^d$ of each classified element. This output can be expressed as a matrix $U \in \mathbb{R}^{d \times N}$, with each row corresponding to u_i. The matrix U is obtained as follows:

1. Consider the kernel matrix K such that

$$K = U^\top U \in \mathbb{R}^{N \times N} \tag{B.9}$$

   is satisfied.
2. Obtain the kernel matrix K from the distance matrix D using the relation

$$K = -\frac{1}{2} HSH, \tag{B.10}$$

   where

$$H \equiv \mathbb{I} - \frac{1}{N} \mathbb{1} \tag{B.11}$$

   ($\mathbb{I}$: identity matrix, $\mathbb{1}$: matrix with all elements equal to 1) is called the centering matrix, which centers the data. This operation is called double centering.
3. Decompose the kernel matrix K by eigenvalue decomposition as

$$K = V \Lambda V^\top, \tag{B.12}$$

   where V is the matrix of eigenvectors $\vec{v}_1, \ldots, \vec{v}_N$ of K, and Λ is the diagonal matrix with eigenvalues $\lambda_1, \ldots, \lambda_N$. Using these, we obtain

$$U = \Lambda^{\frac{1}{2}} V^\top. \tag{B.13}$$

4. The matrix U is of size $\mathbb{R}^{N \times N}$ by default. To reduce the dimension to d, select the first d largest eigenvalues $\lambda_1, \ldots, \lambda_d$ of K and the corresponding eigenvectors $\vec{v}_1, \ldots, \vec{v}_d$.

Classical MDS is used in the `Isomap` algorithm discussed in the main text.

B.3.2 Isomap

The `Isomap` algorithm consists of three steps:

1. **Nearest neighbor search**
   Take an integer K or $\varepsilon > 0$. Calculate the distances

   $$d_{ij}^{\mathscr{X}} \equiv d^{\mathscr{X}}(x_i, x_j) = \|x - x'\|_{\mathscr{X}} \tag{B.14}$$

   between all pairs of data points $x_i, x_j \in \mathscr{X}$ $(i, j = 1, \ldots, n)$ in the feature space $\mathscr{X}$. Typically, Euclidean distance is used. Determine neighboring points on $\mathscr{M}$ by connecting all points up to the K-nearest neighbors or within a ball of radius ε. The performance of `Isomap` depends on the choice of K or ε. `Isomap` uses `sklearn.neighbors.BallTree` for efficient neighbor search.

2. **Calculation of graph distances**
   Construct a weighted neighborhood graph $\mathscr{G} = \mathscr{G}(\mathscr{V}, \mathscr{E})$ for the input data points $\{x_i\}$ $(i = 1, \ldots, N)$. The vertex set $\mathscr{V}$ of the graph represents the data points $\{x_1, \ldots, x_N\}$, and the edge set $\mathscr{E}$ represents the edges e_{ij} indicating neighborhood relationships between the data. Each edge e_{ij} is assigned a weight w_{ij} corresponding to the distance $d_{ij}^{\mathscr{X}}$ between the two points. If two points x_i, x_j are not directly connected by an edge, the weight is 0.
   Estimate the geodesic distances $\{d_{ij}^{\mathscr{X}}\}$ on $\mathscr{M}$ using the graph distances $d_{ij}^{\mathscr{G}}$ on the graph $\mathscr{G}$. The graph distance $d_{ij}^{\mathscr{G}}$ is defined as the length of the shortest path between two points on the graph $\mathscr{G}$. Points that are not neighbors are connected by the shortest path through the nearest neighbors, and the path length is given by the sum of the weights. This length approximates the geodesic distance between distant points.
   Assuming that the data points are sampled from a probability distribution defined on the manifold $\mathscr{M}$, if the manifold is flat, the graph distances $d^{\mathscr{G}}$ converge to the geodesic distances $d^{\mathscr{M}}$ as $N \longrightarrow \infty$ (Bernstein et al., 2001). The most efficient algorithms for this are the Floyd–Warshall algorithm (Floyd, 1962; Warshall, 1962) and Dijkstra's algorithm (Dijkstra, 1959). The former is effective for dense graphs, while the latter is known to be effective for sparse graphs (e.g., Ma & Fu, 2012).

3. **Spectral embedding using MDS**
   Consider the distance matrix $\mathsf{D}^{\mathscr{G}} \equiv \left(d_{ij}^{\mathscr{G}}\right)$ (an $N \times N$ symmetric matrix). Apply classical MDS to the matrix $\mathsf{D}^{\mathscr{G}}$ to reconstruct a d-dimensional space $\mathscr{Y}$ in which the geodesic distances between data points on the manifold $\mathscr{M}$ are as preserved as possible.

   a. Define $\mathsf{S}^{\mathscr{G}} \equiv ((d_{ij}^{\mathscr{G}})^2)$ as the $N \times N$ symmetric matrix with squared graph distances as its elements. Double center it by:

   $$\mathsf{K}_N^{\mathscr{G}} = -\frac{1}{2} \mathsf{H} \mathsf{S}^{\mathscr{G}} \mathsf{H}, \tag{B.15}$$

   $$\mathsf{H} \equiv \mathbb{I}_N - \frac{1}{N} \mathbb{1}_N. \tag{B.16}$$

Here, $\mathbb{I}_N$ is the $(N \times N)$ identity matrix, and $\mathbb{1}_N$ is the $(N \times N)$ symmetric matrix with all elements equal to 1.

4. Choose the embedding vectors $\{\hat{y}_i\}$ to minimize $\|K_N^{\mathscr{G}} - K_N^{\mathscr{Y}}\|$. Here,

$$K_N^{\mathscr{Y}} = -\frac{1}{2}HS^{\mathscr{Y}}H. \tag{B.17}$$

$S^{\mathscr{Y}} = \left((d_{ij}^{\mathscr{Y}})^2\right)$, $d_{ij}^{\mathscr{Y}} = \|y_i - y_j\|$ is the Euclidean distance between y_i and y_j. Decompose $K_N^{\mathscr{G}}$ into its eigenvalues $\Lambda = \mathrm{diag}(\lambda_1, \ldots, \lambda_N)$ and eigenvectors $V = (v_1, \ldots, v_N)$ by eigenvalue decomposition:

$$K_N^{\mathscr{G}} = V\Lambda V^\top. \tag{B.18}$$

The optimal solution is obtained from the d largest eigenvalues $\lambda_1 \geq \cdots \geq \lambda_d$ and the corresponding eigenvectors $v_1, \ldots, v_d$ of $K_N^{\mathscr{G}}$.

5. Embed the graph $\mathscr{G}$ into the d-dimensional subspace $\mathscr{Y}$ as the $d \times N$ matrix

$$Y \equiv (\hat{y}_1, \ldots, \hat{y}_N) = \left(\lambda_1^{\frac{1}{2}} v_1, \ldots, \lambda_d^{\frac{1}{2}} v_d\right). \tag{B.19}$$

B.3.3 UNIFORM MANIFOLD APPROXIMATION AND PROJECTION (UMAP)

Here, we describe the UMAP algorithm in more detail. As briefly introduced in the main text, the UMAP algorithm consists of the following three steps:

1. **Estimation of the Riemannian manifold**

   In UMAP, it is assumed that the data is uniformly distributed on the manifold $\mathscr{M}$. The Riemannian manifold $(\mathscr{M}, g)$ on which the data points lie is estimated. The estimation of the manifold is done using a K-nearest neighbor graph, similar to Isomap. Assuming that the data is given as points in $\mathbb{R}^n$, $\mathscr{M}$ can be considered as embedded in $\mathbb{R}^n$, but the metric g is estimated separately and provided. The following lemma is used for metric estimation (McInnes et al., 2018).

   Lemma 1 ((McInnes et al., 2018)).
   Consider a Riemannian manifold $(\mathscr{M}, g)$ embedded in $\mathbb{R}^n$, and take a point $p \in \mathscr{M}$. For a sufficiently small neighborhood U around the point $p \in \mathscr{M}$, it is assumed that the metric g is a diagonal matrix with $g_{ij} = \text{constant} \times \delta_{ij}$ within U. Here, δ_{ij} represents the Kronecker delta. Taking an open ball $B \subset U$ with radius r centered at point p in $\mathbb{R}^n$, the volume of B on $\mathscr{M}$ can be expressed using the metric g as

$$Vol(B) = \int_B \sqrt{\det(g)}\,dx_1 \ldots dx_n = \sqrt{\det(g)} \int_B dx_1 \ldots dx_n \tag{B.20}$$

(the second equality follows from the fact that g is constant). The rightmost side of equation (B.20) represents the volume of a ball in $\mathbb{R}^n$ with radius r,

$$Vol(B) = \sqrt{\det(g)}\,\frac{\pi^{\frac{n}{2}} r^n}{\Gamma\left(\dfrac{n}{2}+1\right)} \tag{B.21}$$

Suppose the radius r is adjusted so that $\det(g) = r^{-2n}$. Using this radius r, since the metric g is assumed to be expressed as a constant times the identity matrix locally, it can be expressed as $g_{ij} = r^{-2}\delta_{ij}$. Therefore, the distance in g on $\mathcal{M}$ is r^{-2} times the distance in $\mathbb{R}^n$. That is, the distance between two points $p,q \in \mathcal{M}$ is given by

$$d^{\mathcal{M}}(p,q) = r^{-1} d^{\mathbb{R}^n}(p,q) \tag{B.22}$$

This lemma asserts that the distance between a point p and another point $q \in B$ near p can be calculated from the Euclidean distance by appropriately choosing the open ball B. For example, suppose a sequence of points $Y = \{y_1,\ldots,y_N \in \mathbb{R}^n\}$ is uniformly sampled from a Riemannian manifold $\mathcal{M} \in \mathbb{R}^n$. If the metric g of $\mathcal{M}$ is locally expressed as a constant multiple of the identity matrix, then the number of points in Y contained within a ball of radius 1 in the metric g centered at each point y_i will be constant, regardless of the center point. Therefore, by fixing a constant K, and drawing a Euclidean ball around y_i containing K points, the radius of this ball can be defined as the distance 1 in the neighborhood metric of y_i. The distance determined by this method is $d^{\mathcal{M}}$.

2. **Representation of distance spaces with fuzzy topology**

   In UMAP, data points are represented using a method called fuzzy topological representation (Spivak, 2009). This is essentially a representation using weighted graphs, where the edge weights represent the likelihood of a connection between two points. In this section, we aim for an informal introduction, referencing McInnes et al. (2020) without pursuing a rigorous description.

   a. **Fuzzy Set**

      Consider the data structure of a set. Typically, a membership function of a set $\mathscr{S}$ is a function that determines whether an element x belongs to $\mathscr{S}$.

      $$\mu_{\mathscr{S}} \equiv \begin{cases} 1 & x \in \mathscr{S} \\ 0 & x \notin \mathscr{S}. \end{cases} \tag{B.23}$$

      If we denote the entire set of possible x as "Any," then $\mu_{\mathscr{S}} : \text{Any} \longrightarrow \{0,1\}$. Conversely, given a membership function μ, the set itself can be constructed as all x for which $\mu(x) = 1$.

      $$\mathscr{S} = \{x \in \text{Any} \mid \mu(x) = 1\} \tag{B.24}$$

Thus, if the underlying set "Any" is defined, the set $\mathscr{S}$ can be represented as the set μ.

A fuzzy set is an extension of the membership function values in Equation (B.23) to the interval $(0,1]$, where the membership function $\mu_{\mathscr{S}}^F$ returns a real number in $(0,1]$ called the membership strength.

$$\mu(x) = \text{membership strength}(x \in \mathscr{S}).\qquad(\text{B.25})$$

This can also be interpreted as the probability that x belongs to the set $\mathscr{S}$. Given the membership function μ and the set of possible values "Any," the pair (Any, μ) is called a fuzzy set. As with ordinary sets, a set form can be derived from the membership function μ. For a given real number a, the fuzzy set returns all values that belong with a membership strength of at least a:

$$\mathscr{S}(a) = \{x \in \text{Any} \mid \mu(x) \geq a\}\qquad(\text{B.26})$$

Thus, for $a \leq b$, it holds that $\mathscr{S}(a) \supseteq \mathscr{S}(b)$. Such a function $\mathscr{S}$ is also called a fuzzy set.

b. Simplicial set

A simplicial set is a concept in category theory that redefines a simplicial complex. Here, we simply define a simplicial set $\mathscr{S}$ as the set of all subcomplexes of dimension m or less, denoted as $\mathscr{S}_m$. For example, S_0 is the set of vertices of $\mathscr{S}$, and $\mathscr{S}_1$ is the set of edges added to X_0.

c. Fuzzy simplicial set

A simplicial set $\mathscr{S}$ is defined by giving the set $\mathscr{S}_m$ for each m. A fuzzy simplicial set is a fuzzy version of this simplicial set. Given a fuzzy simplicial set $\mathscr{S}$, the set of all sub-simplices of dimension m or less with strength a or greater is denoted as $\mathscr{S}_m(a)$. Let Δ_m be the set of all m-simplices. If the membership function of $\mathscr{S}_m$ is μ_m, then

$$\mathscr{S}_m(a) = \{\sigma \in \bigcup_{i=1}^{m} \Delta_i \mid \mu_m(\sigma) \geq a\}\qquad(\text{B.27})$$

is the expression.

d. FinEPMet, FinReal, and FinSing

Definition C12 (Extended-pseudo-metric space)).

An extended-pseudo-metric space (X,d) is defined by a function $d : X \times X \longrightarrow \mathbb{R}_{\geq 0} \cup \{\infty\}$ that satisfies the following three axioms:

   i. $d(x,y) \geq 0,\ d(x,x) = 0,$

   ii. $d(x,y) = d(y,x)$,

   iii. $d(x,z) \le d(x,y) + d(y,z)$.

Note that infinite distances are allowed, and it is also allowed that the distance between two distinct points can be zero. $\mathcal{M}, g^{\mathcal{M}}$ is an EPMet. Furthermore, when X is a finite space, the EPMet is called FinEPMet. This is used when considering spaces consisting only of input point sequences.

FinReal converts fuzzy m-simplices into FinEPMet. If we denote an m-simplex with strength $a \in (0,1]$ as $\Delta_m(a)$, then

$$\text{FinReal}(\Delta_m(a)) = (\{x_1,\ldots,x_m\}, d_a), \tag{B.28}$$

$$d_a(x_i, x_j) \equiv \begin{cases} -\log a & i \ne j \\ 0 & \text{otherwise} \end{cases}. \tag{B.29}$$

The right-hand side defines the EPMet for a single m-simplex. The base of the distance space is a set of $m+1$ points, corresponding to the $m+1$ vertices of $\Delta_m(a)$.

FinSing assigns a fuzzy simplicial set to FinEPMet. In other words, it defines $\mathcal{S} = \text{FinSing}(Y,d)$. Here, $\mathcal{S}$ gives the set of simplices $\mathcal{S}_m(a)$ for each $m \in \mathbb{N}$ and $a \in (0,1]$, defined as follows:

$$\mathcal{S}_m(a) = \{\overline{f(x_0)\cdots f(x_n)} \mid f \in \text{Hom}(\text{FinReal}(\Delta_m(a), Y)\}. \tag{B.30}$$

Here, $\text{Hom}(A,B)$ denotes the set of all functions $f : A \to B$ for two EPMet (A, d_a) and (B, d_b), which are non-expansive:

$$\forall x,y \in A, d_b(f(x), f(y)) \le d_a(x,y) \tag{B.31}$$

$f \in \text{Hom}(\text{FinReal}(\Delta_m(a)), Y)$ implies that for $i = 1,\ldots,m+1$, $f(x_i) \in Y$, and

$$i \ne j \implies d(f(x_i), f(x_j)) \le -\log(a) \tag{B.32}$$

Also, $\overline{f(x_0)\cdots f(x_m)}$ denotes a simplex (of dimension $\le m$) with $f(x_0)\cdots f(x_m)$ as its vertices.

If the data X can be transformed into a metric space by Lemma 1, it can be mapped to a fuzzy simplicial set using FinSing. The distance defined by Lemma 1 only gives the distance around $x_i \in X$ to x_i, so it is necessary to define the distances between other points. Assuming that points around x_i are only connected to x_i, McInnes et al. (2018) hypothesized:

- Two different points seen from x_i are sufficiently separated, i.e., the distance between them is infinite.
- The distance to the nearest point from x_i is zero.

Thus, the distance d_i observed around x_i, using the distance $d^{\mathcal{M}}$ defined on $\mathcal{M}$ by Lemma 1, can be expressed as:

$$d_i(x_j, x_k) = \begin{cases} 0 & \text{if } j = k \\ d^{\mathcal{M}}(x_i, x_k) - \rho & \text{if } j = i \\ \infty & \text{otherwise} \end{cases} \tag{B.33}$$

Here, ρ represents the value of $d^{\mathcal{M}}$ with the nearest neighbor to x_i.

Given the data $X = \{x_1, \ldots, x_N\} \in \mathbb{R}^n$, we provide the family of EPMet $\{(X, d_i)\}\, i = 1, \ldots, N$. The fuzzy topological representation of X is:

$$\bigcup_{i=1}^{N} \mathrm{FinSing}(X, d_i). \tag{B.34}$$

This is the representation of the data set in UMAP.

3. **Dimensionality reduction**

Represent the original data and the mapped points on the manifold using fuzzy topology. Estimate the data $Y = \{y_1, \ldots, y_N\} \subset \mathbb{R}^d$ $(d \ll n)$ on the low-dimensional manifold corresponding to the data X. It is assumed that Y is uniformly distributed in $\mathbb{R}^d$, considering $\mathbb{R}^d$ itself as a manifold. Therefore, distances are also given by Euclidean distance. When there are two representations X and Y, the ideal distance between them is defined to be as short as possible. To do this, convert the data into membership strengths a and define the mutual entropy as the distance. The membership function of a fuzzy set $\mathcal{S}$ is given by

$$\mu(x) = \sup\{a \in (0, 1] \mid x \in \mathcal{S}(a)\} \tag{B.35}$$

When there are membership functions μ and κ (with domain A) for two representations, their distance is defined as follows:

$$C(\mu, \kappa) = \sum_{a \in A} \mu(a) \log \frac{\mu(a)}{\kappa(a)} + [1 - \mu(a)] \log \frac{1 - \mu(a)}{1 - \kappa(a)} \tag{B.36}$$

Given the data X, find Y that minimizes C. In practice, methods such as stochastic gradient descent are used.

C Physical Constants and Units

C.1 PHYSICAL CONSTANTS

Table C.1

Physical constants.

Constant	Symbol and value
Light speed	$c = 2.998 \times 10^{10} \ [\mathrm{km\,s^{-1}}]$
Planck constant	$h_\mathrm{P} = 6.626 \times 10^{-27} \ [\mathrm{cm^2 g s^{-1}}]$
Reduced Planck constant	$\hbar = 1.05457266 \times 10^{-27} \ [\mathrm{cm^2 g s^{-1}}]$
Gravitational constant	$G = 6.673 \times 10^{-8} \ [\mathrm{cm^3 g^{-1} s^{-2}}]$
Electron charge	$e = 4.8032 \times 10^{-10} \ [\mathrm{esu}]$
Electron mass	$m_e = 9.109 \times 10^{-28} \ [\mathrm{g}]$
Proton mass	$m_p 1.6726 \times 10^{-24} \ [\mathrm{g}]$
Hydrogen mass	$m_\mathrm{H} = 1.6733 \times 10^{-24} \ [\mathrm{g}]$
Atomic mass unit	$\mathrm{amu} = 1.6605 \times 10^{-24} \ [\mathrm{g}]$
Avogadro number	$N_\mathrm{A} = 6.0221 \times 10^{23} \ [\mathrm{cm^{-3}}]$
Boltzmann constant	$k_\mathrm{B} = 1.381 \times 10^{-16} \ [\mathrm{erg K^{-1}}]$
Stefan–Boltzmann constant	$\sigma_\mathrm{SB} = 5.67051 0^{-5} \ [\mathrm{erg\ cm^{-2} K^{-4} s^{-1}}]$
Radiation density constant	$a = 7.5646 \times 10^{-1} \ [\mathrm{erg\ cm^{-3} K^{-4}}]$
Fine structure constant	$\alpha = 7.29735 \times 10^{-3}]$
Rydberg constant	$R_\infty = 2.17991 \times 10^{-11} \ [[erg]$
Bohr radius	$a_0 = 5.29177 \times 10^{-9} \ [\mathrm{cm}]$
Classical electron radius	$r_0 = 2.8179 \times 10^{-13} \ [\mathrm{cm}]$
Gas constant	$R = 8.31434(35) \times 10^{7} \ [\mathrm{erg\ K^{-1} mol^{-1}}]$
Thomson scattering cross section	$\sigma_\mathrm{T} = 6.652 \times 10^{-25} \ [\mathrm{cm^2}]$
Bolometric solar luminosity	$L_\odot = 3.9 \times 10^{33} \ [\mathrm{erg\ s^{-1}}]$
Solar mass	$\mathcal{M}_\odot = 1.99 \times 10^{33} \ [\mathrm{g}]$
Solar radius	$R_\odot = 6.96 \times 10^{10} \ [\mathrm{cm}]$
Solar Temperature	$T_\odot = 5.780 \times 10^{3} \ [\mathrm{K}]$

DOI: 10.1201/9781003104315-C

C.2 UNITS

Table C.2
Physical units.

Quantity	Corresponding standard unit
Electron volt	$1\,\text{eV} = 1.6022 \times 10^{-12}\,[\text{erg}]$
Astronomical unit	$1\,\text{AU} = 1.496 \times 10^{13}\,[\text{cm}]$
Parsec	$1\,[\text{pc}] = 3.086 \times 10^{18}\,[\text{cm}]$
Light year	$1 \times [\text{ly}] = 9.463 \times 10^{17}\,[\text{cm}]$
Year	$1\,[\text{yr}] = 3.156 \times 10^{7}\,[\text{s}]$

Bibliography

Aas, K. 2016, Econometrics, 4, 43

Aas, K., Czado, C., Frigessi, A., & Bakken, H. 2009, Insurance: Mathematics and Economics, 44, 182

Abazajian, K. N., Adelman-McCarthy, J. K., Agüeros, M. A., et al. 2009, Astrophysical Journal Supplement Series, 182, 543, doi: 10.1088/0067-0049/182/2/543

Abbott, B. P., Abbott, R., Abbott, T. D., et al. 2017a, Astrophysical Journal, 848, L12, doi: 10.3847/2041-8213/aa91c9

—. 2017b, Astrophysical Journal, 848, L13, doi: 10.3847/2041-8213/aa920c

Abell, G. O. 1958, The Astrophysical Journal Supplement Series, 3, 211, doi: 10.1086/190036

Acar, E. F., Czado, C., & Lysy, M. 2019, Econometrics and Statistics, 12, 181

Ahn, J., Marron, J. S., Muller, K. M., & Chi, Y.-Y. 2007, Biometrika, 94, 760. http://www.jstor.org/stable/20441411

Akaike, H. 1974a, IEEE Trans. Autom. Contr., AC-19, 716

—. 1974b, IEEE Transactions on Automatic Control, 19, 716

Albrecht, A., & Steinhardt, P. J. 1982, Physical Review Letters, 48, 1220, doi: 10.1103/PhysRevLett.48.1220

Alidoost, F., Su, Z., & Stein, A. 2019, Weather and Climate Extremes, 26, 100227

Allen, D. E., McAleer, M., & Singh, A. K. 2017, Sustainability, 9, 1762

Almeida, C., Czado, C., & Manner, H. 2016, Applied Stochastic Models in Business and Industry, 32, 62

Ando, R., Nakanishi, K., Kohno, K., et al. 2017, Astrophysical Journal, 849, 81, doi: 10.3847/1538-4357/aa8fd4

Aoshima, M., Shen, D., Shen, H., et al. 2018, Australian and New Zealand Journal of Statistics, 60, 4, doi: 10.1111/anzs.12212

Aoshima, M., & Yata, K. 2011, Sequential Analysis, 30, 356, doi: 10.1080/07474946.2011.619088

—. 2014, Annals of the Institute of Statistical Mathematics, 983, doi: 10.1007/s10463-013-0435-8

—. 2015, Sequential Analysis, 34, 279, doi: 10.1080/07474946.2015.1063256

—. 2017, Sugaku Expositions, 30, 137, doi: https://doi.org/10.1090/suga/421

—. 2018, Statistica Sinica, 28, 43

—. 2019, Methodology and Computing in Applied Probability, 21, 663, doi: https://doi.org/10.1007/s11009-018-9646-z

Appleby, S., Chingangbam, P., Park, C., Yogendran, K. P., & Joby, P. K. 2018, Astrophysical Journal, 863, 200 (18pp.)

Ata, M., et al. 2018, Monthly Notices of the Royal Astronomical Society, 473, 4773

Audouze, J., & Tinsley, B. M. 1976, Annual Review of the Astronomy & Astrophysics, 14, 43, doi: 10.1146/annurev.aa.14.090176.000355

Babul, A., & Rees, M. J. 1992, Monthly Notices of the Royal Astronomical Society, 255, 346, doi: 10.1093/mnras/255.2.346

Bag, S., Mondal, R., Sarkar, P., et al. 2019, Monthly Notices of the Royal Astronomical Society, 485, 2235

Bag, S., Mondal, R., Sarkar, P., Bharadwaj, S., & Sahni, V. 2018, Monthly Notices of the Royal Astronomical Society, 477, 1984

Baik, J., Ben Arous, G., & Péché, S. 2005, Annals of Probability, 33, 1643, doi: 10.1214/009117905000000233

Baik, J., & Silverstein, J. W. 2006, Journal of Multivariate Analysis, 97, 1382, doi: 10.1016/j.jmva.2005.08.003

Baldi, R. D., Capetti, A., & Massaro, F. 2018, Astronomy & Astrophysics, 609, A1, doi: 10.1051/0004-6361/201731333

Bardeen, J. M. 1980, Physical Review D, 22, 1882, doi: 10.1103/PhysRevD.22.1882

Bardeen, J. M., Steinhardt, P. J., & Turner, M. S. 1983, Physical Review D, 28, 679, doi: 10.1103/PhysRevD.28.679

Bauer, U., Kerber, M., & Reininghaus, J. 2012, PHAT, a software library for persistent homology. https://bitbucket.org/phat-code/phat/src/master/

Bedford, T., & Cooke, R. M. 2001, Annals of Mathematics and Artificial Intelligence, 32, 245

—. 2002, Annals of Statistics, 30, 1031

Behroozi, P., Wechsler, R. H., Hearin, A. P., et al. 2019, Monthly Notices of the Royal Astronomical Society, 488, 3143, doi: 10.1093/mnras/stz1182

Beisbart, C., Dahlke, R., Mecke, K., & Wagner, H. 2002, Vector- and Tensor-Valued Descriptors for Spatial Patterns, Vol. 600 (New York: Springer), 238–260

Bell, E. F. 2003, The Astrophysical Journal, 586, 794, doi: 10.1086/367829

Benabed, K., Cardoso, J.-F., Prunet, S., & Hivon, E. 2009, Monthly Notices of the Royal Astronomical Society, 400, 219

Bender, R., Burstein, D., & Faber, S. M. 1992, Astrophysical Journal, 399, 462, doi: 10.1086/171940

—. 1993, Astrophysical Journal, 411, 153, doi: 10.1086/172815

Bendo, G. J., Beswick, R. J., D'Cruze, M. J., et al. 2015, Monthly Notices of the Royal Astronomical Society, 450, L80, doi: 10.1093/mnrasl/slv053

Bernardeau, F. 2007, Cosmologie: des fondements théoriques aux observations (Paris: EDP Sciences)

Bernardeau, F., Colombi, S., Gaztanaga, E., & Scoccimarro, R. 2002, Physics Reports, 367, 1

Bernstein, M., Silva, V. D., Langford, J. C., & Tenenbaum, J. B. 2001, Graph Approximations to Geodesics on Embedded Manifolds, Stanford University

Best, P. N., & Heckman, T. M. 2012, Monthly Notices of the Royal Astronomical Society, 421, 1569, doi: 10.1111/j.1365-2966.2012.20414.x

Bhavsar, S. P. 1981, Astrophysical Journall, 246, L5, doi: 10.1086/183542

Biagetti, M., Cole, A., & Shiu, G. 2021, Journal of Cosmology and Astroparticle Physics, 2021, 061 (33pp.)

Bialy, S., & Sternberg, A. 2016, Astrophysical Journal, 822, 83, doi: 10.3847/0004-637X/822/2/83

Bica, E., Alloin, D., & Schmidt, A. 1990, Monthly Notices of the Royal Astronomical Society, 242, 241, doi: 10.1093/mnras/242.2.241

Bigiel, F., Leroy, A., Walter, F., et al. 2008, Astronomical Journal, 136, 2846, doi: 10.1088/0004-6256/136/6/2846

Binggeli, B., Sandage, A., & Tammann, G. A. 1988, Annual Review of the Astronomy & Astrophysics, 26, 509, doi: 10.1146/annurev.aa.26.090188.002453

Binney, J., & Tremaine, S. 2008, Galactic Dynamics: Second Edition (Princeton: Princeton University Press)

Biviano, A. 2000, in Constructing the Universe with Clusters of Galaxies, ed. F. Durret & D. Gerbal, 1, doi: 10.48550/arXiv.astro-ph/0010409

Blanton, M. R. 2006, Astrophysical Journal, 648, 268, doi: 10.1086/505628

Blanton, M. R., et al. 2017, Astronomical Journal, 154, 28 (35pp.)

Boerner, G., & Mo, H. 1989, Astronomy and Astrophysics, 224, 1

Boissier, S. 2013, in Planets, Stars and Stellar Systems. Volume 6: Extragalactic Astronomy and Cosmology, ed. T. D. Oswalt & W. C. Keel, Vol. 6 (Dordrecht: Springer), 141, doi: 10.1007/978-94-007-5609-0_3

Boissier, S. 2017, Star Formation Laws; threshold, resolution effects, star formation laws, state of the art of observations, gas measurements. http://cosmoschool2017.oa.uj.edu.pl/BoissierLecture2SF.pdf

Boissier, S., Monnier Ragaigne, D., Prantzos, N., et al. 2003, Monthly Notices of the Royal Astronomical Society, 343, 653, doi: 10.1046/j.1365-8711.2003.06703.x

Boissier, S., Prantzos, N., Boselli, A., et al. 2003, Monthly Notices of the Royal Astronomical Society, 346, 1215, doi: 10.1111/j.1365-2966.2003.07170.x

Boissonnat, J.-D., Chazal, F., & Yvinec, M. 2018, Geometric and Topological Inference, Cambridge Texts in Applied Mathematics (Cambridge: Cambridge University Press)

Bolatto, A. D., Warren, S. R., Leroy, A. K., et al. 2013, Nature, 499, 450, doi: 10.1038/nature12351

Boquien, M., Kennicutt, R., Calzetti, D., et al. 2016, Astronomy & Astrophysics, 591, A6, doi: 10.1051/0004-6361/201527759

Boselli, A. 2011, A Panchromatic View of Galaxies (John Wiley & Sons, Inc., Berlin)

Boselli, A., Cortese, L., & Boquien, M. 2014a, Astronomy and Astrophysics, 564, A65

Boselli, A., Cortese, L., Boquien, M., et al. 2014b, Astronomy and Astrophysics, 564, A66

Boselli, A., Gavazzi, G., Donas, J., et al. 2001, The Astronomical Journal, 121, 753, doi: 10.1086/318734

Boselli, A., Gavazzi, G., Lequeux, J., et al. 2002, Astronomy & Astrophysics, 385, 454, doi: 10.1051/0004-6361:20020156

Boselli, A., Lequeux, J., & Gavazzi, G. 2004, Astronomy & Astrophysics, 428, 409, doi: 10.1051/0004-6361:20041316

Boselli, A., Eales, S., Cortese, L., et al. 2010, Publication of the Astronomical Society of the Pacific, 122, 261, doi: 10.1086/651535

Boselli, A., Boissier, S., Voyer, E., et al. 2016, Astronomy & Astrophysics, 585, A2, doi: 10.1051/0004-6361/201526915

Bothun, G. D., Impey, C. D., Malin, D. F., & Mould, J. R. 1987, Astronomical Journal, 94, 23, doi: 10.1086/114443

Bouveyron, C., & Brunet, C. 2013, Computational Statistics and Data Analysis, 71, 52, doi: 10.1016/j.csda.2012.12.008

Brauher, J. R., Dale, D. A., & Helou, G. 2008, The Astrophysical Journal Supplement Series, 178, 280, doi: 10.1086/590249

Brechmann, E. C., & Schepsmeier, U. 2013, Journal of Statistical Software, 52, 1

Brinchmann, J., Charlot, S., White, S. D. M., et al. 2004, Monthly Notices of the Royal Astronomical Society, 351, 1151, doi: 10.1111/j.1365-2966.2004.07881.x

Brosche, P. 1973, Astronomy and Astrophysics, 23, 259

Brout, R., Englert, F., & Gunzig, E. 1978, Annals of Physics, 115, 78, doi: 10.1016/0003-4916(78)90176-8

Bryan, G. L., & Norman, M. L. 1998, Astrophysical Journal, 495, 80, doi: 10.1086/305262

Buat, V., & et al. 2007a, Astrophysical Journal Supplement, 173, 404

Buat, V., Giovannoli, E., Takeuchi, T. T., et al. 2011, Astronomy & Astrophysics, 529, A22, doi: 10.1051/0004-6361/201015944

Buat, V., Iglesias-Páramo, J., Seibert, M., et al. 2005, The Astrophysical Journal Letters, 619, L51, doi: 10.1086/423241

Buat, V., Marcillac, D., Burgarella, D., et al. 2007b, Astronomy and Astrophysics, 469, 19

Buat, V., & Xu, C. 1996, Astronomy & Astrophysics, 306, 61

Buat, V., Takeuchi, T. T., Iglesias-Páramo, J., et al. 2007, Astrophysical Journal Supplement Series, 173, 404, doi: 10.1086/516645

Bundy, K., Ellis, R. S., & Conselice, C. J. 2005, Astrophysical Journal, 625, 621, doi: 10.1086/429549

Burgarella, D., Buat, V., Gruppioni, C., et al. 2013, Astronomy & Astrophysics, 554, A70, doi: 10.1051/0004-6361/201321651

Burrows, A., & Vartanyan, D. 2021, Nature, 589, 29, doi: 10.1038/s41586-020-03059-w

Calette, A. R., Avila-Reese, V., Rodríguez-Puebla, A., Hernández-Toledo, H., & Papastergis, E. 2018, Revista Mexicana de Astronomía y Astrofísica, 54, 443

Calistro Rivera, G., Williams, W. L., Hardcastle, M. J., et al. 2017, Monthly Notices of the Royal Astronomical Society, 469, 3468, doi: 10.1093/mnras/stx1040

Callau Poduje, A. C., & Haberlandt, U. 2018, Water, 10, 862

Calzetti, D., Armus, L., Bohlin, R. C., et al. 2000, The Astrophysical Journal, 533, 682, doi: 10.1086/308692

Calzetti, D., Kinney, A. L., & Storchi-Bergmann, T. 1994, The Astrophysical Journal, 429, 582, doi: 10.1086/174346

Calzetti, D., Kennicutt, R. C., Engelbracht, C. W., et al. 2007, Astrophysical Journal, 666, 870, doi: 10.1086/520082

Calzetti, D., Wu, S. Y., Hong, S., et al. 2010, Astrophysical Journal, 714, 1256, doi: 10.1088/0004-637X/714/2/1256

Cambanis, S. 1977, Journal of Multivariate Analysis, 7, 551

Capetti, A., Massaro, F., & Baldi, R. D. 2017a, Astronomy & Astrophysics, 598, A49, doi: 10.1051/0004-6361/201629287

—. 2017b, Astronomy & Astrophysics, 601, A81, doi: 10.1051/0004-6361/201630247

Caplar, N., Lilly, S. J., & Trakhtenbrot, B. 2018, Astrophysical Journal, 867, 148, doi: 10.3847/1538-4357/aadd98

Cardelli, J. A., Clayton, G. C., & Mathis, J. S. 1989, Astrophysical Journal, 345, 245, doi: 10.1086/167900

Carroll, S. M., Press, W. H., & Turner, E. L. 1992, Annual Review of the Astronomy & Astrophysics, 30, 499, doi: 10.1146/annurev.aa.30.090192.002435

Casasola, V., Cassarà, L. P., Bianchi, S., et al. 2017, Astronomy & Astrophysics, 605, A18, doi: 10.1051/0004-6361/201731020

Casella, G., & Hwang, J. T. 1982, Canadian Journal of Statistics, 10, 305, doi: https://doi.org/10.2307/3556196

Catinella, B., Schiminovich, D., Cortese, L., et al. 2013, Monthly Notices of the Royal Astronomical Society, 436, 34, doi: 10.1093/mnras/stt1417

Chabrier, G. 2003, Publications of the Astronomical Society of the Pacific, 115, 763, doi: 10.1086/376392

Chandrasekhar, S. 1943, Reviews of Modern Physics, 15, 1, doi: 10.1103/RevModPhys.15.1

Chazal, F., Cohen-Steiner, D., Glisse, M., Guibas, L. J., & Oudot, S. 2008, Proximity of Persistence Modules and their Diagrams, Research Report RR-6568, INRIA

Chen, Z., Xu, Y., Wang, Y., & Chen, X. 2019, Astrophysical Journal, 885, 23 (14pp.)

Cheung, C. C. 2007, Astronomical Journal, 133, 2097, doi: 10.1086/513095

Chilingarian, I. V., & Zolotukhin, I. Y. 2012, Monthly Notices of the Royal Astronomical Society, 419, 1727, doi: 10.1111/j.1365-2966.2011.19837.x

Chilingarian, I. V., Zolotukhin, I. Y., Katkov, I. Y., et al. 2017, Astrophysical Journal Supplement Series, 228, 14, doi: 10.3847/1538-4365/228/2/14

Chincarini, G. L., Giovanelli, R., & Haynes, M. P. 1983a, Astronomy and Astrophysics, 121, 5

—. 1983b, The Astrophysical Journal, 269, 13, doi: 10.1086/161014

Chingangbam, P., Ganesan, V., Yogendran, K. P., & Park, C. 2017, Physics Letters B, 771, 67

Cohen-Steiner, D., Edelsbrunner, H., & Harer, J. 2007, Discrete & Computational Geometry, 37, 103–120

Cole, A., & Shiu, G. 2018, Journal of Cosmology and Astroparticle Physics, 2018, 025 (21pp.)

Cole, S., Norberg, P., Baugh, C. M., et al. 2001, Monthly Notices of the Royal Astronomical Society, 326, 255, doi: 10.1046/j.1365-8711.2001.04591.x

Collaboration, P., Aghanim, N., Akrami, Y., et al. 2020, Astronomy & Astrophysics, 641, A6, doi: 10.1051/0004-6361/201833910

Colless, M., Dalton, G., Maddox, S., et al. 2001, Monthly Notices of the Royal Astronomical Society, 328, 1039, doi: 10.1046/j.1365-8711.2001.04902.x

Condon, J. J. 1992, Annual Review of Astronomy and Astrophysics, 30, 575, doi: 10.1146/annurev.aa.30.090192.003043

Condon, J. J., & Yin, Q. F. 1990, The Astrophysical Journal, 357, 97, doi: 10.1086/168894

Conroy, C. 2013, Annual Review of Astronomy and Astrophysics, 51, 393, doi: 10.1146/annurev-astro-082812-141017

Cooray, S., Takeuchi, T. T., Kashino, D., et al. 2023, Monthly Notices of the Royal Astronomical Society, 524, 4976, doi: 10.1093/mnras/stad2129

Cormier, D., Madden, S. C., Hony, S., et al. 2010, Astronomy & Astrophysics, 518, L57, doi: 10.1051/0004-6361/201014699

Cortese, L., Boselli, A., Franzetti, P., et al. 2008, Monthly Notices of the Royal Astronomical Society, 386, 1157, doi: 10.1111/j.1365-2966.2008.13118.x

Cousin, M., Buat, V., Lagache, G., et al. 2019, Astronomy & Astrophysics, 627, A132, doi: 10.1051/0004-6361/201834674

Crocce, M., Pueblas, S., & Scoccimarro, R. 2006, Monthly Notices of the Royal Astronomical Society, 373, 369

Cross, N., Driver, S. P., Couch, W., et al. 2001, Monthly Notices of the Royal Astronomical Society, 324, 825, doi: 10.1046/j.1365-8711.2001.04254.x

Cucciati, O., Tresse, L., Ilbert, O., et al. 2012, Astronomy & Astrophysics, 539, A31, doi: 10.1051/0004-6361/201118010

Cutri, R. M., Wright, E. L., Conrow, T., et al. 2011, Explanatory Supplement to the WISE Preliminary Data Release Products. http://wise2.ipac.caltech.edu/docs/release/prelim/expsup/wise{_}prelrel{_}toc.html

da Costa, L. N., Pellegrini, P. S., Davis, M., et al. 1991, The Astrophysical Journal Supplement Series, 75, 935, doi: 10.1086/191555

da Costa, L. N., Willmer, C. N. A., Pellegrini, P. S., et al. 1998, The Astronomical Journal, 116, 1, doi: 10.1086/300410

Dale, D. A., & Helou, G. 2002, The Astrophysical Journal, 576, 159, doi: 10.1086/341632

Dale, D. A., Helou, G., Contursi, A., et al. 2001, The Astrophysical Journal, 549, 215, doi: 10.1086/319077

Daley, D. J., & Vere-Jones, D. 2003, An introduction to the theory of point processes. Vol. I, 2nd edn., Probability and its Applications (New York) (Berlin: Springer-Verlag)

—. 2006, An Introduction to the Theory of Point Processes: Volume I: Elementary Theory and Methods (New York: Springer)

Davies, J. J., Pontzen, A., & Crain, R. A. 2022, Monthly Notices of the Royal Astronomical Society, 515, 1430, doi: 10.1093/mnras/stac1742

Dawson, K. S., et al. 2016, Astronomical Journal, 151, 44 (34pp.)

de Lapparent, V. 2003, Astronomy & Astrophysics, 408, 845, doi: 10.1051/0004-6361:20030898

de Lapparent, V., Galaz, G., Bardelli, S., & Arnouts, S. 2003, Astronomy & Astrophysics, 404, 831, doi: 10.1051/0004-6361:20030451

de Lapparent, V., Geller, M. J., & Huchra, J. P. 1986, The Astrophysical Journal Letters, 302, L1, doi: 10.1086/184625

De Looze, I., Baes, M., Bendo, G. J., Cortese, L., & Fritz, J. 2011, Monthly Notices of the Royal Astronomical Society, 416, 2712, doi: 10.1111/j.1365-2966.2011.19223.x

De Looze, I., Cormier, D., Lebouteiller, V., et al. 2014, Astronomy & Astrophysics, 568, A62, doi: 10.1051/0004-6361/201322489

de Souza, R. E., Capelato, H. V., Arakaki, L., & Logullo, C. 1982, Astrophysical Journal, 263, 557, doi: 10.1086/160526

Debye, P. 1909, Annalen der Physik, 335, 57

Dekel, A., Birnboim, Y., Engel, G., & et al. 2009, Nature, 457, 451, doi: 10.1038/nature07648

Dekel, A., & Silk, J. 1986, Astrophysical Journal, 303, 39, doi: 10.1086/164050

Dempster, A. P., Laird, N. M., & Rubin, D. B. 1977, Journal of the Royal Statistical Society. Series B (Methodological), 39, 1. http://www.jstor.org/stable/2984875

D'Este, G. M. 1981, Biometrika, 68, 339

Dijkstra, E. W. 1959, Numerische Mathematik, 1, 269

Djorgovski, S. 1992, in Morphological and Physical Classification of Galaxies, ed. G. Longo, M. Capaccioli, & G. Busarello (Dordrecht: Springer Netherlands), 337–356

Djorgovski, S., & Davis, M. 1987, Astrophysical Journal, 313, 59, doi: 10.1086/164948

Draine, B. T. 2011, Physics of the Interstellar and Intergalactic Medium (Princeton: Princeton University Press)

Draine, B. T., & Anderson, N. 1985, The Astrophysical Journal, 292, 494, doi: 10.1086/163181

Draine, B. T., & Li, A. 2001, The Astrophysical Journal, 551, 807, doi: 10.1086/320227

Dressler, A. 1980, Astrophysical Journal, 236, 351, doi: 10.1086/157753

Dressler, A., Lynden-Bell, D., Burstein, D., et al. 1987, Astrophysical Journal, 313, 42, doi: 10.1086/164947

Dressler, A., Oemler, Augustus, J., Couch, W. J., et al. 1997, Astrophysical Journal, 490, 577, doi: 10.1086/304890

Driver, S. 2004, Publication of the Astronomical Society of Australia, 21, 344, doi: 10.1071/AS04035

Driver, S. P., Allen, P. D., Graham, A. W., et al. 2006, Monthly Notices of the Royal Astronomical Society, 368, 414, doi: 10.1111/j.1365-2966.2006.10126.x

D'Souza, R., Vegetti, S., & Kauffmann, G. 2015, Monthly Notices of the Royal Astronomical Society, 454, 4027

Dunlop, J. S., & Peacock, J. A. 1990, Monthly Notices of the Royal Astronomical Society, 247, 19

Dutta, S., Khandai, N., & Dey, B. 2020, Monthly Notices of the Royal Astronomical Society, 494, 2664

Eckert, K. D., Kannappan, S. J., Stark, D. V., et al. 2016, The Astrophysical Journal, 824, 124, doi: 10.3847/0004-637X/824/2/124

Edelsbrunner, H., & Harer, J. 2010, Computational Topology - an Introduction (Providence, USA: American Mathematical Society). http://www.ams.org/bookstore-getitem/item=MBK-69

Edelsbrunner, H., Letscher, D., & Zomorodian, A. 2002, Discrete & Computational Geometry, 28, 511–533

Efstathiou, G., & Silk, J. 1983, Fundamentals of Cosmic Physics, 9, 1

Eisenstein, D. J., & Hu, W. 1998, Astrophysical Journal, 496, 605

Eisenstein, D. J., Seo, H.-J., Sirko, E., & Spergel, D. N. 2007, Astrophysical Journal, 664, 675

Eisenstein, D. J., et al. 2005, Astrophysical Journal, 633, 560

Elbaz, D., Daddi, E., Le Borgne, D., et al. 2007, Astronomy & Astrophysics, 468, 33, doi: 10.1051/0004-6361:20077525

Elbers, W., & van de Weygaert, R. 2019, Monthly Notices of the Royal Astronomical Society, 486, 1523

Ellis, R. S. 1997, Annual Review of Astronomy and Astrophysics, 35, 389, doi: 10.1146/annurev.astro.35.1.389

Elmegreen, B. G. 1994, Astrophysical Journal, 425, L73, doi: 10.1086/187313

Engelbracht, C. W., Gordon, K. D., Rieke, G. H., et al. 2005, The Astrophysical Journal Letters, 628, L29, doi: 10.1086/432613

Engelbracht, C. W., Rieke, G. H., Gordon, K. D., et al. 2008, The Astrophysical Journal, 678, 804, doi: 10.1086/529513

Eufrasio, R. T., Lehmer, B. D., Zezas, A., et al. 2017, The Astrophysical Journal, 851, 10, doi: 10.3847/1538-4357/aa9569

Ewen, H. I., & Purcell, E. M. 1951, Nature, 168, 356, doi: 10.1038/168356a0

Fabbiano, G. 2006, Annual Review of Astronomy and Astrophysics, 44, 323, doi: 10.1146/annurev.astro.44.051905.092519

Faber, S. M., & Jackson, R. E. 1976, The Astrophysical Journal, 204, 668, doi: 10.1086/154215

Faber, S. M., Willmer, C. N. A., Wolf, C., et al. 2007, The Astrophysical Journal, 665, 265, doi: 10.1086/519294

Fabian, A. C. 1994, Annual Review of Astronomy and Astrophysics, 32, 277, doi: 10.1146/annurev.aa.32.090194.001425

Fanaroff, B. L., & Riley, J. M. 1974, Monthly Notices of the Royal Astronomical Society, 167, 31P, doi: 10.1093/mnras/167.1.31P

Fang, W., Li, B., & Zhao, G.-B. 2017, Physical Review Letters, 118, 181301 (5pp.)

Farlie, D. J. G. 1960, Biometrika, 47, 307

Fasy, B. T., Lecci, F., Rinaldo, A., et al. 2014, The Annals of Statistics, 42, 2301

Feldbrugge, J., van Engelen, M., van de Weygaert, R., Pranav, P., & Vegter, G. 2019, Journal of Cosmology and Astroparticle Physics, 2019, 052 (47pp.)

Fernández-Ontiveros, J. A., Prieto, M. A., & Acosta-Pulido, J. A. 2009, Monthly Notices of the Royal Astronomical Society, 392, L16, doi: 10.1111/j.1745-3933.2008.00575.x

Floyd, R. W. 1962, Communications of the ACM, 5, 345. https://ci.nii.ac.jp/naid/20000944679/

Freeman, K. C. 1970, Astrophysical Journal, 160, 811, doi: 10.1086/150474

Galametz, M., Kennicutt, R. C., Calzetti, D., et al. 2013, Monthly Notices of the Royal Astronomical Society, 431, 1956, doi: 10.1093/mnras/stt313

Galaz, G., & de Lapparent, V. 1998, Astronomy & Astrophysics, 332, 459. https://arxiv.org/abs/astro-ph/9711093

Gallagher, J. S., Bushouse, H., & Hunter, D. A. 1989, The Astronomical Journal, 97, 700, doi: 10.1086/115015

Galliano, F., Dwek, E., & Chanial, P. 2008, The Astrophysical Journal, 672, 214, doi: 10.1086/523621

Ganesan, V., & Chingangbam, P. 2017, Journal of Cosmology and Astroparticle Physics, 2017, 023 (23pp.)

Gay, C., Pichon, C., & Pogosyan, D. 2012, Physical Review D, 85, 023011 (41pp.)

Geller, M. J., & Huchra, J. P. 1989, Science, 246, 897, doi: 10.1126/science. 246.4932.897

Gendre, M. A., Best, P. N., & Wall, J. V. 2010, Monthly Notices of the Royal Astronomical Society, 404, 1719, doi: 10.1111/j.1365-2966.2010.16413.x

Ginolfi, M., Hunt, L. K., Tortora, C., Schneider, R., & Cresci, G. 2020, Astronomy & Astrophysics, 638, A4, doi: 10.1051/0004-6361/201936304

Giovanelli, R., & Haynes, M. P. 1985, The Astronomical Journal, 90, 2445, doi: 10. 1086/113949

Giovanelli, R., & Haynes, M. P. 1989, Astronomical Journal, 97, 633, doi: 10.1086/ 115010

—. 1991, Annual Review of the Astronomy & Astrophysics, 29, 499, doi: 10.1146/ annurev.aa.29.090191.002435

—. 1993, Astronomical Journal, 105, 1271, doi: 10.1086/116508

Giovanelli, R., Haynes, M. P., & Chincarini, G. L. 1986, The Astrophysical Journal, 300, 77, doi: 10.1086/163784

Giovanelli, R., Haynes, M. P., Myers, S. T., & Roth, J. 1986, Astronomical Journal, 92, 250, doi: 10.1086/114156

Giri, S. K., & Mellema, G. 2021, Monthly Notices of the Royal Astronomical Society, 505, 1863

Glover, S. C. O., & Clark, P. C. 2014, Monthly Notices of the Royal Astronomical Society, 437, 9, doi: 10.1093/mnras/stt1809

Gnat, O., & Ferland, G. J. 2012, The Astrophysical Journal Supplement Series, 199, 20, doi: 10.1088/0067-0049/199/1/20

Gnedenko, B. V., & Korolev, V. Y. 1996, *Random Summation* — Limit Theorems and Applications (Florida: CRC Press)

Golub, T. R., Slonim, D. K., Tamayo, P., et al. 1999, Science, 286, 531, doi: 10. 1126/science.286.5439.531

Goodfellow, I., Bengio, Y., & Courville, A. 2016, Deep Learning (MIT Press)

Gordon, K. D., Clayton, G. C., Witt, A. N., et al. 2000, The Astrophysical Journal, 533, 236, doi: 10.1086/308668

Gott, J. Richard, I., Melott, A. L., & Dickinson, M. 1986, Astrophysical Journal, 306, 341

Gott, J. Richard, I., Weinberg, D. H., & Melott, A. L. 1987, Astrophysical Journal, 319, 1

Gott, J. Richard, I., et al. 1989, Astrophysical Journal, 340, 625

Gräler, B. 2014, Spatial Statistics, 10, 102

Gräler, B., & Pebesma, E. 2011, Procedia Environmental Sciences, 7, 206

Griffiths, R. E., & Padovani, P. 1990, The Astrophysical Journal, 360, 483, doi: 10. 1086/169139

Groth, E. J., & Peebles, P. J. E. 1977, Astrophysical Journal, 217, 385, doi: 10. 1086/155588

Gumbel, E. J. 1960, Journal of the American Statistical Association, 55, 698

Gunn, J. E., & Gott, J. R. 1972, Astrophysical Journal, 176, 1, doi: 10.1086/151605

Guth, A. H. 1981, Physical Review D, 23, 347, doi: 10.1103/PhysRevD.23.347

Guth, A. H., & Pi, S.-Y. 1982, Physical Review Letters, 49, 1110, doi: 10.1103/ PhysRevLett.49.1110

Hall, P., Marron, J. S., & Neeman, A. 2005, Journal of the Royal Statistical Society: Series B (Statistical Methodology), 67, 427, doi: 10.1111/j.1467-9868.2005. 00510.x

Hamajima, K., & Tosa, M. 1975, Publications of the Astronomical Society of Japan, 27, 561

Hand, D. J., Daly, F., McConway, K., Lunn, D., & Ostrowski, E. 1994, A Handbook of Small Data Sets, 1st edition (Chapman and Hall, London)

Hao, C.-N., Kennicutt, R. C., Johnson, B. D., et al. 2011, Astrophysical Journal, 741, 124, doi: 10.1088/0004-637X/741/2/124

Harrison, E. R. 1970, Physical Review D, 1, 2726, doi: 10.1103/PhysRevD.1. 2726

Hartwick, F. D. A. 1971, The Astrophysical Journal, 163, 431, doi: 10.1086/ 150785

Hawking, S. W. 1982, Physics Letters B, 115, 295, doi: 10.1016/0370-2693(82) 90373-2

Haynes, M. P., Giovanelli, R., Martin, A. M., et al. 2011, The Astronomical Journal, 142, 170, doi: 10.1088/0004-6256/142/5/170

Helou, G., Soifer, B. T., & Rowan-Robinson, M. 1985, The Astrophysical Journal Letters, 298, L7, doi: 10.1086/184556

Hettmansperger, T. P. 1984, Statistical Inference Based on Ranks (New York: John Wiley & Sons)

Hikage, C., et al. 2003, Publications of the Astronomical Society of Japan, 55, 911

Hirashita, H., Buat, V., & Inoue, A. K. 2003, Astronomy & Astrophysics, 410, 83, doi: 10.1051/0004-6361:20031144

Hirashita, H., Kaneda, H., Onaka, T., et al. 2008, Publications of the Astronomical Society of Japan, 60, S477, doi: 10.1093/pasj/60.sp2.S477

Hoel, P. G. 1954, Introduction to Mathematical Statistics, 4th edition (Wiley, New york)

Hollander, M., & Wolfe, D. A. 1999, Nonparametric Statistical Methods, 2nd ed. (New York: John Wiley & Sons)

Hollenbach, D. J., & Tielens, A. G. G. M. 1999, Reviews of Modern Physics, 71, 173, doi: 10.1103/RevModPhys.71.173

Houghton, R. C. W. 2015, Monthly Notices of the Royal Astronomical Society, 451, 3427, doi: 10.1093/mnras/stv1113

Huang, J. S., & Kotz, S. 1984, Biometrika, 71, 633

Huchra, J., Davis, M., Latham, D., & Tonry, J. 1983, The Astrophysical Journal Supplement Series, 52, 89, doi: 10.1086/190860

Hunt, L., Magrini, L., Galli, D., et al. 2012, Monthly Notices of the Royal Astronomical Society, 427, 906, doi: 10.1111/j.1365-2966.2012.21761.x

Hunter, D. A., Elmegreen, B. G., & Baker, A. L. 1998, Astrophysical Journal, 493, 595, doi: 10.1086/305158

Hunter, D. A., Kaufman, M., Hollenbach, D. J., et al. 2001, The Astrophysical Journal, 553, 121, doi: 10.1086/320654

Iglesias-Páramo, J., Buat, V., Takeuchi, T. T., et al. 2006, The Astrophysical Journal Supplement Series, 164, 38, doi: 10.1086/502628

Illian, J., Penttinen, A., Stoyan, H., & Stoyan, D. 2008, Statistical Analysis and Modelling of Spatial Point Patterns (New York: John Wiley & Sons)

Ishii, A., Yata, K., & Aoshima, M. 2016, Journal of Statistical Planning and Inference, 170, 186, doi: https://doi.org/10.1016/j.jspi.2015.10.007

Isobe, T., Fiegelson, Akritas, M., & Babu, G. 1990, Astrophysical Journal, 363, 104

Israel, F. P., & Maloney, P. R. 2011, Astronomy & Astrophysics, 531, A19, doi: 10.1051/0004-6361/201016336

Jäger, N., & Morales Nápoles, O. 2017, ASCE-ASME Journal of Risk and Uncertainty in Engineering Systems, Part A: Civil Engineering, 3, 04017014

James, P. A., Shane, N. S., Beckman, J. E., et al. 2004, Astronomy & Astrophysics, 414, 23, doi: 10.1051/0004-6361:20031568

Jiang, I.-G., Yeh, L.-C., Chang, Y.-C., & Hung, W.-L. 2009, Astronomical Journal, 137, 329

Joby, P. K., Chingangbam, P., Ghosh, T., Ganesan, V., & Ravikumar, C. D. 2019, Journal of Cosmology and Astroparticle Physics, 2019, 009 (17pp.)

Johnson, N. L., & Kotz, S. 1977, Communications in Statistics - Theory and Methods, 6, 485

Johnston, R. 2011, Astronomy and Astrophysics Review, 19, 41

Johnstone, I. M. 2001, Annals of Statistics, 29, 295, doi: 10.1214/aos/1009210544

Jones, M. G., Haynes, M. P., Giovanelli, R., & Moorman, C. 2018, Monthly Notices of the Royal Astronomical Society, 477, 2

Junaid, M., & Pogosyan, D. 2015, Physical Review D, 92, 043505 (15pp.)

Jung, S., & Marron, J. S. 2009, Ann. Statist., 37, 4104, doi: 10.1214/09-AOS709

Kaiser, N. 1984, Astrophysical Journal Letters, 284, L9, doi: 10.1086/184341

—. 1987, Monthly Notices of the Royal Astronomical Society, 227, 1, doi: 10.1093/mnras/227.1.1

Kamphuis, P., Józsa, G. I. G., Oh, S. . H., et al. 2015, Monthly Notices of the Royal Astronomical Society, 452, 3139, doi: 10.1093/mnras/stv1480

Kazanas, D. 1980, Astrophysical Journal Letters, 241, L59, doi: 10.1086/183361

Kennicutt, R. C. 1989, Astrophysical Journal, 344, 685, doi: 10.1086/167834

Kennicutt, R. C. 1992, The Astrophysical Journal, 388, 310, doi: 10.1086/171154

Kennicutt, R. C. 1998, Annual Review of the Astronomy & Astrophysics, 36, 189, doi: 10.1146/annurev.astro.36.1.189

Kennicutt, R. C. 1998, The Astrophysical Journal, 498, 541, doi: 10.1086/305588

Kennicutt, R. C., & Evans, N. J. 2012, Annual Review of Astronomy and Astrophysics, 50, 531, doi: 10.1146/annurev-astro-081811-125610

Kennicutt, R. C., Tamblyn, P., & Congdon, C. E. 1994, Astrophysical Journal, 435, 22, doi: 10.1086/174790

Kennicutt, R. C., Calzetti, D., Walter, F., et al. 2007, Astrophysical Journal, 671, 333, doi: 10.1086/522300

Kennicutt, R. C., Hao, C.-N., Calzetti, D., et al. 2009, Astrophysical Journal, 1672, doi: 10.1088/0004-637X/703/2/1672

Kerber, M., Morozov, D., & Nigmetov, A. 2017, Journal of Experimental Algorithmics, 22, 1.4 (20pp.)

Kereš, D., Yun, M. S., & Young, J. S. 2003, The Astrophysical Journal, 582, 659, doi: 10.1086/344820

Kerscher, M., & Tikhonov, A. 2010, Astronomy and Astrophysics, 509, A57 (9pp.)

Kerscher, M., et al. 1997, Monthly Notices of the Royal Astronomical Society, 284, 73

Keto, E., Hora, J. L., Fazio, G. G., Hoffmann, W., & Deutsch, L. 1999, Astrophysical Journal, 518, 183, doi: 10.1086/307246

Khuntia, S. R., Rueda, J. L., & van der Meijden, M. A. M. M. 2019, Wind Energy, 1, 1

Kitaura, F.-S., & Angulo, R. E. 2012, Monthly Notices of the Royal Astronomical Society, 425, 2443

Kitayama, T. 2014, Progress of Theoretical and Experimental Physics, 2014, 06B111, doi: 10.1093/ptep/ptu055

Kitayama, T., & Suto, Y. 1996, Monthly Notices of the Royal Astronomical Society, 280, 638, doi: 10.1093/mnras/280.3.638

Kloubert, M.-L. 2020, Energies, 13, 1727

Kodama, H., & Sasaki, M. 1984, Progress of Theoretical Physics Supplement, 78, 1, doi: 10.1143/PTPS.78.1

Koen, C. 2009, Monthly Notices of the Royal Astronomical Society, 393, 1370, doi: 10.1111/j.1365-2966.2008.14116.x

Komugi, S., Sofue, Y., & Egusa, F. 2006, Publications of the Astronomical Society of Japan, 58, 793, doi: 10.1093/pasj/58.5.793

Kong, X., Charlot, S., Brinchmann, J., et al. 2004, Monthly Notices of the Royal Astronomical Society, 349, 769, doi: 10.1111/j.1365-2966.2004.07556.x

Kono, K. T., Takeuchi, T. T., Cooray, S., Nishizawa, A. J., & Murakami, K. 2020, arXiv:2006.02905, 29pp.

Koo, D. C., & Kron, R. G. 1992, Annual Review of Astronomy and Astrophysics, 30, 613, doi: 10.1146/annurev.aa.30.090192.003145

Koprowski, M. P., Coppin, K. E. K., Geach, J. E., et al. 2020, Monthly Notices of the Royal Astronomical Society, 492, 4927, doi: 10.1093/mnras/staa160

Kotz, S., Balakrishnan, N., & Johnson, N. L. 2000, Continuous Multivariate Distributions, Volume 1: Models and Applications, 2nd edn. (New York: John Wiley & Sons), 51–62

Kozlov, D. 2008, Combinatorial Algebraic Topology (Berlin: Springer)

Krieger, N., Bolatto, A. D., Walter, F., et al. 2019, Astrophysical Journal, 881, 43, doi: 10.3847/1538-4357/ab2d9c

Kroupa, P. 2002, Science, 295, 82, doi: 10.1126/science.1067524

Krumholz, M. R. 2015, Star Formation in Molecular Clouds

Krumholz, M. R., & McKee, C. F. 2005, The Astrophysical Journal, 630, 250, doi: 10.1086/431734

Kullback, S., & Leibler, R. A. 1951, Ann. Math. Statist., 22, 79

Kurowicka, D., & Joe, H. 2011, Dependence Modeling: A Vine Copula Handbook (Singapore: World Scientific)

Lacey, C., & Cole, S. 1993, Monthly Notices of the Royal Astronomical Society, 262, 627, doi: 10.1093/mnras/262.3.627

Lacey, C. G., Baugh, C. M., Frenk, C. S., et al. 2016, Monthly Notices of the Royal Astronomical Society, 462, 3854, doi: 10.1093/mnras/stw1888

Lahav, O., Lilje, P. B., Primack, J. R., & Rees, M. J. 1991, Monthly Notices of the Royal Astronomical Society, 251, 128, doi: 10.1093/mnras/251.1.128

Larson, R. B. 1974, Monthly Notices of the Royal Astronomical Society, 169, 229, doi: 10.1093/mnras/169.2.229

—. 1981, Monthly Notices of the Royal Astronomical Society, 194, 809, doi: 10.1093/mnras/194.4.809

Le Floc'h, E., Papovich, C., Dole, H., et al. 2005, The Astrophysical Journal, 632, 169, doi: 10.1086/432789

Lee, H., Skillman, E. D., Cannon, J. M., & et al. 2006, Astrophysical Journal, 647, 970, doi: 10.1086/505573

Lesgourgues, J. 2011, arXiv:1104.2932, 19pp.

Li, A., & Draine, B. T. 2001, The Astrophysical Journal, 554, 778, doi: 10.1086/323147

Li, Y., Mac Low, M.-M., & Klessen, R. S. 2006, Astrophysical Journal, 639, 879, doi: 10.1086/499350

Libeskind, N. I., et al. 2018, Monthly Notices of the Royal Astronomical Society, 473, 1195

Lightman, A. P., & Schechter, P. L. 1990, Astrophysical Journal Supplement Series, 74, 831, doi: 10.1086/191521

Linde, A. D. 1982, Physics Letters B, 108, 389, doi: 10.1016/0370-2693(82) 91219-9

—. 1983, Physics Letters B, 129, 177, doi: 10.1016/0370-2693(83)90837-7

Ling, C., Wang, Q., Li, R., et al. 2015, Physical Review D, 92, 064024 (7pp.)

Lisenfeld, U., Espada, D., Verdes-Montenegro, L., et al. 2011, Astronomy & Astrophysics, 534, A102, doi: 10.1051/0004-6361/201117056

Liu, J., Bird, S., Zorrilla Matilla, J. M., et al. 2018, Journal of Cosmology and Astroparticle Physics, 2018, 049 (25pp.)

Liu, S., Maljovec, D., Wang, B., Bremer, P.-T., & Pascucci, V. 2017, IEEE Transactions on Visualization and Computer Graphics, 23, 1249, doi: 10.1109/TVCG. 2016.2640960

Loveday, J. 2002, Contemporary Physics, 43, 437, doi: 10.1080/ 0010751021000019166

Luhman, M. L., Satyapal, S., Fischer, J., et al. 2003, The Astrophysical Journal, 594, 758, doi: 10.1086/376965

Lyth, D. H., & Liddle, A. R. 2009, The Primordial Density Perturbation (Cambridge: Cambridge University Press)

Ma, C.-P., & Bertschinger, E. 1995, Astrophysical Journal, 455, 7, doi: 10.1086/ 176550

Ma, C.-P., & Bertschinger, E. 1995, Astrophysical Journal, 455, 7

Ma, Y., & Fu, Y. 2012, Manifold Learning Theory and Applications (CRC Press)

Ma, Z., Xu, H., Zhu, J., et al. 2019, Astrophysical Journal Supplement Series, 240, 34, doi: 10.3847/1538-4365/aaf9a2

Machalski, J., & Godlowski, W. 2000, Astronomy & Astrophysics, 360, 463

Madau, P., & Dickinson, M. 2014, Annual Review of Astronomy and Astrophysics, 52, 415, doi: 10.1146/annurev-astro-081811-125615

Madden, S. C., Poglitsch, A., Geis, N., et al. 1997, The Astrophysical Journal, 483, 200, doi: 10.1086/304247

Madore, B. F., van den Bergh, S., & Rogstad, D. H. 1974, The Astrophysical Journal, 191, 317, doi: 10.1086/152970

Maggiore, M., & Riotto, A. 2010a, Astrophysical Journal, 717, 515, doi: 10.1088/ 0004-637X/717/1/515

—. 2010b, Astrophysical Journal, 717, 526, doi: 10.1088/0004-637X/717/1/526

—. 2010c, Astrophysical Journal, 711, 907, doi: 10.1088/0004-637X/711/2/907

Magnelli, B., Ivison, R. J., Lutz, D., et al. 2015, Astronomy & Astrophysics, 573, A45, doi: 10.1051/0004-6361/201424937

Martin, D. C., et al. 2005, Astrophysical Journal, 619, L59

Martin, D. C., Fanson, J., Schiminovich, D., et al. 2005, Astrophysical Journal, 619, L1, doi: 10.1086/426387

Martín, S., Mangum, J. G., Harada, N., et al. 2021, Astronomy & Astrophysics, 656, A46, doi: 10.1051/0004-6361/202141567

Martinez, V., & Saar, E. 2001, Statistics of the Galaxy Distribution (London: CRC Press)

Marzke, R. O., & da Costa, L. N. 1997, The Astronomical Journal, 113, 185, doi: 10.1086/118243

Marzke, R. O., da Costa, L. N., Pellegrini, P. S., et al. 1998, The Astrophysical Journal, 503, 617, doi: 10.1086/306011

Mashian, N., Oesch, P. A., & Loeb, A. 2016, Monthly Notices of the Royal Astronomical Society, 455, 2101

Mateo, M. L. 1998, Annual Review of Astronomy and Astrophysics, 36, 435, doi: 10.1146/annurev.astro.36.1.435

Mathis, J. S., Rumpl, W., & Nordsieck, K. H. 1977, The Astrophysical Journal, 217, 425, doi: 10.1086/155591

Matsubara, T. 1994, Astrophysical Journal, 434, L43

Matsubara, T. 1995, The Astrophysical Journal Supplement Series, 101, 1, doi: 10.1086/192231

Matsubara, T. 1996, Astrophysical Journal, 457, 13

—. 2003, Astrophysical Journal, 584, 1

—. 2010, Physical Review D, 81, 083505 (9pp.)

Matsubara, T., Hikage, C., & Kuriki, S. 2022, Physical Review D, 105, 023527 (12pp.)

Matsubara, T., & Suto, Y. 1996, Astrophysical Journal, 460, 51

Matsubara, T., & Yokoyama, J. 1996, Astrophysical Journal, 463, 409

Matsubayashi, K., Sugai, H., Hattori, T., et al. 2009, Astrophysical Journal, 701, 1636, doi: 10.1088/0004-637X/701/2/1636

Matthews, T. A., Morgan, W. W., & Schmidt, M. 1964, Astrophysical Journal, 140, 35, doi: 10.1086/147890

Mauch, T., & Sadler, E. M. 2007, Monthly Notices of the Royal Astronomical Society, 375, 931, doi: 10.1111/j.1365-2966.2006.11353.x

McGaugh, S. S. 2005, The Astrophysical Journal, 632, 859, doi: 10.1086/432968

McGaugh, S. S., Schombert, J. M., Bothun, G. D., et al. 2000, The Astrophysical Journal Letters, 533, L99, doi: 10.1086/312628

McInnes, L., Healy, J., & Melville, J. 2020, UMAP: Uniform Manifold Approximation and Projection for Dimension Reduction. https://arxiv.org/abs/1802.03426

McInnes, L., Healy, J., Saul, N., & Groberger, L. 2018, Journal of Open Source Software, 3, 861, doi: 10.21105/joss.00861

McKee, C. F., & Ostriker, J. P. 1977, The Astrophysical Journal, 218, 148, doi: 10.1086/155667

Mecke, K. R., Buchert, T., & Wagner, H. 1994, Astronomy and Astrophysics, 288, 697

Mejdoub, H., & Ben Arab, M. 2018, Research in International Business and Finance, 45, 208

Merritt, D. 1987, Astrophysical Journal, 313, 121, doi: 10.1086/164953

Merz, G., Rezaie, M., Seo, H.-J., et al. 2021, Monthly Notices of the Royal Astronomical Society, 506, 2503, doi: 10.1093/mnras/stab1887

Meurer, G. R., Heckman, T. M., & Calzetti, D. 1999, The Astrophysical Journal, 521, 64, doi: 10.1086/307523

Mie, G. 1908, Annalen der Physik, 330, 377, doi: 10.1002/andp.19083300302

Milne, E. A. 1935, Relativity, Gravitation and World Structure (Oxford: Clarendon Press)

Minamidani, T., Nishimura, A., Miyamoto, Y., et al. 2016, in SPIE 9914, Millimeter, Submillimeter, Far-Infrared Detect. Instrum. Astron. VIII

Miyaji, T., Hasinger, G., Salvato, M., et al. 2015, The Astrophysical Journal, 804, 104, doi: 10.1088/0004-637X/804/2/104

Momose, R., Okumura, S. K., Koda, J., & Sawada, T. 2010, Astrophysical Journal, 721, 383, doi: 10.1088/0004-637X/721/1/383

Momose, R., Koda, J., Kennicutt, R. C., et al. 2013, Astrophysical Journal, 772, 0, doi: 10.1088/2041-8205/772/1/L13

Morgan, W. W., & Lesh, J. R. 1965, Astrophysical Journal, 142, 1364, doi: 10.1086/148422

Morgenstern, D. 1956, Mitteilungen der Mathematischen Statistik, 8, 234

Morozov, D. 2007, Dionysus, a C++ library for computing persistent homolog. https://www.mrzv.org/software/dionysus/

Morrissey, P., Conrow, T., Barlow, T. A., et al. 2007, Astrophysical Journal Supplement Series, 173, 682, doi: 10.1086/520512

Moustakas, J., Kennicutt, R. C., & Tremonti, C. A. 2006, The Astrophysical Journal, 642, 775, doi: 10.1086/500964

Mukhanov, V. F. 1985, Soviet Journal of Experimental and Theoretical Physics Letters, 41, 493

Mukhanov, V. F., & Chibisov, G. V. 1981, Soviet Journal of Experimental and Theoretical Physics Letters, 33, 532

—. 1982, Zhurnal Eksperimentalnoi i Teoreticheskoi Fiziki, 83, 475

Murphy, E. J., Condon, J. J., Schinnerer, E., et al. 2011, Astrophysical Journal, 737, 67, doi: 10.1088/0004-637X/737/2/67

Myers, A. D., et al. 2015, Astrophysical Journal Supplement Series, 221, 27 (24pp.)

Nagler, T., Bumann, C., & Czado, C. 2019, Journal of Multivariate Analysis, 172, 180

Nelsen, R. B. 2006, An Introduction to Copulae, 2nd edn. (New York: Springer)

Newton, K. 1980, Monthly Notices of the Royal Astronomical Society, 190, 689, doi: 10.1093/mnras/190.4.689

Noeske, K. G., Weiner, B. J., Faber, S. M., et al. 2007, The Astrophysical Journal Letters, 660, L43, doi: 10.1086/517926

Obayashi, I., Hiraoka, Y., & Kimura, M. 2018, Journal of Applied and Computational Topology, 1, 421

Oke, J. B., & Sandage, A. 1968, The Astrophysical Journal, 154, 21, doi: 10.1086/149737

Onodera, S., Kuno, N., Tosaki, T., et al. 2010, Astrophysical Journal, 722, L127, doi: 10.1088/2041-8205/722/2/L127

Osterbrock, D. E., & Ferland, G. J. 2006, Astrophysics of Gaseous Nebulae and Active Galactic Nuclei, 2nd edn. (California: University Science Books)

Overzier, R. A., Heckman, T. M., Wang, J., et al. 2011, The Astrophysical Journal Letters, 726, L7, doi: 10.1088/2041-8205/726/1/L7

Pace, Z. J., Tremonti, C., Chen, Y., et al. 2019, Astrophysical Journal, 883, 82, doi: 10.3847/1538-4357/ab3723

Pannella, M., Elbaz, D., Daddi, E., et al. 2015, The Astrophysical Journal, 807, 141, doi: 10.1088/0004-637X/807/2/141

Parroni, C., Cardone, V. F., Maoli, R., & Scaramella, R. 2020, Astronomy and Astrophysics, 633, A71 (15pp.)

Paul, D. 2007, Statistica Sinica, 17, 1617

Peacock, J. A., Cole, S., Norberg, P., et al. 2001, Nature, 410, 169, doi: 10.1038/35065528

Pedregosa, F., Varoquaux, G., Gramfort, A., et al. 2011, Journal of Machine Learning Research, 12, 2825

Peebles, P. J. E. 1980, The large-scale structure of the universe (Princeton: Princeton University Press)

—. 1993, Principles of Physical Cosmology (Princeton: Princeton University Press), doi: 10.1515/9780691206721

Peebles, P. J. E., & Hauser, M. G. 1974, The Astrophysical Journal Supplement Series, 28, 19, doi: 10.1086/190308

Peebles, P. J. E., & Yu, J. T. 1970, Astrophysical Journal, 162, 815, doi: 10.1086/150713

Pei, Y. C. 1992, Astrophysical Journal, 395, 130, doi: 10.1086/171637

Penzias, A. A., Jefferts, K. B., & Wilson, R. W. 1971, Astrophysical Journal, 165, 229, doi: 10.1086/150893

Peters, G. W., Dong, A. X. D., & Kohn, R. 2014, Insurance: Mathematics and Economics, 59, 258

Phillipps, S. 2006, "The Structure and Evolution of Galaxies" (Chichester: John Wiley and Sons, Inc.)

Planck Collaboration, Ade, P. A. R., et al. 2016, Astronomy and Astrophysics, 594, A17 (66pp.)

Poggianti, B. M. 1997, Astronomy and Astrophysics Supplement Series, 122, 399, doi: 10.1051/aas:1997142

Poglitsch, A., Herrmann, F., Genzel, R., et al. 1996, The Astrophysical Journal Letters, 462, L43, doi: 10.1086/310025

Poglitsch, A., Krabbe, A., Madden, S. C., et al. 1995, The Astrophysical Journal, 454, 293, doi: 10.1086/176482

Pogosyan, D., Gay, C., & Pichon, C. 2009, Physical Review D, 80, 081301 (4pp.)

Pollo, A., Guzzo, L., & Le Fèvre, O., e. a. 2006, Astronomy and Astrophysics, 451, 409, doi: 10.1051/0004-6361:20054705

Portillo, S. K. N., Parejko, J. K., Vergara, J. R., & Connolly, A. J. 2020, Astronomical Journal, 160, 45, doi: 10.3847/1538-3881/ab9644

Pranav, P., et al. 2017, Monthly Notices of the Royal Astronomical Society, 465, 4281

—. 2019a, Monthly Notices of the Royal Astronomical Society, 485, 4167

—. 2019b, Astronomy and Astrophysics, 627, A163 (30pp.)

Press, W. H., & Schechter, P. 1974, Astrophysical Journal, 187, 425, doi: 10.1086/152650

Proctor, D. D. 2011, Astrophysical Journal Supplement Series, 194, 31, doi: 10.1088/0067-0049/194/2/31

Pérez-González, P. G., Gallego, J., Zamorano, J., et al. 2003, The Astrophysical Journal Letters, 587, L27, doi: 10.1086/375123

Rahmani, S., Lianou, S., & Barmby, P. 2016, Monthly Notices of the Royal Astronomical Society, 456, 4128, doi: 10.1093/mnras/stv2951

Rampazzo, R., Ciroi, S., Mazzei, P., et al. 2020, Astronomy & Astrophysics, 643, A176, doi: 10.1051/0004-6361/202038996

Ranalli, P., Comastri, A., & Setti, G. 2003, Astronomy & Astrophysics, 399, 39, doi: 10.1051/0004-6361:20021600

—. 2005, Astronomy & Astrophysics, 440, 23, doi: 10.1051/0004-6361:20042598

Rao, C. R. 1965, *Linear Statistical Inference and its Applications* (New York: John Wiley & Sons)

Rees, M. J., & Ostriker, J. P. 1977, Monthly Notices of the Royal Astronomical Society, 179, 541, doi: 10.1093/mnras/179.4.541

Rekola, R., Richer, M. G., McCall, M. L., et al. 2005, Monthly Notices of the Royal Astronomical Society, 361, 330, doi: 10.1111/j.1365-2966.2005.09166.x

Revaz, Y., Jablonka, P., Sawala, T., et al. 2009, Astronomy & Astrophysics, 501, 189, doi: 10.1051/0004-6361/200911734

Rieke, G. H., Lebofsky, M. J., Thompson, R. I., Low, F. J., & Tokunaga, A. T. 1980, Astrophysical Journal, 238, 24, doi: 10.1086/157954

Ritter, C., Côté, B., Herwig, F., Navarro, J. F., & Fryer, C. L. 2018, Astrophysical Journal Supplement Series, 237, 42, doi: 10.3847/1538-4365/aad691

Roberts, M. S. 1963, Annual Review of Astronomy and Astrophysics, 1, 149, doi: 10.1146/annurev.aa.01.090163.001053

Roberts, M. S., & Haynes, M. P. 1994, Annual Review of Astronomy and Astrophysics, 32, 115, doi: 10.1146/annurev.aa.32.090194.000555

Rodighiero, G., Daddi, E., Baronchelli, I., et al. 2011, Astrophysical Journal, 739, L40, doi: 10.1088/2041-8205/739/2/L40

Rodríguez-Puebla, A., Calette, A. R., Avila-Reese, V., Rodriguez-Gomez, V., & Huertas-Company, M. 2020, Publications of the Astronomical Society of Australia, 37, e024

Romano, D., & Matteucci, F. 2007, Monthly Notices of the Royal Astronomical Society, 378, L59, doi: 10.1111/j.1745-3933.2007.00320.x

Ronen, S., Aragon-Salamanca, A., & Lahav, O. 1999, Monthly Notices of the Royal Astronomical Society, 303, 284, doi: 10.1046/j.1365-8711.1999.02222.x

Rosa-González, D., Terlevich, E., & Terlevich, R. 2002, Monthly Notices of the Royal Astronomical Society, 332, 283, doi: 10.1046/j.1365-8711.2002.05285.x

Rosati, P., Borgani, S., & Norman, C. 2002, Annual Review of the Astronomy & Astrophysics, 40, 539, doi: 10.1146/annurev.astro.40.120401.150547

Roweis, S. T., & Saul, L. K. 2000, Science, 290, 2323, doi: 10.1126/science.290.5500.2323

Rubin, R. H. 1968, The Astrophysical Journal, 154, 391, doi: 10.1086/149766

Rudin, W. 1987, *Real and Complex Analysis*, 3rd ed. (New York: McGraw–Hill)

Rybicki, G. B., & Lightman, A. P. 1985, Radiative Processes in Astrophysics (New York, NY: Wiley), doi: 10.1002/9783527618170

Sage, L. J. 1993, Astronomy & Astrophysicss, 100, 537

Saintonge, A., Kauffmann, G., Kramer, C., et al. 2011, Monthly Notices of the Royal Astronomical Society, 415, 32, doi: 10.1111/j.1365-2966.2011.18677.x

Sakamoto, K., Mao, R.-Q., Matsushita, S., et al. 2011, Astrophysical Journal, 735, 19, doi: 10.1088/0004-637X/735/1/19

Salpeter, E. E. 1955, ApJ, 121, 161

Sandage, A. 1986, Astronomy & Astrophysics, 161, 89

Sanders, D. B., & Mirabel, I. F. 1996, Annual Review of Astronomy and Astrophysics, 34, 749, doi: 10.1146/annurev.astro.34.1.749

Sanduleak, N. 1969, The Astronomical Journal, 74, 47, doi: 10.1086/110773

Santini, P., Maiolino, R., Magnelli, B., et al. 2014, Astronomy & Astrophysics, 562, A30, doi: 10.1051/0004-6361/201322835

Santos, L., Wang, K., & Zhao, W. 2016, Journal of Cosmology and Astroparticle Physics, 2016, 029 (29pp.)

Sargsyan, L., Lebouteiller, V., Weedman, D., et al. 2012, The Astrophysical Journal, 755, 171, doi: 10.1088/0004-637X/755/2/171

Sato, K. 1981, Monthly Notices of the Royal Astronomical Society, 195, 467, doi: 10.1093/mnras/195.3.467

Saunders, W., et al. 2000, Monthly Notices of the Royal Astronomical Society, 317, 55

Saunders, W., Rowan-Robinson, M., Lawrence, A., et al. 1990a, Monthly Notices of the Royal Astronomical Society, 242, 318

—. 1990b, Monthly Notices of the Royal Astronomical Society, 242, 318, doi: 10.1093/mnras/242.3.318

Sawada, T., Ikeda, N., Sunada, K., et al. 2008, Publications of the Astronomical Society of Japan, 60, 445, doi: 10.1093/pasj/60.3.445

Sazonova, E., Alatalo, K., Lotz, J., et al. 2020, The Astrophysical Journal, 899, 85, doi: 10.3847/1538-4357/aba42f

Scalo, J. M. 1986, Fundamentals of Cosmic Physics, 11, 1

Schaye, J. 2004, The Astrophysical Journal, 609, 667, doi: 10.1086/421232

Schechter, P. 1976a, The Astrophysical Journal, 203, 297, doi: 10.1086/154079

Schechter, P. L. 1976b, Astrophysical Journal, 203, 297

Scherrer, R. J., Berlind, A. A., Mao, Q., & McBride, C. K. 2010, Astrophysical Journal, 708, L9, doi: 10.1088/2041-8205/708/1/L9

Schlafly, E. F., & Finkbeiner, D. P. 2011, Astrophysical Journal, 737, 103, doi: 10.1088/0004-637X/737/2/103

Schmalzing, J., & Buchert, T. 1997, Astrophysical Journal, 482, L1

Schmalzing, J., & Gorski, K. M. 1998, Monthly Notices of the Royal Astronomical Society, 297, 355

Schmidt, M. 1959, The Astrophysical Journal, 129, 243, doi: 10.1086/146614

Schruba, A., Bialy, S., & Sternberg, A. 2018, Astrophysical Journal, 862, 110, doi: 10.3847/1538-4357/aac6c5

Schucany, W. R., Parr, W. C., & Boyer, J. E. 1978, Biometrika, 65, 650

Schuster, K. F., Kramer, C., Hitschfeld, M., et al. 2007, Astronomy & Astrophysics, 461, 143, doi: 10.1051/0004-6361:20065579

Schwarz, G. 1978, Annals of Statistics, 6, 461

Searle, L., & Zinn, R. 1978, Astrophysical Journal, 225, 357, doi: 10.1086/156499

Serra, P., Oosterloo, T., Morganti, R., et al. 2012, Monthly Notices of the Royal Astronomical Society, 422, 1835, doi: 10.1111/j.1365-2966.2012.20219.x

Seyfert, C. K. 1943, Astrophysical Journal, 97, 28, doi: 10.1086/144488

Shandarin, S. F., & Zeldovich, I. B. 1983, Comments on Astrophysics, 10, 33

Shane, C. D., & Wirtanen, C. A. 1967, A Survey of the Distribution of Galaxies, Vol. 22 (Lick Observatory Publications), 1

Shapley, H. 1933, Proceedings of the National Academy of Science, 19, 389, doi: 10.1073/pnas.19.4.389

—. 1937, Proceedings of the National Academy of Science, 23, 449, doi: 10.1073/pnas.23.8.449

Shapley, H., & Ames, A. 1932, Annals of Harvard College Observatory, 88, 41

Sheth, R. K., & van de Weygaert, R. 2004, Monthly Notices of the Royal Astronomical Society, 350, 517, doi: 10.1111/j.1365-2966.2004.07661.x

Shi, P., & Yang, L. 2018, Journal of the American Statistical Association, 113, 122

Shiraishi, M., Hikage, C., Namba, R., Namikawa, T., & Hazumi, M. 2016, Physical Review D, 94, 043506 (20pp.)

Shiryaev, A. N. 1996, *Probability*, 2nd ed. (New York: Springer-Verlag)

Silk, J. 1997, Astrophysical Journal, 481, 703, doi: 10.1086/304073

Silva, L., Granato, G. L., Bressan, A., et al. 1998, The Astrophysical Journal, 509, 103, doi: 10.1086/306476

Siudek, M., Małek, K., Pollo, A., et al. 2018, Astronomy & Astrophysics, 617, A70, doi: 10.1051/0004-6361/201832784

Sklar, A. 1959, Publications de l'Institut de Statistique de l'Université de Paris, 8, 229

Smith, D. J. B., Haskell, P., Gürkan, G., et al. 2021, Astronomy & Astrophysics, 648, A6, doi: 10.1051/0004-6361/202039343

Smith, S. 1936, Astrophysical Journal, 83, 23, doi: 10.1086/143697

Solarz, A., Takeuchi, T. T., & Pollo, A. 2016, Astronomy & Astrophysics, 592, A155, doi: 10.1051/0004-6361/201629062

Somerville, R. S., & Primack, J. R. 1999, Monthly Notices of the Royal Astronomical Society, 310, 1087, doi: 10.1046/j.1365-8711.1999.03032.x

Sorai, K., Kuno, N., Muraoka, K., et al. 2019, Publications of the Astronomical Society of Japan, 71, S14, doi: 10.1093/pasj/psz115

Spivak, D. I. 2009, Preprint, 4

Springel, V. 2005, Monthly Notices of the Royal Astronomical Society, 364, 1105

Spurio Mancini, A., et al. 2018, Physical Review D, 98, 103507 (21pp.)

Sriboonchitta, S., Liu, J., Kreinovich, V., & Nguyen, H. T. 2014., Modeling Dependence in Econometrics, 1st edn., ed. S. S. Huynh V. N., Kreinovich V., Advances in Intelligent Systems and Computing, (Cham: Springer International Publishing)

Stacey, G. J., Geis, N., Genzel, R., et al. 1991, The Astrophysical Journal, 373, 423, doi: 10.1086/170062

Stacey, G. J., Hailey-Dunsheath, S., Ferkinhoff, C., et al. 2010, The Astrophysical Journal, 724, 957, doi: 10.1088/0004-637X/724/2/957

Starobinsky, A. A. 1980, Physics Letters B, 91, 99, doi: 10.1016/0370-2693(80)90670-X

—. 1982, Physics Letters B, 117, 175, doi: 10.1016/0370-2693(82)90541-X

Stoyan, D., & Stoyan, H. 1994, Fractals, Random Shapes and Point Fields: Methods of Geometrical Statistics (John Wiley & Sons)

Strauss, M. A., & Willick, J. A. 1995, Physics Reports, 261, 271, doi: 10.1016/0370-1573(95)00013-7

Stuart, A., & Ord, K. 1994, Kendall's Advanced Theory of Statistics, Vol. 1: Distribution Theory, 6th edn. (London: Arnold), 275–276

Stuart, A., Ord, K., & Arnold, S. 1999, *Kendall's Advanced Theory of Statistics*, 6th ed. Vol. 2A, Classical Inference and the Linear Model (London: Arnold)

Sullivan, J. M., Wiegand, A., & Eisenstein, D. J. 2019, Monthly Notices of the Royal Astronomical Society, 485, 1708

Sunyaev, R. A., & Zeldovich, Y. B. 1970, Astrophysics and Space Science, 7, 3

Sutherland, R. S., & Dopita, M. A. 1993, The Astrophysical Journal Supplement Series, 88, 253, doi: 10.1086/191823

Takahashi, R., Sato, M., Nishimichi, T., Taruya, A., & Oguri, M. 2012, Astrophysical Journal, 761, 152 (10pp.)

Takeuchi, T. T. 2000, Astrophysics and Space Science, 271, 213

Takeuchi, T. T. 2010, Monthly Notices of the Royal Astronomical Society, 406, 1830, doi: 10.1111/j.1365-2966.2010.16778.x

Takeuchi, T. T. 2010, Monthly Notices of the Royal Astronomical Society, 406, 1830

Takeuchi, T. T., Buat, V., & Burgarella, D. 2005a, Astronomy & Astrophysics, 440, L17, doi: 10.1051/0004-6361:200500158

Takeuchi, T. T., Buat, V., Heinis, S., et al. 2010, Astronomy & Astrophysics, 514, A4, doi: 10.1051/0004-6361/200913476

Takeuchi, T. T., Buat, V., Iglesias-Páramo, J., Boselli, A., & Burgarella, D. 2005b, Astronomy & Astrophysics, 432, 423, doi: 10.1051/0004-6361:20042189

Takeuchi, T. T., Hirashita, H., Ishii, T. T., Hunt, L. K., & Ferrara, A. 2003a, Monthly Notices of the Royal Astronomical Society, 343, 839, doi: 10.1046/j.1365-8711.2003.06726.x

Takeuchi, T. T., Ishii, T. T., Nozawa, T., Kozasa, T., & Hirashita, H. 2005c, Monthly Notices of the Royal Astronomical Society, 362, 592, doi: 10.1111/j.1365-2966.2005.09337.x

Takeuchi, T. T., & Kono, K. T. 2020, Monthly Notices of the Royal Astronomical Society, 498, 4365, doi: 10.1093/mnras/staa2558

Takeuchi, T. T., Kono, K. T., Cooray, S., et al. 2023, Proceedings of the Institute of Statistical Mathematics, 71, 159

Takeuchi, T. T., Yoshikawa, K., & Ishii, T. T. 2000, Astrophysical Journal Supplement Series, 129, 1, doi: 10.1086/313409

Takeuchi, T. T., Yoshikawa, K., & Ishii, T. T. 2003b, Astrophysical Journall, 587, L89, doi: 10.1086/375181

Takeuchi, T. T., Yoshikawa, K., & Ishii, T. T. 2004, Astrophysical Journal, 606, L171, doi: 10.1086/421346

Takeuchi, T. T., Yuan, F.-T., Ikeyama, A., Murata, K. L., & Inoue, A. K. 2012, Astrophysical Journal, 755, 144, doi: 10.1088/0004-637X/755/2/144

Takeuchi, T. T., Yata, K., Egashira, K., et al. 2024, Astrophysical Journals, 271, 44, doi: 10.3847/1538-4365/ad2517

Tan, J. C. 2000, The Astrophysical Journal, 536, 173, doi: 10.1086/308905

Tenenbaum, J. B., Silva, V. d., & Langford, J. C. 2000, Science, 290, 2319, doi: 10.1126/science.290.5500.2319

The GUDHI Project. 2015, GUDHI User and Reference Manual, GUDHI Editorial Board. http://gudhi.gforge.inria.fr/doc/latest/

Tinsley, B. M. 1980, Fundamentals of the Cosmic Physics, 5, 287

Tomita, A. 2010, Galaxies in Being Alive: Introduction to Galactic Astronomy

Tomita, K. 1969, Progress of Theoretical Physics, 42, 9, doi: 10.1143/PTP.42.9

Toomre, A. 1964, The Astrophysical Journal, 139, 1217, doi: 10.1086/147861

Torabi, S., Dourandish, A., Daneshvar Kakhki, M., Kianirad, A., & Mohammadi, H. 2020, in The Economics of Agriculture and Natural Resources: Perspectives on Development in the Middle East and North Africa (MENA) Region, ed. M. Rashidghalam (Singapore: Springer), 99–110

Tosa, M., & Hamajima, K. 1975, Publications of the Astronomical Society of Japan, 27, 501

Totsuji, H., & Kihara, T. 1969, Publications of the Astronomical Society of Japan, 21, 221

Tremonti, C. A., Heckman, T. M., Kauffmann, G., et al. 2004, Astrophysical Journal, 613, 898, doi: 10.1086/423264

Trivedi, P. R., & Zimmer, D. M. 2005, Foundations and Trends in Econometrics, 1, 1

Tully, R. B., & Fisher, J. R. 1977, Astronomy & Astrophysics, 54, 661

Uzeirbegovic, E., Geach, J. E., & Kaviraj, S. 2020, Monthly Notices of the Royal Astronomical Society, 498, 4021, doi: 10.1093/mnras/staa2651

Vallini, L., Gruppioni, C., Pozzi, F., Vignali, C., & Zamorani, G. 2016, Monthly Notices of the Royal Astronomical Society, 456, L40

Valz, P. D., & McLeod, A. I. 1990, American Statistician, 44, 39

Vicinanza, M., Cardone, V. F., Maoli, R., et al. 2019, Physical Review D, 99, 043534 (16pp.)

Villa-Vélez, J. A., Buat, V., Theulé, P., et al. 2021, arXiv preprint arXiv:2108.13321

Vishwanath, S., Sriperumbudur, B. K., Fukumizu, K., & Kuriki, S. 2022, arXiv:2206.01795

Visvanathan, N., & Griersmith, D. 1977, Astronomy & Astrophysics, 59, 317

Visvanathan, N., & Sandage, A. 1977, The Astrophysical Journal, 216, 214, doi: 10.1086/155464

Walcher, J., Groves, B., Budavári, T., et al. 2011, Astrophysics and Space Science, 331, 1, doi: 10.1007/s10509-010-0458-z

Walter, F., Bolatto, A. D., Leroy, A. K., et al. 2017, Astrophysical Journal, 835, 265, doi: 10.3847/1538-4357/835/2/265

Wang, B., & Silk, J. 1994, The Astrophysical Journal, 427, 759, doi: 10.1086/174182

Wang, L., Farrah, D., Connolly, B., et al. 2011, Monthly Notices of the Royal Astronomical Society, 411, 1809, doi: 10.1111/j.1365-2966.2010.17811.x

Warshall, S. 1962, Journal of the ACM, 9, 11

Wasserman, L. 2018, Annual Review of Statistics and Its Application, 5, 501

Watanabe, S. 1969, *Knowing and Guessing* (New York: John Wiley & Sons)

Weinberg, S. 1972, Gravitation and Cosmology: Principles and Applications of the General Theory of Relativity (New York: Wiley-VCH)

Weingartner, J. C., & Draine, B. T. 2001, Astrophysical Journal, 548, 296, doi: 10.1086/318651

Whitaker, K. E., van Dokkum, P. G., Brammer, G., & Franx, M. 2012, Astrophysical Journal, 754, L29, doi: 10.1088/2041-8205/754/2/L29

Whitney, H. 1936, Annals of Mathematics, 37, 645. http://www.jstor.org/stable/1968482

Wiegand, A., Buchert, T., & Ostermann, M. 2014, Monthly Notices of the Royal Astronomical Society, 443, 241

Wiegand, A., & Eisenstein, D. J. 2017, Monthly Notices of the Royal Astronomical Society, 467, 3361

Wilding, G., & otehrs. 2021, Monthly Notices of the Royal Astronomical Society, 507, 2968

Witt, A. N., & Gordon, K. D. 2000, Astrophysical Journal, 528, 799, doi: 10.1086/308197

Wolfire, M. G., Hollenbach, D., & McKee, C. F. 2010, The Astrophysical Journal, 716, 1191, doi: 10.1088/0004-637X/716/2/1191

Wong, T., & Blitz, L. 2002, Astrophysical Journal, 569, 157, doi: 10.1086/339287

Wright, E. L., Eisenhardt, P. R. M., Mainzer, A. K., et al. 2010, Astronomical Journal, 140, 1868, doi: 10.1088/0004-6256/140/6/1868

Wu, C., Wong, O. I., Rudnick, L., et al. 2019, Monthly Notices of the Royal Astronomical Society, 482, 1211, doi: 10.1093/mnras/sty2646

Wyder, T. K., et al. 2005, Astrophysical Journal, 619, L15

Wyder, T. K., Treyer, M. A., Milliard, B., et al. 2005, The Astrophysical Journal Letters, 619, L15, doi: 10.1086/424735

Xu, D., Wei, Q., Elsayed, E. A., Chen, Y., & Kang, R. 2017a, Quality and Reliability Engineering International, 33, 803

Xu, M., Hua, L., & Xu, S. 2017b, Technometrics, 59, 508

Xu, X., Cisewski-Kehe, J., Green, S. B., & Nagai, D. 2019, Astronomy and Computing, 27, 34

Yan, Z., Jerabkova, T., Kroupa, P., & Vazdekis, A. 2019, Astronomy & Astrophysics, 629, A93, doi: 10.1051/0004-6361/201936029

Yata, K., & Aoshima, M. 2009, Communications in Statistics - Theory and Methods, 38, 2634, doi: 10.1080/03610910902936083

—. 2010a, Comm. Statist. Theory Methods, Special Issue Honoring M. Akahira, (ed. M. Aoshima), 39, 1511

—. 2010b, Journal of multivariate analysis, 101, 2060. https://ci.nii.ac.jp/naid/120002381162/

—. 2012, Journal of Multivariate Analysis, 105, 193 , doi: https://doi.org/10.1016/j.jmva.2011.09.002

—. 2013, Journal of Multivariate Analysis, 122, 334 , doi: https://doi.org/10.1016/j.jmva.2013.08.003

—. 2022, To appear in Statistica Sinica (arXiv 2209.14891). https://arxiv.org/abs/2209.14891

Yoshiura, S., Shimabukuro, H., Takahashi, K., & Matsubara, T. 2017, Monthly Notices of the Royal Astronomical Society, 465, 394

Young, J. S., Allen, L., Kenney, J. D. P., Lesser, A., & Rownd, B. 1996, Astronomical Journal, 112, 1903, doi: 10.1086/118152

Young, L. M., Bureau, M., Davis, T. A., et al. 2011, Monthly Notices of the Royal Astronomical Society, 414, 940, doi: 10.1111/j.1365-2966.2011.18561.x

Yuan, F. T., Takeuchi, T. T., Buat, V., & Burgarella, D. 2012, Publication of Korean Astronomical Society, 27, 345, doi: 10.5303/PKAS.2012.27.4.345

Yuan, F.-T., Takeuchi, T. T., Buat, V., et al. 2011, Publications of the Astronomical Society of Japan, 63, 1207, doi: 10.1093/pasj/63.6.1207

Zehavi, I., Zheng, Z., & Weinberg, D. H. e. a. 2005, The Astrophysical Journal, 630, 1, doi: 10.1086/431891

Zeldovich, Y. B. 1972, Monthly Notices of the Royal Astronomical Society, 160, 1P, doi: 10.1093/mnras/160.1.1P

Zhang, H., & Zaritsky, D. 2016, Monthly Notices of the Royal Astronomical Society, 455, 1364, doi: 10.1093/mnras/stv2413

Zomorodian, A., & Carlsson, G. 2004, in Proceedings of the Twentieth Annual Symposium on Computational Geometry, SCG'04 (New York: Association for Computing Machinery), 347–356

Zwicky, F. 1933, Helvetica Physica Acta, 6, 110

—. 1937, Astrophysical Journal, 86, 217, doi: 10.1086/143864

Zwicky, F., Herzog, E., & Wild, P. 1963, Catalogue of Galaxies and of Clusters of Galaxies, Vol. 2 (Pasadena: California Institute of Technology (CIT))

Zwicky, F., Herzog, E., Wild, P., Karpowicz, M., & Kowal, C. T. 1961, Catalogue of Galaxies and of Clusters of Galaxies, Vol. 1 (Pasadena: California Institute of Technology (CIT))

Index

H-volume, 179
K-correction, 347
Q parameter (Toomre's Q parameter), 43
Q parameter: gas disk, 44
Q parameter: stellar disk, 43
Q_A parameter, 44
β-decay, 72
Čech complex, 299
Hα, 90
n-th central moment, 154
p–p chain, 59
[CII] 158 μm line, 105
[OI] 146 μm line, 38
[OI] 63 μm line, 38
[OI] 63 μm line, 105
[OII] λ3727 line, 104
[OIII] 88 μm line, 105
[OIII] λ5007 line, 104
Isomap, 284, 353
UMAP, 284, 355
HII region, 35
2-increasing, 179
21-cm line, 27
2170 bump, 51

[CII] 158 μm line, 38

absorption: dust, 51
absorption: free-free, 59
activation function, 250
active galactic nuclei (AGN), 20
active galactic nucleus (AGN), 7, 144, 317
adiabatic relation, 41
ALMA, 227
alpha complex, 301
angular diameter distance, 344
artificial general intelligence (AGI), 247
artificial intelligence (AI), 247
asymptotic giant branch (AGB), 67
attenuation, 49, 54, 89, 96
autoencoder, 255

axial twist, 12
axis concentration, 218

baryon acoustic oscillation (BAO), 134, 315
baryon Tully–Fisher relation, 25
bias, 142
biased galaxy formation, 112
Big Bang, 122
binding energy, 149
birthrate parameter, 108
bispectrum, 120, 121
black hole, 62, 69
blue loop, 66
Boltzmann distribution, 27
bottleneck distance, 306
boxy, 12
Bremsstrahlung (free-free) emission, 106
bulge, 3

Calzetti curve, 54
carbon monoxide (CO), 33
cascade, 100
Case B, 90
catch-up, 131
censored, 193
Center for Astrophysics (CfA) Galaxy Redshift Survey , 112
Cepheid, 65
Chandrasekhar limit, 68
chemical evolution, 74
closed-box model, 79
cloud-cloud collision, 47
cluster of galaxies, 110
cluster: Coma, 111
cluster: Virgo, 111
clustering, 112
CNO cycle, 59
CO, 207
CO-to-H$_2$ conversion factor, 34
collision, 29
color, 14

color excess, 51
color–magnitude relation, 25
COMING, 172
comoving coordinate, 125
comoving volume, 17
concentration on a sphere, 218
consistent estimator, 155
continuity equation, 39, 125
convective core, 59
convolutional layer, 250
convolutional neural network (CNN), 318
cooling, 26
cooling function, 27, 146
cooling: dust, 57
cooling: grain, 39
cooling: photoelectric, 50
copula, 178
copula: Farlie–Gumbel–Morgenstern, 178, 184
copula: Gaussian, 187
copula: pair, 199
copula: vine, 195
correlation coefficient, 177, 178, 181, 184
correlation function: N-point, 119, 293
correlation function: 2-point, 114, 117, 134, 293
correlation function: 3-point, 119, 121
correlation function: 4-point, 119, 121
Cosmic Microwave Background (CMB), 119, 123, 296
cosmological constant, 338
cosmological parameter, 338
cosmological redshift, 341
Cramér–Rao inequality, 156
cross section: extinction, 50
cross validation (CV), 329
curvature-dominated, 340

dark halo, 137, 142, 145
dark matter, 3, 110
dark matter: cold (CDM), 123
data augmentation, 322
de-excitation: collisional, 38

Debye model, 55
decoupling, 134
deep learning, 248
degeneracy pressure, 64
Delaunay complex, 301
density contrast, 116, 121
density field, 116
density parameter: matter, 338
density parameter: radiation, 338
differential geometry, 159, 336
differential rotation, 44
dimensionality reduction, 352
disk scale length, 12
disky, 12
dispersion, 154
dispersion relation, 42
distribution function, 153
Doppler shift, 229
dredg-up: first, 63
dredg-up: second, 64
dust, 3, 26, 45
dust emission, 91
dust-to-metal ratio, 44
dust: carbonaceous, 51
dust: emission, 55
dust: extinction, 33
dust: formation, 53
dust: stochastic heating, 55
dusty starburst, 95

Eddington luminosity, 61
eigenspectrum, 230
Einstein A coefficient, 30
Einstein B coefficient, 31
Einstein C coefficient, 29
Einstein's equation, 337
Einstein-de Sitter Universe, 129
electron degeneracy, 63
emission line: forbidden, 103
emission line: submillimeter, 229
emission line: vibro-rotational, 38
energy conservation equation, 40
energy flux, 40
ensemble average, 114
epicyclic frequency, 43

epoch, 249
equation of state, 39, 61
equivalent width (EW), 109
Ergodic hypothesis, 114
ergodic hypothesis, 292
estimation, 155
Euler characteristic, 294
Euler equation, 39, 125
evolutionary synthesis, 84
excitation: collisional, 38
excursion set approach, 138
expectation-maximization (EM) algorithm, 165
expectation-maximization (EM) algorithm: Fisher (FEM), 278
extended Baryon Oscillation Spectroscopic Survey (eBOSS), 312
extended Press–Schechter (EPS) formalism, 139
extinction, 49
extinction: Galactic, 173
extinction: LMC, 56
extinction: Milky Way, 56
extinction: SMC, 56

Faber–Jackson relation, 21
Faint Images of the Radio Sky at Twenty-cm (VLAFIRST), 319
Fanaroff–Riley (FR) classification, 317
far infrared (FIR)–radio correlation, 108
far-infrared (FIR), 39
far-ultraviolet (FUV), 37
feedback: active galactic nucleus (AGN), 147
feedback: stellar, 147
fermion, 63
fine-structure line, 104
Fréchet-Hoeffding lower bound, 179
Fréchet-Hoeffding upper bound, 179
free-fall time, 47
Friedmann equations, 337
fundamental plane relation, 22

G-dwarf problem, 80
galactic wind, 147

galaxy bias, 144
galaxy formation, 144
galaxy manifold, 278, 280
galaxy: cD, 11
galaxy: dwarf, 148
galaxy: dwarf elliptical, 11
galaxy: dwarf spheroidal (dSph), 21
galaxy: elliptical, 5
galaxy: Fanaroff–Riley I (FR-I), 317
galaxy: Fanaroff–Riley II (FR-II), 318
galaxy: irregular, 5
galaxy: low surface brightness (LSB), 8
galaxy: luminous infrared (LIRG), 20
galaxy: morphology, 5
galaxy: peculiar, 7
galaxy: radio, 317
galaxy: spiral, 5
galaxy: starburst, 7
Gamma function, 185
gas consumption timescale, 109
gas ejectation rate, 76
gas infall rate, 76
gas outflow rate, 76
gas-to-dust mass ratio, 26
Gaussian curvature, 336
Gaussian mixture model, 166
Gaussian random field, 122
genus, 294
genus curve, 294
geometric representation, 216
giant molecular cloud (GMC), 47
graph, 196
gravitational collapse, 44, 69
gravitational lensing, 110
gravitational potential, 125
gravitational wave, 73

halo bias, 144
halo mass function, 17
Harrison–Zel'dovich spectrum, 122
Hayashi track, 66
HCN, 243
heat capacity, 55
heating, 26
heating: cosmic-ray, 38

heating: dust, 55
heating: photoelectric, 38, 50
Heaviside step function, 192
heavy element, 26, 71
Heisenberg's uncertainty principle, 31
helium burning, 66
helium core, 63
helium core burning, 64
helium flash, 64
helium shell burning, 66
Hertzsprung gap, 66
Hertzsprung–Russell (HR) diagram, 65
high dimension low sample size (HDLSS), 214
high-dimensional statistical analysis, 213
HNC, 243
homogeneous background, 127
homology, 301
homology group, 302
homotopy, 349
homotopy equivalence, 350
horizon, 125
Hubble parameter, 338
Hubble sequence, 13
hydrogen recombination line, 14
hydrogen: atomic (HI), 26, 27, 206
hydrogen: ionized (HII), 35
hydrogen: molecular (H_2), 33, 45, 165, 207
hyperparameter, 329
hypothesis testing, 155

incomplete data, 164
infall model, 82
inflation, 122
information criterion: Akaike (AIC), 162, 204
information criterion: Bayesian (BIC), 204
information entropy, 157
information matrix (Fisher matrix), 159, 160
infrared (IR), 91
infrared (IR) luminosity, 91
infrared excess (IRX), 94

initial mass function (IMF), 74, 84
initial mass function: Kroupa, 76
initial mass function: Salpeter, 76, 90
instability: Jeans (gravitational), 40, 109, 128
instantaneous recycling approximation, 77, 109
internal energy, 40
interstellar medium (ISM), 3, 26, 74, 76, 215
interstellar medium (ISM): multiphase, 26
intracluster medium (ICM), 110
inverse analysis, 307
IRX–β relation, 98
isodensity surface, 294
isophote, 11

Jeans criterion, 42
Jeans length, 129
Jeans mass, 42, 129
Jeans radius, 42
Jeans wavelength, 42

Kennicutt–Schmidt (K-S) law, 47, 165
Kullback–Leibler information, 158

Large Magellanic Cloud (LMC), 51
large-scale structure, 14, 113
least-square method, 166
likelihood function, 157
line broadening, 31
line broadening: Doppler, 31
line broadening: natural, 31
line element, 336
linearization, 41
local thermodynamic equilibrium (LTE), 30
log-likelihood function, 167
luminosity distance, 344
luminosity function, 17, 138, 178
luminosity function: bivariate, 178
luminosity function: double power-law, 20
luminosity function: Saunders, 19, 189

luminosity function: Schechter, 17, 185, 190
Lyman photon, 91

M31, 46
M33, 46
Mészáros effect, 130
machine learning, 247
machine learning: supervised, 247
machine learning: unsupervised, 247
main sequence: star-forming galaxy, 25
manifold learning, 281, 283
mass ejection, 54
mass function: cluster, 111
mass function: multivariate, 204
mass function: Press–Schechter, 112, 138
mass-to-light ratio, 13, 111, 147
matter-dominated (MD), 340
matter-radiation equality, 130
maximum likelihood estimator, 157
Maxwell–Boltzmann distribution, 63
mean free path, 47
merging, 141
metal abundance, 77
metallicity, 45, 74, 77
metric, 336
metric: Friedmann-Lemaître-Robertson-Walker metric, 336
mid-infrared (MIR), 38
Mie theory, 51
Milky Way, 3
millimeter (mm), 33
minimum variance, 156
Minkowski functional, 294
model selection, 202
molecular cloud, 33, 109, 165
moment, 121
moment measure, 114
morphology, 252
morphology–density relation, 14
morphology: radio galaxy, 318
MRN distribution, 53
multidimensional scaling (MDS), 285, 352

nebula, 112
negative absorption, 31
neon burning, 66
neural network, 249
neural network: convolutional (CNN), 249
neutron star, 62, 68, 69
neutron star merger, 73
Newtonian fluid approximation, 125
NGC 253, 227
Nobeyama Radio Observatory (NRO), 172
nonparametric, 156
nova, 54

onion-like structure, 68
optical thickness, 50
optimization function, 250
outflow, 227, 240
overdense region, 135
overdensity, 116
oxygen-neon-magnesium core, 66

particle horizon, 346
peak model, 142
peculiar velocity, 126
peculiar velocity field, 131
persistence diagram, 304
persistent homology, 295
persistent homology group, 303
perturbation, 40
phase transition, 44
photodisintegration, 70
photodissociation region (PDR), 36
photoionization, 35
photon-dominated region, 36
Planck function (blackbody), 58
point cloud, 116
Poisson equation, 39, 125
polycyclic aromatic hydrocarbon (PAH), 38, 51, 57
polyspectrum, 120
population mean, 153
population synthesis, 84
potential energy, 40

power spectrum, 112, 117
Press–Schechter formalism, 135
pressure: electron degeneracy, 68
principal component analysis (PCA), 220, 248, 254
principal component analysis: automatic sparse (A-SPCA), 226
principal component analysis: noise-reduction (NRPCA), 224
probability density function, 153, 167
profile: de Vaucouleurs, 11
profile: exponential, 12
profile: Sérsic, 13
profile: surface brightness, 11

r-process, 72
radiation-dominated (RD), 339
radiation-dominated era, 130
radiative pressure, 62
radiative transfer: H$\,$I line, 29
radio continuum, 106
Radio Galaxy Zoo, 333
rank correlation: Kendall, 182
rank correlation: Spearman, 182
recombination, 35, 131
recommendation system, 248
reddening, 51
Reference Catalog of galaxy SEDs (RCSED), 253
reflection nebula, 3
reinforcement learning, 248
Riemannian manifold, 351
Roberts time, 109
rotational invariance, 121

s-process, 72
Sérsic index, 13
Sérsic profile, 254
sample mean, 153
scale factor, 336
scaling law (scaling relation), 21
scattering: dust, 51
scattering: Thomson, 60
Schechter function, 111
Schmidt law, 46, 80, 166

selection effect, 192
selective extinction, 49
self-shielding, 38
shell burning: helium, 64
shell burning: hydrogen, 63
shell flash, 66
shock wave, 145
simple stellar population (SSP), 84
simplex, 298
simplicial complex, 298
Sklar's theorem, 181
SLoan Digital Sky Survey (SDSS), 312
Sloan Digital Sky Survey (SDSS), 113, 253, 281
Small Magellanic Cloud (SMC), 46
Small Magellanic Clould (SMC), 51
smoothing scale, 139
sound speed, 41
Southern Sky Redshift Survey, 113
specific star formation rate (SSFR), 108, 273
spectral energy distribution (SED), 84
spectral energy distribution (SED) fitting, 89
spectral synthesis, 84, 89
spherical collapse, 135
spheroid: oblate, 12
spheroid: prolate, 12
spheroid: triaxial, 12
spin, 27
spontaneous emission, 28, 30
stability analysis, 40
stagspansion, 130, 131
star, 3
star formation, 43
star formation efficiency (SFE), 109
star formation rate (SFR), 14, 76, 166
star formation rate (SFR) indicator, 89
star: asymptotic giant branch (AGB), 64
star: horizontal branch (HB), 64
star: main sequence, 59
star: Population I, 3
star: Population II, 3
star: red giant branch (RGB), 63
star: Wolf-Rayet (WR), 54, 66

steady-state solution, 40
stellar evolution, 59
stellar remnants, 78
stellar wind, 66
stimulated emission, 31
stochastic field, 121
stochastic process, 139
Strömgren radius, 36
strong inconsistency, 216
structure formation, 123
submillimeter (submm), 33
supernova, 62, 68, 69
supernova shock, 27
supernova: core-collapse, 70
supernova: thermonuclear runaway, 70
supernova: Type Ia, 54, 69
supernova: Type Ib, 54, 70
supernova: Type Ic, 54, 70
supernova: Type II, 54, 69, 147
surface brightness, 8, 147
survival analysis, 193
survival time, 142
synchrotron emission, 107

thermal pulse, 66
thermodynamic equilibrium, 32
topological data analysis (TDA), 295
topological space, 350
training data, 249
transfer function, 123
tree, 196
triple-alpha reaction, 64, 65

trispectrum, 120, 121
truncated, 193
Tully–Fisher relation, 22
tuning-fork diagram, 5
turbulence, 48
turn-around, 135
Two-Degree Field (2dF) Redshift
    Survey, 113

ultraviolet (UV) continuum, 90
ultraviolet (UV) radiation, 165
ultraviolet (UV) spectral slope, 95
unbiased estimator, 156

validation data, 249
variational autoencoder (VAE), 248, 258
vine, 197
vine: canonical (C-vine), 199
vine: drawable (D-vine), 198
vine: regular, 197
virial equilibrium, 111
virial temperature, 145
virial theorem, 136
void, 296
volume average, 114

wave equation, 42
white dwarf, 62

X-ray luminosity, 106

yield: metal, 77
yield: stellar, 71

For Product Safety Concerns and Information please contact our EU
representative GPSR@taylorandfrancis.com
Taylor & Francis Verlag GmbH, Kaufingerstraße 24, 80331 München, Germany

www.ingramcontent.com/pod-product-compliance
Ingram Content Group UK Ltd.
Pitfield, Milton Keynes, MK11 3LW, UK
UKHW022318100726
473146UK00009B/534